Archibald Sandeman

Pelicotetics

The Science of Quantity, an elementary treatise on algebra and its groundwork, arithmetic

Archibald Sandeman

Pelicotetics
The Science of Quantity, an elementary treatise on algebra and its groundwork, arithmetic

ISBN/EAN: 9783337189952

Printed in Europe, USA, Canada, Australia, Japan

Cover: Foto ©berggeist007 / pixelio.de

More available books at **www.hansebooks.com**

PELICOTETICS

OR

THE SCIENCE OF Q

AN ELEMENTARY

ON

ALGEB

AND

ITS GROUNDWO

ARITHMET

BY

ARCHIBALD SANDEMAN

"....δύο μεγεθῶν ὁμογενῶν ἡ κατὰ πη
ἄλληλα ποιὰ σχέσις."

Cambridge
DEIGHTON BELL AND CO.
LONDON: BELL AND DALDY
1868

PREFACE

THIS book seeks to make Arithmetic and Algebra a science,—a piece of knowledge to wit everywhere reasoned out in an orderly way from principles expressly laid down—, and toward that end has to run wide of the track of the common books.

In the arithmetic of whole numbers these books at starting are so altogether taken up with numerical notation and notational processes as with the help of sundry ellipses and ambiguities of language to becloud the first notion of pure number and to miss the operations that are the very heart and life of arithmetic's every part that therefore abide ever the same under all systems of notation and that therefore before aught else have a right to the name of arithmetical operations. Yet the notation itself marks numbers only as results of certain of those operations and so can only be understood by knowing what the operations are. Besides the properties of, and the processes that have to do with, the notation spring wholly from, and so can only be known through, the laws of the operations. Hence the notation's exact meaning is mistaken the processes themselves are not made out and of the operational laws the more striking alone get any handling whatever though of the roughest kind while the rest are deemed unworthy of even so much as the barest hint.

Passing on to the arithmetic of fractions the mist and darkness settles down thickens and spreads. Because forsooth multiplying one whole number by another yields the same product as multiplying this other by the first the two operations become henceforth absolutely undistinguishable,—as if seemingly operations to be the same have only to give the same results—, and thus it comes about that the one guiding principle throughout fractional operations is marred in itself and maimed and crippled for its work that these operations are first misunderstood and then neither traced back to their springs nor truly performed at all and particularly that in both whole and fractional arithmetic it is held the same operation from the product and the multiplicand of a multiplication to get the multiplier as from the product and the multiplier to get the multiplicand. To such a length indeed are distinctions cast aside of things that differ as to lead to the saying that "every multiplication is a division and every division a multiplication" without the smallest misgiving in the sayer.

The third and last part of arithmetic where the numerical relations of magnitude fall to be dealt with is worse than left out. For not only is the phenomenon of Incommensurability, through which alone arises any need of ratio either the thing or the name, given the go by to but everything touching ratio is put as no more than fractional relationship. Operations with incommensurable numerical quantities are therefore far too much to look for but that books claiming of all things intense practicality should pass over the arithmetic of approximates is startling. Nay as if to go wrong of set purpose to the furthermost the quantities which come straightest from and have most of all to do with ratio are the ones named irrational. The confusions too of multiplication and division could hardly flourish anywhere ranklier than in what goes by the name of the Rule of Three.

Arithmetic which as to matter is either whole fractional or incommensurable is as to manner of handling either pure notational or symbolic. In symbolic arithmetic the books with all the mistakes confusions and falsities above spoken of thick upon them plunge headlong into downright contradictions. For instance the sum got by the addition to $2a+3b$ of $4a+b$ is first said to be $2a+3b+(4a+b)$ but afterwards $6a+4b$ and the process of making out that $2a+3b+(4a+b)=6a+4b$ is called an operation of addition. Again the product of $a-b$ and $a+b$ is said to be $(a-b)(a+b)$,—whether $a-b$ or $a+b$ is the multiplier either is not stated or if stated is soon after either forgotten unheeded or contradicted—, but shortly the same product is said to be a^2-b^2 and then the method of reaching this result is called multiplication. Likewise although at first $5\surd a-2\surd a$ is said to be the remainder got by subtracting $2\surd a$ from $5\surd a$ yet after a while $3\surd a$ is said to be so and the quotient got from dividing $\surd a$ by $\surd b$ though at first symbolized by $\frac{\surd a}{\surd b}$ turns out nevertheless to be $\sqrt{\frac{a}{b}}$. No symbol perhaps fares so ill at the hands of book writers as the fraction symbol which, or rather what from likeness of shape can only be meant for it, is sometimes found with a number of specified units written in the numerator's place and a number of either specified or unspecified units in the denominator's and sometimes with symbols written termwise of two magnitudes neither expressed numerically. To use this last symbol is to cut away the very ground for taking the fraction shaped symbol at all to be the numerical representative of a ratio, inasmuch as what fits it for becoming so is that a fraction's terms express magnitudes numerically in reference to a common unit magnitude, and to use the other is either to set at naught or not to see that settling what unit a fraction refers to settles at once what common unit the fraction's terms refer to. Moreover so utterly slighted is anything like principle in symbolic language that $\surd a\surd b$ or $\surd a.\surd b$ is taken to mean the product by $\surd a$ of $\surd b$ instead of the second root of $a\surd b$ that $\surd a(b+c)$ is put in the stead of $(\surd a)(b+c)$ and that $\sin A \sin B$ is made to stand for what

$(\sin A)\sin B$ rightly stands for. It is too not only heedlessness of principle but even a flat contradiction as well of the symbol's definition to use $ab\times c$ or $ab.c$ for the product of ab into c instead of for the product of a into $b\times c$ or $b.c$ and $n+1.n+2.n+3$ for what $(n+1)(n+2)(n+3)$ rightly symbolizes. In ways such as these the symbolic language which might could and with greater delicacy of touch and skill in handling would cut out sharply the nicest distinctions and lay bare and open up the knottiest entanglements has its fine edge so turned and blunted and hacked that at last it becomes unfit for any but the coarsest work and that on the whole there is nothing which it more does than witness to the roughness of hand and the sightlessness of eye of its many users.

The algebra of the books is quite of a piece with the arithmetic. Indeed since it is Arithmetic alone with its operations laws theorems and processes that first brings into being and afterwards everywhere underlies shapes moulds and gives the cue to Algebra there can be no sound true and thorough algebra without a sound true and thorough arithmetic. No wonder therefore that the principle of all algebraic extension of meaning and use to arithmetical symbols should be loosely grasped lightly prized and even run clean counter to, that there should be no clear well marked boundary between what is matter of definition and principle and what is matter of demonstration, that algebraic operations and laws of operation should be unnoticed undistinguished undefined unproved and even utterly confounded with one another, and that in particular such things should lie far away out of sight and be never dreamt of as the meaning member of an operational equivalence being always the stepping stone whence as from sure footing meaning is laid down for the unmeaning member this meaning becoming in turn if need be a stepping stone in like manner to still other meanings and the need there may be of generalized laws for generalized operations. But notwithstanding all the strayings tumbles and flounderings of algebra writers some conclusions reached by them are nowhit less than wonderful. For instance $\frac{1}{1+x}$ which never ceases to have meaning is said to be equal in some sense to $1-x+x^2-x^3+\cdots$ (with the successive operations symbolized understood to be endless) even when this has no meaning whatever, to wit when x has an absolute value not less than 1; in particular to make all sure it is said that $\frac{1}{2}=1-1+1-1+\cdots$ that $\frac{1}{3}=1-2+2^2-2^3+\cdots$ and the like. This astounding result is got by saying that the product of the multiplication by $1+x$ of $1-x+x^2-x^3+\cdots$ is $1-x+x^2-x^3+\cdots+x-x^2+x^3-\cdots$ which is said to be the sum of the addition to $1-x+x^2-x^3+\cdots$ of $x-x^2+x^3-\cdots$ and this sum again is said to be 1 by actual addition. The process in full so far as one can guess,—for long ago all distinctions between operations and operational laws have been thrown to the winds and everything has been made the sheerest ciphering and symbol grinding—, seems to be the following, $(1+x)(1-x+x^2-x^3+\cdots)$

$$=1\times(1-x+x^2-\cdots)+x(1-x+x^2-\cdots)=1-x+x^2-\cdots+(x-x^2+\cdots)$$
$$=1-x+x^2-x^3+\cdots+x-x^2+x^3-\cdots$$
$$=1+x-x+x^2-x^2+x^3-x^3+\cdots$$
$$=1-(x-x)+(x^2-x^2)-(x^3-x^3)+\cdots$$

Now of the six expressionlike things here written as if asserted to be equal to one another the first five are empty of all meaning so that either the first five are equal in being all equally meaningless and in this sense it is untrue that any of them is equal to the sixth or the whole set of statements is arrant nonsense in the strict sense of the word and inasmuch as nothing whatever is said there is nothing whatever to be gainsaid nothing to be proved nothing to be disproved. Algebraers thus run away with by their overmastering symbols are at last driven to the strange and wild shift and outrageously overtowering extravagance and absurdity of finding and raising high as a principle that a chain of reasoning to be strong and good need not have meaning in every link that in other words the conclusiveness of an argument has nothing to do with the intelligibility of its several steps or that things may be thoroughly made out true for reasons nowise to be understood.

Small need then to say as a wind up that arithmetic and algebra in their wonted setting forth cannot but be educationally bad and mischievous scientifically misleading bewildering unhelping balking stunning deadening and killing and philosophically worthless.

QUEENS' COLLEGE CAMBRIDGE
7 *September* 1867

CONTENTS

I CHAP. Number in Itself

Numerical and Symbolic Wholes Multiples and Submultiples

III CHAP. Magnitude in Relation to Number

IV CHAP. THE PASSAGE FROM ARITHMETIC TO ALGEBRA

V CHAP. Equations of the First and Second Degrees

VI CHAP. PARANOMIC PROCESSES

VII CHAP. DITENSIVE GENERALIZATIONS

PELICOTETICS

PART FIRST

ARITHMETIC

CHAPTER I

NUMBER IN ITSELF.

1. IF any thing and any other thing be put together, and to the group thus made another thing be put, and to the new group of things then made another thing be put, and so on, other groups being made successively in the same way by putting to each group made another thing to make the next following group; and if the things that make up the several groups be viewed only as distinct individual members of the groups, leaving utterly unheeded what the things are, how they are arranged in the groups, and all else: still the groups differ from one another, and from the things that make them up, as to what is called the NUMBER of things in each of them. Accordingly groups so viewed are spoken of as DIFFERENT NUMBERS OF THINGS or as DIFFERENT NUMBERS simply.

ARITHMETIC is the Science of Number.

2. *Definition.*

Each thing in any group of things is called ONE of the things or a UNIT or UNITY.

A group made by putting to a thing another thing is called TWO things or units.

A group made by putting to a group of *two* things another thing is called THREE things.

— — — — — — — — *three* — — — — — — FOUR — .

— — — — — — — — *four* — — — — — — FIVE — .

And so on, the groups following next in order being severally called SIX, SEVEN, EIGHT, NINE, TEN.

Since the groups that can be thus made each from the next before are endless the definitions of different specific numbers must likewise be endless. The names however (hereafter to be dealt with) of all but a few numbers are so chosen as to show the order of successive formation of the numbers and hence serve to define as well as to express them.

3. *Def.* A number is said to be EQUAL to, or the SAME as, another number when to each unit of the one there is a unit of the other.

This is (art. 2) just to say that those numbers are equal to one another which have the same name.

4. *Def.* A number is said to be GREATER than another number when a number equal to the former is among those numbers that can be made by putting units in succession to the latter and to the numbers successively made. Also the latter number is then said to be LESS than the former.

5. *Def.* That made by barely putting things together is called their AGGREGATE.

6. AXIOM. *Equals to the same are equal to one another.*

Hence *Equals to equals are equal to one another.*

For an equal to one of two equals is, by the axiom, equal to the other and therefore, again by the axiom, is equal to an equal to that other.

7. AX. *A whole is equal to the aggregate of its parts.*

8. AX. *A whole is greater than its own part.*

9. AX. *An equal to, and much more a greater than, the greater of two unequals is greater than an equal to, and much more a less than, the less.*

10. AX. *If to equals equals be severally put the wholes are equal.*

11. AX. *If to unequals equals be severally put, or to equals unequals, the whole with the greater unequal is greater than the whole with the less.*

12. AX. *A whole with parts as many as, and severally greater than, the parts of another whole is greater than that other.*

13. Hence *If from equals equals be severally taken the remainders are equal.*

Else by putting to unequal remainders the equals taken the equal wholes would (art. 11) be unequal.

14. Also *If from unequals equals be severally taken the remainder of the greater unequal is greater than the remainder of the less.*

Else by putting to the remainders the equals taken the greater of the unequal wholes would be either (art. 10) equal to, or (art. 11) less than, the less.

15. Again *If from equals unequals be severally taken the remainder over above the greater unequal is less than the remainder over above the less.*

Else by putting to the remainders the unequals taken the equal wholes would (either art. 11 or art. 12) be unequal.

16. Much more *If from the greater and the less of two unequals the less and the greater of two unequals be severally taken the first remainder is greater than the other.*

17. The foregoing axioms and axiomatic consequences are shared by Arithmetic in common with other sciences. The following postulates belong to Arithmetic alone.

POSTULATE. *Let it be granted that the units which make up a number are all equal to one another.*

Things may be numbered together as units however unlike they are to one another. Inasmuch then as they are deemed all equally separate individual things in any group of things so are they EQUAL units of the number of things in the group.

18. POST. *Let it be granted that any numbers make up as parts some number as a whole.*

Groups of things cannot be grouped without making a single group of the things. Several aggregates of units cannot but make up one whole aggregate of those units.

19. *Def.* The number which as a whole is made up of numbers as parts is called the SUM of those numbers.

20. POST. *Let it be granted that any number greater than* ONE *may as a whole be made up of any number less than itself and some other number as parts, hence this other likewise if greater than* ONE *of any number less than itself and some other, and so on; moreover that the number may as a whole be made up of all those less numbers and the last come to of those other numbers as parts.*

21. PROPOSITION. *To find the sum of two given numbers.*

The sum of every two numbers has to be found singly. If either of the two be *one* the sum is given (art. 2) as a matter of definition. If neither be *one* the way to find the sum is always as follows:—

Let *four* and *six* be the numbers whose sum is sought.

The aggregate of *four* and *one* is (art. 7) equal to their sum which (art. 2) is *five.* To each of these equals put *one* and (art. 10) the aggregate of *four, one,* and *one,* is equal to the aggregate of *five* and *one.* But because (art. 2) the aggregate of *one* and *one* is *two,* putting to *four* each, (art. 10) the aggregate of *four, one,* and *one,* is equal to the aggregate of *four* and *two.* Also the aggregate of *five* and *one* is (art. 7) equal to their sum which (art. 2) is *six.* Therefore (art. 6) the aggregate of *four* and *two* is equal to *six.*

Again to these equals severally put *one* and (art. 10) the aggregate of *four, two,* and *one,* is equal to the aggregate of *six* and *one.* But the aggregate of *two* and *one* being (art. 7) equal to their sum to wit (art. 2) *three,* to *four* put each and (art. 10) the aggregate of *four, two,* and *one,* is equal to the aggregate of *four* and *three.* Also (art. 7) the aggregate is equal to the sum of *six* and *one* to wit (art. 2) *seven.* Therefore (art. 6) the aggregate of *four* and *three* is equal to *seven.*

In like manner may it be shown turn by turn that the aggregate of *four* and *four* is equal to *eight,* of *four* and *five* to *nine,* and at length of *four* and *six* to *ten.*

But the aggregate of *four* and *six* is (art. 7) also equal to their sum. Therefore (art. 6) the sum of *four* and *six* is equal to *ten.*

Hence further (art. 9) the sum of *four* and *six* is greater than any number less than *ten* and is less than any number greater.

Therefore (arts. 3, 4) the sum of *four* and *six* is *ten.*

22. *Def.* The operation just gone through in art. 21 is called the ADDITION TO A NUMBER OF A NUMBER.

The instance above taken shows that the sum got by the addition to *four* of *six* is *ten*.

23. *Def.* The following marks, called DIGITS, FIGURES, or CIPHERS, have the meanings severally written under them,

1	2	3	4	5	6	7	8	9
one	two	three	four	five	six	seven	eight	nine.

24. *Def.* Letters of the alphabet, either alone or marked with accents, dashes, suffixes, or other marks, being taken for symbols of any numbers, the symbol $a+b$, read "a PLUS b", stands for the sum got by the addition to a of b.

Thus $1+1$ is the symbol of what is denoted by 2, $2+1$ symbolizes what 3 denotes, $3+1$ stands for what 4 represents, $4+1$ expresses the same as 5, and so on.

25. *Def.* The aggregate of any things is expressed symbolically by writing in any order the symbols of the things with a comma between each adjoining two and shutting all in by a pair of like shaped brackets such as ( ), { }, [ ].

26. *Def.* The symbol $=$ stands for *is equal to*, *equals*, or such other part of the verb *to be equal to* as its manner of use may call for.

27. *Def.* The symbol $\therefore$ means *hence* or *therefore*, and the symbol $\because$ *since* or *because*.

28. PROP. *If to equal numbers equal numbers be severally added the sums are equal.*

For if a, a', b, b', be any numbers such that $a=a'$ and $b=b'$,

$$\text{(art. 10)} \quad (a,\ b) = (a',\ b').$$

$$\text{But (art. 7)} \quad a+b=(a,\ b) \quad \text{and} \quad a'+b'=(a',\ b').$$

$$\therefore \text{(art. 6)} \quad a+b=a'+b'.$$

29. PROP. *The sum got by adding to one number another is the same as the sum got by adding to this other the first.*

For if a, b, be any numbers, (art. 7)

$$a+b=(a,\ b) \quad \text{and} \quad b+a=(a,\ b).$$

$$\therefore \text{(art. 6)} \quad a+b=b+a.$$

Wherefore (art. 3) these are the very same number.

30. PROP. *To find the sum of more than two given numbers.*

First to any one of three given numbers add (art. 21) another and to the sum found add the third. The sum last found is the sum of the three numbers.

For the aggregate of the first two numbers being (art. 7) equal to their sum, to each put the third number and (art. 10) the aggregate of the three numbers is equal to the aggregate of that sum and the third number. But (art. 7) these aggregates are equal severally to the sums of the numbers aggregated. Therefore (art. 6) the sum of the three numbers is equal to, and hence (arts. 9, 3, 4) can be no other than, the sum got by adding to the sum of the first two numbers the third number.

Next the sum of four given numbers is the sum got by adding (art. 21) to the sum of any three of them the fourth. For the former is (art. 7) equal to the aggregate of the four numbers. And, because the sum is (art. 7) equal to the aggregate of the three numbers, (art. 10) the latter is also equal to the same.

In like manner the sum of five given numbers is the sum got by adding to the sum of any four of them the fifth. And so on for six or more given numbers.

31. *Def.* The sum got by adding to $a+b$ c is symbolized by $a+b+c$, the sum got by adding to $a+b+c$ d by $a+b+c+d$, and so on.

32. *Def.* Shutting within a pair of brackets the symbolic expression of the result of one or more operations marks that the expression is to be understood just as a letter expressing the result would be understood in the same place.

Thus $a+(b+c)$ stands for the sum got by the addition to a of $b+c$, $a+(b+c+d)$ for the sum of the addition to a of $b+c+d$, $v+(w+x)+\{d+(e+f)\}$ for the sum of the addition to $v+(w+x)$ of $d+(e+f)$, and the like.

33. PROP. *If the letters stand for any numbers,*

$$x+(a+b+c+\cdots+f+g+h)=x+a+b+c+\cdots+f+g+h.$$

First of all, if a, b, c, be any three numbers, $a+(b+c)=a+b+c$.

For (art. 7) $b+c=(b, c)$, to a put each and (art. 10) $(a, b+c)=(a, b, c)$. But (art. 7) $a+(b+c)=(a, b+c)$. $\therefore$ (art. 6) $a+(b+c)=(a, b, c)$.

Again $a+b=(a, b)$, to each put c and $(a+b, c)=(a, b, c)$. But $a+b+c=(a+b, c)$. $\therefore$ $a+b+c=(a, b, c)$.

$$\therefore \text{(art. 6) } a+(b+c)=a+b+c.$$

Hence $x+(a+b+c+\cdots+f+g+h)=x+(a+b+c+\cdots+f+g)+h$, by what has just been proved, and ∴ again (arts. 28, 6)

$$=x+(a+b+c+\cdots+f)+g+h,$$

∴ again in like manner $=x+(a+b+\cdots+c)+f+g+h,$

$$= \cdot \quad \cdot \quad \cdot \quad \cdot \quad \cdot \quad \cdot \quad \cdot \quad \cdot \quad \cdot$$

at length ∴ $=x+(a+b)+c+\cdots+f+g+h,$

and ∴ at last $=x+a+b+c+\cdots+f+g+h.$

34. PROP. *The final sum got by adding all but one of any numbers in succession to that one and to the sums successively got is the same in whatever order the numbers are taken.*

It has already (art. 29) been shown that for any two numbers a, b, $a+b=b+a$. And the general proposition may be proved in like manner from art. 30. For $a, b, c, \ldots f, g, h$, being any numbers it is there shown that $a+b+c+\cdots+f+g+h=(a, b, c, \ldots f, g, h)$, and also, taking the numbers in any other order, that $g+c+h+\cdots+b+f+a =(a, b, c, \ldots f, g, h)$.

$$\therefore\ a+b+c+\cdots+f+g+h=g+c+h+\cdots+b+f+a.$$

But the proposition may be better viewed as a consequence of the propositions laid down in arts. 29, 33.

Let a, b, c, be any three numbers.

∵ (art. 29) $a+b=b+a$, to each add c and (art. 28) $a+b+c=b+a+c$.
So likewise $a+c+b=c+a+b$ and $b+c+a=c+b+a$.
And the orders in which a, b, c, are written in these expressions are all the orders that can be; for the orders in the last two expressions are the only ones with a last, those in the two next before the only ones with b last, and those in the first two the only ones with c last.

Now (art. 33) $a+b+c=a+(b+c)$ and ∴ (arts. 29, 28) $=a+(c+b)$,
∴ too $=b+c+a$ (art. 29), $=a+c+b$ (art. 33).

$$\therefore\ a+b+c=b+a+c=a+c+b=c+a+b=b+c+a=c+b+a.$$

Again let a, b, c, d, be any four numbers.

By what has just now been shown and art. 28,

$a+b+c+d$	$a+b+d+c$	$a+c+d+b$	$b+c+d+a$
$=b+a+c+d$	$=b+a+d+c$	$=c+a+d+b$	$=c+b+d+a$
$=a+c+b+d$	$=a+d+b+c$	$=a+d+c+b$	$=b+d+c+a$
$=c+a+b+d$	$=d+a+b+c$	$=d+a+c+b$	$=d+b+c+a$
$=b+c+a+d$	$=b+d+a+c$	$=c+d+a+b$	$=c+d+b+a$
$=c+b+a+d$,	$=d+b+a+c$,	$=d+c+a+b$,	$=d+c+b+a$.

Also (art. 33)
$a+b+c+d=a+(b+c+d)$ and ∴ by the above $=a+(c+d+b)=a+(b+d+c)$,
and ∴ $=b+c+d+a$ (art. 29), (art. 33) $=a+c+d+b$ $=a+b+d+c$.

Hence (art. 6) all the expressions in the four groups above of six each are equal.

The proposition thus proved for any four numbers may next be proved in like manner for any five, then in like manner for any six, and so on.

35. *Def.* Different operations or sets of operations are said to be EQUIVALENT which when performed with the same numbers give the same result whatever those numbers be. And the symbolically expressed results of equivalent operations or sets of operations are said to be OPERATIONALLY EQUIVALENT.

Thus (art. 29) $a+b=b+a$ and (art. 33) $a+(b+c)=a+b+c$ whatever numbers a, b, c, stand for. $a+b$ then is operationally equivalent to $b+a$ and $a+(b+c)$ to $a+b+c$. Equalities like these have to do only with the very nature of the operations and not at all with the particular numbers operated with.

36. *Def.* The proposition of arts. 29, 34, is called the LAW OF THE COMMUTATION OF ADDITIONS and that of art. 33 the LAW OF THE DISTRIBUTION OF ADDITION OVER ADDITIONS.

The proposition of art. 29 and the first case of the proposition of art. 33 are not only the simplest cases severally of the general propositions of arts. 34 and 33 but also together the groundwork whereon these have been built, and must therefore be held to be FUNDAMENTAL LAWS OF OPERATIONAL EQUIVALENCE.

37. *Def.* The operation of finding the sum of a given number of numbers each equal to a given number is called MULTIPLICATION; the former given number is called the MULTIPLIER, the latter the MULTIPLICAND, and the sum the PRODUCT OF THE MULTIPLICATION BY THE MULTIPLIER OF THE MULTIPLICAND.

This operation is a special case of the general operation of finding the sum of any numbers (arts. 21, 30).

The units that make up, or the units of, the multiplier are each equal to the multiplicand and therefore (art. 6) all equal to one another. These units then are equal to one another not only in the sense in which (art. 17) all the units of a number are equal but also in the further sense of being equal numbers (art. 3). But (art. 28) equal numbers enter into operations of addition all in the very same way and so far are not to be distinguished from one another. Hence the equal unit numbers of the multiplier, though distinguishable both as individual units of the multiplier and as individual parts of the product which as a whole they make up, are not distinguishable as numbers. That therefore about them alone need be marked which their common name expresses, and hence instead of saying that the units of the multiplier are numbers bearing the name of the multiplicand it is enough to say that the multiplier refers to the multiplicand as unit.

Moreover although it is the name only of the multiplier that expresses how many numbers each having the same name as the multiplicand make up the product yet it may be said for shortness that the multiplier itself expresses this.

For example in finding what number seven threes make up no two of these threes are the same three. Yet in adding to three three adding to the sum got three and so on till the sum of all the seven threes is at length found, what is done at any stage with any one three is just what would be done at that stage with any other three; there is nothing to distinguish the threes but the order in which they are taken. It is enough simply that each of the seven units is three or that the multiplier seven refers to the multiplicand three as unit. Also the number of the threes that make up the product is seven or *seven* expresses how many threes make up the product.

38. *Def.* The symbol $b\times a$, $b.a$, or ba, read "b INTO a", stands for the product of the multiplication by b of a; $c\times b\times a$, $c.b.a$, or cba, for the product of the multiplication by c of ba; $d\times c\times b\times a$, $d.c.b.a$, or $dcba$, for the product of the multiplication by d of cba; and so on.

The multiplication sign $\times$, or the dot ., may be left out only when the symbol cannot reasonably be confounded with the symbol of anything else. Thus the product of 2 into 3 must not be symbolized by 23, for this (as will soon be seen) is made to stand for something else; even the symbol 2.3 may be mistaken for 2·3 a symbol afterwards to be used for quite another thing.

The symbol 2×7 stands for the very same as $7+7$, 3×4 for the same as $4+4+4$, and generally ax for the same as $x+x+x+\cdots+x$ with x written a times, or as this may be written

$$\underbrace{x+x+x+\cdots+x+x}_{a\ x\text{s}}.$$

It is clear that $a\times 1=a=1\times a$ whatever number a stand for.

Of course $(a+b)c$ stands for the product of $a+b$ into c, $a(b+c)$

for that of a into $b+c$, $(abc)d$ for that of abc into d, and the like (art. 32). Also $1+2\times3$ stands for the sum of the addition to 1 of 2×3, $1+2\times3+4\times5\times6$ for the sum of the addition to that sum of $4\times5\times6$, and the like (arts. 24, 31).

39. Prop. *The product made by multiplying by a sum of numbers a number is the same as the sum of the products made by multiplying by each of those numbers that number.*

First for any three numbers a, b, c.

$(a+b)c$ is the same as $\underbrace{c+c+c+\cdots+c}_{a\ cs}+\underbrace{c+c+c+\cdots+c}_{b\ c\text{s}}$

and $\therefore$ (art. 33) $=\underbrace{c+c+c+\cdots+c}_{a\ cs}+\underbrace{(c+c+c+\cdots+c)}_{b\ cs}$ which is $ac+bc$.

Hence for any numbers $a, b, c, \ldots f, g, h, x$,

$$(a+b+c+\cdots+f+g+h)x = (a+b+c+\cdots+f+g)x+hx \text{ by the above,}$$
$$= (a+b+c+\cdots+f)x+gx+hx \text{ by the same,}$$
$$= \cdot\ \cdot\ \cdot\ \cdot\ \cdot\ \cdot\ \cdot\ \cdot\ \cdot\ \cdot\ \cdot$$
$$= (a+b)x+cx+\cdots+fx+gx+hx,$$
$$= ax+bx+cx+\cdots+fx+gx+hx.$$

40. *Def.* The proposition just proved (art. 39) is called the Law of the Distribution of Multiplication Of over Additions.

41. Prop. *If the letters stand for any numbers,*

$$abc\ldots mnpqr = (abc\ldots mnpq)r.$$

For (art. 39) if $q_1, q_2, q_3, \ldots q_p$, be any p numbers,

$$q_1r+q_2r+q_3r+\cdots+q_pr = (q_1+q_2+q_3+\cdots+q_p)r.$$

If therefore in particular $q_1=q_2=q_3=\cdots=q_p$ and each $=q$,

$$pqr = (pq)r.$$

Hence, making use of this again and again,

$$abc\ldots mnpqr = abc\ldots mn(pq)r$$
$$= abc\ldots m(npq)r$$
$$= \cdot\ \cdot\ \cdot\ \cdot\ \cdot$$
$$= a(bc\ldots mnpq)r,$$

and at last $\qquad = (abc\ldots mnpq)r.$

42. *Def.* This proposition (art. 41) is called the Law of the Colligation of Multiplications.

43. *Def.* The symbol a^2 stands for aa, a^3 for aaa, and generally a^n for $aa\ldots aaa$ with a written n times or as it may be written

$$\underbrace{aa\ldots aaa}_{n\ as}.$$

That symbolized by a^n is called the *n*TH POWER of the number symbolized by a and this power is said to be of the *n*TH DEGREE; also the number symbolized by n is called the INDEX or EXPONENT of the power.

In this system of symbolization a^1 must stand for a, and hence the first power of a number is just the number itself.

44. PROP. *If the letters stand for any numbers,*

$$a^w a^v a^u \ldots a^q a^p a^n a^m = a^{m+n+p+q+\cdots+u+v+w}.$$

For first $a^n a^m$ is ${}_{|}(aa\ldots aaa)_{|}aa\ldots aaa_{|}$,
$\overleftarrow{\;n \text{ as }}\;\ast\;\overrightarrow{\;m \text{ as }}$

and $\therefore$ (art. 41) $= {}^{|}aa\ldots\ldots aaa^{|}aa\ldots aaa^{|}$, which is a^{m+n}.

Hence

$$\begin{aligned} a^w a^v a^u \ldots a^q a^p a^n a^m &= a^w a^v a^u \ldots a^q a^p a^{m+n} \\ &= a^w a^v a^u \ldots a^q a^{m+n+p} \\ &= \cdot \quad \cdot \quad \cdot \quad \cdot \quad \cdot \\ &= a^w a^{m+n+p+q+\cdots+u+v} \\ &= a^{m+n+p+q+\cdots+u+v+w}. \end{aligned}$$

45. PROP. *If the letters stand for any numbers,*

$$(a^n)^x = a^{xn}.$$

For (art. 44) $a^w a^v \ldots a^p a^n a^m = a^{m+n+p+\cdots+v+w}$, whatever numbers $a, m, n, \ldots w$, be. If then in particular $m = n = p = \cdots = v = w$, and there be x of them,

$$(a^n)^x = a^{xn}.$$

46. *Def.* The mark 0, called ZERO, stands for *naught, nothing*, or *not any*.

Clearly $0x = 0$ and $x \times 0 = 0$ whatever number x be.

47. PROP. *To explain how numbers are denoted decimally.*

Every number less than *ten* is denoted by a digit (art. 23).

For the purpose of explaining how every other number is denoted let upright strokes be drawn side by side so as to make a column between each adjoining two, and let the columns thus made be named severally, in order from right to left, $A, B, C, \ldots$

From not fewer than ten things ten may be counted off, and should any be left over, from them if not fewer than ten ten more may be counted off, and should any be again left over, from them in turn if not fewer than ten other ten may be counted off, and so on. But since by counting off the things singly there would at length be none of them left over, much sooner by counting them off by tens must there at length be left over either none or fewer of them than ten. Hence any number not less than ten may be made up of one or more groups of ten units each either alone or with some number of ungrouped units less than ten.

If there be any ungrouped units their number being less than ten may be recorded by writing in the column *A* the digit which expresses that number. If the number of the groups be less than ten it may be recorded by writing in the column *B* the digit which expresses it, and this digit which of itself expresses only the number of the groups may when placed in the column *B* be taken further to denote the number of the units in those groups. Moreover if there be any ungrouped units the digit in the column *B* and the digit in the column *A* denoting severally the number of grouped and the number of ungrouped units may be written side by side and taken jointly to denote the number made up of those numbers. In this way then may any number be denoted which is not less than ten but less than the number of things in ten groups of ten things each, that is than the product made by multiplying by ten ten.

5	2
8	

But if the groups, which for distinction may be called groups of the first rank, be not fewer than ten they like the original things may be counted off by tens into one or more groups of a second rank until there be left over either none of them or fewer than ten. If any first rank groups be left over ungrouped the number of them being less than ten may as before be recorded by writing in the column *B* the digit which expresses that number and the digit so written may be taken as before to denote the number of the units in those groups. Now if the second rank groups be fewer than ten their number may be recorded by writing in the column *C* the digit which expresses this number, and this digit expressing the number of second rank groups may be taken to denote the number of the units in these groups. Moreover, should any units or any first rank groups be left over ungrouped, the digit in *C* denoting the number of units in the second rank groups and either or both of the digits in *B*, *A*, denoting severally the number of units in the first rank groups and the number of the ungrouped units may be written on a level all in a row and taken jointly to denote the number made up of, and consequently determined by, those numbers. Since ten units make a first rank group and ten first rank groups make a

6	4	1
9		
3		8
2	7	

second rank group, there may in this way be denoted any number not less than the product made by multiplying by ten ten but less than the product made by multiplying by ten that product.

Again the second rank groups if not fewer than ten may in turn be told off by tens into groups of a third rank until none or fewer than ten of them be left over, these groups of the third rank if not fewer than ten may be told off by tens in like manner into groups of a fourth rank, and so on. But since all numbers are (arts. 2, 22) the sums got by adding in unbroken succession *one* to *one* and to the successive sums got, and the sum got by adding *one* to any number is (arts. 4, 11, 12) less than the sum of *ten* numbers each equal to that number, the final sum of any successive additions of *one* to *one* and to the successive sums got is less than the final product of as many successive multiplications by *ten* of *one* and of the successive products made. Wherefore in telling off by tens to the utmost the units of any given number into first rank groups, these first rank groups into second rank groups, and so on, groups of some rank must be come to at length which are fewer than ten and out of which consequently no group of a higher rank can be made. Hence by first writing severally in the columns *A*, *B*, *C*, ... and all in a row the digits which express such numbers each less than ten of units, first rank groups, second rank groups,... as may be in turn left over ungrouped and at last the number of the groups that yield no further group; by then taking the digit which expresses the number of groups of any rank to denote the number of units in those groups; and by lastly taking the whole row of digits to denote the number made up of the numbers severally denoted by the digits; it follows that any number whatever may be denoted.

In this system of notation then the numbers to which as units the digits written in the several columns in order from right to left are understood to refer are severally *one* and the products made in succession by continued multiplications by *ten* of *one* and of the products successively made, each digit is understood to denote the product made by multiplying by the number which it expresses the number to which as unit it refers, and a row of digits in different columns is understood to denote the final sum got by adding in succession the numbers denoted by all but the first of the digits taken in order from left to right to the number denoted by the first and to the sums successively got. If t be made to stand for *ten*, 1

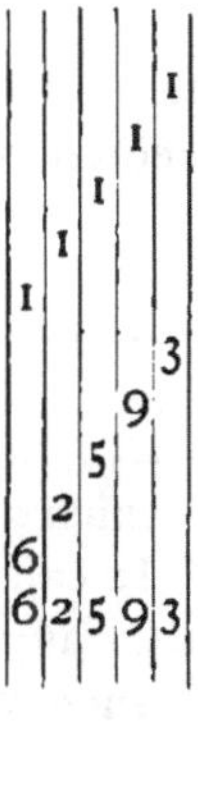

written in the columns *B*, *C*, *D*, ... denotes severally t, t^2, t^3, ... The digits 3, 9, 5, 2, 6, written in the several columns *A*, *B*, *C*, *D*, *E*, denote severally 3, $9t$, $5t^2$, $2t^3$, $6t^4$; and a row of the same digits so written denotes $6t^4+2t^3+5t^2+9t+3$.

When a number is denoted in the way now shown the only use of the columns is to mark the ranks of the groups to which as units the several digits refer. But if a digit were in every column to the right of any column wherein a digit is the bare order of succession of the digits would do this. Hence by using zero as a digit to fill up empty columns the columns themselves may be done without. When columns are done away with by this use of zero as a digit to fill up gaps the other digits are called *significant*.

Thus in a row of digits without columns the numbers to which as units the digits in backward order, that is from right to left, severally refer are *one* and the successive products made by continuedly multiplying by *ten one* and the products successively made, which therefore are severally denoted by

1, 10, 100, 1000, 10000, ...;

each digit denotes the product made by multiplying by the number which it, primarily or apart by itself, expresses the number to which as unit, from its place in the row, it refers; and the whole row denotes the last sum got by adding successively the numbers denoted by all but the first of the digits in order, that is from left to right, to the number denoted by the first and to the successive sums got. For example the row of digits

83000704

denotes the last sum of successive additions of all but the first of the following numbers in order to the first and to the sums got in succession, to wit

the number denoted by	80000000	or the product made by multiplying by the number denoted by	8	the number denoted by	10000000
....	3000000	....	3	....	1000000
....	700	....	7	....	100
....	4	....	4	....	1;

of which also the symbolic expression is $8t^7+3t^6+7t^2+4$.

It can make no change in the meaning of a row of digits to write any number of zeros at the beginning of the row. For instance, 750, 0750, 00750, ... all denote the same.

A number as above denoted is viewed as the sum of the products of certain multiplications of which the multipliers, each less than t, are alone written but so written as to show by their order the several multiplicands, to wit, 1, t, t^2, t^3, ... in backward order.

In dealing with decimally denoted numbers the use of stroke bounded columns may help to clearness whenever the ranks of the units referred to by the digits are at all likely to be mistaken.

48. *The Principle of* DIGIT KNITTING *in the notation of numbers.*

This principle has two parts and may be shown from a particular instance. If t mean *ten*,

$$4950638002 \text{ is } 4t^9+9t^8+5t^7+6t^5+3t^4+8t^3+2 \text{ and } \therefore \text{(art. 33)}$$
$$=4t^9+(9t^8+5t^7+6t^5)+3t^4+(8t^3+2).$$

But $9t^8+5t^7+6t^5$ is 950600000 and $8t^3+2$ is 8002. If then the sum of the numbers denoted by the digits in a knot of consecutive digits taken anywhere in a row of digits, that is the number which would be denoted by the knot were every other digit in the row 0, be called shortly *the number denoted by the knot*, it thus follows that the number denoted by a row of digits is the same as the sum of the numbers denoted severally by any knots of digits or by any knots of digits and single digits into which the row may be broken up.

Again $5t^7$ is (art. 43) $5ttt^5$ and $\therefore$ (art. 41) $=(5tt)t^5$ or $(5t^2)t^5$. So likewise $9t^8=(9t^3)t^5$.

$$\therefore\ 9t^8+5t^7+6t^5=(9t^3)t^5+(5t^2)t^5+6t^5 \text{ and } \therefore \text{(art. 39)}$$
$$=(9t^3+5t^2+6)t^5 \text{ or } 9506\times 100000.$$

It thus follows that the number denoted by a knot of consecutive digits in a row is the same as the product made by multiplying by the number which the knot were it a row apart by itself would denote the number to which as unit the last digit in the knot from its place in the row refers.

Hence a row of digits may be broken up any how into knots of consecutive digits and taken to mean, in a secondary sense equivalent operationally to the primary, either (1) the sum of the numbers denoted severally by the knots, or (2) the sum of the products made by multiplying by the number which each knot viewed as an independent row denotes the number to which as unit its last digit refers. For example the row 47729006380 1025 may be cut into 47, 72900, 6, 3801, 025, and taken to be the final sum got by adding successively all but the first of the numbers in order in either of the two following sets to the first and to the successive sums got :—

470 000000 000000	The product by			47 of	10 000000 000000
7 290000 000000	...	...	...	72900 ...	100 000000
60 000000	...	...	...	6 ...	10 000000
3 801000	...	...	...	3801 ...	1000
25.	...	...	...	25 ...	1.

Any of the knots into which a row is thus broken up and through which it is given a secondary meaning may when viewed as an independent row be itself in like manner broken up into knots and through them in its turn be given a like secondary meaning.

When the digits of a row are knit regularly in backward order into knots of equal numbers, by first writing the rows which denote the numbers referred to as units by the last digits of all the knots but the last and then cutting atwo each of them where the next following knot ends the foregoing principle shows that these numbers in backward order are severally the products of continued multiplications by the last of them of *one* and of the successive products made. The independent proof of this is that if n be the number of digits in each knot the numbers to which as units the last digits of the knots in backward order severally refer are (art. 33) the same as 1, t^n, t^{2n}, t^{3n},... which (art. 45) are operational equivalents of 1, t^n, $(t^n)^2$, $(t^n)^3$,... severally.

3,74,95,18,62
100
100,00
100,0000
100,000000

49. PROP. *To explain how numbers are named.*

The name of any number less than *ten* is given in art. 2. Numbers wholly made up of groups of the first rank are named in order

10	20	30	40	50	60	70	80	90
ten	*twenty*	*thirty*	*forty*	*fifty*	*sixty*	*seventy*	*eighty*	*ninety.*

Other numbers denoted by two digits have names compounded of the names of the numbers denoted severally by the digits; 11 is named *eleven* and 12 *twelve;* then follow

13	14	15	16	17	18	19
thirteen	*fourteen*	*fifteen*	*sixteen*	*seventeen*	*eighteen*	*nineteen;*

and the rest are all named after a uniform law, as

21	22	35	47	86	99
twenty-one	*twenty-two*	*thirty-five*	*forty-seven*	*eighty-six*	*ninety-nine.*

A group of the second rank is called a HUNDRED, and numbers made up only of second rank groups are called in order

100	200	300	. . . . .	900
one hundred	*two hundred*	*three hundred*	. . .	*nine hundred.*

Any other number denoted by a row of three digits is viewed through the principle of digit knitting (art. 48) as the sum got by adding to the number denoted by the first of the digits the number denoted by the knot of the other two and named accordingly, as for instance 108 *one hundred and eight*, 511 *five hundred and eleven*, 694 *six hundred and ninety-four*.

A group of the third rank is called a THOUSAND; and any number denoted by more than three digits but by not more than twice three or six is viewed (art. 48) as the sum got by adding to the number denoted by the digit or knot of digits standing before the last three digits in the row the number denoted by the knot of the last three, also the former of these when denoted by a knot is viewed (art. 48) as the product made by multiplying by the number which the knot as a separate row denotes the number to which as unit its last digit refers. Thus the numbers denoted by the following rows are named severally

8000 *eight thousand*,
5053 *five thousand and fifty-three*,
10000 *ten thousand*,
12067 *twelve thousand and sixty-seven*,
30200 *thirty thousand two hundred*,
100000 *one hundred thousand*,
300500 *three hundred thousand five hundred*,
207080 *two hundred and seven thousand and eighty*,
570406 *five hundred and seventy thousand four hundred and six*,
999999 *nine hundred and ninety-nine thousand nine hundred and ninety-nine*.

A group of the sixth rank is called a MILLION. When a number is denoted by a row of more than six digits the digits are knit regularly by sixes in backward order, the number is by the principle of digit knitting viewed as the sum of the numbers denoted by the several knots, and the number denoted by each of the knots but the last is by the same principle viewed as a product. In the multiplications that yield the products thus secondarily denoted by the knots the multipliers being each denoted by not more than six digits have already been given names and by reason of the regularity of the knitting the multiplicands in backward order are (art. 48) the successive products of continued multiplications by *one million* of *one* and of the products successively made which after the first are severally called a BILLION, a TRILLION, a QUADRILLION, and so on. Thus the numbers denoted by the following rows have the names severally written against them :—

1,000000	*one million ;*
1000,000000	*one thousand millions ;*
53444,827921	*fifty-three thousand four hundred and forty-four millions, eight hundred and twenty-seven thousand nine hundred and twenty-one ;*
100000,000000	*one hundred thousand millions ;*
9876,543210,123456	*nine thousand eight hundred and seventy-six billions, five hundred and forty-three thousand two hundred and ten millions, one hundred and twenty-three thousand four hundred and fifty-six.*

In a row of many enough digits then the digits are first knit backwards by sixes, then in each knot of six the digits are knit backwards by threes, and lastly in each knot of three the digits are knit backwards by twos.

Of course the digits in a row may be sundrywise knit and the number denoted by the row sundrywise named after the secondary meanings given to the row by the principle of knitting. Thus the row 1623 may be cut into 16, 23, and the number denoted by it named *sixteen hundred and twenty-three;* 1000000 may be taken as 1000,000 and called *one thousand thousand* or knitting by twos the digits of the first knot *ten hundred thousand.*

50. PROP. *To find the decimally denoted sum of given decimally denoted numbers.*

(1) The sum of any two numbers each less than 10 must first be found by pure addition (art. 21). In the accompanying table the sum got by adding to any number in the top row any not greater number in the side column is set down under the former over against the latter, and hence by the law of the commutation of additions (art. 29) the sum got by adding to any number in the side column any not less number in the top row is over against the former under the latter.

	9	8	7	6	5	4	3	2	1
1	10	9	8	7	6	5	4	3	2
2	11	10	9	8	7	6	5	4	
3	12	11	10	9	8	7	6		
4	13	12	11	10	9	8			
5	14	13	12	11	10				
6	15	14	13	12					
7	16	15	14						
8	17	16							
9	18								

The numbers standing against 1 in the row next the top are (art. 2) a matter of definition. And by making use of the law of the distribution of addition over additions (art. 33) the numbers in each of the following rows may be gathered from the numbers in the row next before as a matter of definition.

(2) The sum got by adding to any number not less than 10 and not greater than 90 any number less than 10 is next to be found as in that one of the two following instances which may be fit. Putting t for 10,

$32+7$ is $3t+2+7$ and $\therefore$ (art. 33) $=3t+(2+7)$ which by the first case above is $3t+9$ or 39.

$$\begin{array}{r} 32 \\ 7 \\ \hline 39 \end{array}$$

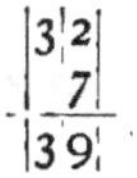

$56+8$ is $5t+6+8=5t+(6+8)$ which by (1) is $5t+14$ or $5t+(t+4)$ and $\therefore$ (art. 33) $=5t+t+4$ and this by the meaning of the symbols (art. 38) is $6t+4$ or 64.

$$\begin{array}{r} 56 \\ 8 \\ \hline 64 \end{array}$$

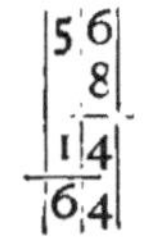

The way shown in the one instance or in the other is to be taken according as adding to the number denoted by the last digit of the first number the other number gives a sum less or not less than 10.

(3) Hence the sum of any ten or fewer numbers may be found as in the following instance.

$386+75+4+2790+507+600666$ is when expressed symbolically

$3t^2+8t+6+(7t+5)+4+(2t^3+7t^2+9t)+(5t^2+7)+(6t^5+6t^2+6t+6)$ and $\therefore$

(art. 33) $=3t^2+8t+6+7t+5+4+2t^3+7t^2+9t+5t^2+7+6t^5+6t^2+6t+6, \therefore$

(art. 34) $=6t^5+2t^3+3t^2+7t^2+5t^2+6t^2+8t+7t+9t+6t+6+5+4+7+6, \therefore$

(art. 33) $=6t^5+ \cdot \cdot \cdot \cdot \cdot \cdot \cdot \cdot \cdot \cdot \cdot +6t+(6+5+4+7+6)$.

But by the case (1) 6+5 is 11, and by the case (2) 11+4 is 15, 15+7 22, 22+6 28; that is 6+5+4+7+6 is 28. And the greatest sum which can here arise in any instance is the sum of ten 9s, that is as found in like manner by (1) and (2) 90. The above expression therefore

$=6t^5+\cdots\cdots+6t+(2t+8)$ and $\therefore$ (art. 33)

$=6t^5+\cdots+6t^2+8t+7t+9t+6t+2t+8, \therefore$ (art. 33)

$=6t^5+\cdots+6t^2+(8t+7t+9t+6t+2t)+8, \therefore$ (art. 39)

$=6t^5+\cdots+6t^2+(8+7+9+6+2)t+8$.

$$\begin{array}{r} 386 \\ 75 \\ 4 \\ 2790 \\ 507 \\ 600666 \\ \hline 604428 \end{array}$$

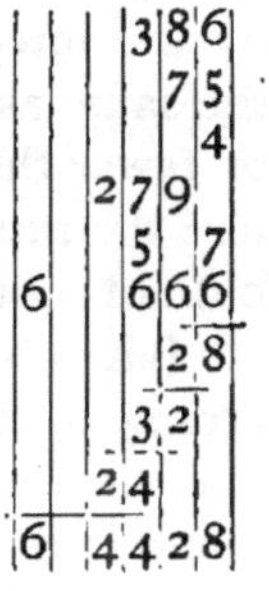

As before 8+7+9+6+2 is 32. And the greatest sum which can in any instance arise at this stage is 99 the sum of eleven 9s, to wit a 9 from each of ten given numbers and another 9 from 90 the greatest sum of the numbers denoted by the last digits of those numbers. The expression therefore

$=6t^5+\cdots+6t^2+(3t+2)t+8$ and $\therefore$ (art. 39)

$=6t^5+\cdots+6t^2+\{(3t)t+2t\}+8$, $\therefore$ (art. 41)

$=6t^5+\cdots+6t^2+(3t^2+2t)+8$, $\therefore$ again (art. 33)

$=6t^5+2t^3+3t^2+7t^2+5t^2+6t^2+3t^2+2t+8$, $\therefore$ by the same steps as before

$=6t^5+2t^3+(3+7+5+6+3)t^2+2t+8.$

Again as before $3+7+5+6+3$ is 24. Also the greatest sum of the kind which can in any instance arise either at this or at any after stage is still 99, the sum to wit of a 9 from every one of ten rows and another 9 the first of the 9s in 99 got from the sum of the numbers expressed by those digits that refer to the next lower power of t as unit. The expression got therefore

$=6t^5+2t^3+(2t+4)t^2+2t+8$ and $\therefore$ as before

$=6t^5+2t^3+2t^3+4t^2+2t+8$

$=6t^5+(2+2)t^3+4t^2+2t+8.$

$\therefore$ $386+75+4+2790+507+600666$ is 604428.

When the bare result of this process is all that is sought those steps which show only the nature and proof of the process may of course be left out. It is enough then to write in order the rows denoting the numbers each of them but the first under the one before with digits which refer to groups of the same rank as units in the same up and down column, to draw a cross stroke under the last row, and to write under the stroke the digits of the row denoting the sum as they are got in backward order each in the column of digits whence it arises.

(4) Lastly the given numbers if more than ten may be arranged in sets of ten or fewer each and the sum of each set found by the foregoing cases. The sum of the sums so found is (art. 33) the same as the sum of the given numbers. When the sets are ten or fewer the sum of their sums may be found as before and therefore the sum of any given numbers of which there are more than t but not more than t^2.

But when the sets are more than ten their sums may be arranged in sets of ten or fewer each, the sum of each set of sums may be found as before, and as before the sum of the sums so found is the same as the sum of the former sums and therefore the same as the sum of the given numbers. Hence when the new sums are ten or fewer and their sum may therefore be found the sum may thus be found of any given numbers of which there are more than t^2 but not more than t^3.

In like manner by arranging the last sums when more than ten into sets of a third kind of ten or fewer each the sum of any given numbers may next be found of which there are more than t^3 but not more than t^4. And so on.

Hence the sum of any given numbers of which the number is not greater than some power of t may be found in this way. But as was shown in art. 47 there is a power of t greater than any given number. Therefore the sum of any given numbers may be so found.

The way now shown of finding the sum of more than ten given numbers is perhaps the best for shutting out mistakes of ciphering. Yet this fourth case may often be more easily dealt with in the same way as the third case. Of course the sum of any given numbers may always be found from the first case by means of the laws used in the second and third cases.

51. *Def.* The process of the last art. is called the NOTATIONAL ADDITIVE PROCESS.

This process is to be carefully distinguished from the Pure Operation of Addition (arts. 21, 30). Although the first great end of a numerical notation is to represent by means of a few marks all numbers whatever yet it serves the hardly less end of finding from the results of some few operations the results of all other operations without actually performing these other through the nature of the notation and the general laws of operational equivalence. Thus from the thirty-six sums of additions given in the Table of the (1) case of art. 50 the sums of all other additions are found by the notational additive process. NUMERICAL OPERATIONS are the same whatever system of notation be used and have nothing to do with NOTATIONAL PROCESSES.

52. PROP. *The product made by multiplying by a number a sum of numbers is the same as the sum of the products made by multiplying by that number each of those numbers.*

For a, b, c, being any three numbers,

$a(b+c)$ means $\underbrace{b+c+(b+c)+(b+c)+\cdots+(b+c)}_{a\ (b+c)\text{s}}$ and $\therefore$ (art. 33)

$=b+c+b+c+b+c+\cdots+b+c$, $\therefore$ (art 34)

$=b+b+b+\cdots+b+c+c+c+\cdots+c$, $\therefore$ (art. 33)

$=b+b+b+\cdots+b+(c+c+c+\cdots+c)$ which means $ab+ac$.

Hence $x, a, b, c, \ldots f, g, h$, being any numbers,

$$x(a+b+c+\cdots+f+g+h)=x(a+b+c+\cdots+f+g)+xh \text{ and } \therefore$$
$$=x(a+b+c+\cdots+f)+xg+xh$$
$$= \quad \cdot \quad \cdot \quad \cdot \quad \cdot \quad \cdot \quad \cdot \quad \cdot \quad \cdot$$
$$=xa+xb+xc+\cdots+xf+xg+xh.$$

53. *Def.* The proposition of the last art. is called the LAW OF THE DISTRIBUTION OF MULIPLICATION BY OVER ADDITIONS.

54. PROP. *The final product made by multiplying by all but one of any numbers in succession that one and the successive products made is the same in whatever order the numbers are taken.*

First if a, b, be any two numbers.

ab is $a_{|}(1+1+1+\cdots+1)_{|}$ which (art. 52)
$\leftarrow$ - - b is - - $\rightarrow$

$=a\times 1+a\times 1+a\times 1+\cdots+a\times 1$ or ${}_{|}a+a+a+\cdots+a_{|}$ which is ba.
$\leftarrow$ - - b as - $\rightarrow$.

Next if a, b, c, be any three numbers. Because by what has just been shown $bc=cb$, (art. 28) $2bc=2cb$, $\therefore$ again (art. 28) $3bc=3cb$, and so on; $\therefore$ $abc=acb$. So $bac=bca$ and $cab=cba$.

But $abc=(ab)c$ (art. 41) and $\therefore$ by the case above $=(ba)c$
$=cab$ by the case above $=bac$ (art. 41).
$\therefore$ $abc=acb=bac=bca=cab=cba$.

Now let the proposition thus proved for any two and for any three numbers be taken as true for any n numbers. Then if a, b, c, $\ldots f$, g, h, i, be any $n+1$ numbers,

$abc\ldots fghi=(abc\ldots fgh)i$ (art. 41) and $\therefore$ by the hypothesis
$=(bac\ldots fgh)i=(cab\ldots fgh)i=\quad\cdot\quad\cdot\quad\cdot\quad=(habc\ldots fg)i$
$=iabc\ldots fgh$ by the first case, and (art. 41)
$=bac\ldots fghi=cab\ldots fghi=\quad\cdot\quad\cdot\quad\cdot\quad=habc\ldots fgi$.

But because by hypothesis $bc\ldots fghi$ is not changed in result if the order of the n letters b, c, $\ldots f$, g, h, i, be changed $abc\ldots fghi$ gives the same result as if the letters after a were written in any other order. So $bac\ldots fghi$ gives the same result as if the letters after b were written in any other order. And the like holds for each of the expressions $cab\ldots fghi$, . . . $habc\ldots fgi$, $iabc\ldots fgh$. Hence the proposition if true for any n numbers is true for any $n+1$. But it is true for any 3 numbers. It is therefore true for any $3+1$ or 4 numbers. Therefore it is true for any $4+1$ or 5 numbers. And so on.

55. *Def.* The proposition of the last art. is called the LAW OF THE COMMUTATION OF MULTIPLICATIONS.

56. *Def.* The common result of multiplying in whatever order one of any numbers by another the product made by a third and so on to the end is called the PRODUCT OF THE MULTIPLICATION TOGETHER OF THE NUM-

BERS or shortly the PRODUCT OF THE NUMBERS, and the numbers are called the FACTORS OF THE PRODUCT.

57. PROP. *To find the decimally denoted product of the multiplication by one given decimally denoted number of another.*

(1) When each of the two numbers is less than 10 the product must be found either by the pure operation (art. 37) or by the notational additive process (art. 50). In the accompanying table the product made by multiplying by any number in the side column any not less number in the top row is set down over against the former under the latter, and hence by the law of the commutation of multiplications (art. 54) the product made by multiplying by any number in the top row any not greater number in the side column is under the former over against the latter.

1	9	8	7	6	5	4	3	2	1
2	18	16	14	12	10	8	6	4	
3	27	24	21	18	15	12	9		
4	36	32	28	24	20	16			
5	45	40	35	30	25				
6	54	48	42	36					
7	63	56	49						
8	72	64							
9	81								

Each number in any row but the first is of course the sum got by adding to the number next above it the number at the top of the column.

(2) When each of the two numbers is denoted by a single significant digit and one at least of them is not less than 10 let at^m be the multiplier and bt^n the multiplicand, t standing for 10.

$(at^m)bt^n = at^m bt^n$ (art. 41), ∴ (art. 54) $= abt^m t^n$ and ∴ (arts. 41, 44) $= (ab)t^{n+m}$.

Now ab is known by the first case and is denoted by either one or two digits, also the column is known in which a digit refers to t^{n+m} as unit. Therefore, and by the principle of digit knitting (art. 48) if need be, the product $(at^m)bt^n$ is denoted either by the single digit denoting ab written in that column or by the second of the two digits denoting ab written in that column and the first in the next column before.

In making use of this process, even though the result only is sought, thus much at least of its steps ought to be borne in mind,—first that the number to which as unit the significant digit of the multiplicand refers is multiplied by the number to which as unit the significant digit of the multiplier refers, secondly that the number primarily expressed by the former digit is multiplied by the number primarily expressed by the latter, and thirdly that the product of the first of those multiplications is multiplied by the product of the other. Hence if the multiplicand be first written, then below it the multiplier, so that digits which refer to groups of the same rank as units may be in the same column, and below the multiplier a stroke be drawn across the page, all that is needed for writing below the stroke the product in like manner is to mark that the product made by multiplying the number to which as unit the multiplicand's significant digit refers by the number to which as unit the multiplier's significant digit refers is by the principle of digit knitting denoted by the digit 1 written in the column as many columns to the left of the column in which the multiplicand's significant digit is as the multiplier's significant digit is columns to the left of its last digit, and therefore to write either in the column so fixed the single digit, or in this column and the next column to the left severally the second and the first of the two digits, denoting the product made by multiplying the number expressed by the multiplicand's significant digit by the number expressed by the multiplier's.

$$\begin{array}{r} 600 \\ 8000 \\ \hline 4800000 \end{array}$$

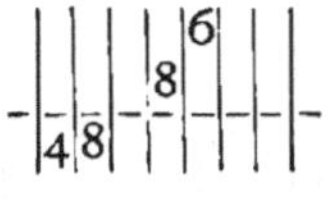

If the multiplier be less than 10 $abt^n = (ab)t^n$, if the multiplicand $(at^m)b = (ab)t^m$, and all else is as before. These are at once brought under the general case by making the unmeaning t^0 mean 1.

(3) When one of the two numbers is denoted by a single significant digit and the other by a row of digits two or more of them significant, if the one denoted by a single significant digit be the multiplicand the product is by the law of the distribution of multiplication of over additions (art. 39) the same as the sum of the products made by multiplying the multiplicand by each of the numbers denoted by the several digits of the multiplier. These products may be found by the first and second cases above and their sum may be found by the notational additive process (art. 50). By finding the products in backward order the multiplicative and the ad-

$$\begin{array}{r} 70 \\ 4906 \\ \hline 343420 \end{array} \qquad \begin{array}{r} 70 \\ 4906 \\ \hline 280000 \\ 63000 \\ 420 \\ \hline 343420 \end{array}$$

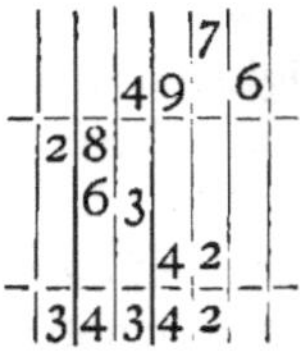

ditive parts of the process may be carried on together and then the result of the whole process need only be written.

But if the multiplier be the one denoted by a single significant

```
      600'850
       60'000
36'051'000'000
```

```
   600'850
    60'000
36'000'000'000
    48'000'000
     3'000'000
36'051'000'000
```

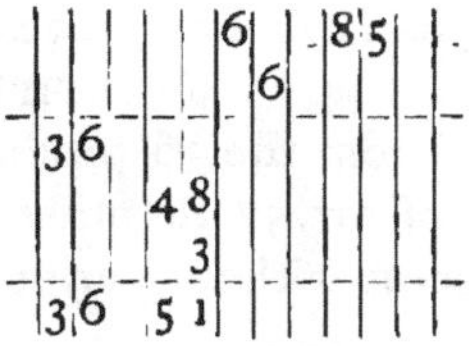

digit the product is by the law of the distribution of multiplication by over additions (art. 52) the same as the sum of the products made by multiplying by the multiplier each of the numbers denoted severally by the digits of the multiplicand. And these products and this sum may be found as before.

(4) When lastly each of the two numbers is denoted by a row of digits two or more of them significant, the product is made by multiplying the multiplicand by the sum of the numbers denoted by the several digits of the multiplier and therefore (art. 39) is the same as the sum of the products made by multiplying the multiplicand by each of those numbers. These products may be found by the third case above and their sum may be found by the notational additive process (art. 50). In doing this it will tend to clearness to write in order the multiplicand, the multiplier, the partial products, and the whole product, so that digits which refer to groups of the same rank as units may be all in the same column, and to draw two strokes the one between the multiplier and the first partial product the other between the last partial product and the whole product. The columns being well marked to which the digits belong the zero digits at the ends of the rows denoting the partial products are needless. The better to mark the digit columns and their order of succession dotted lines may be drawn up and down between neighbouring columns at equal intervals.

```
          81'027
         300'349
24'308'1
    24'308'1
     3'241'08
        729'243
24'336'378'423
```

The result in this case may be got in another way. For the product is made by multiplying by the multiplier the sum of the numbers denoted severally by the digits of the multiplicand and therefore (art. 52) is the same as the sum of the products made by

```
          81'027
         300'349
24'027'92
    300'349
     6'006'98
      2'102'443
24'336'378'423
```

multiplying by the multiplier each of those numbers. These products may be found by the foregoing third case and their sum may then be found as before.

58. *Def.* The process of the last art. is called the NOTATIONAL MULTIPLICATIVE PROCESS.

From the 36 products of multiplications tabulated in the case (1) of art. 57 all other products of multiplications are got by this process without performing (art. 37) the multiplications.

59. PROP. *Having given a number and a part of it to find the other part.*

This has to be done separately for every given number and part but may always be done as follows:—

Let *eleven* be a given whole and *four* a given part of that whole.

Eleven is (art. 2) the sum and therefore (art. 7) is equal to the aggregate of *ten* and *one*. But since *ten* is the sum and therefore equal to the aggregate of *nine* and *one*, to each put *one* and (art. 10) the aggregate of *ten* and *one* is equal to the aggregate of *nine*, *one*, and *one*. Therefore (art. 6) *eleven* is equal to the aggregate of *nine*, *one*, and *one*. But since the aggregate of *one* and *one* is *two*, to *nine* put each and the aggregate of *nine*, *one*, and *one*, is equal to the aggregate of *nine* and *two*. Therefore *eleven* is equal to the aggregate of *nine* and *two*.

Again since *nine* is the sum and therefore equal to the aggregate of *eight* and *one*, to each put *two* and the aggregate of *nine* and *two* is equal to the aggregate of *eight*, *one*, and *two*. Therefore *eleven* is equal to the aggregate of *eight*, *one*, and *two*. But since the aggregate of *one* and *two* is equal to the sum of *one* and *two* which is *three*, to *eight* put each and the aggregate of *eight*, *one*, and *two*, is equal to the aggregate of *eight* and *three*. Therefore *eleven* is equal to the aggregate of *eight* and *three*.

In like manner may it be shown that *eleven* is equal to the aggregate of *seven* and *four*.

But (arts. 20, 7) *eleven* is equal to the aggregate of the part sought of it and *four*. Therefore the aggregate of *seven* and *four* is equal to the aggregate of the part of *eleven* sought and *four*. From each of these equals take *four* and (art. 13) *seven* is equal to the part sought. Hence (art. 9) any number greater than *seven* is greater than the part sought and any number less than *seven* is less.

Therefore (arts. 3, 4) the part sought of *eleven* is *seven*.

This result may also be got if instead of stopping in the path above taken as soon as *eleven* is shown to be equal to the aggregate of *seven* and *four* it be gone on further to show in like manner that *eleven* is equal to the aggregate of *six* and *five*, of *five* and *six*, and at length of *four* and *seven*.

The result may be got in yet a third way by showing as in art. 21 that the aggregate of *four* and *one* is equal to *five*, of *four* and *two* to *six*, ... and at last of *four* and *seven* to *eleven*.

60. *Def.* The operation gone through in the last article is called the SUBTRACTION FROM A NUMBER OF A NUMBER; the former number is called the MINUEND, the latter the SUBTRAHEND, and the result of the operation the REMAINDER, EXCESS, or DIFFERENCE.

The instance taken above shows that the remainder got by the subtraction from *eleven* of *four* is *seven*.

When from a given number of things a given number of them is taken it is by this operation that the number left is found. If the number taken be the whole of the things and therefore the number of them left none at all, however little of operation there then is, and however little a number can be rightly said to be made up of two parts when by one of these parts is meant the number itself and by the other naught, yet this from its close ties with other subtractions must be held to be a subtraction too.

Although the manner in which things are arranged has nothing to do with the number of them still any particular number of things greater than *one* can be known and dealt with only by passing through all less numbers one by one in ascending order of greatness up to it and so climbing from number to number by unit steps. When therefore one part of a given number is given the other part must be found by taking the given part to be made up of one or more unit steps either at the beginning or at the end of this upward passage. In art. 59 the first way shown of subtracting from *eleven four* was by going down four steps, first from *eleven* to *ten*, secondly from *ten* to *nine*, thirdly from *nine* to *eight*, and fourthly from *eight* to *seven;* the second way was by going down so many steps from *eleven* that *four* upward steps might be left; and the third was by going up so many steps from *four* that *eleven* might be reached. Thus in the first of these ways there is a downwardly, and in each of the others an upwardly, taking away of *four* from the natural ascent to *eleven*. Hence subtraction is of two kinds, DESUBTRACTION or subtraction downward and SURSUBTRACTION or subtraction upward; the term *remainder* most fitly applies to a result of the first kind of subtraction and *excess* or *difference* to a result of the other.

61. *Def.* The symbol $a-b$, read "a MINUS b", stands for the remainder got by the subtraction from a of b, $a-b-c$ for the remainder got by the subtraction from $a-b$ of c, $a-b-c-d$ for the remainder got by the subtraction from $a-b-c$ of d, and so on.

From what is said in art. 60 $a-a$ is 0 whatever number a stands for.

The kind of subtraction may be symbolically specified when needful by using $a \llcorner b$ for the remainder left by the desubtraction from a of b and reading it "a MINUS DEORSUM b" and using $a \neg b$ for the excess of a over b, or the difference between a and a not greater number b, and reading this "a MINUS SURSUM b"; the small stroke drawn downwards to, at the a end of, the minus stroke in the first symbol and upwards to, at the b end of, that stroke in the other serving to mark severally the downward course of operation from a and the upward course of operation from b. For example though by the very nature of subtraction (art. 59) $(a-b, b)=a$ and $\therefore$ (arts. 7, 21) $a-b+b=a$, $b+(a-b)=a$, yet $a \llcorner b+b=a$ more immediately than $a \neg b+b=a$ and $b+(a \neg b)=a$ more immediately than $b+(a \llcorner b)=a$. Indeed $a \llcorner b$, $a \neg b$, may be defined as the numbers severally such that $a \llcorner b+b=a$ and $b+(a \neg b)=a$.

62. PROP. *If from equal numbers equal numbers be severally subtracted the remainders are equal.*

For if a, a', b, b', be any numbers such that $a=a'$, $b=b'$, a is not less than b, and hence (arts. 6, 9) a' is not less than b'; then (art. 59)

$$(a-b, b)=a \quad \text{and} \quad (a'-b', b')=a'.$$

$$\therefore \text{ (art. 6) } (a-b, b)=(a'-b', b').$$

Hence from these equals taking severally the equals b, b', (art. 13)

$$a-b=a'-b'.$$

63. PROP. *If the letters stand for any numbers that give meaning to the statement,*

$$a-b=a+c-(b+c).$$

$\because$ (art. 59) $(a-b, b)=a$ to each put c and (art. 10) $(a-b, b, c)=(a, c)$.

But $\because$ (art. 7) $b+c=(b, c)$ to $a-b$ putting each $(a-b, b+c)=(a-b, b, c)$. And $a+c=(a, c)$. $\therefore$ (art. 6)

$$(a-b, b+c)=a+c.$$

$$\text{But (art. 59) } \{a+c-(b+c), b+c\}=a+c.$$

$$\therefore (a-b, b+c)=\{a+c-(b+c), b+c\}.$$

From each of these equals take $b+c$ and (art. 13) $a-b=a+c-(b+c)$.

64. *Def.* The proposition of the last art. is called the LAW OF RELATIVITY IN SUBTRACTION.

65. PROP. *If the letters be any numbers that give meaning to the statement,*

$$x-(a+b+c+\cdots+f+g+h)=x-h-g-f-\cdots-c-b-a.$$

$$\text{For (art. 59) } \{a-(b+c), b+c\}=a.$$

But $\because$ $(a-c-b, b)=a-c$ to each putting c (art. 10) $(a-c-b, b, c)$ $=(a-c, c)$ and $\therefore$ (art. 59) $=a$. Moreover $\because$ (art. 7) $b+c=(b, c)$ to $a-c-b$ putting each $(a-c-b, b+c)=(a-c-b, b, c)$. $\therefore$

$$(a-c-b, b+c)=a.$$

$\therefore$ $\{a-(b+c), b+c\}=(a-c-b, b+c)$. And from each taking $b+c$

$$a-(b+c)=a-c-b.$$

Hence making use time after time of this first and simplest case of the proposition,

$$\begin{aligned} x-(a+b+c+\cdots+f+g+h) &= x-h-(a+b+c+\cdots+f+g) \\ &= x-h-g-(a+b+c+\cdots+f) \\ &= \cdot\ \cdot\ \cdot\ \cdot\ \cdot\ \cdot\ \cdot\ \cdot\ \cdot \\ &= x-h-g-f-\cdots-c-b-a. \end{aligned}$$

66. *Def.* The proposition of art. 65 is called the LAW OF THE DISTRIBUTION OF SUBTRACTION OVER ADDITIONS.

67. PROP. *The last remainder got by subtracting numbers in succession from a number and from the successive remainders got is the same in whatever order the numbers are taken.*

Let $a, b, c, \ldots f, g, h$, be numbers that can be severally subtracted from a number x and the remainders successively got, and let those numbers be taken in any other order say $h, a, g, \ldots b, f, c$.

$$\begin{aligned} x-a-b-c-\cdots-f-g-h &= x-(h+g+f+\cdots+c+b+a) \text{ (art. 65) and } \therefore \text{ (art. 34)} \\ &= x-(c+f+b+\cdots+g+a+h), \therefore \text{ (art. 65)} \\ &= x-h-a-g-\cdots-b-f-c. \end{aligned}$$

68. *Def.* The last article's proposition is called the LAW OF THE COMMUTATION OF SUBTRACTIONS.

69. PROP. *Of successive additions and subtractions when the number either first added to or first subtracted from and all the numbers added are one set of numbers and all the numbers subtracted one set the result is the same whatever be the order of succession.*

First of all $a+b-c=a-c+b$ if $a, b, c,$ be any numbers that give meaning to the statement.

For (art. 59) $(a+b-c, c)=a+b$ and $\therefore$ (art. 7) $=(a, b)$.

But $\because (a-c, c)=a$ to each put b and (art. 10) $(a-c, c, b)=(a, b)$.

$$\therefore (a+b-c, c)=(a-c, c, b).$$

From each of these equals take c and (art. 13)

$$a+b-c=(a-c, b) \text{ and } \therefore \text{(art. 7)} =a-c+b.$$

Secondly it hence follows that

$$\begin{aligned} a-p-q-r-\cdots-u-v-w+b+c+\cdots+f+g+h \\ &=a-p-\cdots-u-v+b-w+c+\cdots+g+h \\ &=a-p-\cdots-u+b-v-w+c+\cdots+g+h \\ &= \cdot\ \cdot\ \cdot\ \cdot\ \cdot\ \cdot\ \cdot\ \cdot\ \cdot\ \cdot\ \cdot \\ &=a+b-p-q-\cdots-u-v-w+c+\cdots+g+h; \end{aligned}$$

so in like manner $=a+b+c-p-q-\cdots-v-w+d+\cdots+g+h$

$= \cdot\ \cdot\ \cdot\ \cdot\ \cdot\ \cdot\ \cdot\ \cdot\ \cdot\ \cdot\ \cdot$

and at last $=a+b+c+\cdots+f+g+h-p-q-r-\cdots-u-v-w.$

Thirdly and lastly therefore, taking for instance as a result of successive additions and subtractions

$a_1+a_2+a_3-b_1-b_2-b_3-b_4+a_4+a_5+a_6-b_5-b_6+a_7-b_7+\cdots$; this

$=a_1+a_2+a_3+a_4+a_5+a_6-b_1-b_2-b_3-b_4-b_5-b_6+a_7-b_7+\cdots$ by the second case,

$=a_1+a_2+a_3+a_4+a_5+a_6+a_7-b_1-b_2-b_3-b_4-b_5-b_6-b_7+\cdots$ by the same,

$= \cdot\ \cdot\ \cdot\ \cdot\ \cdot\ \cdot\ \cdot\ \cdot\ \cdot\ \cdot$

$$=a_1+a_2+a_3+\cdots-b_1-b_2-b_3-\cdots$$

But taking the operations in any other possible order so that the number either first added to or first subtracted from and all the numbers added are $a_1, a_2, a_3, \ldots$ and all the numbers subtracted $b_1, b_2, b_3, \ldots$ as for instance

$$a_7+a_4-b_6-b_1-b_7+a_1+a_6+a_3-b_2-b_5+a_2+a_5-b_4-b_3+\cdots;$$

this as before $=a_7+a_4+a_1+a_6+\cdots-b_6-b_1-b_7-b_2-\cdots$

which by the commutative law of additions (art. 34) and the commutative law of subtractions (art. 67)

$$=a_1+a_2+a_3+\cdots-b_1-b_2-b_3-\cdots$$

Hence the two orders of succession of the operations lead to one result.

70. *Def.* The last article's proposition is called the LAW OF THE COMMUTATION OF ADDITIONS AND SUBTRACTIONS.

71. PROP. *If the letters be any numbers that give the statement meaning,*

$$x+\left(\begin{array}{l}a_1+a_2+\cdots+a_\alpha-m_1-m_2-\cdots-m_\mu+b_1+\cdots+b_\beta-n_1-\cdots-n_\nu+ \\ \qquad\qquad\qquad\qquad\qquad \cdots\cdot+k_1+\cdots+k_\kappa-t_1-\cdots-t_\tau\end{array}\right)$$
$$=x+a_1+a_2+\cdots+a_\alpha-m_1-m_2-\cdots-m_\mu$$
$$+b_1+b_2+\cdots+b_\beta-n_1-n_2-\cdots-n_\nu+\cdots\cdot+k_1+\cdots+k_\kappa-t_1-\cdots-t_\tau.$$

First if a, b, c, be any three numbers of which b is not less than c,

$$a+(b-c)=a+b-c.$$

For (art. 7) $a+(b-c)=(a,\ b-c)$. To each put c and (art. 10) $\{a+(b-c),\ c\}=(a,\ b-c,\ c)$. But $\because$ $(b-c,\ c)=b$ (art. 59), to a putting each $(a,\ b-c,\ c)=(a,\ b)$ and $\therefore=a+b$. $\therefore$ $\{a+(b-c),\ c\}=a+b$.

But $(a+b-c,\ c)=a+b$.

$\therefore$ $\{a+(b-c),\ c\}=(a+b-c,\ c)$. And taking c from each (art. 13) $a+(b-c)=a+b-c$.

Hence secondly

$$x+(a-p-q-r-\cdots-u-v-w)=x+(a-p-q-r-\cdots-u-v)-w$$
$$=x+(a-p-q-r-\cdots-u)-v-w$$
$$=\cdot\quad\cdot\quad\cdot\quad\cdot\quad\cdot\quad\cdot\quad\cdot\quad\cdot\quad\cdot$$
$$=x+a-p-q-r-\cdots-u-v-w.$$

Hence lastly

$$x+\left(\begin{array}{l}a_1+a_2+\cdots+a_\alpha-m_1-m_2-\cdots-m_\mu+b_1+\cdots+b_\beta-n_1-\cdots-n_\nu+ \\ \qquad\qquad\qquad\qquad\qquad \cdots\cdot+k_1+\cdots+k_\kappa-t_1-\cdots-t_\tau\end{array}\right)$$
$$=x+(a_1+\cdots\cdots+k_\kappa)-t_1-t_2-\cdots-t_\tau,\ \therefore\ \text{(art. 33)}$$
$$=x+(a_1+\cdots\cdots-s_\sigma)+k_1+k_2+\cdots+k_\kappa-t_1-t_2-\cdots-t_\tau,$$
$$=\cdot\quad\cdot\quad\cdot\quad\cdot\quad\cdot\quad\cdot\quad\cdot\quad\cdot\quad\cdot$$
$$=x+a_1+a_2+\cdots+a_\alpha-m_1-m_2-\cdots-m_\mu+b_1+\cdots+b_\beta-n_1-\cdots-n_\nu+$$
$$\cdots\cdot+k_1+\cdots+k_\kappa-t_1-\cdots-t_\tau.$$

72. *Def.* The last article's proposition is called the LAW OF THE DISTRIBUTION OF ADDITION OVER ADDITIONS AND SUBTRACTIONS.

73. PROP. *If the letters be any numbers that give the statement meaning,*

$$x-\left(\begin{array}{l}a_1+a_2+\cdots+a_\alpha-m_1-m_2-\cdots-m_\mu+b_1+\cdots+b_\beta-n_1-\cdots-n_\nu+ \\ \qquad\qquad\qquad\qquad\qquad \cdots\cdot+k_1+\cdots+k_\kappa-t_1-\cdots-t_\tau\end{array}\right)$$
$$=x+t_\tau+\cdots+t_2+t_1-k_\kappa-\cdots-k_2-k_1+$$
$$\cdots\cdot+n_\nu+\cdots+n_1-b_\beta-\cdots-b_1+m_\mu+\cdots+m_1-a_\alpha-\cdots-a_1.$$

In the first place if a, b, c, be any three numbers such that b is not less than c and not greater than $a+c$,

$$a-(b-c)=a+c-b.$$

$\because$ (art. 59) $\{a-(b-c),\ b-c\}=a$ to each put c and (art. 10) $\{a-(b-c),\ b-c,\ c\}=(a,\ c)$. But $\because$ $(b-c,\ c)=b$ to $a-(b-c)$ putting each $\{a-(b-c),\ b-c,\ c\}=\{a-(b-c),\ b\}$. And (art. 7) $(a,\ c)=a+c$. $\therefore$ (art. 6) $\{a-(b-c),\ b\}=a+c$.

But $(a+c-b,\ b)=a+c$.

$\therefore$ $\{a-(b-c),\ b\}=(a+c-b,\ b)$. From each taking b $\therefore$ (art. 13) $a-(b-c)=a+c-b$.

In the second place hence

$$\begin{aligned} x-(a-p-q-r-\cdots-u-v-w) &= x+w-(a-p-q-r-\cdots-u-v) \\ &= x+w+v-(a-p-q-r-\cdots-u) \\ &= \ \cdot\ \cdot\ \cdot\ \cdot\ \cdot\ \cdot\ \cdot\ \cdot\ \cdot \\ &= x+w+v+u+\cdots+r+q+p-a. \end{aligned}$$

Hence in the third and last place

$$x-\left(\begin{matrix}a_1+a_2+\cdots+a_\alpha-m_1-m_2-\cdots-m_\mu+b_1+\cdots+b_\beta-n_1-\cdots-n_\nu+ \\ \cdots\cdots+k_1+\cdots+k_\kappa-t_1-\cdots-t_\tau\end{matrix}\right)$$

$$=x+t_\tau+t_{\tau-1}+\cdots+t_2+t_1-(a_1+\cdots\cdots\cdots+k_\kappa),\ \therefore \text{(art. 65)}$$

$$=x+t_\tau+t_{\tau-1}+\cdots+t_2+t_1-k_\kappa-k_{\kappa-1}-\cdots-k_2-k_1-(a_1+\cdots\cdots-s_\sigma)$$

and by repetition of the like steps at length

$$\begin{aligned} &=x+t_\tau+\cdots+t_2+t_1-k_\kappa-\cdots-k_2-k_1+ \\ &\qquad\cdots\cdots+n_\nu+\cdots+n_1-b_\beta-\cdots-b_1+m_\mu+\cdots+m_1-a_\alpha-\cdots-a_1. \end{aligned}$$

74. *Def.* Art. 73's proposition is called the LAW OF THE DISTRIBUTION OF SUBTRACTION OVER ADDITIONS AND SUBTRACTIONS.

75. PROP. *If the letters be any numbers that give the statement meaning,*

$$\left(\begin{matrix}a_1+a_2+\cdots+a_\alpha-m_1-m_2-\cdots-m_\mu+b_1+\cdots+b_\beta-n_1-\cdots-n_\nu+ \\ \cdots\cdots+k_1+\cdots+k_\kappa-t_1-\cdots-t_\tau\end{matrix}\right)x$$

$$\begin{aligned} &=a_1x+a_2x+\cdots+a_\alpha x-m_1x-m_2x-\cdots-m_\mu x+ \\ &\qquad\cdots\cdots+k_1x+\cdots+k_\kappa x-t_1x-\cdots-t_\tau x. \end{aligned}$$

First if a, b, c, be any three numbers of which a is not less than b,

$$(a-b)c=ac-bc.$$

For $(a-b)c$ is $\underbrace{c+c+c+\cdots+c}_{(a-b)\ cs}$ which (art. 60) is the same as

$$\underbrace{c+c+c+\cdots+c}_{(a-b)\ cs}+\{\underbrace{c+c+c+\cdots+c}_{b\ cs}-(\underbrace{c+c+c+\cdots+c}_{b\ cs})\} \text{ and } \therefore \text{(art. 71)}$$

$$=\underbrace{c+c+c+\cdots+c+c+c+c+\cdots+c}_{a\ cs}-(\underbrace{c+c+c+\cdots+c}_{b\ cs}) \text{ or } ac-bc.$$

Wherefore secondly

$$\begin{aligned}(a-p-q-r-\cdots-u-v-w)x &= (a-p-q-r-\cdots-u-v)x-wx\\ &= (a-p-q-r-\cdots-u)x-vx-wx\\ &= \cdot\ \cdot\ \cdot\ \cdot\ \cdot\ \cdot\ \cdot\ \cdot\ \cdot\\ &= ax-px-qx-rx-\cdots-ux-vx-wx.\end{aligned}$$

Therefore thirdly and lastly

$$\begin{pmatrix}a_1+a_2+\cdots+a_\alpha-m_1-m_2-\cdots-m_\mu+b_1+\cdots+b_\beta-n_1-\cdots-n_\nu+\\ \cdots+k_1+\cdots+k_\kappa-t_1-\cdots-t_\tau\end{pmatrix}x$$

$$\begin{aligned}&=(a_1+\cdots\cdots\cdots+k_\kappa)x-t_1x-t_2x-\cdots-t_\tau x,\ \therefore \text{(art. 39)}\\ &=(a_1+\cdots\cdots-s_\sigma)x+k_1x+k_2x+\cdots+k_\kappa x-t_1x-t_2x-\cdots-t_\tau x\\ &= \cdot\ \cdot\ \cdot\ \cdot\ \cdot\ \cdot\ \cdot\ \cdot\ \cdot\ \cdot\ \cdot\ \cdot\ \cdot\ \cdot\\ &=a_1x+a_2x+\cdots+a_\alpha x-m_1x-m_2x-\cdots-m_\mu x+b_1x+\cdots+b_\beta x-n_1x-\cdots-n_\nu x+\\ &\qquad\cdots+k_1x+\cdots+k_\kappa x-t_1x-\cdots-t_\tau x.\end{aligned}$$

76. *Def.* Art. 75's proposition is called the LAW OF THE DISTRIBUTION OF MULTIPLICATION OF OVER ADDITIONS AND SUBTRACTIONS.

77. PROP. *If the letters be any numbers that give the statement meaning,*

$$x\begin{pmatrix}a_1+a_2+\cdots+a_\alpha-m_1-m_2-\cdots-m_\mu+b_1+\cdots+b_\beta-n_1-\cdots-n_\nu+\\ \cdots+k_1+\cdots+k_\kappa-t_1-\cdots-t_\tau\end{pmatrix}$$

$$\begin{aligned}&=xa_1+xa_2+\cdots+xa_\alpha-xm_1-xm_2-\cdots-xm_\mu+\\ &\qquad\cdots\cdots+xk_1+\cdots+xk_\kappa-xt_1-\cdots-xt_\tau.\end{aligned}$$

This follows at once from the last proposition by the commutative law of Multiplications (art. 54). It may also be proved independently thus:—

First $a(b-c)$ is $\underbrace{b-c+(b-c)+(b-c)+\cdots+(b-c)}_{a(b-c)\text{s}}$ and $\therefore$ (art. 71)

$$\begin{aligned}&=b-c+b-c+b-c+\cdots+b-c,\ \therefore \text{(art. 69)}\\ &=b+b+b+\cdots+b-c-c-c-\cdots-c,\ \therefore \text{(art. 65)}\\ &=b+b+b+\cdots+b-(c+\cdots+c+c+c)\ \text{which is } ab-ac.\end{aligned}$$

Next $x(a-p-q-r-\cdots-u-v-w)$

$$\begin{aligned}&=x(a-p-q-r-\cdots-u-v)-xw\ \text{ by the above,}\\ &=x(a-p-q-r-\cdots-u)-xv-xw\ \text{ by the same,}\\ &= \cdot\ \cdot\ \cdot\ \cdot\ \cdot\ \cdot\ \cdot\ \cdot\ \cdot\\ &=xa-xp-xq-xr-\cdots-xu-xv-xw.\end{aligned}$$

Lastly $\therefore$ $x\begin{pmatrix}a_1+a_2+\cdots+a_\alpha-m_1-m_2-\cdots-m_\mu+b_1+\cdots+b_\beta-n_1-\cdots-n_\nu+\\ \cdots+k_1+\cdots+k_\kappa-t_1-\cdots-t_\tau\end{pmatrix}$

$$\begin{aligned}&=x(a_1+\cdots\cdots\cdots+k_\kappa)-xt_1-xt_2-\cdots-xt_\tau,\ \therefore \text{(art. 52)}\\ &=x(a_1+\cdots\cdots-s_\sigma)+xk_1+xk_2+\cdots+xk_\kappa-xt_1-xt_2-\cdots-xt_\tau,\\ &= \cdot\ \cdot\ \cdot\ \cdot\ \cdot\ \cdot\ \cdot\ \cdot\ \cdot\ \cdot\ \cdot\ \cdot\ \cdot\end{aligned}$$

$=xa_1+xa_2+\cdots+xa_\alpha-xm_1-xm_2-\cdots-xm_\mu+xb_1+\cdots+xb_\beta-xn_1-\cdots-xn_\nu+$
$\cdots\cdots-xt_1-\cdots-xt_\tau.$

78. *Def.* Art. 77's proposition is called the LAW OF THE DISTRIBUTION OF MULTIPLICATION BY OVER ADDITIONS AND SUBTRACTIONS.

79. PROP. *To find the decimally denoted remainder of the subtraction from the greater of two given decimally denoted numbers of the less.*

If a be the greater of two given numbers and b the less $(a-b, b)$ $=a$ (art. 59). But if x be any number greater than $a-b$ and y any number less (x, b) is greater than $(a-b, b)$ and (y, b) is less (art. 11). Wherefore (arts. 7, 6, 9) $a-b+b=a$ whereas $x+b$ is greater than a and $y+b$ is less. $a-b$ then is the only number to which if b be added the sum is a. Hence from the additive process of art. 50 a kind of counter process may be drawn for finding the decimally denoted remainder by simply marking what the digits in backward order of a number must be so that the sum got by adding thereto the subtrahend may be the minuend. For example let the minuend be 327640 and the subtrahend 82653. Beginning with the last digits, What number denoted by one digit is that which by adding to it 3 gives a sum denoted by two digits with the second of them 0? Since (art. 50) the greatest sum of two numbers each denoted by one digit is 18 the sum can only be 10 and therefore the number sought can only be 7. Passing on to the digits next before and not forgetting the 1 brought (art. 50) from the last digits, What number is z when $z+5+1$, or (art. 33) its equivalent $z+(5+1)$ to wit $z+6$, is a number denoted by two digits of which the second is 4. The greatest number which can ever arise at this, or any after, stage is $9+9+1$ or 19; therefore $z+6$ is 14 and therefore z is 8. In like manner going on to the other digits, What number and 7 make 16? just 9. What and 3 7? 4. What and 8 12? 4. What and 1 3? 2. Hence the digits in backward order of the remainder are 7, 8, 9, 4, 4, 2, and the remainder itself is 244987.

$$\begin{array}{r} 327640 \\ 82653 \\ \hline 244987 \end{array}$$

But the remainder may be found independently of the notational additive process.

All those remainders must first be found by the pure operation (art. 59) which can be got by subtracting either from a number not greater than 10 a less number or from a number greater than 10 but not greater than 19 a number less than 10 but not less than the number denoted by the second of the two digits denoting the number. In the accompanying table each remainder of the kind is written under the minuend in the top row over against the subtrahend in the side column.

	19	18	17	16	15	14	13	12	11	10	9	8	7	6	5	4	3	2
1									10	9	8	7	6	5	4	3	2	1
2								10	9	8	7	6	5	4	3	2	1	
3							10	9	8	7	6	5	4	3	2	1		
4						10	9	8	7	6	5	4	3	2	1			
5					10	9	8	7	6	5	4	3	2	1				
6				10	9	8	7	6	5	4	3	2	1					
7			10	9	8	7	6	5	4	3	2	1						
8		10	9	8	7	6	5	4	3	2	1							
9	10	9	8	7	6	5	4	3	2	1								

Then if the numbers expressed by the digits in backward order of the minuend be severally not less than the numbers expressed by the digits in backward order of the subtrahend the remainder may be found as in the following instance. Let t stand for 10.

47186−2163 is $4t^4+7t^3+t^2+8t+6-(2t^3+t^2+6t+3)$ and ∴ (art. 65)

$=4t^4+7t^3+t^2+8t+6-3-6t-t^2-2t^3$, ∴ (art. 69)

$=4t^4+7t^3-2t^3+t^2-t^2+8t-6t+6-3$, ∴ (art. 71)

$=4t^4+(7t^3-2t^3)+(t^2-t^2)+(8t-6t)+(6-3)$, ∴ (art. 75)

$=4t^4+(7-2)t^3+(8-6)t+(6-3)$ which by the above is $4t^4+5t^3+2t+3$ or 45023.

$$\begin{array}{r} 47186 \\ 2163 \\ \hline 45023 \end{array}$$

And every other case may be treated as follows:—

66350428−9160472

is $6t^7+6t^6+3t^5+5t^4+4t^2+2t+8-(9t^6+t^5+6t^4+4t^2+7t+2)$ and ∴ (art. 65)

$=6t^7+6t^6+3t^5+5t^4+4t^2+2t+8-2-7t-4t^2-6t^4-t^5-9t^6$, ∴ (art. 69)

$=6t^7+6t^6-9t^6+3t^5-t^5+5t^4-6t^4+4t^2-4t^2+2t-7t+8-2$.

Now $6t^7+6t^6-9t^6$ is $5t^7+t^7+6t^6-9t^6$ and ∴ (art. 71) $=5t^7+(tt^6+6t^6-9t^6)$, ∴ (art. 75) $=5t^7+(t+6-9)t^6$ which by the first case above is $5t^7+7t^6$. Hence (arts. 28, 62) $6t^7+6t^6-9t^6+3t^5-t^5=5t^7+7t^6+3t^5-t^5$, ∴ (art. 71) $=5t^7+7t^6+(3t^5-t^5)$ and ∴ (art. 75) $=5t^7+7t^6+(3-1)t^5$ or by the first case $5t^7+7t^6+2t^5$. Hence again in like manner

$$\begin{array}{r} 66350428 \\ 9160472 \\ \hline 57189956 \end{array}$$

6	6	3	5		4	2	8
	9	1	6		4	7	2
6							
5	7	2					
		1	9				
			8	9	9	5	6
5	7	1	8	9	9	5	6

$6t^7+6t^6-9t^6+3t^5-t^5+5t^4-6t^4=5t^7+7t^6+(t^5+t^5)+5t^4-6t^4$ and $\therefore$ (art. 33)
$=5t^7+7t^6+t^5+t^5+5t^4-6t^4$, $\therefore$ as before
$=5t^7+7t^6+t^5+(t+5-6)t^4$ or $5t^7+7t^6+t^5+9t^4$.

Hence by like steps (art. 6) 66350428–9160472

$$=5t^7+7t^6+t^5+(8t^4+t^4)+4t^2-4t^2+2t-7t+8-2$$
$$=5t^7+7t^6+t^5+8t^4+(9t^3+t^3)+4t^2-4t^2+2t-7t+8-2$$
$$= \cdot\ \cdot\ \cdot\ \cdot\ +8t^4+9t^3+(t+4-4)t^2+2t-7t+8-2$$
$$= \cdot\ \cdot\ \cdot\ \cdot\ +8t^4+9t^3+(9t^2+t^2)+2t-7t+8-2$$
$$= \cdot\ \cdot\ \cdot\ \cdot\ \cdot\ \cdot\ +9t^3+9t^2+(t+2-7)t+(8-2)$$
$$=5t^7+7t^6+t^5+8t^4+9t^3+9t^2+5t+6 \text{ or } 57189956.$$

When nothing but the result of this process is sought it is enough to write in any order the minuend and the subtrahend with digits that refer to groups of the same rank as units in the same column and drawing a stroke to write the remainder in like manner below, finding its digits in order as follows:—Taking the example above 66350428–9160472, say 6 less 0 gives 6 but as in the next column 6 is less than 9 write 5 and carry 1 to the next column; pass to the next column and say 16 less 9 gives 7 and as in the column next following 3 is greater than 1 write 7; as to the next following column 3 less 1 gives 2 and as before write 1 and carry 1; as to the next in order 15 less 6 9 but casting the eye on the columns to the right the first with the minuend's digit not the same as the subtrahend's is one with the former's 2 and the latter's 7, therefore write 8 and carry 1; next 10 less 0 10, 9 and carry 1; so 14 less 4 10, 9 and carry 1; 12 less 7 5; and lastly 8 less 2 6.

80. *Def.* Either of the processes of art. 79 is called the NOTATIONAL SUBTRACTIVE PROCESS.

81. PROP. *To find the number of numbers each equal to a given number which either by themselves or along with some number to be found less than that number make up as parts a given number as a whole.*

Let *a*, *b*, be any two given numbers.

If *a* be not greater than *b* it is itself either less than *b* or equal to *b*.

If *a* be greater than *b* it may (arts. 20, 59) be made up of two parts $a-b$ and *b*. If $a-b$ be less than *b* *a* is made up of this part less than *b* and one part equal to *b*. If $a-b=b$ *a* is made up of two parts each equal to *b*.

But if $a-b$ be greater than *b* it may be made up of two parts $a-b-b$ and *b*, and then (art. 20) *a* is made up of three parts $a-b-b$, *b*, and *b*. If $a-b-b$ be less than *b* *a* is made up of this part less than *b* and two parts each equal to *b*. But if $a-b-b=b$ *a* is wholly made up of three parts each equal to *b*.

Again if $a-b-b$ be greater than *b* it may be made up of $a-b-b-b$ and *b*, and then (art. 20) *a* is made up of $a-b-b-b$, *b*, *b*, *b*; and so on.

In thus finding the successive parts $a-b$, $a-b-b$, $a-b-b-b$,...... there must sooner or later arise one that is not greater than *b*. For if *b* were 1 this would happen before *a* of them were found and if *b* were anything else much sooner (arts. 59, 15).

Hence *a* if not less than *b* may be wholly made up either of one or more parts each equal to *b* or of these and another part less than *b*, moreover the number of the parts each equal to *b* and the part less than *b* may be found in this way.

82. *Def.* The operation gone through in art. 81 is called the DIVISION OF A NUMBER BY A NUMBER; the former number is called the DIVIDEND, the latter the DIVISOR, the number of the numbers each the same as the divisor which either alone or together with a number less than the divisor make up the dividend the QUOTIENT, and this number less than the divisor the REMAINDER OF THE DIVISION.

When the dividend is made up only of parts each equal to the divisor, or in other words when there is no remainder, the division is said to be EXACT. In an exact division then (art. 37) the dividend is equal to the product made by multiplying by the quotient the divisor and the quotient expresses how many numbers each equal to the divisor make up the dividend.

83. PROP. *To find the decimally denoted quotient and remainder of the division of one decimally denoted number by another.*

The decimally denoted dividend and divisor being severally *a* and *b*, the successive remainders $a-b$, $a-b-b$, $a-b-b-b$,... may be found by the notational subtractive process (art. 79) instead of by pure subtraction (art. 59), and as in pure division (art. 81) the number of the subtractive processes gone through when at length either no remainder is left or a remainder less than *b* is the same as the quotient.

If *q* be the quotient or the number of the successive subtractions in the division and r be the remainder when there is one of the division or 0 when there is none (which may be taken as a remainder 0).

$$r = a\underbrace{-b-b-\cdots-b}_{q\ bs} \text{ and } \therefore \text{ (art. 65)} = a-(b+\cdots+b+b) \text{ or } a-qb.$$

Hence if of the products b, $2b$, $3b$,... the one xb can be found that makes $a-xb$ less than b x is the quotient and $a-xb$ is the remainder. Of course the notational multiplicative process (art. 57) may be used for finding this product.

Again if $e, f, g, \ldots l, m, n$, be any numbers such that

$$q = e+f+g+\cdots+l+m+n$$

$$r = a\underbrace{-b-b-\cdots-b}_{e\ bs}\underbrace{-b-b-\cdots-b}_{f\ bs}-\cdots\cdots\underbrace{-b-b-\cdots-b}_{n\ bs} \text{ and } \therefore \text{ (art. 65)}$$

$$= a-(b+\cdots+b+b)-(b+\cdots+b+b)-\cdots\cdots-(b+\cdots+b+b)$$

$$\text{or } a-eb-fb-\cdots\cdots-nb$$

$$\therefore \text{ again (art. 65)} = a-(nb+mb+\cdots\cdots+gb+fb+eb), \therefore \text{ (art. 39)}$$

$$= a-(n+m+l+\cdots+g+f+e)b \text{ and } \therefore \text{ also (art. 34)}$$

$$= a-(e+f+g+\cdots+l+m+n)b.$$

Hence if products eb, fb, gb,...lb, mb, nb, can be found such that $a-eb-fb-\cdots-mb-nb$ is less than b this is the same as the remainder of the division and $e+f+\cdots+m+n$ is the same as the quotient.

The readiness with which products of multiplications can be found by the notational process (art. 57) when the multipliers are denoted by single significant digits points out multipliers of this kind as the fittest to be used for e, f, g,...n. These multipliers may further be so chosen that their significant digits refer to groups of different ranks as units and therefore that their sum is got without any additive process by barely writing those digits in a row. Here as in other notational processes it will tend to clearness to write in the same column all digits that refer to groups of the same rank as units. To keep in mind how the process is linked to the multiplicative process of art. 57 as a kind of counter process it is well to write the divisor first, to draw a stroke a little way below so as to leave room between for the quotient, to write the dividend close below the stroke, and then to carry on the after work below. For example let the divisor be 47692 and the dividend 327790485. It is readily seen that 10000×47692 is greater than the dividend and that 1000×47692

```
      |47692
      |
 ___________
 3277|90485
```

is less. Trying then 1000×47692, 2000×47692,... 9000×47692, the dividend is found to be greater than 6000×47692 but less than 7000×47692; therefore writing the row which denotes 6000×47692, to wit 286152000, or such of its digits as may be needed to mark it, the remainder of the subtraction from the dividend of this product is found to be 41638485. Dealing with this remainder in the same way it is found to be greater than 800×47692 but less than 900×47692; therefore writing the needful digits of the former which is 38153600 the new remainder is 3484885. In like manner the subtraction from this of 70×47692 or 3338440 gives the remainder 146445, and at length the subtraction from this last of 3×47692 or 143076 gives for remainder 3369 which is less than 47692. Hence 6873 is the quotient and 3369 is the remainder of the division of 327790485 by 47692.

```
    47692
     6873
---------
327790485
286152
---------
 41638
 381536
 ------
  34848
  333844
  ------
   14644
   143076
   ------
     3369
```

84. *Def.* The process of the last article is called the NOTATIONAL DIVISIVE PROCESS.

CHAPTER II

NUMBER IN RELATION TO MAGNITUDE

85. *Def.* A MAGNITUDE is whatever can be viewed as a whole made up of parts anywise like the whole.

Thus a straight line is a magnitude because made up of parts each of them itself a straight line. So is a plane surface because made up of parts each a plane surface, and also an angle because made up of parts each an angle. But a man's body though made up as a whole of head, arms, and the rest, as parts is not a magnitude because the head is not like the whole a man's body, neither are the arms, and so on. Still the bulk of a man's body is a magnitude being made up of the bulks of the several parts, and likewise the weight being made up of the weights of the parts.

86. *Def.* A magnitude is OF THE SAME KIND AS, or OF A DIFFERENT KIND FROM, another magnitude according as it can, or cannot, be thought to be either greater than, equal to, or less than, that other.

Thus all straight lines are magnitudes of the same kind since one of any two straight lines must be thought either greater than, equal to, or less than, the other. So likewise are all weights. But a straight line and a weight are so utterly different in kind that the one cannot be thought either greater than, equal to, or less than, the other.

It belongs to the science which has to do with a particular kind of magnitude to settle precisely when one magnitude of the kind is equal to, and when greater than, another. But first of all there must be that about the magnitudes and common to them which gives rise to this way of dealing with them.

87. *Def.* A magnitude either equal to, or wholly made up of parts each equal to, another magnitude is called a MULTIPLE of that other; and this other is called a SUBMULTIPLE or MEASURE of the first.

The parts being each equal to the same are (art. 6) all equal to one another.

A magnitude wholly made up of *n* parts each equal to another magnitude is said shortly to be *n* times that other and the number *n* is said to express, although it is indeed the bare name of the number which expresses, what multiple the first magnitude is of the other. Hence the product of a multiplication is the multiple of the multiplicand expressed by the multiplier, and the quotient of an exact division expresses what multiple the dividend is of the divisor.

88. *Equimultiples of equals are equal.*

For if to equals equals be severally put the wholes are equal (art. 10), and if to the equal wholes equals be severally put the new wholes are equal, and so on. Hence a whole with parts as many as, and severally equal to, the parts of another whole is equal to that other. In particular when the parts of each whole are equal to the same equimultiples of equals are equal.

89. *A multiple of the greater of two unequals is greater than the same multiple of the less.*

For (art. 9) this is a case of a whole with parts as many as, and severally greater than, the parts of another whole, to wit when of each whole the parts are equal to the same, and therefore (art. 12) the proposition holds.

90. *Equisubmultiples of equals are equal.*

Else equimultiples of unequals would be equal and (art. 89) this cannot be.

91. *A submultiple of the greater of two unequals is greater than the same submultiple of the less.*

Else either equimultiples of equals would be unequal against art. 88, or a multiple of the less of two unequals would be greater than the same multiple of the greater against art. 89.

92. *Def.* Any multiple of any submultiple of a magnitude is called a FRACTION of the magnitude.

A magnitude is (art. 8) greater than each of any equal parts into which it may be wholly cut; also if a magnitude be wholly cut first into a number of equal parts and then into a greater number (art. 4) of equal parts so many of the latter parts as there are of the former make up only a part of, and therefore (art. 8) what is less than, the magnitude and hence (art. 91) each of the former parts is greater than each of the latter. Hence those submultiples of a magnitude that are equal to one another are precisely those of each of which the magnitude is the multiple expressed by some one and the same number.

Def. The number that expresses what multiple a magnitude is of a submultiple of it is called the DENOMINATOR of all fractions of the magnitude that are multiples of this submultiple, and the number that expresses what multiple one of the fractions is of the submultiple so determined is called the NUMERATOR of that one. The numerator and denominator of a fraction are called its TERMS.

From what is here said and art. 89, of two fractions having the same numerator but different denominators that is the greater which has the less denominator.

93. *Def.* A fraction with numerator m and denominator n is called m nth parts or shortly m-$n$ths and is written $\frac{m}{n}$.

The fractions $\frac{1}{2}, \frac{2}{2}, \frac{3}{2}, \ldots$ are also named severally *one-half, two-halves, three-halves, ...* and $\frac{1}{4}, \frac{2}{4}, \frac{3}{4}, \ldots$ severally *one-quarter, two-quarters, three-quarters, ...*

94. *Def.* A fraction is said to be PROPER or IMPROPER according as its numerator is less or not less than its denominator.

An improper fraction of a magnitude either is a multiple of the magnitude, to wit when the numerator is exactly divisible (art. 82) by the denominator and the quotient then expresses what multiple, or is made up of two parts one that multiple of the magnitude which is expressed by the quotient of the division of the numerator by the denominator and the other that proper fraction of the magnitude which has the same denominator as the fraction and for numerator the remainder of the inexact division. Thus (art. 83) $\frac{327790485}{47692}$ of a magnitude is made up of 6873 times the magnitude and $\frac{3369}{47692}$ of the magnitude and this is written $6873\frac{3369}{47692}$, and read "6873 and 3369-47692ths".

95. Multiples of a magnitude are one kind of fractions of it whether expressed as 1, 2, 3, ... times the magnitude, taking

(art. 87) an equal to a magnitude to be a submultiple of it, or as $\frac{n}{n}$, $\frac{2n}{n}$, $\frac{3n}{n}$, ... of the magnitude (art. 94), n being any number. But multiples of a magnitude are also numbers having for units magnitudes that are equal to one another by being each equal to the magnitude. Hence under the general class of things called fractions are embraced as a special class all such numbers as have equal magnitudes for units. From the common practice in language then of making the word used for a part of a class stand for the whole there springs the

Def. With the understanding that the unit is a magnitude and that the units are equal magnitudes the term NUMBER is widened in meaning so as to apply to any fraction of a magnitude as well as to any multiple. The magnitude of which a number in this sense is a multiple or fraction is called the UNIT or UNIT MAGNITUDE TO WHICH THE NUMBER REFERS. Numbers that are multiples of the unit magnitude are called WHOLE NUMBERS and all others FRACTIONAL NUMBERS.

When magnitudes are in this way viewed as numbers the numbers are spoken of as expressing what multiples or fractions the magnitudes severally are of the unit magnitude although as in arts. 37, 87, it is only the names of the numbers that truly do so.

Def. That submultiple of the unit magnitude which the denominator of a fraction determines and in reference to which as unit subordinate to the unit magnitude the fraction is a whole number is called the SUBUNIT of the fraction.

96. *Def.* A CONTINUOUS magnitude is one of which every part is a magnitude.

Magnitudes are always understood to be continuous unless expressly stated not to be.

97. POST. *Let it be granted that a magnitude may be wholly cut into any assigned number of equal parts.*

98. PROP. *A fraction may be expressed in terms of the products made by multiplying by its corresponding terms any whole number.*

Let $\frac{m}{n}$ be any fraction and x any whole number.

Cut into x equal parts (art. 97) each of the m subunits that make up the fraction and also each of the n subunits of the same denomination that make up the unit to which the fraction refers. All these parts are (art. 90) equal to one another. Also (art. 37) the fraction is made up of mx of them and the unit magnitude of nx. Therefore (arts. 87, 92) the fraction is $\frac{mx}{nx}$.

This is simply taking for the subunit of the fraction a submultiple of the subunit given. It is the one principle by which all operations with fractions are performed and is called the PRINCIPLE OF FRACTIONAL SUBUNIT CHANGE.

99. PROP. *To express fractions of which particular terms differ so as to have the corresponding particular terms the same.*

Let $\frac{a}{a'}$, $\frac{b}{b'}$, $\frac{c}{c'}$, ... $\frac{r}{r'}$, $\frac{s}{s'}$, $\frac{t}{t'}$, be fractions of which the terms a, b', c, ... r', s, t', are different numbers. Whatever whole numbers $\alpha, \beta, \gamma, \ldots \rho, \sigma, \tau$, are these fractions may (art. 98) be expressed as

$$\frac{a\alpha}{a'\alpha},\ \frac{b\beta}{b'\beta},\ \frac{c\gamma}{c'\gamma}, \ldots\ldots \frac{r\rho}{r'\rho},\ \frac{s\sigma}{s'\sigma},\ \frac{t\tau}{t'\tau}.$$

And to have the corresponding terms the same as one another all that is needed is to choose $\alpha, \beta, \ldots \tau$, so that

$$a\alpha = b'\beta = c\gamma = \cdot \quad \cdot \quad \cdot = r'\rho = s\sigma = t'\tau.$$

This (arts. 54, 56) may be done by taking for α the product of all but a of the numbers a, b', c, ... r', s, t', for β the product of all but b' of them, . . . for τ the product of all but t'. The same would be done also by taking for $\alpha, \beta, \ldots \tau$, severally the products of those products each into any the same whole number. It may often be done (as will afterwards be seen) by taking for $\alpha, \beta, \ldots \tau$, numbers severally less than $b'c..r'st'$, $ac..r'st'$, ... $ab'c..r's$.

100. PROP. *To find the order of greatness of given fractions.*

Fractions with the same denominator are (arts. 88, 20, 8) in the same order of greatness as their numerators and with the same numerator (art. 92) in the reverse order of greatness to their denominators. And fractions having not may (art. 99) be so expressed as to have either the same denominator or the same numerator.

101. PROP. *To find the fraction which as a whole is made up of given fractions as parts.*

Given fractions either have or (art. 99) may be expressed so as to have a common denominator and are then the whole numbers expressed by the several numerators in reference to that submultiple of the common unit magnitude as common unit which the common denominator determines. Hence (art. 19) the whole which they as parts make up is that multiple of the submultiple so determined which the sum of the numerators expresses and hence (art. 92) is the fraction whose numerator is this sum and denominator the common denominator.

The sum of the numerators can be got either immediately (arts. 21, 30) or mediately (art. 50) only by adding to one of the numerators another adding to the sum got another and so on to the end.

102. By widening the meaning of terms as in art. 95 there arises the

Def. The fraction which as a whole is made up of fractions as parts is called the SUM of those fractions. The operation of finding a fraction whose numerator is the sum got by adding to the numerator of one the numerator of the other of two fractions expressed so as to have a common denominator and whose denominator is this common denominator is called the ADDITION TO THE FIRST FRACTION OF THE OTHER. If the letters stand for any numbers whole or fractional $a+b$ symbolizes the sum got by adding to a b, $a+b+c$ the sum got by adding to that sum c, and so on.

103. PROP. *If a magnitude and a part of it be given fractions of a common unit magnitude to find what fraction of the unit magnitude the other part is.*

A whole magnitude and a part of it that are given fractions of a unit either are or (art. 99) may be expressed as whole numbers of the equal subunits determined by some common denominator, and then (art. 59) the other part is the whole number of those same subunits got as the remainder of the subtraction from the one of the other. Hence this other part is that fraction of the unit magnitude whose numerator is this remainder and denominator the common denominator.

104. Widening the meaning of the terms and symbols of arts. 60, 61, after the manner of arts. 95, 102, gives the

Def. The operation of art. 103 is called the SUBTRACTION FROM A FRACTION OF A FRACTION; the former fraction is called the MINUEND, the latter the SUBTRA-

HEND, and the result the REMAINDER, EXCESS, or DIFFERENCE. If a, b, be any two numbers whole or fractional of which a is not less than b, $a-b$ stands for the remainder of the subtraction from a of b, if c be any number whole or fractional not greater than $a-b$ $a-b-c$ stands for the remainder of the subtraction from $a-b$ of c, and so on.

As this subtraction springs out of, and rests on, the subtraction of art. 59 all that is said of the latter in art. 60 touching its kinds holds of the former.

105. PROP. *The laws of operational equivalence that have to do only with additions and subtractions are the same for fractional as for whole numbers.*

Additions and subtractions of fractions are (arts. 101, 103) additions and subtractions of whole numbers of such equal submultiples of the unit magnitude as some common denominator determines. And what is true of whole numbers in general is true of whole numbers in particular of which the units are equal magnitudes. Therefore the propositions of arts. 28, 62, the laws of commutation,—of additions (arts. 29, 34), of subtractions (art. 67), and of additions and subtractions (art. 69)—, the laws of distribution of addition and of subtraction,—over additions (arts. 33, 66), and over either subtractions or additions and subtractions (arts. 71, 73)—, and the law of relativity in subtraction (art. 63), are all true no less for fractional numbers than for whole.

106. PROP. *If a magnitude be a given fraction of another magnitude and this other a given fraction of a third magnitude, to find what fraction the first magnitude is of the third.*

Let a magnitude be $\frac{m}{n}$ of another magnitude and this other $\frac{r}{s}$ of a third magnitude.

If u, v, be any two whole numbers, (art. 98) the first magnitude is $\frac{mu}{nu}$ of the second and the second $\frac{rv}{sv}$ of the third. If further u, v, be (art. 99) so chosen that $nu = rv$ and the common result be called x, the first magnitude is $\frac{mu}{x}$ of the second and the second $\frac{x}{sv}$ of the third. The second magnitude then is at once x times each of the mu submultiples that make up the first and x times each of the

sv submultiples that make up the third. Therefore (art. 90) each of the former submultiples is equal to each of the latter and therefore (art. 92) the first magnitude is $\frac{mu}{sv}$ of the third.

Although it will always do to take u r and v n (art. 54) so that the first magnitude is $\frac{mr}{sn}$ of the third yet u, v, may often be taken less numbers. Partly for this reason but chiefly to keep in clear view the nature of the operation the method here given is rather to be borne in mind than the result.

107. Finding what fraction the first of three magnitudes is of the third when the first is a given fraction of the second and the second a given fraction of the third is a general operation of which a particular kind is finding what multiple the first of three magnitudes is of the third when the first is a given multiple of the second and the second a given multiple of the third. But (art. 37) the latter operation is the multiplication by one whole number of another when the units are equal magnitudes. Here as in arts. 95, 102, 104, instead of making a new name for the general operation the old name of the particular operation is used in a generalized sense.

Def. The operation of finding what fraction of the unit magnitude a magnitude is which is a given fraction of a given fraction of the unit magnitude is called the MULTIPLICATION BY THE FORMER GIVEN FRACTION OF THE LATTER; the former given fraction is called the MULTIPLIER, the latter the MULTIPLICAND, and the result of the operation the PRODUCT. If $a, b, c, \ldots$ be any numbers whole or fractional ba, $b.a$, or $b \times a$, stands for the product made by multiplying a by b, cba, $c.b.a$, or $c \times b \times a$, for the product made by multiplying that product by c, and so on.

108. PROP. *The laws of operational equivalence that have to do with multiplications are the same for fractional as for whole numbers.*

The laws of multiplicational equivalence are given in arts. 39, 41, 52, 54, 75, 77, and by means only of propositions that relate to additions and subtractions are there proved from the following simplest cases of them:—

(1) $ab = ba$. (2) $abc = (ab)c$.

(3) $\begin{cases}(a+b)c = ac+bc. \\ c(a+b) = ca+cb.\end{cases}$ (4) $\begin{cases}(a-b)c = ac-bc. \\ c(a-b) = ca-cb.\end{cases}$

All but one of those propositions about additions and subtractions have (art. 105) been shown still true with the new meanings given to addition and subtraction. The one not shown true is that—The products made by multiplying by equal numbers equal numbers are equal. It is not shown because multiplication is no longer always successive additions as in art. 37. It may however be proved thus:—

Equal multipliers either are, or (art. 99) may be expressed as, fractions that have a common denominator and then (art. 100) their numerators are equal. Of equal multiplicands the equisubmultiples determined by the common denominator are (art. 90) equal and therefore (art. 88) of those equal equisubmultiples the equimultiples expressed by the equal numerators are equal, that is the products.

Hence if the above simplest cases of the laws be proved in the sense now given to them the further proof of the laws becomes the same as before but with the language understood in the wider meaning now borne by it. Writing then for *a*, *b*, *c*, symbols explicitly fractional

(1) $\frac{m}{n}\frac{r}{s} = \frac{mr}{sn}$ (art. 106) and $\therefore$ (art. 54) $= \frac{rm}{ns} = \frac{r}{s}\frac{m}{n}$ (art. 106).

(2) $\frac{m}{n}\frac{r}{s}\frac{u}{v} = \frac{m}{n}\frac{ru}{vs} = \frac{mru}{(vs)n}$ (art. 106) and $\therefore$ (art. 41)

$$= \frac{(mr)u}{vsn} = \frac{mr}{sn}\frac{u}{v} = \left(\frac{m}{n}\frac{r}{s}\right)\frac{u}{v}.$$

(3) $\left(\frac{m}{n}+\frac{r}{s}\right)\frac{u}{v} = \left(\frac{ms}{ns}+\frac{rn}{sn}\right)\frac{u}{v} = \frac{ms+rn}{ns}\frac{u}{v}$ (art. 101) and $\therefore$ (art. 106)

$$= \frac{(ms+rn)u}{vns}, \therefore \text{(art. 39)} = \frac{(ms)u+(rn)u}{vns},$$

$\therefore$ (art. 101) $= \frac{(ms)u}{vns}+\frac{(rn)u}{vsn}$ and $\therefore$ (art. 106) $= \frac{ms}{ns}\frac{u}{v}+\frac{rn}{sn}\frac{u}{v}$ or $\frac{m}{n}\frac{u}{v}+\frac{r}{s}\frac{u}{v}$.

Either in the same way as this (using art. 52 instead of art. 39) or from this by (1)

$$\frac{u}{v}\left(\frac{m}{n}+\frac{r}{s}\right) = \frac{u}{v}\frac{m}{n}+\frac{u}{v}\frac{r}{s}.$$

(4) $\frac{u}{v}\left(\frac{m}{n}-\frac{r}{s}\right) = \frac{u}{v}\frac{ms-rn}{ns}$ (art. 103) and $\therefore$ (art. 106) $= \frac{u(ms-rn)}{(ns)v}$,

$\therefore$ (art. 77) $= \frac{ums-urn}{(ns)v} = \frac{ums}{(ns)v}-\frac{urn}{(sn)v}$ (art. 103)

and $\therefore$ (art. 106) $= \frac{u}{v}\frac{ms}{ns} - \frac{u}{v}\frac{rn}{sn}$ or $\frac{u}{v}\frac{m}{n} - \frac{u}{v}\frac{r}{s}$.

Either in the same way as this (using art. 75 instead of art. 77) or from this by (1)

$$\left(\frac{m}{n} - \frac{r}{s}\right)\frac{u}{v} = \frac{m}{n}\frac{u}{v} - \frac{r}{s}\frac{u}{v}.$$

109. PROP. *If two magnitudes be given fractions of a third magnitude to find what fraction one of the two is of the other.*

Let one magnitude be $\frac{m}{n}$ and another $\frac{r}{s}$ of a third magnitude.

If u, v, be any two whole numbers the first magnitude is $\frac{mu}{nu}$ and the second $\frac{rv}{sv}$ of the third (art. 98). If further u, v, be taken so that $nu = sv$ and this common product be called x the first is $\frac{mu}{x}$ and the second $\frac{rv}{x}$ of the third. The first is then mu times and the second rv times $\frac{1}{x}$ of the third. Therefore (art. 92) the first is $\frac{mu}{rv}$ of the second.

By making u s and v n (art. 54) the first magnitude is $\frac{ms}{rn}$ of the second. But here as in art. 106 the method is rather to be kept in mind than the result.

Since the second magnitude is $\frac{r}{s}$ of the third magnitude the third (art. 92) is $\frac{s}{r}$ of the second. The first magnitude being then $\frac{m}{n}$ of the third and the third $\frac{s}{r}$ of the second the first (art. 107) is $\frac{m}{n}\frac{s}{r}$ of the second. This (art. 106) gives the same result as before.

110. The finding what fraction one of two magnitudes is of the other when each is a given fraction of a third magnitude is a general operation under which there falls as a particular the finding what multiple, if any, one of two magnitudes is of the other when each is a given multiple of a third magnitude. But this particular operation (art. 82) is the exact division of one whole number by another when the units are equal magnitudes. Generalizing then as before in arts. 95, 102, 104, 107, the name of the particular operation,

Def. The operation of finding what fraction one of two given fractions of the unit magnitude is of the other is called the DIVISION OF THE FORMER GIVEN FRACTION BY THE LATTER; the former given fraction is called the DIVIDEND, the latter the DIVISOR, and the result of the operation the QUOTIENT.

The division of a whole number m by a whole number n gives the quotient $\frac{m}{n}$. If m be exactly divisible by n this (art.98) is only another way of expressing the quotient found by art.81. But if m be not exactly divisible by n let m (art.81) be made up of q parts each equal to n and a remainder r less than n so that (arts. 31, 38) $m=qn+r$, then

$$\frac{m}{n} \text{ is } \frac{qn+r}{n} \text{ and } \therefore \text{ (art.101)} = \frac{qn}{n}+\frac{r}{n} \text{ or (art.98) } q+\frac{r}{n}.$$

Hence the division of art.81 unless exact must be held to be an incomplete operation inasmuch as it cannot take account in the quotient of the remainder.

111. *Def.* The quotient got by dividing the unit magnitude by any fraction of the unit magnitude is called the RECIPROCAL of that fraction.

Hence (art. 109 or more immediately art. 92 used as in the latter part of art. 109) the reciprocal of a fraction is a fraction of which the denominator is the numerator and the numerator the denominator of the fraction.

112. Since a magnitude of which the unit magnitude is a given fraction is (art. 111) the fraction of the unit magnitude expressed by the given fraction's reciprocal magnitudes of which the unit magnitude is the same given fraction are (art. 100) all equal to one another. As then in art. 92 a magnitude is expressed numerically in reference to another magnitude as unit by means of the multiples which they severally are of some one magnitude so more generally may a magnitude be expressed numerically in reference to another magnitude as unit by means of the fractions which they severally are of some one magnitude. And as in that art. a magnitude m times a magnitude is in reference to a unit magnitude n times the same magnitude called $\frac{m}{n}$ so by a widened use of language spoken and written there comes the

Def. A magnitude which is a certain fraction $\frac{m}{n}$ of a magnitude of which the magnitude taken for unit is a certain fraction $\frac{r}{s}$ is called a COMPLEX FRACTION of the unit magnitude and is written $\frac{\left(\frac{m}{n}\right)}{\left(\frac{r}{s}\right)}$; the former fraction $\frac{m}{n}$ is called the NUMERATOR and the latter $\frac{r}{s}$ the DENOMINATOR of the complex fraction, also the numerator and denominator of a complex fraction are called its TERMS.

A fraction with its terms whole numbers is called a SIMPLE fraction when distinction is needed.

Since (art. 110) the quotient of a division expresses what multiple or fraction the dividend is of the divisor, or in other words is the numerical expression of the dividend in reference to the divisor as unit, the quotient of the division of $\frac{m}{n}$ of a magnitude by $\frac{r}{s}$ of the same magnitude may be expressed by $\frac{\left(\frac{m}{n}\right)}{\left(\frac{r}{s}\right)}$.

The complex fraction $\frac{\left(\frac{m}{n}\right)}{\left(\frac{r}{s}\right)}$ may (art. 98) be expressed as $\frac{\left(\frac{mu}{nu}\right)}{\left(\frac{rv}{sv}\right)}$, u v being any whole numbers, or, making $nu = sv$ (art. 99) and calling each x, as $\frac{\left(\frac{mu}{x}\right)}{\left(\frac{rv}{x}\right)}$. This, being $\frac{mu}{x}$ of a magnitude of which the unit magnitude is $\frac{rv}{x}$, and therefore mu times a magnitude $\left(\frac{1}{x}\right.$ of the other$\left.\right)$ rv times which the unit magnitude is, is $\frac{mu}{rv}$ and therefore so also is the complex fraction.

Again $\frac{\left(\frac{m}{n}\right)}{\left(\frac{r}{s}\right)}$, being $\frac{m}{n}$ of a magnitude of which the unit magnitude is $\frac{r}{s}$ and therefore (art. 92) $\frac{m}{n}$ of a magnitude which is $\frac{s}{r}$ of the unit magnitude, is (art. 107) $\frac{m}{n}\frac{s}{r}$.

As a complex fraction is made after the shape of a simple fraction by taking one or each term a fraction instead of a whole number so may a more complex fraction be made by taking a complex fraction as term. And by taking for term a fraction of any complexity a fraction may be made of still greater complexity. By expressing in the way just shown as a simple fraction each fraction of the first degree of complexity that enters into a fraction of any degree, doing the same with the resulting fraction of lower degree, and so on, any fraction however complex may be expressed as a simple fraction.

113. PROP. *Products made by multiplying equal multiplicands are in the same order of greatness as the multipliers.*

For multipliers either are, or (arts. 99, 112) may be expressed as, simple fractions having a common denominator and then (art. 100) are in the same order of greatness as the numerators. Now of equal multiplicands the equisubmultiples determined by the common denominator are (art. 90) equal and therefore (arts. 88, 20, 8) of those equal equisubmultiples the multiples expressed by the numerators, that is the products, are in the same order of greatness as the numerators. Therefore the products are in the same order of greatness as the multipliers.

114. The product of a multiplication (art. 107) expresses numerically in reference to a unit magnitude a magnitude expressed numerically by the multiplier in reference to another magnitude as unit expressed numerically by the multiplicand in reference to the unit magnitude.

The quotient of a division (art. 110) expresses numerically one of two magnitudes in reference to the other as unit the former of which is expressed numerically by the dividend and the latter by the divisor each in reference to a common unit magnitude.

Now if *a*, *b*, be any two numbers whole or fractional (arts. 93, 112) the symbol $\frac{a}{b}$ stands for the numerical expression in reference to some understood unit magnitude of a magnitude expressed numerically by *a* in reference to a magnitude as unit in reference to which as unit the unit magnitude is expressed numerically by b. This symbol then is as in arts. 110, 112, an expression of the quotient of the division of a by b.

Also, making $\frac{a}{b}$ refer to a magnitude as unit expressed numerically by b in reference to the unit magnitude, in the symbolic expression $\frac{a}{b}b$ the numbers a, b, b, all refer to the same unit magnitude. For, as shown in arts. 92, 112, and there made the very hinge on which all fractional relationship simple or complex turns, magnitudes in reference to which as units the same magnitude is expressed by the same whole or fractional number are all equal to one another. Therefore $\frac{a}{b}b=a$. But by the nature of multiplication and division, if q be the quotient of the division of a by b, $qb=a$. Therefore $\frac{a}{b}b=qb$. Therefore (art. 113) q is equal to $\frac{a}{b}$, any number greater than q is greater than $\frac{a}{b}$, and any number less than q is less than $\frac{a}{b}$.

Hence any other symbol than $\frac{a}{b}$ for the quotient of the division of a by b is needless and hence the

Def. The symbol $\frac{a}{b}$, read "a BY b", stands for the quotient of the division of a by b and therefore for the number marked out by this alone that $\frac{a}{b}b=a$.

It is at once clear that $\frac{ab}{b}=a$; for multiplying b by each the products are the same.

115. PROP. *Quotients of divisions by equal divisors are in the same order of greatness as the dividends.*

For (art. 114) in divisions by equal divisors the dividends are equal severally to the products made by multiplying severally by the quotients the equal divisors and therefore (art. 113) are in the same order of greatness as the quotient multipliers.

116. PROP. *If the letters be any whole or fractional numbers,*

$$\frac{a}{b}=\frac{ac}{bc}.$$

$\because$ (art. 114) $\frac{a}{b}b=a$, multiply by each c and (art. 113) $\left(\frac{a}{b}b\right)c=ac$;

$\therefore$ (arts. 41, 108, 6) $\frac{a}{b}bc=ac$. But (art. 114) $\frac{ac}{bc}bc=ac$. $\therefore$ (art. 6)

$\frac{a}{b}bc=\frac{ac}{bc}bc$, and $\therefore$ (art. 113) $\frac{a}{b}=\frac{ac}{bc}$.

117. *Def.* This proposition of art. 116 is called the LAW OF RELATIVITY IN DIVISION.

The principle of subunit change of art. 98 is a particular case of the law.

118. PROP. *If the letters be any whole or fractional numbers,*

$$\frac{\left[\frac{\left[\frac{\left[\frac{\left(\frac{x}{a}\right)}{b}\right]}{c}\right]}{\vdots}\right]}{g}\right]}{h}=\frac{x}{hg\ldots cba}.$$

First $\frac{\left(\frac{a}{b}\right)}{c}=\frac{\frac{a}{b}b}{cb}$ (art. 116) and $\therefore$ (art. 114) $=\frac{a}{cb}$.

Hence $\frac{\left[\frac{\left[\frac{\left[\frac{\left[\frac{\left(\frac{x}{a}\right)}{b}\right]}{c}\right]}{d}\right]}{\vdots}\right]}{h}=\frac{\left[\frac{\left[\frac{\left[\frac{\left(\frac{x}{ba}\right)}{c}\right]}{d}\right]}{\vdots}\right]}{h}$, by the above and art. 115,

$=\frac{\left[\frac{\left[\frac{\left(\frac{x}{cba}\right)}{d}\right]}{\vdots}\right]}{h}$, by the same, $= - - - - =\frac{x}{hg\ldots cba}$.

119. *Def.* Art. 118's proposition is called the LAW OF THE COLLIGATION OF DIVISIONS.

120. PROP. *The last quotient got from dividing by numbers in succession a number and the successive quotients got is the same in whatever order the numbers are taken.*

Let any numbers $a, b, c, \ldots g, h$, be divisors in succession of a number x and the successive quotients got, and let those numbers be taken in any other order of succession say $c, h, a, \ldots b, g$. Then (arts. 118, 54, 108)

$$\frac{\left[\frac{\left[\frac{\left[\frac{\left[\frac{x}{a}\right]}{b}\right]}{c}\right]}{\vdots}\right]}{g}}{h}=\frac{x}{hg\ldots cba}=\frac{x}{gb\ldots ahc}=\frac{\left[\frac{\left[\frac{\left[\frac{\left[\frac{x}{c}\right]}{h}\right]}{a}\right]}{\vdots}\right]}{b}}{g}.$$

121. *Def.* Art. 120's proposition is called the LAW OF THE COMMUTATION OF DIVISIONS.

122. PROP. *Of successive multiplications and divisions when the number either first multiplied or first divided and all the numbers multiplied by are one set of numbers and all the numbers divided by are one set the result is the same whatever be the order of succession.*

First if a, b, c, be any whole or fractional numbers $\frac{ab}{c}=a\frac{b}{c}$.

For (art. 114) $\frac{ab}{c}=\frac{a\frac{b}{c}c}{c}$ $\therefore$ (arts. 41, 108) $=\frac{\left(a\frac{b}{c}\right)c}{c}$ and $\therefore$ (art. 114) $=a\frac{b}{c}$.

Hence secondly if the letters be any whole or fractional numbers

$$hgf\ldots dcb\frac{\left[\frac{\left[\frac{\left[\frac{\left[\frac{a}{m}\right]}{n}\right]}{\vdots}\right]}{n}\right]}{v}}{w}=hgf\ldots dc\frac{b\frac{\left[\frac{\left[\frac{\left[\frac{a}{m}\right]}{n}\right]}{\vdots}\right]}{n}}{v}}{w}$$

$$=hg\ldots dc\frac{\left[b\frac{\left[\frac{\left[\frac{a}{m}\right]}{n}\right]}{\vdots}\right]}{n}\right]}{v}}{w}=\cdots=hgf\ldots dc\frac{\left[\frac{\left[\frac{\left[\frac{\left[\frac{ba}{m}\right]}{n}\right]}{\vdots}\right]}{n}\right]}{v}}{w};$$

$$\text{so also} \therefore = hgf\ldots d\,\frac{\left[\frac{\left[\frac{\left[\frac{\left[\frac{cba}{m}\right]}{n}\right]}{\vdots}\right]}{v}\right]}{w} = \text{- - - -} = \frac{\left[\frac{\left[\frac{\left[\frac{\left[\frac{hgf\ldots cba}{m}\right]}{n}\right]}{\vdots}\right]}{v}\right]}{w}.$$

Hence thirdly and lastly taking as a result of successive multiplications and divisions

$$\ldots\frac{\frac{a_7\frac{\left[a_6a_5a_4\frac{\left[\frac{\left[\frac{\left[\frac{a_3a_2a_1}{v_1}\right]}{v_2}\right]}{v_3}\right]}{v_4}\right]}{v_5}}{v_6}}{\substack{v_7\\ \vdots}}, \text{ this } = \ldots \frac{\frac{a_7\frac{\left[\frac{\left[\frac{\left[\frac{\left[\frac{\left[\frac{a_6a_5a_4a_3a_2a_1}{v_1}\right]}{v_2}\right]}{v_3}\right]}{v_4}\right]}{v_5}\right]}{v_6}}{\substack{v_7\\ \vdots}}$$

$$= \ldots \frac{\left[\frac{\left[\frac{\left[\frac{\left[\frac{\left[\frac{a_7a_6a_5a_4a_3a_2a_1}{v_1}\right]}{v_2}\right]}{v_3}\right]}{v_4}\right]}{v_5}\right]}{v_6}\right]}{\substack{v_7\\ \vdots}} = \text{- - - -} = \frac{\left[\frac{\left[\frac{\ldots a_3a_2a_1}{v_1}\right]}{v_2}\right]}{\substack{v_3\\ \vdots}}.$$

But taking the operations in any other order so that the number either first multiplied or first divided and all the numbers multiplied by are a_1, a_2, a_3, … and all the numbers divided by are v_1, v_2, v_3, … as in

$$\ldots\frac{\left[a_5a_2\frac{\left[a_3a_6a_1\frac{\left[\frac{\left[\frac{a_4a_7}{v_6}\right]}{v_1}\right]}{v_7}\right]}{\substack{v_2\\ v_5}}\right]}{\frac{v_4}{\substack{v_3\\ \vdots}}}; \text{ this as before } = \frac{\left[\frac{\left[\frac{\left[\frac{\ldots a_6a_1a_4a_7}{v_6}\right]}{v_1}\right]}{v_7}\right]}{\substack{v_2\\ \vdots}},$$

$$\text{which (arts. 54, 108, 120)} = \frac{\left[\frac{\left[\frac{\ldots a_3a_2a_1}{v_1}\right]}{v_2}\right]}{\substack{v_3\\ \vdots}}.$$

Therefore the different orders of succession of the operations give the same result.

123. *Def.* Art. 122's proposition is called the LAW OF THE COMMUTATION OF MULTIPLICATIONS AND DIVISIONS.

124. PROP. *If the letters be any numbers whole or fractional,*

$$\frac{ab}{c} = \frac{a}{c}b.$$

For (art. 114) $\frac{ab}{c} = \frac{\left(\frac{a}{c}c\right)b}{c}$ ∴ (arts. 41, 108) $= \frac{\frac{a}{c}cb}{c}$ ∴ (arts. 54, 108) $= \frac{\frac{a}{c}bc}{c}$ ∴ (arts. 41, 108) $= \frac{\left(\frac{a}{c}b\right)c}{c}$ ∴ (art. 114) $= \frac{a}{c}b.$

125. *Def.* Art. 124's proposition is called the LAW OF THE COLLIGATION OF MULTIPLICATION OF AND DIVISION.

126. PROP. *To find an expression with a single quotient symbol operationally equivalent to a product of several explicitly symbolized quotients.*

First $\frac{a}{a'}\frac{b}{b'} = \frac{au}{a'u}\frac{bv}{b'v}$ (art. 116) if u v be any numbers, $= \frac{au}{x}\frac{x}{b'v}$ if u v be taken so that $a'u = bv$ and the common result be called x. This can be done (arts. 54, 108) by making u b and v a'; but it may often be done in a simpler way. Hence (art. 122) $\frac{a}{a'}\frac{b}{b'} = \frac{\frac{au}{x}x}{b'v}$ and ∴ (art. 114) $= \frac{au}{b'v}$. This is a generalization of art. 106's proposition and in particular yields $\frac{a}{a'}\frac{b}{b'} = \frac{ab}{b'a'}$.

Hence $\frac{a}{a'}\frac{b}{b'}\frac{c}{c'}\cdots\frac{f}{f'}\frac{g}{g'}\frac{h}{h'} = \frac{a}{a'}\cdots\frac{f}{f'}\frac{gh}{h'g'} = \frac{a}{a'}\cdots\frac{e}{e'}\frac{fgh}{(h'g')f'}$ and ∴ (arts. 41, 108) $= \frac{a}{a'}\cdots\frac{e}{e'}\frac{fgh}{h'g'f'} = \text{-}\ \text{-}\ \text{-}$ and at length $= \frac{abc\ldots fgh}{h'g'f'\ldots c'b'a'}$.

127. PROP. *If the letters be any numbers whole or fractional,*

$$\frac{ab}{c} = \frac{a}{\left(\frac{c}{b}\right)}.$$

For (art. 114) $\frac{ab}{c} = \frac{ab}{\frac{c}{b}b}$ and ∴ (art. 116) $= \frac{a}{\left(\frac{c}{b}\right)}$.

128. *Def.* The proposition of art. 127 is called the LAW OF THE COLLIGATION OF MULTIPLICATION BY AND DIVISION.

129. PROP. *To find an expression with a single quotient symbol operationally equivalent to the quotient of the division of one explicitly symbolized quotient by another.*

First $\dfrac{\left(\dfrac{a}{a'}\right)}{\left(\dfrac{b}{b'}\right)} = \dfrac{\left(\dfrac{au}{a'u}\right)}{\left(\dfrac{bv}{b'v}\right)}$ (art. 116) if u v be any numbers and $= \dfrac{\left(\dfrac{au}{x}\right)}{\left(\dfrac{bv}{x}\right)}$

if $a'u = b'v$ and each be called x, ∴ (art. 116) $= \dfrac{\dfrac{au}{x}x}{\dfrac{bv}{x}x}$ or (art. 114) $\dfrac{au}{bv}$.

Since (arts. 54, 108) $a'b' = b'a'$ u may be taken b' and v a'; still $a'u$ may be often more simply made equal to $b'v$. Hence $\dfrac{\left(\dfrac{a}{a'}\right)}{\left(\dfrac{b}{b'}\right)} = \dfrac{ab'}{ba'}$.

Hence further $\dfrac{\left\{\dfrac{\left(\dfrac{a}{a'}\right)}{\left(\dfrac{b}{b'}\right)}\right\}}{\left(\dfrac{c}{c'}\right)} = \dfrac{\left(\dfrac{ab'}{ba'}\right)}{\left(\dfrac{c}{c'}\right)} = \dfrac{(ab')c'}{cba'}$ and ∴ (arts. 41, 108) $= \dfrac{ab'c'}{cba'}$.

Likewise $\dfrac{\left[\dfrac{\left\{\dfrac{\left(\dfrac{a}{a'}\right)}{\left(\dfrac{b}{b'}\right)}\right\}}{\left(\dfrac{c}{c'}\right)}\right]}{\left(\dfrac{d}{d'}\right)} = \dfrac{ab'c'd'}{dcba'}$. And so on.

Also $\dfrac{\left(\dfrac{a}{a'}\right)}{\left\{\dfrac{\left(\dfrac{b}{b'}\right)}{\left(\dfrac{c}{c'}\right)}\right\}} = \dfrac{\left(\dfrac{a}{a'}\right)}{\left(\dfrac{bc'}{cb'}\right)} = \dfrac{acb'}{bc'a'}$, $\dfrac{\left(\dfrac{a}{a'}\right)}{\left[\dfrac{\left(\dfrac{b}{b'}\right)}{\left\{\dfrac{\left(\dfrac{c}{c'}\right)}{\left(\dfrac{d}{d'}\right)}\right\}}\right]} = \dfrac{\left(\dfrac{a}{a'}\right)}{\left(\dfrac{bdc'}{cd'b'}\right)} = \dfrac{acd'b'}{bdc'a'}$, and the like.

130. PROP. *If the letters be any whole or fractional numbers that give meaning to the statement,*

$$\frac{a_1+a_2+\cdots+a_\alpha-m_1-m_2-\cdots-m_\mu+b_1+\cdots+b_\beta-n_1-\cdots-n_\nu+\cdots+k_1+\cdots+k_\kappa-t_1-\cdots-t_\tau}{x}$$

$$=\frac{a_1}{x}+\frac{a_2}{x}+\cdots+\frac{a_\alpha}{x}-\frac{m_1}{x}-\frac{m_2}{x}-\cdots-\frac{m_\mu}{x}+\frac{b_1}{x}+\cdots+\frac{b_\beta}{x}-\frac{n_1}{x}-\cdots+\frac{k_\kappa}{x}-\frac{t_1}{x}-\cdots-\frac{t_\tau}{x}.$$

For (art. 114) $\frac{a_1+\cdots+a_\alpha-m_1-\cdots-m_\mu+b_1+\cdots\cdots-t_\tau}{x}$

$$=\frac{\frac{a_1}{x}x+\cdots+\frac{a_\alpha}{x}x-\frac{m_1}{x}x-\cdots-\frac{m_\mu}{x}x+\cdots\cdots-\frac{t_\tau}{x}x}{x}$$

$$\therefore \text{(arts. 75, 108)} = \frac{\left(\frac{a_1}{x}+\cdots+\frac{a_\alpha}{x}-\frac{m_1}{x}-\cdots-\frac{m_\mu}{x}+\cdots\cdots-\frac{t_\tau}{x}\right)x}{x}$$

$$\text{and } \therefore \text{(art. 114)} = \frac{a_1}{x}+\cdots+\frac{a_\alpha}{x}-\frac{m_1}{x}-\cdots-\frac{m_\mu}{x}+\cdots\cdots-\frac{t_\tau}{x}.$$

131. *Def.* Art. 130's proposition is called the LAW OF THE DISTRIBUTION OF DIVISION OVER ADDITIONS AND SUBTRACTIONS.

132. PROP. *To find an expression with a single quotient symbol operationally equivalent to the result of successive additions and subtractions of explicitly symbolized quotients.*

Let $\frac{a'}{a}+\frac{b'}{b}+\frac{c'}{c}-\frac{m'}{m}-\frac{n'}{n}-\frac{p'}{p}+\frac{d'}{d}+\frac{e'}{e}-\frac{q'}{q}$ be taken for instance of a result of successive additions and subtractions of quotients. If $\alpha, \beta, \gamma, \mu, \nu, \varpi, \delta, \epsilon, \chi$, be any numbers whole or fractional that result (art. 116)

$$=\frac{a'\alpha}{a\alpha}+\frac{b'\beta}{b\beta}+\frac{c'\gamma}{c\gamma}-\frac{m'\mu}{m\mu}-\frac{n'\nu}{n\nu}-\frac{p'\varpi}{p\varpi}+\frac{d'\delta}{d\delta}+\frac{e'\epsilon}{e\epsilon}-\frac{q'\chi}{q\chi}.$$

And if $\alpha, \beta, \ldots \chi$, can be so taken that $a\alpha=b\beta=\cdots=q\chi$ and each of these be called x this (art. 130)

$$=\frac{a'\alpha+b'\beta+c'\gamma-m'\mu-n'\nu-p'\varpi+d'\delta+e'\epsilon-q'\chi}{x}.$$

By taking α the product of all but a of the numbers $a, b, c, m, n, p, d, e, q$, β the product of all of them but b, - - - - χ the product of all of them but q, (arts. 54, 108) $a\alpha=b\beta=\cdots=q\chi$; and the same may often be brought about by taking $\alpha, \beta, \ldots \chi$, simpler products than these.

133. POST. *Let it be granted that of any given magnitude however small a multiple may be taken greater than any given magnitude of the same kind however great.*

134. PROP. *To explain the decimal notation of fractional numbers.*

Any multiple of a magnitude is denoted decimally either by a digit or by a row of digits as in art. 47 when the magnitude is made the unit. The single digit or the last digit in a row of digits denoting a multiple of the unit magnitude being the digit which refers only to the unit magnitude is called the UNIT'S DIGIT and the numbers to which as units the other digits in backward order of a row severally refer are spoken of not only as groups of units of the first, second, third, - - - ranks severally but also as multiples of the unit magnitude of those several ranks. Now all that is needed further for decimally denoting fractional numbers is to make digits stretch away in a row to the right of the unit's digit no less than to the left and to understand universally that that to which as unit any digit refers is ten times that to which as unit the next digit to the right refers or the same thing is one-tenth of that to which as unit the next digit to the left refers.

But the unit's digit being no longer the last digit in the row must have some other mark set on it and to this end a dot, called the DECIMAL DOT or POINT, is put midway between it and the next digit to the right, or when stroke bounded columns are used the column with the unit's digit is separated from the next column to the right by a double stroke. Hence if t be *ten* the numbers or multiples of the first, second, third, - - - ranks to which as units the digits to the left of the unit's digit in backward order severally refer are t, t^2, t^3,... and are therefore severally denoted by 10, 100, 1000,... ; the numbers or submultiples of the unit magnitude of the first, second, third, - - - ranks to which as units the digits to the right of the unit's digit in order severally refer are $\frac{1}{t}$, $\left(\frac{1}{t}\right)^2$, $\left(\frac{1}{t}\right)^3$,... and are therefore severally denoted by 0·1, 0·01, 0·001,...; each digit denotes the product made by multiplying by the number which it of itself expresses the number to which as unit it refers; and a row of digits denotes the final sum got by adding successively the numbers denoted by all but the first of the digits in order to the number denoted by the first and to the successive sums got. For example 837·2546 denotes $8t^2+3t+7+2\frac{1}{t}+5\left(\frac{1}{t}\right)^2+4\left(\frac{1}{t}\right)^3+6\left(\frac{1}{t}\right)^4$. The digit next to the unit's digit on the right is said to be IN THE FIRST PLACE OF DECIMALS, the digit next in order to this on the right IN THE SECOND, and so on for the rest.

A fraction of a magnitude (art. 94) is either a multiple of the magnitude, a proper fraction of the magnitude, or made up of a multiple and a proper fraction. Again a proper fraction of the unit magnitude (art. 109) is some fraction of $\frac{1}{t}$ (*one-tenth*) of the unit and therefore (art. 94) is either a multiple of $\frac{1}{t}$, a proper fraction of $\frac{1}{t}$, or made up of a multiple of $\frac{1}{t}$ and a proper fraction of $\frac{1}{t}$; the multiple too of $\frac{1}{t}$ must in each case be less than t since the fraction of the unit magnitude is proper. In like manner a proper fraction of $\frac{1}{t}$ of the unit is some fraction of $\left(\frac{1}{t}\right)^2$ (*one-tenth of one-tenth*) of the unit and therefore as before is either a multiple less than t of $\left(\frac{1}{t}\right)^2$, a proper fraction of $\left(\frac{1}{t}\right)^2$, or made up of a multiple less than t of $\left(\frac{1}{t}\right)^2$ and a proper fraction of $\left(\frac{1}{t}\right)^2$. And so on for the submultiples of higher ranks $\left(\frac{1}{t}\right)^3$, $\left(\frac{1}{t}\right)^4$, ... $\left(\frac{1}{t}\right)^i$, the last of these being the submultiple of the unit magnitude of any rank whatever the ith. Wherefore a fraction is either a multiple of 1, made up of one or more multiples each less than t of $\frac{1}{t}$, $\left(\frac{1}{t}\right)^2$, $\left(\frac{1}{t}\right)^3$, ... $\left(\frac{1}{t}\right)^i$, a proper fraction of $\left(\frac{1}{t}\right)^i$, or made up of several of these. But any multiple of 1 (art. 47) and any multiple less than t of any of the submultiples $\frac{1}{t}$, $\left(\frac{1}{t}\right)^2$, $\left(\frac{1}{t}\right)^3$, ... $\left(\frac{1}{t}\right)^i$, may be denoted. When therefore any fraction of the unit is given there may be denoted decimally either this fraction itself or a fraction differing from it by a proper fraction of a submultiple of the unit of any assigned rank.

Moreover of any given magnitude however small of the same kind as the unit magnitude a multiple may (art. 133) be taken greater than the unit magnitude, and as shown in art. 47 by taking i great enough t^i may be made greater than the number expressing what multiple is so taken. That submultiple therefore (arts. 91, 92) of the unit of which the unit is the multiple then expressed by t^i is less than the given magnitude and therefore (art. 9) much more so is a proper fraction of that submultiple. But a submultiple of the unit of any rank being t times the submultiple of the next higher rank and the unit t times the submultiple of the first rank the unit is t^i times the submultiple $\left(\frac{1}{t}\right)^i$. Hence if a fraction cannot be decimally denoted another fraction at least can be differing therefrom by less than any given magnitude however small.

135. PROP. *In the widest meaning yet given to the symbols,*

$$\left.\begin{matrix}\left(\frac{1}{a}\right)^n a^m \\ a^m\left(\frac{1}{a}\right)^n\end{matrix}\right\} = \left\{\begin{matrix} a^{m-n} \text{ if } m \text{ be not less than } n, \\ \left(\frac{1}{a}\right)^{n-m} \text{ if } m \text{ be not greater than } n.\end{matrix}\right.$$

The word *power* must (art. 43) have a meaning answering to the meaning of *multiplication*. But the index laws $a^n a^m = a^{m+n}$, $(a^n)^m = a^{mn}$, of arts. 44, 45, are still true now that a may be a fractional number since the proofs of these laws (art. 108) still hold.

$\left(\frac{1}{a}\right)^n a^m$ is (art. 43)

$$\underbrace{\left(\frac{1}{a}\frac{1}{a}\cdots\frac{1}{a}\frac{1}{a}\right)}_{n\ \frac{1}{a}\text{s}}\underbrace{aa\ldots\ldots aa}_{m\ a\text{s}} \text{ and } \therefore \text{ (arts. 41, 108)} = \underbrace{\frac{1}{a}\frac{1}{a}\cdots\frac{1}{a}\frac{1}{a}}_{n\ \frac{1}{a}\text{s}}\underbrace{aa\ldots\ldots aa}_{m\ a\text{s}}$$

$$\therefore \text{ if } m \text{ be greater than } n = \underbrace{\frac{1}{a}\frac{1}{a}\cdots\frac{1}{a}\frac{1}{a}}_{n\ \frac{1}{a}\text{s}}\underbrace{aa\ldots\ldots aa}_{n\ a\text{s}}\underbrace{aa\ldots\ldots\ldots aa}_{(m-n)\ a\text{s}}$$

$$\therefore \text{ (arts. 41, 108)} = {}^{n}\!\left(\frac{1}{a}\,{}^{n-1}\!\left(\frac{1}{a}\cdots{}^{2}\!\left(\frac{1}{a}\,{}^{1}\!\left(\frac{1}{a}a\right)_1 a\right)_2\cdots a\right)_{n-1} a\right)_n a^{m-n},$$

the ith pair of brackets being marked ${}^i(\ )_i$;

$$\text{and if } m \text{ be less than } n = \underbrace{\frac{1}{a}\frac{1}{a}\cdots\cdots\frac{1}{a}\frac{1}{a}}_{(n-m)\ \frac{1}{a}\text{s}}\underbrace{\frac{1}{a}\frac{1}{a}\cdots\frac{1}{a}\frac{1}{a}}_{m\ \frac{1}{a}\text{s}}\underbrace{aa\ldots aa}_{m\ a\text{s}}$$

$$= \underbrace{\frac{1}{a}\frac{1}{a}\cdots\cdots\frac{1}{a}\frac{1}{a}\frac{1}{a}}_{(n-m)\ \frac{1}{a}\text{s}}\,{}^{m-1}\!\left(\frac{1}{a}\cdots{}^{2}\!\left(\frac{1}{a}\,{}^{1}\!\left(\frac{1}{a}a\right)_1 a\right)_2\cdots a\right)_{m-1} a.$$

But (art. 114) $\frac{1}{a}a = 1$ and (arts. 37, 106) $1 \times a$ is a. Therefore the former result is a^{m-n} and the latter $\left(\frac{1}{a}\right)^{n-m}$.

$a^m\left(\frac{1}{a}\right)^n$ being (arts. 54, 108) $= \left(\frac{1}{a}\right)^n a^m$ is therefore also operationally equivalent to one or other of these. Yet this may be shown by simpler laws as follows:—

$$a^m\left(\frac{1}{a}\right)^n \text{ is } \underbrace{(aa\ldots\ldots aa)}_{m \text{ as}}\underbrace{\frac{1}{a}\frac{1}{a}\cdots\cdots\frac{1}{a}\frac{1}{a}}_{n\ \frac{1}{a}\text{s}} \text{ and } \therefore = \underbrace{aa\ldots\ldots aa}_{m \text{ as}}\underbrace{\frac{1}{a}\frac{1}{a}\cdots\cdots\frac{1}{a}\frac{1}{a}}_{n\ \frac{1}{a}\text{s}}$$

$$\therefore \text{ if } m \text{ be greater than } n = \underbrace{aa\ldots\ldots\ldots aa}_{(m-n) \text{ as}}\underbrace{aa\ldots\ldots aa}_{n \text{ as}}\underbrace{\frac{1}{a}\frac{1}{a}\cdots\cdots\frac{1}{a}\frac{1}{a}}_{n\ \frac{1}{a}\text{s}}$$

$$\text{and } \therefore = \underbrace{aa\ldots\ldots\ldots aa}_{(m-n) \text{ as}}\,{}^{n-1}\!\left(a\ldots{}^{2}\!\left(a\,{}^{1}\!\left(a\frac{1}{a}\right)_1\frac{1}{a}\right)_2\cdots\frac{1}{a}\right)_{n-1}\frac{1}{a},$$

$$\text{but if } m \text{ be less than } n = \underbrace{aa\ldots aa}_{m \text{ as}}\underbrace{\frac{1}{a}\frac{1}{a}\cdots\cdots\frac{1}{a}\frac{1}{a}}_{m\ \frac{1}{a}\text{s}}\underbrace{\frac{1}{a}\frac{1}{a}\cdots\cdots\cdots\frac{1}{a}\frac{1}{a}}_{(n-m)\ \frac{1}{a}\text{s}}$$

$$\text{and } \therefore = {}^{m}\!\left(a\,{}^{m-1}\!\left(a\ldots{}^{2}\!\left(a\,{}^{1}\!\left(a\frac{1}{a}\right)_1\frac{1}{a}\right)_2\cdots\frac{1}{a}\right)_{m-1}\frac{1}{a}\right)_m\left(\frac{1}{a}\right)^{n-m}.$$

Now because the magnitude expressed by $a\frac{1}{a}$ in reference to the unit magnitude is (art. 107) expressed by a in reference to a magnitude as unit which is expressed by $\frac{1}{a}$ in reference to the unit magnitude and in reference to which as unit therefore (art. 114) the unit magnitude is expressed by a, $a\frac{1}{a} = 1$. Therefore the result in the former case is a^{m-n} and in the latter $\left(\frac{1}{a}\right)^{n-m}$.

When m is greater than n the more immediate operational equivalences (arts. 60, 61) are $\left(\frac{1}{a}\right)^n a^m = a^{m \llcorner n}$, $a^m\left(\frac{1}{a}\right)^n = a^{m \urcorner n}$, and when m is less than n $\left(\frac{1}{a}\right)^n a^m = \left(\frac{1}{a}\right)^{n \urcorner m}$, $a^m\left(\frac{1}{a}\right)^n = \left(\frac{1}{a}\right)^{n \llcorner m}$.

The case when $m = n$ is brought under the law by giving to a^0 or $\left(\frac{1}{a}\right)^0$ the meaning 1 as in art. 57.

136. *The* PRINCIPLE OF DIGIT KNITTING *holds in the decimal notation of fractions.*

For taking an instance and putting t for 10, 7208310·5940062 is

$7t^6+2t^5+8t^3+3t^2+t+5\frac{1}{t}+9\left(\frac{1}{t}\right)^2+4\left(\frac{1}{t}\right)^3+6\left(\frac{1}{t}\right)^6+2\left(\frac{1}{t}\right)^7$ and ∴ (arts. 33, 105)

$$=7t^6+2t^5+8t^3+\left\{3t^2+t+5\frac{1}{t}+9\left(\frac{1}{t}\right)^2+4\left(\frac{1}{t}\right)^3\right\}+\left\{6\left(\frac{1}{t}\right)^6+2\left(\frac{1}{t}\right)^7\right\}$$

or 7208000+310·594+0·0000062.

Further (arts. 135, 41, 108) $9\left(\frac{1}{t}\right)^2=9t\left(\frac{1}{t}\right)^3=(9t)\left(\frac{1}{t}\right)^3$, $5\frac{1}{t}=5t^2\left(\frac{1}{t}\right)^3$

$=(5t^2)\left(\frac{1}{t}\right)^3$, $t=t^4\left(\frac{1}{t}\right)^3$, $3t^2=3t^5\left(\frac{1}{t}\right)^3=(3t^5)\left(\frac{1}{t}\right)^3$,

$\therefore 3t^2+t+5\frac{1}{t}+9\left(\frac{1}{t}\right)^2+4\left(\frac{1}{t}\right)^3=(3t^5)\left(\frac{1}{t}\right)^3+t^4\left(\frac{1}{t}\right)^3+(5t^2)\left(\frac{1}{t}\right)^3+(9t)\left(\frac{1}{t}\right)^3+4\left(\frac{1}{t}\right)^3$

and ∴ (arts. 39, 108) $=(3t^5+t^4+5t^2+9t+4)\left(\frac{1}{t}\right)^3$.

So $6\left(\frac{1}{t}\right)^6+2\left(\frac{1}{t}\right)^7=(6t^2+2t)\left(\frac{1}{t}\right)^8$, and (arts. 44, 41) $7t^6+2t^5+8t^3$

$=(7t^3)t^3+(2t^2)t^3+8t^3$ and ∴ (art. 39) $=(7t^3+2t^2+8)t^3$.

∴ 7208310·5940062

$$=(7t^3+2t^2+8)t^3+(3t^5+t^4+5t^2+9t+4)\left(\frac{1}{t}\right)^3+(6t^2+2t)\left(\frac{1}{t}\right)^8,$$

that is 7208×1000+310594×0·001+620×0·00000001.

Hence the number denoted by a row of digits is equivalent operationally (1) to the sum of the numbers denoted by any knots of consecutive digits into which the row may be cut,—meaning by *the number denoted by a knot* the sum of the numbers severally denoted by the digits of the knot—, and (2) to the sum of the products made by multiplying by the whole number which each knot were it a separate row with its last digit the unit's digit would denote the number to which as unit its last digit refers.

137. A decimally denoted fractional number is usually named by first naming as in art. 49 the number denoted by all the digits to the left of the decimal dot and then naming in order the numbers primarily expressed by the several digits to the right. Thus 4012·53006 is named *four thousand and twelve, dot, five three naught naught six.*

But by help of the principle of digit knitting a fractional number decimally denoted may also be named in the same way as a whole one. First taking the unit's digit for the first digit in a knot of digits and any digit to the right for the last it is found what multiple the unit magnitude is of the submultiple referred to as unit by the latter digit and hence what submultiple of the unit magnitude this submultiple is. Then taking any digit to the right of the unit's digit for the last digit in another knot of digits it is found what multiple the number denoted by the knot is of the submultiple referred to as unit by the digit. Thus 1 = 10×0·1, 1 = 100×0·01, 1 = 1000×0·001, - - - and hence 0·1 is *one-tenth*, 0·01 is *one hundredth*, 0·001 is *one thousandth*, - - -; 4012·53006 may be named *four thousand and twelve, and fifty-three thousand and six hundredthousandths*, or *forty hundred, one thousand two hundred and fifty-three hundredths, and sixty millionths* or in yet other ways; telling off the digits by sixes to the right of the decimal dot as well as to the left 90,182007·300064,005508,610000 may be read *ninety millions, one hundred and eighty-two thousand and seven, three hundred thousand and sixty-four millionths, five thousand five hundred and eight billionths, six hundred and ten thousand trillionths.*

Hence too may a decimally denoted fraction be expressed in the primary, necessary, general, common, or universal, form of a fraction by taking for numerator and denominator the numbers found as above that express what multiples the fraction and the unit are severally of the submultiple referred to as unit by the last digit in the row denoting the fraction. Thus 80·637 may be expressed as $\frac{80637}{1000}$ or $\frac{806370}{10000}$ or $\frac{8063700}{100000}$ or - - - .

138. PROP. *To find the decimally denoted results of operations with decimally denoted fractional numbers.*

The notational processes for finding results of operations with decimally denoted fractional numbers are simply those extensions of the processes of arts. 50, 57, 79, 83, that naturally spring from the extensions of the operations. And as the laws of operational equivalence on which these processes rest are (arts. 105, 108) the same for the extended as for the unextended operations the extended processes are so far the same as the unextended. The use of the index law of art. 135 is indeed all they differ in.

In the ADDITIVE PROCESS the first two cases of art. 50 stand just as before, the third case undergoes the slight change shown in the instance following, and then (arts. 102, 105) the fourth case still holds in the wide sense now given to the language. Taking t for 10,

59·63+0·95048+2000·7002+0·861+6·7+372+0·0705 is

$$5t+9+6\frac{1}{t}+3\left(\frac{1}{t}\right)^2$$
$$+\left\{9\frac{1}{t}+5\left(\frac{1}{t}\right)^2+4\left(\frac{1}{t}\right)^4+8\left(\frac{1}{t}\right)^5\right\}$$
$$+\left\{2t^3+7\frac{1}{t}+2\left(\frac{1}{t}\right)^4\right\}$$
$$+\left\{8\frac{1}{t}+6\left(\frac{1}{t}\right)^2+\left(\frac{1}{t}\right)^3\right\}$$
$$+\left(6+7\frac{1}{t}\right)+(3t+7t+2)+\left\{7\left(\frac{1}{t}\right)^2+5\left(\frac{1}{t}\right)^4\right\}$$

	59·63	
	0·950	48
2	000·700	2
	0·861	
	6·7	
	372	
	0·070	5
2	440·912	18

		5	9	6	3			
				9	5		4	8
2				7			2	
				8	6	1		
			6	7				
	3	7	2					
					7		5	
2	4	4		9	1	2	1	8

and ∴ as in art. 50 (arts. 33, 34, 105)

$$=2t^3+3t^2+5t+7t+9+6+2+6\frac{1}{t}+9\frac{1}{t}+7\frac{1}{t}+8\frac{1}{t}+7\frac{1}{t}$$
$$+3\left(\frac{1}{t}\right)^2+5\left(\frac{1}{t}\right)^2+6\left(\frac{1}{t}\right)^2+7\left(\frac{1}{t}\right)^2+\left(\frac{1}{t}\right)^3+4\left(\frac{1}{t}\right)^4+2\left(\frac{1}{t}\right)^4+5\left(\frac{1}{t}\right)^4+8\left(\frac{1}{t}\right)^5$$

∴ (arts. 33, 105, 39, 108) $\quad =2t^3+\cdots+\left(\frac{1}{t}\right)^3+(4+2+5)\left(\frac{1}{t}\right)^4+8\left(\frac{1}{t}\right)^5$

∴ (arts. 108, 105), ∵ (art. 50) $4+2+5$ is $t+1$,

$$=2t^3+\cdots+\left(\frac{1}{t}\right)^3+t\left(\frac{1}{t}\right)^4+\left(\frac{1}{t}\right)^4+8\left(\frac{1}{t}\right)^5$$

∴ likewise, ∵ (art. 135) $t\left(\frac{1}{t}\right)^4=\left(\frac{1}{t}\right)^3$,

$$=2t^3+\cdots+7\frac{1}{t}+(3+5+6+7)\left(\frac{1}{t}\right)^2+\left\{\left(\frac{1}{t}\right)^3+\left(\frac{1}{t}\right)^3\right\}+\left(\frac{1}{t}\right)^4+8\left(\frac{1}{t}\right)^5$$

∴ again, ∵ $3+5+6+7$ is $2t+1$ and $(2t)\left(\frac{1}{t}\right)^2=2t\left(\frac{1}{t}\right)^2=2\frac{1}{t}$,

$$=2t^3+\cdots+2+(6+9+7+8+7+2)\frac{1}{t}+\left(\frac{1}{t}\right)^2+2\left(\frac{1}{t}\right)^3+\left(\frac{1}{t}\right)^4+8\left(\frac{1}{t}\right)^5$$

and ∴, as $(6+9+7+8+7+2)\frac{1}{t}$ is $(3t+9)\frac{1}{t}=(3t)\frac{1}{t}+9\frac{1}{t}=3t\frac{1}{t}+9\frac{1}{t}=3+9\frac{1}{t}$,

$$=2t^3+3t^2+5t+7t+(9+6+2+3)+9\frac{1}{t}+\left(\frac{1}{t}\right)^2+2\left(\frac{1}{t}\right)^3+\left(\frac{1}{t}\right)^4+8\left(\frac{1}{t}\right)^5$$

and ∴ in the same way as in art. 50 = 2440·91218.

Here as in art. 50 when the result barely is sought the rows denoting the numbers may be written in order with each after the first close under the one before it and with digits that refer to multiples or submultiples of the same rank as units in the same column.

The SUBTRACTIVE PROCESS is the same as the process of art. 79 in all but the last case where the change must be wrought shown in the instance below.

5370·940381−4675·600383006 is

$$5t^3+3t^2+7t+9\frac{1}{t}+4\left(\frac{1}{t}\right)^2+3\left(\frac{1}{t}\right)^4+8\left(\frac{1}{t}\right)^5+\left(\frac{1}{t}\right)^6$$

$$-\left\{4t^3+6t^2+7t+5+6\frac{1}{t}+3\left(\frac{1}{t}\right)^4+8\left(\frac{1}{t}\right)^5+3\left(\frac{1}{t}\right)^6+6\left(\frac{1}{t}\right)^9\right\}$$

and ∴ as in art. 79 (arts. 65, 69, 105)

$$=5t^3-4t^3+3t^2-6t^2+7t-7t-5+9\frac{1}{t}-6\frac{1}{t}+4\left(\frac{1}{t}\right)^2$$

$$+3\left(\frac{1}{t}\right)^4-3\left(\frac{1}{t}\right)^4+8\left(\frac{1}{t}\right)^5-8\left(\frac{1}{t}\right)^5+\left(\frac{1}{t}\right)^6-3\left(\frac{1}{t}\right)^6-6\left(\frac{1}{t}\right)^9$$

$$\begin{array}{r} 5370·940381 \\ 4675·600383006 \\ \hline 695·339997994 \end{array}$$

∴ by the same steps as in art. 79 $=6t^2+9t+5+9\frac{1}{t}-6\frac{1}{t}+4\left(\frac{1}{t}\right)^2+\cdots\cdot$

∴ (arts. 71, 105) $=6t^2+9t+5+\left(9\frac{1}{t}-6\frac{1}{t}\right)+4\left(\frac{1}{t}\right)^2+\cdots\cdot$

∴ (art. 108) $=6t^2+9t+5+(9-6)\frac{1}{t}+4\left(\frac{1}{t}\right)^2+\cdots\cdot$

∴ (art. 79) $=6t^2+9t+5+3\frac{1}{t}+\left\{3\left(\frac{1}{t}\right)^2+\left(\frac{1}{t}\right)^2\right\}+3\left(\frac{1}{t}\right)^4-3\left(\frac{1}{t}\right)^4+\cdots\cdot$

∴ (arts. 33, 105, 135)

$$=6t^2+9t+5+3\frac{1}{t}+3\left(\frac{1}{t}\right)^2+(9+1)\left(\frac{1}{t}\right)^3+3\left(\frac{1}{t}\right)^4-3\left(\frac{1}{t}\right)^4+\cdots\cdot$$

∴ (art. 108) $=\cdots+3\left(\frac{1}{t}\right)^2+\left\{9\left(\frac{1}{t}\right)^3+\left(\frac{1}{t}\right)^3\right\}+3\left(\frac{1}{t}\right)^4-3\left(\frac{1}{t}\right)^4+\cdots$

∴ (arts. 33, 105) $=\cdots+3\left(\frac{1}{t}\right)^2+9\left(\frac{1}{t}\right)^3+\left(\frac{1}{t}\right)^3+\cdots$

∴ (arts. 135, 71, 105) $=\cdots+9\left(\frac{1}{t}\right)^3+\left\{t\left(\frac{1}{t}\right)^4+3\left(\frac{1}{t}\right)^4-3\left(\frac{1}{t}\right)^4\right\}+8\left(\frac{1}{t}\right)^5-\cdots$

∴ (arts. 75, 108) $=\cdot\cdot+9\left(\frac{1}{t}\right)^3+(t+3-3)\left(\frac{1}{t}\right)^4+8\left(\frac{1}{t}\right)^5-\cdot\cdot$

∴ by like steps $=\cdot\cdot+9\left(\frac{1}{t}\right)^3+9\left(\frac{1}{t}\right)^4+(t+8-8)\left(\frac{1}{t}\right)^5+\left(\frac{1}{t}\right)^6-\cdot\cdot$

so also $=\cdot\cdot+9\left(\frac{1}{t}\right)^4+9\left(\frac{1}{t}\right)^5+(t+1-3)\left(\frac{1}{t}\right)^6-6\left(\frac{1}{t}\right)^9$

$$=\cdot\cdot+9\left(\frac{1}{t}\right)^5+(7+1)\left(\frac{1}{t}\right)^6-6\left(\frac{1}{t}\right)^9=\cdot\cdot+9\left(\frac{1}{t}\right)^5+7\left(\frac{1}{t}\right)^6+(9+1)\left(\frac{1}{t}\right)^7-6\left(\frac{1}{t}\right)^9$$

$$=\cdot\cdot+7\left(\frac{1}{t}\right)^6+9\left(\frac{1}{t}\right)^7+(9+1)\left(\frac{1}{t}\right)^8-6\left(\frac{1}{t}\right)^9$$

and at length $=6t^2+9t+5+3\frac{1}{t}+3\left(\frac{1}{t}\right)^2+9\left(\frac{1}{t}\right)^3+9\left(\frac{1}{t}\right)^4+9\left(\frac{1}{t}\right)^5$

$+7\left(\frac{1}{t}\right)^6+9\left(\frac{1}{t}\right)^7+9\left(\frac{1}{t}\right)^8+(t-6)\left(\frac{1}{t}\right)^9$ or 695·339997994.

The digits in order of the result may hence be got in the same way as in art. 79.

In the MULTIPLICATIVE PROCESS the first case of art. 57 remains unchanged. The second case now is when of the multiplier and multiplicand each is denoted by a single significant digit but not in the unit's place for both, and here the multiplier may be $a\left(\frac{1}{t}\right)^m$ instead of at^m and the multiplicand $b\left(\frac{1}{t}\right)^n$ instead of bt^n. Hence the product instead of being $(at^m)bt^n$ which $=(ab)t^{n+m}$ may be one or other of these:—

$$\left\{a\left(\frac{1}{t}\right)^m\right\}bt^n = a\left(\frac{1}{t}\right)^m bt^n \text{ (arts. 41, 108)}, \ \therefore \text{ (arts. 54, 108)} = ab\left(\frac{1}{t}\right)^m t^n,$$

$$\therefore \text{ (arts. 135, 41, 108)} = (ab)\left(\frac{1}{t}\right)^{m-n} \text{ or } (ab)t^{n-m}$$

according as m is not less or not greater than n and $=ab$ if $m=n$;

$$(at^m)b\left(\frac{1}{t}\right)^n = at^m b\left(\frac{1}{t}\right)^n = abt^m\left(\frac{1}{t}\right)^n = (ab)t^{m-n},\ (ab)\left(\frac{1}{t}\right)^{n-m}, \text{ or } ab,$$

in like manner as before;

$$\left\{a\left(\frac{1}{t}\right)^m\right\} b\left(\frac{1}{t}\right)^n = a\left(\frac{1}{t}\right)^m b\left(\frac{1}{t}\right)^n = ab\left(\frac{1}{t}\right)^m\left(\frac{1}{t}\right)^n = (ab)\left(\frac{1}{t}\right)^{n+m}$$

as in art. 57 (art. 108).

In all these as well as in the corresponding product $(ab)t^{n+m}$ of art. 57 the handiest way of finding the number to which as unit ab refers is to mark that (arts. 41, 108) the product of a multiplication by t^m or by $\left(\frac{1}{t}\right)^m$ of any number is the same as the final product of m successive multiplications by t or by $\frac{1}{t}$ of that number and of the products got in succession and hence from the nature of the notation (art. 134) that the number referred to as unit by ab is the number referred to as unit by the digit m columns to the left or m columns to the right of the digit b. When ab is denoted by two digits the first digit of the two must by the principle of digit knitting (arts. 48, 136) go into the column next to the left of the column with the second.

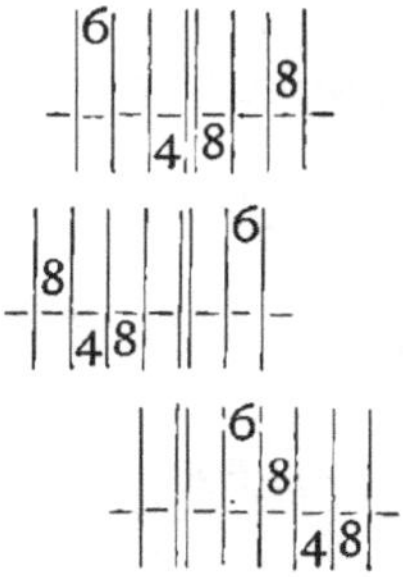

When the second case of art. 57 is thus changed the other two cases hold just as they are if only the language be understood in all the wideness of meaning now borne by it. Thus the product of the multiplication by 0·03 of 72·00605 (arts. 52, 108) is the sum of the products of the multiplications by 0·03 of each of the numbers 70, 2, 0·006, 0·00005; and these products may be found by what goes next before and therefore their sum by the notational additive process above. So too may the products be found of the multiplications by each of the numbers 0·008, 0·0004, 0·0000009, of that same 72·00605. Hence further may the sum of these products by 0·03, 0·008, 0·0004, 0·0000009, severally of 72·00605 be found which (arts. 39, 108) is the same as the product of the multiplication by 0·0384009 of 72·00605. Or 0·0384009×72·00605 may be found equally well by finding in the same way first 0·03×70+0·008×70+0·0004×70+0·0000009×70 which (arts. 39, 108)=0·0384009×70 and likewise the products by 0·0384009 of each of the numbers 2, 0·006, 0·00005, and then the sum of the products so found which (arts. 52, 108) is the same as the product sought.

```
72·006 05
 0·038 400 9
2 160 181 5
  576 048 4
   28 802 42
        64 805 445
2·765 097 125 445

72·006 05
 0·038 400 9
2 688 063
   76 801 8
     230 405 4
         1 920 045
2·765 097 125 445
```

The DIVISIVE PROCESS is the last of the processes in art. 83 changed twowise (1) in understanding that (arts. 105, 108)

$$a-eb-fb-gb-\cdots-lb-mb-nb=a-(e+f+g+\cdots+l+m+n)b$$

for any numbers whole or fractional and (2) in then seeking thereby the result of none other than what is (art. 110) a full, complete, thorough, division. If in finding the successive remainders $a-eb$, $a-eb-fb$, $a-eb-fb-gb$, - - - one that is 0 be never come to the quotient cannot be denoted decimally; there can then (art. 134) be denoted decimally only what differs from the quotient by less than any given magnitude however small of the same kind as the magnitude expressed by it. Thus to find the decimally denoted quotient of the division of 2·765097125445 by 72·00605 the former number is greater than 0·03×72·00605 but less than 0·04×72·00605 and is greater than 0·03×72·00605 by 0·604915625445, this last is greater than 0·008×72·00605 but less than 0·009×72·00605 and exceeds 0·008×72·00605 by 0·028867225445, this last again exceeds

```
72·006 05
 0·038 400 9
2·765 097 125 445
2 160 181 5
  604 915 6
  576 048 4
   28 867 2
   28 802 42
        64 80
        64 805 445
```

0·0004×72·00605 by 0·000064805445, and this last is 0·0000009×72·00605; so that $\frac{2·765097125445}{72·00605}$ is decimally denoted by 0·0384009. Likewise taking up the example of art. 83 it is found that 327790485 is greater than 6873×47692 by 3369 than 6873·07×47692 by 30·56 than 6873·0706×47692 by 1·9448 than 6873·07064×47692 by 0·03712 than 6873·0706407×47692 by 0·0037356 and so on. Therefore among other things $\frac{327790485}{47692}$ is greater than 6873·0706 by less than 0·00005, is greater than 6873·07064 by less than 0·0000008, and is less than 6873·070641 by less than 0·0000003; so that this quotient is better expressed to the fourth decimal place by 6873·0706 than by 6873·0707 but to the sixth decimal place is more nearly 6873·070641 than 6873·070640.

```
          47692
           6873·0706407
---------------------
   327790485
        3369
        333844
          3056
          286152
           19448
           190768
             3712
             333844
              37356
```

By this divisive process a fraction expressed in the common or general fractional form (usually called a VULGAR FRACTION) may be expressed either exactly or to any required degree of nearness as a decimally denoted fraction (usually called a DECIMAL FRACTION).

If a, b, q, be decimally denoted numbers such that $a = qb$ and e, f, ... m, n, be the numbers severally denoted by the digits in order of q the divisive process finds e, f, ... m, n, and through them q by breaking up a into eb, fb, ... mb, nb, and is in this manner a counter process to the multiplicative process which finds a by making it up of eb, fb, ... mb, nb. Now if r, s, ... y, z be the numbers severally denoted by the digits of b the other multiplicative process which finds a by making it up of qr, qs, ... qy, qz, has in like manner a counter process by which when a and q are given r, s, ... y, z, and through them b can be found by breaking up a into qr, qs, ... qy, qz. For (arts. 65, 105)

$$a-qr-qs-\cdots-qy-qz = a-(qz+qy+\cdots+qs+qr), \therefore \text{ (arts. 52, 105)}$$
$$= a-q(z+y+\cdots+s+r) \text{ and } \therefore \text{ (arts. 34, 105)}$$
$$= a-q(r+s+\cdots+y+z).$$

Thus to find what decimally denoted number it is that multiplied by 0·0384009 gives as product 2·765097125445; writing 0·0384009 with room above where to write the digits of the number sought as they are found and with a stroke drawn close below, then writing

```
  72·006|05   |
  0·0384|009  |
 ---------------------
  2·765|097 125|445
  2 688|063    |
 -----------
     77|034    |
     76|801 8  |
     --------
       |232 3  |
       |230 405|4
       -----------
           1|920 0
           1|920 045
           -----------
```

2·765097125445 under the stroke, and doing all the other writing needed below, so that everywhere digits referring to multiples or submultiples of the same rank as units are in the same column, the greatest number denoted by a single significant digit which multiplied by 0·0384009 gives a product not greater than 2·765097125445 is after slight trial found to be 70 and 2·765097125445–0·0384009×70 is by the foregoing multiplicative and subtractive processes found to be 0·077034125445. In like manner 2 is found the greatest number denoted by a single significant digit which multiplied by 0·0384009 gives a product not greater than 0·077034125445, and so on. And so at length the number sought is found to be 72·00605.

Here as in the divisive process it may happen that no decimally denoted number when multiplied by a given decimally denoted number gives another given decimally denoted number for product. All that can then be done is to find a decimally denoted number differing from the number sought by less than any given magnitude of the same kind as the magnitude expressed thereby.

That number which multiplied by q gives the product a is symbolized by $\frac{1}{q}a$. For this (art. 107) is the numerical expression in reference to the unit magnitude of a magnitude which in reference to another magnitude as unit expressed numerically in reference to the unit magnitude by a is expressed numerically by $\frac{1}{q}$ and in reference to which therefore as unit (art. 114) that other is expressed numerically by q. Since then the number that multiplied by q gives the product a is the product made by multiplying a by the reciprocal of q the notational process for finding it when a and q are given decimally denoted numbers is called the RECIPROCAL MULTIPLICATIVE PROCESS.

139. PROP. *In the widest meaning yet given to the symbols,*

$$\left.\begin{array}{r}\frac{1}{a^n}a^m\\[6pt] \dfrac{a^m}{a^n}\\[6pt] a^m\frac{1}{a^n}\end{array}\right\} = \begin{cases} a^{m-n} & \text{if } m \text{ be not less than } n,\\[4pt] \dfrac{1}{a^{n-m}} & \text{if } m \text{ be not greater than } n.\end{cases}$$

For if m be greater than n a^m is

both $|aa\ldots aa|aa\ldots\ldots aa|$ and $|aa\ldots\ldots aa|aa\ldots aa|$,
$\quad$ n as ✱ $(m-n)$ as $\qquad$ $(m-n)$ as ✱ n as

and ∴ (arts. 41, 108) $= a^n a^{m-n}$ $\qquad = a^{m-n}a^n$.

Hence $\frac{1}{a^n}a^m = \frac{1}{a^n}a^n a^{m-n}$ ∴ (arts. 41, 108) $= \left(\frac{1}{a^n}a^n\right)a^{m-n}$ which (art. 114) is a^{m-n},

$$\frac{a^m}{a^n} = \frac{a^{m-n}a^n}{a^n} \text{ and } \therefore \text{ (art. 114)} = a^{m-n},$$

$$a^m\frac{1}{a^n} = (a^{m-n}a^n)\frac{1}{a^n} \therefore \text{ (arts. 41, 108)} = a^{m-n}a^n\frac{1}{a^n} \text{ which (arts. 107, 114)}$$

is a^{m-n}.

When m is less than n $a^n = a^m a^{n-m}$ and also $= a^{n-m}a^m$ as before. Hence

$$\frac{1}{a^n}a^m = \frac{1}{a^m a^{n-m}}a^m \therefore \text{ (art. 126)} = \left(\frac{1}{a^{n-m}}\frac{1}{a^m}\right)a^m \text{ and } \therefore \text{ as before}$$

$$= \frac{1}{a^{n-m}}\frac{1}{a^m}a^m \text{ which is } \frac{1}{a^{n-m}},$$

$$\frac{a^m}{a^n} = \frac{a^m}{a^{n-m}a^m} \text{ and } \therefore \text{ (art. 116)} = \frac{1}{a^{n-m}},$$

$$a^m\frac{1}{a^n} = a^m\frac{1}{a^{n-m}a^m} = a^m\frac{1}{a^m}\frac{1}{a^{n-m}} = \left(a^m\frac{1}{a^m}\right)\frac{1}{a^{n-m}} \text{ which is } \frac{1}{a^{n-m}}.$$

More immediately

$$\frac{1}{a^n}a^m = a^{m\llcorner n} \text{ or } \frac{1}{a^{n\llcorner m}} \text{ while both } \frac{a^m}{a^n} \text{ and } a^m\frac{1}{a^n} = a^{m\urcorner n} \text{ or } \frac{1}{a^{n\urcorner m}}.$$

When $m = n$ the law holds by making a^0 mean 1 as in arts. 57, 135.

140. Prop. $a^n b^n c^n \ldots f^n g^n h^n = (abc\ldots fgh)^n$.

First $a^n b^n$ is

$|(aa\ldots aa)|bb\ldots bb|$ ∴ (arts. 41, 108) $= aa\ldots aabb\ldots bb$ ∴ (arts. 54, 108)
n as ✱ n bs

$= abab\ldots abab$ ∴ (arts. 41, 108) $= (ab)(ab)\ldots(ab)ab$ that is $(ab)^n$.

Hence $a^n b^n c^n \ldots f^n g^n h^n$

$$= a^n b^n c^n \ldots f^n (gh)^n = a^n b^n \ldots e^n (fgh)^n = \text{- - - -} = (abc\ldots fgh)^n.$$

141. Prop. $\frac{a^n}{b^n} = \left(\frac{a}{b}\right)^n$.

For this is what the proposition $\frac{a_1 a_2 \ldots a_{n-1} a_n}{b_n b_{n-1} \ldots b_2 b_1} = \frac{a_1}{b_1}\frac{a_2}{b_2}\cdots\frac{a_{n-1}}{b_{n-1}}\frac{a_n}{b_n}$ in art. 126 becomes when each of the numbers $a_1, a_2, \ldots a_n$, is a and each of the numbers $b_1, b_2, \ldots b_n$, is b.

142. Before passing on to sundry uses made of what has so far been done it may be well to set forth here in one view arranged under their several heads the laws laid down with the symbolic statement either of each law itself or of its simplest and fundamental case :—

THE FUNDAMENTAL LAWS OF OPERATIONAL EQUIVALENCE

I. *Laws of Commutation*

(i) of Additions and Subtractions; (1) $a+b=b+a$ (arts. 29, 34, 105), (2) $a-b-c=a-c-b$ (arts. 67, 105), (3) $a+b-c=a-c+b$ (arts. 69, 105).

(ii) of Multiplications and Divisions; (1) $ab=ba$ (arts. 54, 108), (2) $\frac{\left(\frac{a}{b}\right)}{c}=\frac{\left(\frac{a}{c}\right)}{b}$ (art. 120), (3) $\frac{ab}{c}=a\frac{b}{c}$ (art. 122).

II. *Laws of Distribution*

(i) of Addition; (1) $a+(b+c)=a+b+c$ (arts. 33, 105), (2) $a+(b-c)=a+b-c$ (arts. 71, 105).

(ii) of Subtraction; (1) $a-(b+c)=a-c-b$ (arts. 65, 105), (2) $a-(b-c)=a+c-b$ (arts. 73, 105).

(iii) of Multiplication; (1) $\begin{cases}(a+b)c=ac+bc \text{ (arts. 39, 108)},\\ c(a+b)=ca+cb \text{ (arts. 52, 108)},\end{cases}$

(2) $\begin{cases}(a-b)c=ac-bc \text{ (arts. 75, 108)},\\ c(a-b)=ca-cb \text{ (arts. 77, 108)}.\end{cases}$

(iv) of Division; (1) $\frac{a+b}{c}=\frac{a}{c}+\frac{b}{c}$ (art. 130), (2) $\frac{a-b}{c}=\frac{a}{c}-\frac{b}{c}$ (art. 130).

III. *Laws of Colligation*

(i) of Multiplications; $abc=(ab)c$ (arts. 41, 108).

(ii) of Divisions; $\frac{\left(\frac{a}{b}\right)}{c}=\frac{a}{cb}$ (art. 118).

(iii) of Multiplication and Division; (1) $\frac{ab}{c}=\frac{a}{c}b$ (art. 124),

(2) $\frac{ab}{c}=\frac{a}{\left(\frac{c}{b}\right)}$ (art. 127).

IV. *Laws of Relativity*

(i) in Subtraction; $a-b=a+c-(b+c)$ (arts. 63, 105).

(ii) in Division; $\frac{a}{b}=\frac{ac}{bc}$ (art. 116).

From these come what may be arranged as follows under the head of

Index Laws

(1) $a^n a^m = a^{m+n}$ (arts. 44, 108), (2) $(a^n)^m = a^{mn}$ (arts. 45, 108),

(3) $$\left.\begin{array}{c} \left(\frac{1}{a}\right)^n a^m \\ \frac{1}{a^n}a^m \\ \frac{a^m}{a^n} \\ a^m\frac{1}{a^n} \\ a^m\left(\frac{1}{a}\right)^n \end{array}\right\} = a^{m-n} \text{ if } m \text{ be not less than } n \text{ (arts. 135, 139)}$$

but if m be not greater than n $$\left\{\begin{array}{l} \left.\begin{array}{c} \left(\frac{1}{a}\right)^n a^m \\ a^m\left(\frac{1}{a}\right)^n \end{array}\right\} = \left(\frac{1}{a}\right)^{n-m} \text{ (art. 135)}, \\ \left.\begin{array}{c} \frac{1}{a^n}a^m \\ \frac{a^m}{a^n} \\ a^m\frac{1}{a^n} \end{array}\right\} = \frac{1}{a^{n-m}} \text{ (art. 139)}, \end{array}\right.$$

(4) $a^n b^n = (ab)^n$ (art. 140), (5) $\frac{a^n}{b^n} = \left(\frac{a}{b}\right)^n$ (art. 141).

143. INDIRECT ARITHMETICAL QUESTIONS.

When numbers are sought linked to given numbers in given ways not, or not at once seen to be, those in which results of operations, or sets of operations, of addition subtraction multiplication and division (arts. 21, 30, 37, 59, 81, 101, 103, 106, 109) are linked to the numbers operated with, the first thing to be found out is of which, or of what sets, of these operations with the given numbers are the sought numbers the results. This is done by means of the laws of operational equivalence and the following propositions.

Numbers are equal precisely when

(i) *the sums are equal got by severally adding either to them, or them to, equal numbers*

(ii) *the remainders are equal got by severally subtracting either from them, or them from, equal numbers* (iij) *the products are equal got by severally multiplying either them by, or by them, equal numbers of which none is* 0

(iv) *the quotients are equal got by severally dividing either them by, or by them, equal numbers of which none is* 0.

Let a, a', be any numbers and b, b', any numbers that are equal to one another.

It is precisely when $a=a'$ (arts. 10, 11) that $(a, b)=(a', b')$ and therefore (arts. 6, 9) that either $a+b=a'+b'$ or $b+a=b'+a'$.

Next if a be not less than b and a' than b' it is (arts. 6, 9) precisely when $a=a'$ that $(a-b, b)=(a'-b', b')$ and therefore (arts. 13, 14) that $a-b=a'-b'$. Besides if a be not greater than b and a' than b' (art. 6) $(b-a, a)=(b'-a', a')$ and therefore (arts. 13, 15) it is precisely when $a=a'$ that $b-a=b'-a'$.

Again b, b', if not 0s either are, or (art. 99) may be expressed as, fractions that have a common denominator and they have then also (art. 100) a common numerator. Hence (arts. 90, 91, 88, 89) it is precisely when $a=a'$ that $ba=b'a'$. It has been already shown (art. 113) that $a=a'$ precisely when $ab=a'b'$.

Lastly by what has just been shown it is precisely when $a=a'$ that $b'a=ba'$ and therefore (arts. 6, 9) that $\left(\frac{b'}{a'}a'\right)a=\left(\frac{b}{a}a\right)a'$ or that $\frac{b'}{a'}a'a=\frac{b}{a}aa'$ and therefore by what has just been shown ($\because a'a=aa'$) that $\frac{b'}{a'}=\frac{b}{a}$. It has been shown already (art. 115) that $a=a'$ precisely when $\frac{a}{b}=\frac{a'}{b'}$.

(1). What is the height of a house wall in which a window 6 feet high has under it $\frac{1}{3}$ and above it $\frac{1}{2}$ of the whole height?

Let the height sought be x feet. Then under the window there are $\frac{1}{3}x$ feet and above it $\frac{1}{2}x$ feet.

$$\therefore \frac{1}{3}x+6+\frac{1}{2}x=x.$$

This happens precisely when, by subtracting $\frac{1}{2}x$ from each of the numbers $\frac{1}{3}x+6+\frac{1}{2}x$ and x,

$$\frac{1}{3}x+6+\frac{1}{2}x-\frac{1}{2}x=x-\frac{1}{2}x$$

and $\therefore$ (arts. 71, 105) precisely when $\frac{1}{3}x+6+\left(\frac{1}{2}x-\frac{1}{2}x\right)=x-\frac{1}{2}x$,

$\therefore$, subtracting $\frac{1}{3}x$ from each of the numbers $\frac{1}{3}x+6$ and $x-\frac{1}{2}x$, precisely when

$$\frac{1}{3}x+6-\frac{1}{3}x=x-\frac{1}{2}x-\frac{1}{3}x$$

and $\therefore$ (arts. 69, 105) precisely when $\frac{1}{3}x-\frac{1}{3}x+6=x-\frac{1}{2}x-\frac{1}{3}x$

and $\therefore$ (arts. 75, 108) precisely when $6=\left(1-\frac{1}{2}-\frac{1}{3}\right)x$ that is $\frac{1}{6}x$,

$\therefore$, multiplying by 6 each of the numbers 6 and $\frac{1}{6}x$, precisely when

$$6\times 6=6\times\frac{1}{6}x$$

and $\therefore$ (arts. 41, 108) precisely when $6\times 6=\left(6\times\frac{1}{6}\right)x$ that is when $36=x$. And the height can only be 36 feet.

This result may be otherwise got as follows, where each statement after the first holds just when the statement next before it holds and therefore any one statement holds just when any other does.

$$\frac{1}{3}x+6+\frac{1}{2}x=x,\quad 6+\frac{1}{3}x+\frac{1}{2}x=x,\quad 6+\left(\frac{1}{3}x+\frac{1}{2}x\right)=x,$$

$$6+\left(\frac{1}{3}+\frac{1}{2}\right)x-\left(\frac{1}{3}+\frac{1}{2}\right)x=x-\left(\frac{1}{3}+\frac{1}{2}\right)x,$$

$$6+\left\{\left(\frac{1}{3}+\frac{1}{2}\right)x-\left(\frac{1}{3}+\frac{1}{2}\right)x\right\}=\left\{1-\left(\frac{1}{3}+\frac{1}{2}\right)\right\}x,$$

$$6=\frac{1}{6}x,\text{ and as before } 6\times 6=x.$$

(2). How may a debt of 5*l.* be paid with 29 coins some of them crowns and the rest florins?

Let there be x crowns. Then there are $29-x$ florins, the x crowns are worth $x\times 5$ shillings, and the $29-x$ florins are worth $(29-x)\times 2$ shillings.

$$\therefore\ x\times 5+(29-x)\times 2=5\times 20 \text{ or } 100.$$

This happens precisely when any one of the following happens, because each of them happens precisely when the one next before it happens for the reason written against it.

(art. 75) $\qquad x\times 5+(29\times 2-x\times 2)=100,$

(art. 71) $\qquad x\times 5+29\times 2-x\times 2=100,$

(art. 69) $\qquad x\times 5-x\times 2+29\times 2=100,$

(art. 77 and subtracting from each 58) $x(5-2)+58-58=100-58$,
(art. 71) $x\times 3+(58-58)=42$,
(dividing each by 3) $\frac{x\times 3}{3}=\frac{42}{3}$,
(art. 114) $x=14$,
(subtracting each from 29) $29-x=15$.
So that 5*l.* can be paid in the required way only with 14 crowns and 15 florins.

(3). A workman engaging for 36 days to take 3*s.* 6*d.* every day he worked and to pay 1*s.* 6*d.* every day he did not made 2*l.* 16*s.* How many of the 36 days did he work?

If x be the number of days he worked, $36-x$ is the number he did not; by working then he gained $x\times 3\frac{1}{2}$ shillings and by not working lost $(36-x)\times 1\frac{1}{2}$. Hence

$$x\times 3\frac{1}{2}-(36-x)\times 1\frac{1}{2}=2\times 20+16.$$

This happens precisely when any of the following does:—

$$x\times 3\frac{1}{2}-\left(36\times 1\frac{1}{2}-x\times 1\frac{1}{2}\right)=56,$$

$$x\times 3\frac{1}{2}+x\times 1\frac{1}{2}-36\times 1\frac{1}{2}=56 \quad \text{(arts. 73, 105)},$$

$$x\left(3\frac{1}{2}+1\frac{1}{2}\right)-54+54=56+54,$$

$$x\times 5=110 \text{ by the nature of subtraction (arts. 59, 103)},$$

$$\frac{x\times 5}{5}=\frac{110}{5}.$$

$$x=22.$$

And it was therefore 22 out of the 36 days that he worked.

(4). A smuggler with brandy that would bring 9*l.* 18*s.* had after selling 10 gallons $\frac{1}{3}$ of the rest seized and so made only 8*l.* 2*s.* Find the number of gallons and the price of a gallon.

Let there be x gallons. The price of a gallon then is $\frac{1}{x}(9\times 20+18)$ or $\frac{1}{x}\times 198$ shillings and after selling 10 gallons $\frac{1}{3}$ of the remaining $x-10$ gallons would therefore bring $\left\{\frac{1}{3}(x-10)\right\}\frac{1}{x}\times 198$ shillings which by the question is $198-(8\times 20+2)$ or 36 shillings. Hence

$$\left\{\frac{1}{3}(x-10)\right\}\frac{1}{x}\times 198=36.$$

This holds just when any one of these does:—

$$\frac{1}{3}(x-10)\frac{1}{x}\times 198 = 36 \quad \text{(arts. 41, 108)},$$

$$\frac{1}{3}x\frac{1}{x}\times 198-\frac{1}{3}\times 10\frac{1}{x}\times 198 = 36 \quad \text{(arts. 75, 77, 108)},$$

$$\frac{1}{3}\left(x\frac{1}{x}\right)\times 198-\frac{1}{3}\times 10\frac{1}{x}\times 198 = 36,$$

$$66-(66-\frac{1}{3}\times 10\frac{1}{x}\times 198) = 66-36$$

by subtracting from 66 each of the last two numbers,

$$66+\frac{1}{3}\times 10\frac{1}{x}\times 198-66 = 30,$$

$$66-66+\frac{1}{3}\times 10\frac{1}{x}\times 198 = 30,$$

$$\frac{1}{10}\times 3\times\frac{1}{3}\times 10\frac{1}{x}\times 198 = \frac{1}{10}\times 3\times 30,$$

$$\left\{\frac{1}{10}\left(3\times\frac{1}{3}\right)\times 10\right\}\frac{1}{x}\times 198 = 9,$$

$$\frac{198}{\frac{1}{x}\times 198} = \frac{198}{9}$$ dividing 198 by each of the last two numbers,

$$\frac{1}{\left(\frac{1}{x}\right)} = 22 \quad \text{(art. 116)},$$

$$x = 22 \quad \text{since } \frac{1}{\left(\frac{1}{x}\right)} = \frac{x}{\frac{1}{x}x} \text{ (art. 116)} = \frac{x}{1} \text{ (art. 114)}.$$

The number of gallons therefore is 22 and by the fourth line back the price of a gallon is 9 shillings.

These results may also be got thus:—Taking x as before for the number of gallons the 10 gallons and $\frac{2}{3}$ of the rest not seized at $\frac{1}{x}\times 198$ shillings the gallon yield $8\times 20+2$ or 162 shillings.

$$\therefore \left\{10+\frac{2}{3}(x-10)\right\}\frac{1}{x}\times 198 = 162.$$

And this happens just when any one of these does:—

$$\left\{10+\left(\frac{2}{3}x-\frac{2}{3}\times 10\right)\right\}\frac{1}{x}\times 198 = 162, \quad \left(10+\frac{2}{3}x-\frac{20}{3}\right)\frac{1}{x}\times 198 = 162,$$

$$\left(10-\frac{20}{3}+\frac{2}{3}x\right)\frac{1}{x}\times 198 = 162, \quad \frac{10}{3}\frac{1}{x}\times 198+\left(\frac{2}{3}x\right)\frac{1}{x}\times 198 = 162,$$

$$\frac{10}{3}\frac{1}{x}\times198+\frac{2}{3}x\frac{1}{x}\times198=162,\quad \frac{10}{3}\frac{1}{x}\times198+\frac{2}{3}\left(x\frac{1}{x}\right)\times198=162,$$

$$\frac{10}{3}\frac{1}{x}\times198+132-132=162-132,\quad \frac{10}{3}\frac{1}{x}\times198+(132-132)=30,$$

$$\frac{3}{10}\times\frac{10}{3}\frac{1}{x}\times198=\frac{3}{10}\times30,\quad \left(\frac{3}{10}\times\frac{10}{3}\right)\frac{1}{x}\times198=9,$$ and the rest as before.

The same question may still be dealt with in another way. If v shillings be the price of a gallon $\frac{198}{v}$ is the number of gallons and hence v is to be such that

$$\left\{10+\frac{2}{3}\left(\frac{198}{v}-10\right)\right\}v=162.$$

This happens when and never but when any one of the following happens. $\left\{10+\left(\frac{2}{3}\frac{198}{v}-\frac{2}{3}\times10\right)\right\}v=162$, $\left(10+\frac{2}{3}\frac{198}{v}-\frac{2}{3}\times10\right)v=162$, $\left(10-\frac{2}{3}\times10+\frac{2}{3}\frac{198}{v}\right)v=162$, $\frac{10}{3}v+\left(\frac{2}{3}\frac{198}{v}\right)v=162$, $\frac{10}{3}v+\frac{2}{3}\frac{198}{v}v=162$, $\frac{10}{3}v+132-132=162-132$, $\frac{10}{3}v+(132-132)=30$, $\frac{3}{10}\frac{10}{3}v=\frac{3}{10}\times30$, $\left(\frac{3}{10}\frac{10}{3}\right)v=9$ or $v=9$, $\frac{198}{v}=22$.

(5). How much tea at 2*s*. 6*d*. must be mixed with 60lbs. at 4*s*. 4*d*. that the mixture may be at 3*s*. 9*d*.?

Let it be x lbs. Then the 60lbs. is worth $60\times4\frac{1}{3}$ shillings, the x lbs. $x\times2\frac{1}{2}$ shillings, and the mixture of $60+x$ lbs. $(60+x)\times3\frac{3}{4}$ shillings.

$$\therefore\ 60\times4\frac{1}{3}+x\times2\frac{1}{2}=(60+x)\times3\frac{3}{4}.$$

This is the case precisely when any one of these is:—

$$60\times4\frac{1}{3}+x\times2\frac{1}{2}=60\times3\frac{3}{4}+x\times3\frac{3}{4}\quad\text{(arts. 39, 108)},$$

$$60\times4\frac{1}{3}+x\times2\frac{1}{2}-\left(60\times3\frac{3}{4}+x\times2\frac{1}{2}\right)=60\times3\frac{3}{4}+x\times3\frac{3}{4}-\left(60\times3\frac{3}{4}+x\times2\frac{1}{2}\right),$$

$$60\times4\frac{1}{3}+x\times2\frac{1}{2}-x\times2\frac{1}{2}-60\times3\frac{3}{4}$$
$$=60\times3\frac{3}{4}+x\times3\frac{3}{4}-x\times2\frac{1}{2}-60\times3\frac{3}{4}\quad\text{(arts. 65, 105)},$$

$$60\times4\frac{1}{3}+\left(x\times2\frac{1}{2}-x\times2\frac{1}{2}\right)-60\times3\frac{3}{4}$$
$$=60\times3\frac{3}{4}-60\times3\frac{3}{4}+x\times3\frac{3}{4}-x\times2\frac{1}{2}\quad\text{(arts. 71, 69, 105)},$$

$$60\left(4\frac{1}{3}-3\frac{3}{4}\right)=x\left(3\frac{3}{4}-2\frac{1}{2}\right) \text{ (arts. 77, 108)},$$

$$\frac{60\times\frac{7}{12}}{\left(\frac{5}{4}\right)}=\frac{x\times\frac{5}{4}}{\left(\frac{5}{4}\right)},$$

$$28=x.$$

(6). What sums of money have two men when according as the one takes from, or gives to, the other 3*s.* so is the first's sum made equal to, or $\frac{5}{8}$ of, the other's?

If the first man have x shillings he must have $x+3$ shillings after taking 3 from the second and this being what the second then has the second must at first have had $x+3+3$ which (arts. 33, 105) $=x+(3+3)$ or $x+6$ shillings. Hence when the first gives the second 3*s.* the first's sum becomes $x-3$ shillings and the second's $x+6+3$ which $=x+(6+3)$ or $x+9$ shillings. Therefore

$$x-3=\frac{5}{8}(x+9).$$

And this holds precisely when any one of these following does:

$$x-3=\frac{5}{8}x+\frac{5}{8}\times 9 \text{ (arts. 52, 108)},$$

$$x-3+3=\frac{5}{8}x+\frac{5}{8}\times 9+3,$$

$$x-\frac{5}{8}x=\frac{5}{8}x+\frac{5}{8}\times 9+3-\frac{5}{8}x,$$

$$\left(1-\frac{5}{8}\right)x=\frac{5}{8}x-\frac{5}{8}x+\frac{5}{8}\times 9+3,$$

$$\frac{3}{8}x=\frac{69}{8},$$

$$\frac{8}{3}\times\frac{3}{8}x=\frac{8}{3}\times\frac{69}{8},$$

$$\left(\frac{8}{3}\times\frac{3}{8}\right)x, \text{ that is } x, =23,$$

$$x+6=29.$$

Hence the first has 23*s.* and the other 29*s.*

(7). How many sheep has a man who gets 90*l.* 6*s.* by selling $\frac{2}{3}$ of them at 22*s.* each $\frac{1}{4}$ of them at 30*s.* each and the rest at 35*s.* each?

Let there be x sheep. Since after taking away $\frac{2}{3}$ and $\frac{1}{4}$ of the sheep there is $1-\frac{2}{3}-\frac{1}{4}$ or $\frac{1}{12}$ of them left,

$$\left(\frac{2}{3}x\right)\times 1\frac{1}{10}+\left(\frac{1}{4}x\right)\times 1\frac{1}{2}+\left(\frac{1}{12}x\right)\times 1\frac{3}{4}=90\frac{3}{10};$$

which happens precisely when any one of the following happens.

$$\frac{2}{3}x\times\frac{11}{10}+\frac{1}{4}x\times\frac{3}{2}+\frac{1}{12}x\times\frac{7}{4}=\frac{903}{10},$$

$$x\times\frac{2}{3}\times\frac{11}{10}+x\times\frac{1}{4}\times\frac{3}{2}+x\times\frac{1}{12}\times\frac{7}{4}=\frac{903}{10} \quad \text{(arts. 54, 108)},$$

$$x\left(\frac{22}{30}+\frac{3}{8}+\frac{7}{48}\right)=\frac{903}{10},$$

$$x\times\frac{301}{240}=\frac{903}{10},$$

$$\frac{x\times\frac{301}{240}}{\left(\frac{301}{240}\right)}=\frac{\left(\frac{903}{10}\right)}{\left(\frac{301}{240}\right)},$$

$$x=72.$$

(8). How many leaps must a greyhound take to catch a hare 50 hare leaps before it if 2 greyhound leaps be equal to 3 hare leaps and the greyhound take 3 leaps to the hare's 4?

Let x be the number of the greyhound's leaps. Since x greyhound leaps are equal to $\frac{x}{2}\times 3$ hare leaps the hare takes $\frac{x}{2}\times 3-50$ leaps and therefore runs during $\frac{\frac{x}{2}\times 3-50}{4}$ times the time of four of its leaps. Hence

$$\frac{\frac{x}{2}\times 3-50}{4}=\frac{x}{3}.$$

This relation holds exactly when any one of the following relations holds:—

$$\frac{\left(x\times\frac{1}{2}\right)\times 3-50}{4}=x\times\frac{1}{3} \quad \text{(art. 122)},$$

$$\frac{x\times\frac{1}{2}\times 3}{4}-\frac{50}{4}=x\times\frac{1}{3} \quad \text{(arts. 41, 108, 130)},$$

$$x\times\frac{\frac{1}{2}\times 3}{4}-\frac{50}{4}=x\times\frac{1}{3}\quad\text{(art. 122)},$$

$$x\times\frac{3}{8}-\frac{50}{4}+\frac{50}{4}=x\times\frac{1}{3}+\frac{50}{4},$$

$$x\times\frac{3}{8}-x\times\frac{1}{3}=x\times\frac{1}{3}+\frac{50}{4}-x\times\frac{1}{3},$$

$$x\left(\frac{3}{8}-\frac{1}{3}\right)=x\times\frac{1}{3}-x\times\frac{1}{3}+\frac{50}{4},$$

$$\frac{x\times\frac{1}{24}}{\left(\frac{1}{24}\right)}=\frac{\left(\frac{50}{4}\right)}{\left(\frac{1}{24}\right)},$$

$$x=300.$$

The question may be otherwise treated thus:—Let the greyhound catch the hare in t times the time in which either the greyhound takes 3 leaps or the hare 4. Then expressing numerically the distance run by the greyhound in reference to that distance as unit which is equal to 2 greyhound leaps and 3 hare leaps,

$$\frac{t\times 3}{2}=\frac{50+t\times 4}{3}.$$

And this happens just when any of the following does:—

$$t\times\frac{3}{2}=\frac{50}{3}+\frac{t\times 4}{3},\quad t\times\frac{3}{2}-t\times\frac{4}{3}=\frac{50}{3}+t\times\frac{4}{3}-t\times\frac{4}{3},$$

$$t\left(\frac{3}{2}-\frac{4}{3}\right)=\frac{50}{3}+\left(t\times\frac{4}{3}-t\times\frac{4}{3}\right),\quad \frac{t\times\frac{1}{6}}{\left(\frac{1}{6}\right)}=\frac{\left(\frac{50}{3}\right)}{\left(\frac{1}{6}\right)},\quad t=100,$$

and the greyhound's leaps are $t\times 3$ or 300.

144. Into every simple arithmetical operation whether of addition, subtraction, multiplication, or division, two numbers enter of which either may be viewed as the number operated on and then the other becomes the number by which the operation is performed or shortly the operating number. According as the number added to or added, subtracted from or subtracted, multiplied or multiplied by, divided or divided by, is held to be the number operated on so may the corresponding operation be named ADDITION TO or ADDITION OF, SUBTRACTION FROM or SUBTRACTION OF, MULTIPLICATION OF or MULTIPLICATION BY, DIVISION OF or DIVISION BY.

Def. If an operation performed on a number give a certain number for result the operation which performed on the latter number gives back at once the first number for result is called the INVERSE OPERATION.

The inverse operation of an inverse operation is clearly the original operation.

Those steps taken in dealing with the questions of the last article that are neither bare symbolic statements of the questions nor laws of operational equivalence are inversions of operations.

145. Let any number x be taken for a number operated on and any number a for a number operating.

The result of *addition to* is $x+a$ and (arts. 71, 105) $x+a-a$ $=x+(a-a)$ that is x. The inverse operation then of *addition to* is *subtraction from*. The inverse operation of *subtraction from* ought therefore to be *addition to* and accordingly $x-a+a$ (arts. 59, 103) $=x$.

The result of *addition of* is $a+x$ and (arts. 69, 105) $a+x-a$ $=a-a+x$ or x. So that the inverse of *addition of* is *subtraction from*. Accordingly also $a+(x-a)=a+x-a=a-a+x$ or x.

The result of *subtraction of* is $a-x$ and $a-(a-x)=a+x-a$ (arts. 73, 105) $=a-a+x$ or x as before. Therefore the inverse of *subtraction of* is *subtraction of*.

Thus of either *addition to* or *addition of* and *subtraction from*, and also of *subtraction of* and *subtraction of*, each must be held to be the inverse operation of the other. But, distinguishing between the kinds of subtraction (arts. 60, 61, 104), $x+a \llcorner a=x$, $x \llcorner a+a=x$, $a+x \neg a=x$, $a+(x \neg a)=x$, $a \neg (a \llcorner x)=x$, $a \llcorner (a \neg x)=x$, more immediately and directly than the corresponding expressions above and therefore a still closer relationship of inversion is borne to one another by *addition to* and *desubtraction from*, by *addition of* and *sursubtraction from*, and by *desubtraction of* and *sursubtraction of*.

The result of *multiplication of* is ax and (arts. 41, 108) $\frac{1}{a}ax=\left(\frac{1}{a}a\right)x$ which (art. 114) is x. If then the multiplication of a number by the reciprocal of a number be named RECIPROCAL MULTIPLICATION OF or RECIPROCAL MULTIPLICATION BY according as the former number or the latter is made the number operated on the inverse operation of *multiplication of* is *reciprocal multiplication of*. Accordingly too $a\frac{1}{a}x=\left(a\frac{1}{a}\right)x$ or x.

The result of *multiplication by* is xa and (art. 114) $\frac{xa}{a}=x$. Hence the inverse of *multiplication by* is *division of*. Conversely $\frac{x}{a}a=x$.

Lastly the result of *division by* is $\frac{a}{x}$ and $\frac{1}{\left(\frac{a}{x}\right)}a=\frac{x}{\frac{a}{x}x}a$ (art. 116) and ∴ (art. 114) $=\frac{x}{a}a=x$. So that the inverse operation of *division by* is *reciprocal multiplication by*. And accordingly $\frac{a}{\frac{1}{x}a}=\frac{1}{\left(\frac{1}{x}\right)}=x$.

Hence in each of the following pairs of operations each operation is the inverse of the other:—

Addition To	and	*Desubtraction From,*
Addition Of	and	*Sursubtraction From,*
Desubtraction Of	and	*Sursubtraction Of,*
Multiplication Of	and	*Reciprocal multiplication Of,*
Multiplication By	and	*Division Of,*
Division By	and	*Reciprocal multiplication By.*

SYMBOLICALLY GENERALIZED PROBLEMS

146. Arithmetical questions alike in everything but the particular numbers given in them are all treated in the same way and therefore may be brought together under a single treatment by using general symbols for the given numbers. The results are then embraced in one symbolically expressed result of certain operations with those symbolized numbers or in what is called a GENERAL FORMULA.

To find two numbers of which the sum is s and the difference d.

Let x be the greater. Then $x-d$ is the less and

$$x+(x-d)=s,$$

which happens precisely when any of these happens:—

$$x+x-d=s \text{ (arts. 71, 105)},$$

$$2x=s+d$$

because (art. 145) the inverse of *subtraction from* is *addition to*,

$$x=\frac{1}{2}(s+d)$$

because the inverse of *multiplication of* is *reciprocal multiplication of*,

$$x-d=\frac{1}{2}s+\frac{1}{2}d-d=\frac{1}{2}s-\left(d-\frac{1}{2}d\right)=\frac{1}{2}s-\left(1-\frac{1}{2}\right)d=\frac{1}{2}(s-d).$$

The greater number sought therefore is $\frac{1}{2}(s+d)$ and the less $\frac{1}{2}(s-d)$.

Thus at an election where 823 votes were given and the winning candidate had a majority of 211, the winning candidate must have had $\frac{1}{2}(823+211)$ or 517 votes and the other $\frac{1}{2}(823-211)$ or 306.

So a rod 53·27 inches long to be cut into two parts with one greater by 10·44 inches than the other must have the one part $\frac{1}{2}(53\cdot27+10\cdot44)$ or 31·855 inches long and the other $\frac{1}{2}(53\cdot27-10\cdot44)$ or 21·415.

147. To cut a given number a into $n+1$ parts so that
the 2d part may be greater by h_1 than e_1 times the 1st part,
— 3rd — — — — — h_2 — e_2 — — 2d —
— 4th — — — — — h_3 — e_3 — — 3rd —
. . . .
— $(n+1)$th — — — — h_n — e_n — — nth —.

Let the first part be x. The second part is then e_1x+h_1. Therefore the 3rd part is $e_2(e_1x+h_1)+h_2$ and therefore (arts. 52, 108) the same as $e_2e_1x+e_2h_1+h_2$. Therefore again the 4th part is the same as $e_3(e_2e_1x+e_2h_1+h_2)+h_3$ and therefore also as $e_3e_2e_1x+e_3e_2h_1+e_3h_2+h_3$. And so on till at last the $(n+1)$th part is the same as

$$e_n..e_2e_1x+e_n..e_2h_1+e_n..e_3h_2+\cdots\cdots+e_nh_{n-1}+h_n.$$

Hence x is to be such that

$$x+(e_1x+h_1)+(e_2e_1x+e_2h_1+h_2)+\cdots+(e_n..e_2e_1x+e_n..e_3e_2h_1+\cdots+e_nh_{n-1}+h_n)=a.$$

And this happens just when any of the following does:—

$$x+e_1x+h_1+(e_2e_1)x+e_2h_1+h_2+(e_3e_2e_1)x+(e_3e_2)h_1+e_3h_2+h_3$$
$$+(e_4e_3e_2e_1)x+\cdots\cdots\cdots\cdots+h_n=a,$$

$$x+e_1x+(e_2e_1)x+\cdots+(e_n..e_2e_1)x+h_1+e_2h_1+(e_3e_2)h_1+\cdots+(e_n..e_3e_2)h_1+$$
$$\cdots\cdots+h_{n-1}+e_nh_{n-1}+h_n=a,$$

$$x+e_1x+\cdots+(e_n..e_1)x+\{h_1+e_2h_1+\cdots+(e_n..e_2)h_1\}$$
$$+\{h_2+e_3h_2+\cdots+(e_n..e_3)h_2\}+\cdots+(h_{n-1}+e_nh_{n-1})+h_n=a,$$

$$(1+e_1+e_2e_1+e_3e_2e_1+\cdots+e_n..e_3e_2e_1)x+(1+e_2+e_3e_2+e_4e_3e_2+\cdots+e_n..e_4e_3e_2)h_1$$
$$+(1+e_3+\cdots+e_n..e_4e_3)h_2+\cdots\cdots+(1+e_{n-1}+e_ne_{n-1})h_{n-2}+(1+e_n)h_{n-1}+h_n=a,$$

$$(1+e_1+\cdots+e_n..e_1)x$$
$$+\{(1+e_2+\cdots+e_n..e_2)h_1+(1+e_3+\cdots+e_n..e_3)h_2+\cdots\cdots+(1+e_n)h_{n-1}+h_n\}=a$$

or as it may be written for shortness $Ex+H=a$,

$$Ex=a-H$$

the inverse operation of *addition to* being *subtraction from*,

$$x=\frac{1}{E}(a-H).$$

And when the first part is thus found the other parts must be found as above. Calling the *i*th part x_i, $x_{i+1}=e_i x_i+h_i$; and hence x_1 being found the other parts may be found each from the one before it by making i in succession 1, 2, 3, ... n, that is

$$x_2=e_1x_1+h_1,\ x_3=e_2x_2+h_2,\ \ldots\ldots x_{n+1}=e_nx_n+h_n.$$

For example to cut a line of 1858 fathoms into 5 parts each after the first greater by 7 fathoms than 3ce the one next before it the first part must be

$$\frac{1}{1+3+3.3+3.3.3+3.3.3.3}\left[\begin{matrix}1858\\-\{(1+3+3.3+3.3.3).7+(1+3+3.3).7+(1+3).7+7\}\end{matrix}\right]$$

or 12 fathoms and the other parts in order then are 3.12+7 or 43, 3.43+7 or 136, 415, and 1252, fathoms.

Again to deal out 23*l.* among three men so that the second's share may be 6*l.* more than $\frac{1}{2}$ the first's and the third's 5*l.* more than $\frac{1}{3}$ the second's the first's share must be

$$\frac{1}{1+\frac{1}{2}+\frac{1}{3}\times\frac{1}{2}}\left[23-\left\{\left(1+\frac{1}{3}\right)\times 6+5\right\}\right]$$

or 6*l.*; the second's share is therefore $\frac{1}{2}\times 6+6$ or 9*l.* and the third's $\frac{1}{3}\times 9+5$ or 8*l.*

The problem of art. 146 may be brought under this general problem as a particular case by taking it to be either the cutting a number s into two parts so that the second part may be greater by a number d than the first part or the cutting s, with a stretch of language, into any number of parts more than 2 so that the second of them may be greater by d than 1ce the first and every other greater by 0 than 0 times the one before it. Either way gives the first part $\frac{1}{2}(s-d)$ and the second part is therefore $1\times\frac{1}{2}(s-d)+d$

$$=\frac{1}{2}s-\frac{1}{2}d+d=\frac{1}{2}s+d-\frac{1}{2}d=\frac{1}{2}s+\left(d-\frac{1}{2}d\right)=\frac{1}{2}s+\left(1-\frac{1}{2}\right)d=\frac{1}{2}(s+d).$$

The second way of viewing the problem gives the third part

$0 \times \frac{1}{2}(s+d)+0$ which is 0 and every other part $0 \times 0+0$ which is also 0.

Cutting a given number into any assigned number of equal parts is the special case when every part but the first is greater by 0 than 1 times the part next before.

148. To find the capital in a business at first when at the end of n years it becomes a pounds by having
in the 1st year changed at the rate of c_1 pounds the pound and had withdrawn from it d_1 pounds

— 2nd — — — — — c_2 — — — — — — — d_2 —

. . . . . .

— nth — — — — — c_n — — — — — — — d_n —.

If at first the capital be x pounds at the end of the first year it becomes xc_1-d_1 pounds. Therefore at the end of the 2nd year it is $(xc_1-d_1)c_2-d_2$ which $=(xc_1)c_2-d_1c_2-d_2=xc_1c_2-d_1c_2-d_2$. Therefore at the end of the 3rd year it $=(xc_1c_2-d_1c_2-d_2)c_3-d_3=(xc_1c_2)c_3-(d_1c_2)c_3-d_2c_3-d_3$ $=xc_1c_2c_3-d_1c_2c_3-d_2c_3-d_3$. And so on until at length at the end of the nth year

$$xc_1c_2\ldots c_{n-1}c_n-d_1c_2c_3\ldots c_{n-1}c_n-d_2c_3\ldots c_n-\cdots-d_{n-2}c_{n-1}c_n-d_{n-1}c_n-d_n=a.$$

This happens precisely when any of these does: to wit

$$xc_1c_2\ldots c_{n-1}c_n-(d_n+d_{n-1}c_n+d_{n-2}c_{n-1}c_n+\cdots+d_1c_2c_3\ldots c_{n-1}c_n)=a,$$

$$xc_1c_2c_3\ldots c_{n-2}c_{n-1}c_n=a+(d_n+d_{n-1}c_n+d_{n-2}c_{n-1}c_n+\cdots+d_1c_2c_3\ldots c_{n-1}c_n),$$

$$x=\frac{a+(d_n+d_{n-1}c_n+d_{n-2}c_{n-1}c_n+\cdots+d_1c_2c_3\ldots c_{n-2}c_{n-1}c_n)}{c_1c_2c_3\ldots c_{n-2}c_{n-1}c_n}$$

since the inverse operation of *multiplication by* is *division of.*

The result may also be got in a direct manner. The capital at the beginning of the nth year is at the rate of c_n pounds the pound changed into $a+d_n$ pounds and can therefore only be $\frac{a+d_n}{c_n}$ pounds. In the same way therefore the capital at the beginning of the $(n-1)$th year is $\dfrac{\frac{a+d_n}{c_n}+d_{n-1}}{c_{n-1}}$ and $\therefore=\dfrac{\left(\frac{a+d_n}{c_n}+d_{n-1}\right)c_n}{c_{n-1}c_n}=\dfrac{a+d_n+d_{n-1}c_n}{c_{n-1}c_n}$.

Hence in the same way the capital at the beginning of the $(n-2)$th year

$$=\frac{\frac{a+d_n+d_{n-1}c_n}{c_{n-1}c_n}+d_{n-2}}{c_{n-2}}=\frac{\left(\frac{a+d_n+d_{n-1}c_n}{c_{n-1}c_n}+d_{n-2}\right)c_{n-1}c_n}{c_{n-2}c_{n-1}c_n}=\frac{a+d_n+d_{n-1}c_n+d_{n-2}c_{n-1}c_n}{c_{n-2}c_{n-1}c_n}.$$

And so at last the capital at the beginning of the 1st year

$$=\frac{a+d_n+d_{n-1}c_n+d_{n-2}c_{n-1}c_n+\cdots+d_1c_2c_3\ldots c_{n-1}c_n}{c_1c_2\ldots c_{n-2}c_{n-1}c_n}$$

which by the law of the distribution of addition over additions (arts. 33, 105) is the same number as the result got before.

149. To find when the minute hand of a clock is m minute divisions before the hour hand.

The hour and minute hands of a clock are to be taken as ever turning round the same way upon a common spindle each passing through equal spaces in all equal lengths of time but with such speeds that while the minute hand makes a full round of the face every hour marking by the number of minute divisions passed through since the beginning of the hour how many of the hour's 60 minutes have passed the hour hand makes $\frac{1}{12}$ of a round. Moreover the hour and the minute divisions begin together at the 12 or 0 o'clock hour line and the 60 or 0 minute line, so that between every adjoining two hour lines there are $\frac{1}{12}\times 60$ or 5 minute divisions. Hence the minute hand moves at the rate of 1 minute division a minute and the hour hand at the rate of $\frac{1}{60}\times 5$ or $\frac{1}{12}$ minute divisions a minute.

Let it be then at x minutes past h o'clock that the minute hand is m minute divisions before the hour hand on the clock face. In the x minutes taken by the minute hand to move from the 12 o'clock hour line to a place x minute divisions distant therefrom in the direction of motion of the hands the hour hand moves over $x\times\frac{1}{12}$ minute divisions. But at h o'clock the hour hand is distant from the 12 o'clock hour line $h\times 5$ minute divisions in the direction of the hand's motion. Therefore at x minutes past h o'clock the hour hand is $h\times 5+x\times\frac{1}{12}$ minute divisions distant from the 12 o'clock line in the said direction.

$$\therefore\ x=h\times 5+x\times\frac{1}{12}+m.$$

This holds just when any of the following holds:

$$x=h\times 5+m+x\times\frac{1}{12},$$

$$x - x\times\frac{1}{12} = h\times 5+m,$$

$$x\left(1-\frac{1}{12}\right) = h\times 5+m,$$

$$x = \frac{h\times 5+m}{\left(\frac{11}{12}\right)} = (h\times 5+m)\times\frac{12}{11} \text{ (art. 116).}$$

Thus making m 0 and h 0, 1, 2, ..., the hour and minute hands are together at 0 minutes past 0 o'clock that is 12 o'clock, at $5\frac{5}{11}$ minutes past 1 o'clock, at $10\frac{10}{11}$ past 2, $16\frac{4}{11}$ past 3, $21\frac{9}{11}$ past 4, $27\frac{3}{11}$ past 5, $32\frac{8}{11}$ past 6, $38\frac{2}{11}$ past 7, $43\frac{7}{11}$ past 8, $49\frac{1}{11}$ past 9, $54\frac{6}{11}$ past 10, 60 past 11 which is 12, $65\frac{5}{11}$ past 12 which is $5\frac{5}{11}$ past 1, and so on.

The minute hand is 7 minute divisions before the hour hand at $(0\times 5+7)\times 1\frac{1}{11}$ or $7\frac{7}{11}$ past 12, at $(3\times 5+7)\times\frac{12}{11}$ or 24 minutes past 3, and others.

Also the minute hand is 23 minute divisions behind the hour hand just when 60–23 minute divisions before and therefore is so at $(11\times 5+37)\times\frac{12}{11}$ or $100\frac{4}{11}$ minutes past 11 that is $40\frac{4}{11}$ past 12, at $(12\times 5+37)\times\frac{12}{11}$ or $105\frac{9}{11}$ minutes past 12 that is $45\frac{9}{11}$ minutes past 1, at $(4\times 5+37)\times\frac{12}{11}$ or $62\frac{2}{11}$ minutes past 4 that is $2\frac{2}{11}$ past 5, and the rest.

And generally the minute hand is m' minute divisions behind the hour hand at $\{h\times 5+(60-m')\}\frac{12}{11}$ and hence at $(h\times 5+12\times 5-m')\times\frac{12}{11}$ and at $\{(h+12)\times 5-m'\}\times\frac{12}{11}$ minutes past h o'clock and therefore also at $(h'\times 5-m')\times\frac{12}{11}$ minutes past h' o'clock.

150. If in m grains of a mixture of two compounds that substances A, B, form severally with a substance C there be n grains of C and A, B, enter into their compounds with C at the several

rates of a, and of b, grains to the grain, to find how much there is in the mixture of each of the substances A, B.

Let there be x grains of A. There are then $\frac{x}{a}$ grains of the compound of A and C, $\therefore$ $\frac{x}{a}(1-a)$ grains of C compounded with A, $\therefore$ $n-\frac{x}{a}(1-a)$ grains of C compounded with B, $\therefore$ $\frac{n-\frac{x}{a}(1-a)}{1-b}$ grains of the compound of B and C, $\therefore$

$$\frac{n-\frac{x}{a}(1-a)}{1-b}+\frac{x}{a}=m.$$

And this holds just when any of the following does:

$$\frac{n}{1-b}-\frac{\frac{x}{a}(1-a)}{1-b}+\frac{x}{a}=m,$$

$$\frac{n}{1-b}+\frac{x}{a}-\frac{x}{a}\,\frac{1-a}{1-b}=m \text{ (art. 122)},$$

$$\frac{n}{1-b}+\left(\frac{x}{a}-\frac{x}{a}\,\frac{1-a}{1-b}\right) \text{ or } \frac{n}{1-b}-\left(\frac{x}{a}\,\frac{1-a}{1-b}-\frac{x}{a}\right)=m$$

according as $\frac{x}{a}$ is greater or less than $\frac{x}{a}\,\frac{1-a}{1-b}$,

$$\frac{n}{1-b}+\frac{x}{a}\left(1-\frac{1-a}{1-b}\right) \text{ or } \frac{n}{1-b}-\frac{x}{a}\left(\frac{1-a}{1-b}-1\right)=m$$

according as 1 is greater or less than $\frac{1-a}{1-b}$,

$$\frac{x}{a}\left(1-\frac{1-a}{1-b}\right)=m-\frac{n}{1-b} \text{ or } \frac{x}{a}\left(\frac{1-a}{1-b}-1\right)=\frac{n}{1-b}-m \text{ (art. 145)},$$

$$\frac{x}{a} \text{ either}=\frac{m-\frac{n}{1-b}}{1-\frac{1-a}{1-b}}=\frac{\left(m-\frac{n}{1-b}\right)(1-b)}{\left(1-\frac{1-a}{1-b}\right)(1-b)}=\frac{m(1-b)-\frac{n}{1-b}(1-b)}{1-b-(1-a)}=\frac{m(1-b)-n}{1-b+a-1}$$

$$=\frac{m(1-b)-n}{1-1+a-b}=\frac{m(1-b)-n}{a-b} \text{ or }=\frac{\frac{n}{1-b}-m}{\frac{1-a}{1-b}-1}=\frac{\left(\frac{n}{1-b}-m\right)(1-b)}{\left(\frac{1-a}{1-b}-1\right)(1-b)}$$

$$=\frac{\frac{n}{1-b}(1-b)-m(1-b)}{1-a-(1-b)}=\frac{n-m(1-b)}{1-a+b-1}=\frac{n-m(1-b)}{1-1+b-a}=\frac{n-m(1-b)}{b-a},$$

$$x=\frac{m(1-b)-n}{a-b}a \text{ or } \frac{n-m(1-b)}{b-a}a.$$

The number of grains of B in the mixture is $\dfrac{n-\frac{x}{a}(1-a)}{1-b}b$ and may hence be found. But the number of grains of B may be got in the same way as the number of grains of A above by simply putting b where a is and a where b. Therefore by thus interchanging a and b it follows at once that there are in the mixture either $\dfrac{n-m(1-a)}{a-b}b$ or $\dfrac{m(1-a)-n}{b-a}b$ grains of B.

If $a=b$ the substances A and B cannot be distinguished from one another and all that is then known is that either of them or the two together make up $m-n$ grains of the mixture.

151. If two liquids that undergo no change of bulk by mixing weigh at the several rates per gallon of $1+a$ and $1+b$ times the weight of a gallon of water to find how much must be taken of them to make n gallons of a mixture weighing at the rate per gallon of $1+c$ times the weight of a gallon of water, a being greater than c and b less.

Let it be x gallons of the first liquid that must be taken. Then $n-x$ gallons of the other must be taken and since the liquids and their mixture weigh severally $x(1+a)$, $(n-x)(1+b)$, $n(1+c)$, water gallon weights

$$(n-x)(1+b)+x(1+a)=n(1+c)$$

which happens exactly when severally,—

$$n(1+b)-x(1+b)+x(1+a)=n(1+c),$$
$$x(1+a)-x(1+b)+n(1+b)=n(1+c),$$
$$x\{1+a-(1+b)\}=n(1+c)-n(1+b),$$
$$x(1+a-b-1)=n\{1+c-(1+b)\},$$
$$x(1-1+a-b)=n(1+c-b-1)=n(1-1+c-b),$$
$$x=\frac{n(c-b)}{a-b},$$
$$n-x=\frac{n(a-b)}{a-b}-\frac{n(c-b)}{a-b}=\frac{n(a-b)-n(c-b)}{a-b}=\frac{n\{a-b-(c-b)\}}{a-b}$$
$$=\frac{n(a-b+b-c)}{a-b}=\frac{n(a-c)}{a-b}.$$

152. Arithmetical questions of the direct kind may likewise be dealt with in this general symbolic way.

Thus if p pounds be put out at simple interest at the yearly rate of r pounds a pound in n years 1 pound bears nr pounds interest and p

pounds therefore pnr pounds, or p pounds bears in 1 year pr pounds interest and therefore in n years npr pounds. Hence p pounds rises in n years to $p+pnr$ which $=p(1+nr)$ pounds and the sum of money which in n years rises to the amount of a pounds is $\frac{a}{1+nr}$ pounds. A debt of a pounds due n years hence may therefore be paid now with $\frac{a}{1+nr}$ pounds reckoning simple interest at r pounds a pound a year and the discount on the debt therefore is

$$a-\frac{a}{1+nr}=a-a\frac{1}{1+nr}=a\left(1-\frac{1}{1+nr}\right)=a\left(\frac{1+nr}{1+nr}-\frac{1}{1+nr}\right)=a\frac{1+nr-1}{1+nr}$$

$$=a\frac{1-1+nr}{1+nr}=a\frac{nr}{1+nr}=\frac{anr}{1+nr}\text{ pounds.}$$

If the rate of interest be c pounds per cent. r the rate per pound is $\frac{1}{100}c$ or $0{\cdot}01c$.

153. When money is at compound interest at the yearly rate of r pounds a pound a sum of money is changed in a year at the rate of $1+r$ pounds a pound. Therefore 1 pound becomes at the end of 1 year $1+r$ pounds, at the end of 2 years $(1+r)^2$ pounds, at the end of 3 years $(1+r)^2(1+r)$ which (arts. 44, 108) $=(1+r)^{1+2}$ or $(1+r)^3$ pounds, at the end of 4 years $(1+r)^3(1+r)$ which $=(1+r)^{1+3}$ or $(1+r)^4$ pounds, and generally at the end of n years $(1+r)^n$ pounds. Hence in n years a sum of p pounds is changed at the rate of $(1+r)^n$ pounds a pound and therefore becomes $p(1+r)^n$ pounds. Wherefore the present worth of a pounds due n years hence reckoning compound interest at the rate of r pounds a pound a year, that is the sum of money which at this rate of interest would in n years mount up to a pounds, is $\frac{a}{(1+r)^n}$ pounds.

If the interest be due after each $\frac{1}{v}$ of a year instead of after each year but still at the rate of r pounds a pound simple interest a year money changes at the rate of $1+\frac{1}{v}r$ pounds a pound in $\frac{1}{v}$ of a year and therefore as before at the rate of $(1+\frac{1}{v}r)^v$ pounds a pound in 1 year. Then p pounds becomes at the end of n years $p\{(1+\frac{1}{v}r)^v\}^n$ which (arts. 45, 108) $=p(1+\frac{1}{v}r)^{nv}$ pounds.

154. If n men can singly do a work in $d_1, d_2, d_3, \ldots d_n$, days severally, in how many days can they do it together?

Taking each man to do an equal part of the work in every equal part of the time he works the 1st man does $\frac{1}{d_1}$ of the work in a day, the 2nd $\frac{1}{d_2}$, the 3rd $\frac{1}{d_3}$, . . . the nth $\frac{1}{d_n}$. Therefore working altogether they do $\frac{1}{d_1}+\frac{1}{d_2}+\frac{1}{d_3}+\cdots+\frac{1}{d_n}$ of the work in a day and can therefore do the whole work in $\dfrac{1}{\frac{1}{d_1}+\frac{1}{d_2}+\frac{1}{d_3}+\cdots+\frac{1}{d_n}}$ days.

In like manner if two men can together do a work in d days which one of them can do alone in d' days the other can do $\frac{1}{d}-\frac{1}{d'}$ of the work by himself in one day and the whole in $\dfrac{1}{\frac{1}{d}-\frac{1}{d'}}$ days.

155. If m men reap a acres in d days of h hours each, (1) how many acres can m' men reap in d' days of h' hours each? (2) how many men are needed to reap a' acres in d' days of h' hours each? (3) in how many days of h' hours each can m' men reap a' acres? and (4) how many hours a day must m' men work so as to reap a' acres in d' days?

Taking every man to reap in every equal part of every hour of every day an equal part of an acre,

since m	men in	d	days of	h	hours each reap	a	acres,
1	— —	d	— —	h	— — —	$\frac{1}{m}a$	—
1	— —	1	— —	h	— — —	$\frac{1}{d}\frac{1}{m}a$	--
1	— —		— —	1	— — —	$\frac{1}{h}\frac{1}{d}\frac{1}{m}a$	—
1	— —	1	— —	h'	— — —	$h'\frac{1}{h}\frac{1}{d}\frac{1}{m}a$	—
1	— —	d'	— —	h'	— — —	$d'h'\frac{1}{h}\frac{1}{d}\frac{1}{m}a$	—
m'	— —	d'	— —	h'	— — —	$m'd'h'\frac{1}{h}\frac{1}{d}\frac{1}{m}a$	—.

To reap a' acres in d' days of h' hours each $\dfrac{a'}{d'h'\frac{1}{h}\frac{1}{d}\frac{1}{m}a}$ men are needed. Since if m' men reap a' acres each of them reaps $\frac{1}{m'}a'$ acres it is in $\dfrac{\frac{1}{m'}a'}{h'\frac{1}{h}\frac{1}{d}\frac{1}{m}a}$ days of h' hours each that m' men can reap a' acres. And lastly since if m' men reap in d' days a' acres 1 man reaps in 1 day $\frac{1}{d'}\frac{1}{m'}a'$ acres it is $\dfrac{\frac{1}{d'}\frac{1}{m'}a'}{\frac{1}{h}\frac{1}{d}\frac{1}{m}a}$ hours a day that m' men must work for d' days to reap a' acres.

156. At what mean rate in ounces the cubic foot does a framework weigh made with $c_1, c_2, \ldots c_n$, cubic feet severally of n kinds of material weighing at the several mean rates of $u_1, u_2, \ldots u_n$, ounces the cubic foot?

The whole bulk of material being $c_1+c_2+\cdots+c_n$ cubic feet and the whole weight $c_1u_1+c_2u_2+\cdots+c_nu_n$ ounces the framework weighs at the mean rate of $\dfrac{1}{c_1+c_2+\cdots+c_n}(c_1u_1+c_2u_2+\cdots+c_nu_n)$ ounces the cubic foot.

157. How many ounces the cubic foot does a compound weigh composed of $w_1, w_2, \ldots w_n$, ounces severally of n substances weighing severally $u_1, u_2, \ldots u_n$, ounces the cubic foot?

The bulk of w_i ounces of a substance weighing u_i ounces the cubic foot being $\frac{w_i}{u_i}$ cubic feet the bulk of all the n substances is $\frac{w_1}{u_1}+\frac{w_2}{u_2}+\cdots+\frac{w_n}{u_n}$ cubic feet and if when change of bulk happens in composition it be at the rate of k cubic feet the cubic foot the bulk of the compound is $\left(\frac{w_1}{u_1}+\frac{w_2}{u_2}+\cdots+\frac{w_n}{u_n}\right)k$ cubic feet. Also the weight of the compound, being the weight of all the substances composing it, is $w_1+w_2+\cdots+w_n$ ounces. Hence the compound weighs

$$\frac{1}{\left(\frac{w_1}{u_1}+\frac{w_2}{u_2}+\cdots+\frac{w_n}{u_n}\right)k}(w_1+w_2+\cdots+w_n) \text{ ounces the cubic foot.}$$

If no change of bulk happens in forming the compound k is 1.

GENERAL THEOREMS

158. There are theorems in Arithmetic of so general a kind and of such common use as to rank only lower than the very Laws of the science. The conditions fulfilled by the numbers found in art. 146 and by the numbers that s and d there symbolize give two theorems, to wit—(1) The greater of two numbers is half the sum got by adding to, and the less half the remainder got by subtracting from, the sum of the numbers their difference—(2) The greater of two numbers is the sum got by adding to, and the less the remainder got by subtracting from, half the sum of the numbers half their difference. These theorems may be otherwise shown thus, a being any, and b any not greater, number:—

$$a \text{ is } \left(\frac{1}{2}\times 2\right)a \text{ and } \therefore = \frac{1}{2}(b-b+a+a) = \frac{1}{2}(a+b+a-b) = \frac{1}{2}\{a+b+(a-b)\}$$

$$= \frac{1}{2}(a+b)+\frac{1}{2}(a-b). \quad b \text{ is } \left(\frac{1}{2}\times 2\right)b \text{ and } \therefore = \frac{1}{2}(a-a+b+b) = \frac{1}{2}(a+b+b-a)$$

$$= \frac{1}{2}\{a+b-(a-b)\} = \frac{1}{2}(a+b)-\frac{1}{2}(a-b).$$

159. The sum of any number but 1 and its reciprocal is greater than 2 by the product of the differences from 1 of the number and the reciprocal. For if x be any number

$$x+\frac{1}{x} = 2-1-1+x+\frac{1}{x} = \text{both } 2+x-1+\frac{1}{x}-1 \text{ and } 2+\frac{1}{x}-1+x-1$$

and $\therefore$ according as x is not less or not greater than 1

$$= 2+(x-1)-\left(x\frac{1}{x}-\frac{1}{x}\right) \qquad \text{or} \qquad = 2+\left(\frac{1}{x}-1\right)-\left(x\frac{1}{x}-x\right)$$

$$= 2+\left\{x-1-(x-1)\frac{1}{x}\right\} \qquad\qquad = 2+\left\{\frac{1}{x}-1-x\left(\frac{1}{x}-1\right)\right\}$$

$$= 2+(x-1)\left(1-\frac{1}{x}\right) \qquad\qquad = 2+(1-x)\left(\frac{1}{x}-1\right).$$

160. If the excess of any number greater than 1 over its reciprocal be divided by the sum of the number and 1 the quotient is less than 1 by the number's reciprocal. For if x be any number greater than 1

$$\frac{x-\frac{1}{x}}{x+1} = \frac{x+(1-1)-\frac{1}{x}}{x+1} = \frac{x+1-\frac{1}{x}x-\frac{1}{x}}{x+1} = \frac{x+1-\left(\frac{1}{x}+\frac{1}{x}x\right)}{x+1} = \frac{x+1-\frac{1}{x}(1+x)}{x+1}$$

$$= \frac{x+1}{x+1}-\frac{\frac{1}{x}(x+1)}{x+1} = 1-\frac{1}{x}.$$

Also if the excess of any number greater than 1 over its reciprocal be divided by the excess of the number over 1 the quotient is greater than 1 by the number's reciprocal. For

$$\frac{x-\frac{1}{x}}{x-1}=\frac{x-1+1-\frac{1}{x}}{x-1}=\frac{x-1+\left(\frac{1}{x}x-\frac{1}{x}\right)}{x-1}=\frac{x-1+\frac{1}{x}(x-1)}{x-1}$$

$$=\frac{x-1}{x-1}+\frac{\frac{1}{x}(x-1)}{x-1}=1+\frac{1}{x}.$$

161. The difference of the second powers of two numbers is equal to the product of the sum and the difference of the numbers. For if a, b, be any two numbers of which a is not the less

$$a^2-b^2=a^2-ab+ab-b^2=\begin{cases}a^2-ba+(ab-b^2)=(a-b)a+(a-b)b=(a-b)(a+b).\\ a^2-ab+(ba-b^2)=a(a-b)+b(a-b)=(a+b)(a-b).\end{cases}$$

Hence if any number ($2a$) be cut into two equal parts (a, a) and into two unequal parts ($a+b$, $a-b$ (art. 158)) the product of the unequal parts is less than the product of the equal parts by the second power of the difference (b) between each of the equal and either of the unequal parts.

Hence too of any three numbers ($a-b$, a, $a+b$) whereof the first is less than the second and the second than the third by the same difference (b) the product of the greatest and least is less than the second power of the middle one by the second power of the common difference.

Moreover (art. 158) $ab=\left\{\frac{1}{2}(a+b)+\frac{1}{2}(a-b)\right\}\left\{\frac{1}{2}(a+b)-\frac{1}{2}(a-b)\right\}$

and $\therefore=\left\{\frac{1}{2}(a+b)\right\}^2-\left\{\frac{1}{2}(a-b)\right\}^2$.

162. The second power of the sum of two numbers is greater, and the second power of their difference is less, than the sum of their second powers by twice their product. For a, b, being any two numbers such that a is not less than b

$$(a+b)^2=a(a+b)+b(a+b)=a^2+ab+(ba+b^2)=a^2+ab+ab+b^2$$
$$=a^2+b^2+ab+ab=a^2+b^2+(ab+ab)\ \text{ or }\ a^2+b^2+2ab.$$

$$(a-b)^2=a(a-b)-b(a-b)=a^2-ab-(ba-b^2)=a^2-ab+b^2-ab$$
$$=a^2+b^2-ab-ab=a^2+b^2-(ab+ab)\ \text{ or }\ a^2+b^2-2ab.$$

Hence $\frac{1}{2}\{(a+b)^2+(a-b)^2\}=\frac{1}{2}\{a^2+b^2+2ab+(a^2+b^2-2ab)\}=a^2+b^2$

(art. 158). And $\frac{1}{2}\{(a+b)^2-(a-b)^2=\frac{1}{2}\{a^2+b^2+2ab-(a^2+b^2-2ab)\}=2ab.$

$$\therefore\ (a+b)^2+(a-b)^2 = 2(a^2+b^2) = 2a^2+2b^2,$$

$$(a+b)^2-(a-b)^2 = 2\times 2ab = (2\times 2)ab \text{ or } 4ab,$$

$$ab = \frac{1}{4}\{(a+b)^2-(a-b)^2\} = \left(\frac{1}{2}\right)^2(a+b)^2-\left(\frac{1}{2}\right)^2(a-b)^2$$

and $\therefore$ (art. 140) $= \left\{\frac{1}{2}(a+b)\right\}^2 - \left\{\frac{1}{2}(a-b)\right\}^2$ as in art. 161.

If a be α^2 and b β^2 $\left\{\frac{1}{2}(\alpha^2+\beta^2)\right\}^2 - \left\{\frac{1}{2}(\alpha^2-\beta^2)\right\}^2 = \alpha^2\beta^2 = (\alpha\beta)^2$

and $\left\{\frac{1}{2}(\alpha^2+\beta^2)\right\}^2 = (\alpha\beta)^2 + \left\{\frac{1}{2}(\alpha^2-\beta^2)\right\}^2.$

163. If a, b, c, be three numbers of which the sum of any two is not less than the third $(a+b-c)(c+a-b)(b+c-a)(a+b+c)$

$$= 2\{(bc)^2+(ca)^2+(ab)^2\}-\{(a^2)^2+(b^2)^2+(c^2)^2\}.$$

For $(a+b-c)(c+a-b)(b+c-a)(a+b+c)$

$= \{(a+b-c)(a+c-b)\}(b+c-a)(b+c+a)$ and $\therefore =$ either $[\{a+(b-c)\}\{a-(b-c)\}]\{(b+c)^2-a^2\}$ or $[\{a-(c-b)\}\{a+(c-b)\}]\{(b+c)^2-a^2\}$

$= $ either $\{a^2-(b-c)^2\}(b^2+c^2+2bc-a^2)$ or $\{a^2-(c-b)^2\}(b^2+c^2+2bc-a^2)$

$= \{a^2-(b^2+c^2-2bc)\}(b^2+c^2+2bc-a^2) = \{a^2+2bc-(b^2+c^2)\}(b^2+c^2+2bc-a^2)$

$= \{2bc+a^2-(b^2+c^2)\}\{2bc+(b^2+c^2)-a^2\}$

$= $ either $[2bc+\{a^2-(b^2+c^2)\}][2bc-\{a^2-(b^2+c^2)\}]$

or $\{2bc-(b^2+c^2-a^2)\}\{2bc+(b^2+c^2-a^2)\}$

$= $ either $(2bc)^2-\{a^2-(b^2+c^2)\}^2$ or $(2bc)^2-(b^2+c^2-a^2)^2$

$= 2^2b^2c^2-\{(a^2)^2+(b^2+c^2)^2-2a^2(b^2+c^2)\}$

$= 4b^2c^2-[(a^2)^2+\{(b^2)^2+(c^2)^2+2b^2c^2\}-(2a^2b^2+2a^2c^2)]$

$= 4b^2c^2-\{(a^2)^2+(b^2)^2+(c^2)^2+2b^2c^2-2a^2c^2-2a^2b^2\}$

$= 4b^2c^2+2a^2b^2+2c^2a^2-2b^2c^2-\{(a^2)^2+(b^2)^2+(c^2)^2\}$

$= 4b^2c^2-2b^2c^2+2c^2a^2+2a^2b^2-\{(a^2)^2+(b^2)^2+(c^2)^2\}$

$= (4-2)(bc)^2+2(ca)^2+2(ab)^2-\{(a^2)^2+(b^2)^2+(c^2)^2\}$

$= 2\{(bc)^2+(ca)^2+(ab)^2\}-\{(a^2)^2+(b^2)^2+(c^2)^2\}.$

164. *Def.* A set of numbers determined one by one in order after a fixed law is called a SERIES; the several numbers are called the TERMS of the series, the first and last terms the EXTREMES, and all the other terms the MEANS.

165. *Def.* Numbers are said to be in ARITHMETICAL PROGRESSION when either every one but the first is greater or every one but the first is less than the one next before it by the same number called the COMMON DIFFERENCE.

Let a be the first term of a series of numbers in arithmetical progression and b the common difference.

If the numbers go on increasing the 2nd term is $a+b$, the third term is $a+b+b$ and $\therefore = a+(b+b)$ that is $a+2b$, $\therefore$ the 4th term $= a+2b+b = a+(2b+b)$ that is $a+3b$, and if generally after the law of the first 4 terms the ith term be operationally equivalent to $a+(i-1)b$ the next following term the $(i+1)$th $= a+(i-1)b+b$ $= a+\{(i-1)b+b\}$ that is $a+ib$ after the same law. This law $\therefore$ that holds for the first 4 terms if it hold for any term whatever holds for the next term following. It holds $\therefore$ for the $4+1$ or 5th term $\therefore$ for the $5+1$ or 6th term and so on to the end of the series. The ith term $\therefore = a+(i-1)b$.

But if the numbers go on decreasing the 2nd term is $a-b$, the 3rd term is $a-b-b$ and $\therefore = a-(b+b)$ that is $a-2b$, $\therefore$ the 4th $= a-2b-b = a-(b+2b) = a-(2b+b)$ that is $a-3b$, and if generally the ith $= a-(i-1)b$ the $(i+1)$th $= a-(i-1)b-b = a-\{b+(i-1)b\}$ $= a-\{(i-1)b+b\}$ that is $a-ib$ and follows the same law. This law $\therefore$ holds for the 5th term $\therefore$ for the 6th and so on throughout the series.

Hence in particular the n consecutive whole numbers $1, 2, 3, \ldots n$, are severally equivalent operationally to $1, 1+1, 1+2, \ldots 1+(n-1)$, whereof the ith term is $1+(i-1)$ and $= i-1+1$ that is i; and the same numbers in backward order are severally equivalent operationally to $n, n-1, n-2, \ldots n-(n-1)$, whereof the ith term is $n-(i-1)$ and $= n+1-i = n-i+1$. Wherefore if u_i be the ith term of any series of n numbers $u_1, u_2, \ldots u_n$, the ith term counting backwards from the nth or last term as the first is $u_{n-(i-1)}$.

166. Of an arithmetical progression the sum of every two terms equidistant from the beginning and the end is the same number, and the sum of all the terms is equal to half the product made by multiplying that number by the number expressing how many terms there are.

Let a be the first term and b the common difference of a series of n numbers $u_1, u_2, u_3, \ldots u_n$, in arithmetical progression.

If the numbers increase (art. 165)

$$u_i = a+(i-1)b \quad \text{and} \quad u_{n-(i-1)} = a+\{n-(i-1)-1\}b, \ \therefore$$

$$u_i+u_{n-(i-1)} = a+(i-1)b+[a+\{n-(i-1)-1\}b] = a+(i-1)b+a+\{n-1-(i-1)\}b$$
$$= a+a+\{(n-1)b-(i-1)b\}+(i-1)b = 2a+(n-1)b-(i-1)b+(i-1)b$$
$$= 2a+(n-1)b.$$

But if the numbers decrease (art. 165)

$$u_i = a-(i-1)b \quad \text{and} \quad u_{n-(i-1)} = a-\{n-(i-1)-1\}b, \ \therefore$$

$$u_i+u_{n-(i-1)}=a-(i-1)b+[a-\{n-(i-1)-1\}b]=a-(i-1)b+a-\{n-1-(i-1)\}b$$
$$=a+a-(i-1)b-\{(n-1)b-(i-1)b\}=2a-(i-1)b+(i-1)b-(n-1)b$$
$$=2a-(n-1)b.$$

So that $u_i+u_{n-(i-1)}$ is the same whichever of the numbers 1, 2, 3, ... n, the number i is.

Moreover $2(u_1+u_2+u_3+\cdots+u_n)$ is

$$u_1+u_2+u_3+\cdots+u_n+(u_1+u_2+u_3+\cdots+u_n) \text{ and } \therefore$$
$$=u_1+\cdots+u_n+u_1+u_2+\cdots+u_n=u_1+u_n+u_2+u_{n-1}+u_3+u_{n-2}+\cdots\cdots+u_n+u_1$$

by commutation of the additions and taking the us in the order the first of the one set and the last of the other then the 2nd of the one and the 2nd last of the other then the 3rd and the 3rd last and so on, $=u_1+u_n+(u_2+u_{n-1})+(u_3+u_{n-2})+\cdots+(u_n+u_1)$

$$=u_i+u_{n-(i-1)}+(u_i+u_{n-(i-1)})+\cdots+(u_i+u_{n-(i-1)})$$

by what has just been shown, that is $n(u_i+u_{n-(i-1)})$.

$$\therefore\ u_1+u_2+u_3+\cdots+u_n=\frac{1}{2}n(u_i+u_{n-(i-1)}).$$

167. The matter of arts. 165, 166, may be put to sundry uses such for example as the following.

(1). $1+2+3+\cdots+n=\frac{1}{2}n(n+1)$. Thus the sum of 10000 terms of the series 1, 2, 3, ... $=\frac{1}{2}\times 10000\times 10001$ and $\therefore$ is 50005000.

The nth term of the series 1, 3, 5, . . .

$$=1+(n-1)\times 2=1+(n\times 2-2)=n\times 2-2+1=n\times 2+1-2=n\times 2-(2-1)$$

that is $n\times 2-1$. $\therefore$ the sum of n terms $1+3+5+\cdots+(n\times 2-1)$

$$=\frac{1}{2}n(n\times 2-1+1)=2\times\frac{1}{2}n^2=\left(2\times\frac{1}{2}\right)n^2 \text{ that is } n^2.$$

Of the series 3, $4\frac{1}{2}$, 6, . . . the 456th term $=3+(456-1)\times 1\frac{1}{2}$ and is $\therefore$ $685\frac{1}{2}$, also the sum of 456 terms $=\frac{1}{2}\times 456\times\left(3+685\frac{1}{2}\right)$ and is $\therefore$ 156978.

The 223rd term of 611, $608\frac{1}{3}$, $605\frac{2}{3}$, . . . $=611-(223-1)\times 2\frac{2}{3}$ and $\therefore$ is 19, $\therefore$ the sum of the first 223 terms $=\frac{1}{2}\times 223\times(19+611)$ and $\therefore$ is 70245.

(2). To find m arithmetical means between a given number a and a given greater number a'. If b be the common difference of the series of increasing numbers in arithmetical progression of which the terms in order are a m numbers and a', since the last of the m numbers is the $(1+m)$th term and $\therefore$ a' the $(1+m+1)$th,

$a' = a+(1+m+1-1)b$ and $\therefore = a+\{1+m+(1-1)\}b$ that is $a+(1+m)b$.

Wherefore $(1+m)b = a'-a$, $b = \frac{1}{1+m}(a'-a)$. And $\because$ $1+m-1 = 1-1+m$ the several means then $= a+b,\ a+2b, \ldots a+mb$.

Thus to find 10 arithmetical means between 19 and 74; $19+(1+10)b = 74$, $b = \frac{1}{11}(74-19)$ that is 5, and the means $\therefore$ are 24, 29, 34, 39, 44, 49, 54, 59, 64, 69.

The thing may also be done by finding the common difference of the series of decreasing numbers in arithmetical progression with the terms in order a' the means and a. Then $a'-(1+m+1-1)b = a$.

(3). To find the series of numbers in arithmetical progression whose ith term is u and $(i+i')$th $u_{i+i'}$. The series must have a first term a and a common difference b such that

either $\begin{cases} a+(i-1)b = u_i \text{ and} \\ a+(i+i'-1)b = u_{i+i'} \end{cases}$ or $\begin{cases} a-(i-1)b = u_i \text{ and} \\ a-(i+i'-1)b = u_{i+i'} \end{cases}$

according as u_i is less or greater than $u_{i+i'}$. Of these alternative pairs of conditions the former holds just when first $a = u_i-(i-1)b$ and then either $u_{i+i'} = u_i-(i-1)b+(i-1+i')b$ $= u_i-(i-1)b+\{(i-1)b+i'b\} = u_i-(i-1)b+(i-1)b+i'b$, $i'b = u_{i+i'}-u_i$, or $b = \frac{1}{i'}(u_{i+i'}-u_i)$. And the latter alternative pair holds just when first $a = u_i+(i-1)b$ and then either $u_{i+i'} = u_i+(i-1)b-(i'+i-1)b$ $= u_i+(i-1)b-\{i'+(i-1)\}b = u_i+(i-1)b-\{i'b+(i-1)b\}$ $= u_i+(i-1)b-(i-1)b-i'b = u_i+\{(i-1)b-(i-1)b\}-i'b$, $i'b = u_i-u_{i+i'}$, or $b = \frac{1}{i'}(u_i-u_{i+i'})$.

Thus of the series of numbers in arithmetical progression whose 20th term is 64 and 45th 139 the common difference is $\frac{1}{45-20}(139-64)$ or 3 and the first term is $64-(20-1)\times 3$ or 7; hence the series is 7, 10, 13,....

Again the series of numbers in arithmetical progression whose 15th term is 27·28 and 104th 6·81 has its common difference $\left(\frac{1}{104-15}(27{\cdot}28-6{\cdot}81)\right)$ 0·23 its first term $(27{\cdot}28+(15-1)\times 0{\cdot}23)$ 30·5 and is itself 30·5, 30·27, 30·04, ...

168. If v_i stand for $a+ib$

$$v_1v_2v_3\ldots v_{r-2}v_{r-1}v_r+v_2v_3v_4\ldots v_{r-1}v_rv_{r+1}+v_3v_4v_5\ldots v_rv_{r+1}v_{r+2}+\cdots+v_nv_{n+1}v_{n+2}\ldots v_{n+r-3}v_{n+r-2}v_{n+r-1}$$

$$=\frac{v_nv_{n+1}v_{n+2}\ldots v_{n+r-3}v_{n+r-2}v_{n+r-1}v_{n+r}-v_0v_1v_2v_3\ldots v_{r-2}v_{r-1}v_r}{(r+1)b}.$$

Since by the nature of subtraction (arts. 59, 103)

$$\alpha=\alpha-\beta+\beta=\alpha-\beta+\beta-\gamma+\gamma=\alpha-\beta+\beta-\gamma+\gamma-\delta+\delta=\ \cdot\ \cdot\ \cdot$$

when $\alpha, \beta, \gamma, \delta, \ldots$ are any numbers that give the statements meaning

$$v_nv_{n+1}\ldots v_{n+r-1}v_{n+r}-v_0v_1\ldots v_{r-1}v_r$$

$$=v_nv_{n+1}\ldots v_{n+r-1}v_{n+r}-v_{n-1}v_n\ldots v_{n+r-2}v_{n+r-1}+v_{n-1}v_n\ldots v_{n+r-2}v_{n+r-1}$$
$$-v_{n-2}v_{n-1}\ldots v_{n+r-3}v_{n+r-2}+v_{n-2}v_{n-1}\ldots v_{n+r-3}v_{n+r-2}-$$
$$\cdots\cdots-v_1v_2\ldots v_rv_{r+1}+v_1v_2\ldots v_rv_{r+1}-v_0v_1\ldots v_{r-1}v_r,\ \therefore$$

$$=(v_n\ldots v_{n+r-1})v_{n+r}-(v_n\ldots v_{n+r-1})v_{n-1}+\{(v_{n-1}\ldots v_{n+r-2})v_{n+r-1}-(v_{n-1}\ldots v_{n+r-2})v_{n-2}\}+$$
$$\cdots\cdots+\{(v_2\ldots v_{r+1})v_{r+2}-(v_2\ldots v_{r+1})v_1\}+\{(v_1\ldots v_r)v_{r+1}-(v_1\ldots v_r)v_0\}$$
$$=(v_n\ldots v_{n+r-1})(v_{n+r}-v_{n-1})+(v_{n-1}\ldots v_{n+r-2})(v_{n+r-1}-v_{n-2})$$
$$+(v_{n-2}\ldots v_{n+r-3})(v_{n+r-2}-v_{n-3})+\cdots+(v_1\ldots v_r)(v_{r+1}-v_0).$$

But $\because v_{i+r+1}-v_i=a+\{i+(r+1)\}b-(a+ib)=a+\{ib+(r+1)b\}-(a+ib)$ $=a+ib+(r+1)b-(a+ib)=a+ib-(a+ib)+(r+1)b$ that is $(r+1)b$, and $n-k+r+1=n+r+1-k=n+r-(k-1)$ if k be not less than 1,

$$(r+1)b=v_{n+r}-v_{n-1}=v_{n+r-1}-v_{n-2}=v_{n+r-2}-v_{n-3}=\cdots=v_{r+1}-v_0$$

and the above $\therefore$

$$=(v_n\ldots v_{n+r-1}+v_{n-1}\ldots v_{n+r-2}+v_{n-2}\ldots v_{n+r-3}+\cdots+v_1\ldots v_r)(r+1)b.$$

$$\therefore\ \frac{v_n\ldots v_{n+r-1}v_{n+r}-v_0v_1\ldots v_r}{(r+1)b}=v_n\ldots v_{n+r-1}+v_{n-1}\ldots v_{n+r-2}+v_{n-2}\ldots v_{n+r-3}+\cdots+v_1\ldots v_r$$

and $\therefore$ also $=v_1v_2\ldots v_{r-2}v_{r-1}v_r+v_2v_3\ldots v_{r-1}v_rv_{r+1}+v_3v_4\ldots v_rv_{r+1}v_{r+2}+\cdots+v_nv_{n+1}\ldots v_{n+r-3}v_{n+r-2}v_{n+r-1}.$

When r is 1 this gives

$$a+b+(a+2b)+(a+3b)+\cdots+\{a+(n-1)b\}=\frac{\{a+(n-1)b\}(a+nb)-a(a+b)}{2b}$$

and $\therefore$

$$=\frac{\{a+(nb-b)\}a+\{a+(n-1)b\}nb-(a^2+ab)}{2b}$$

$$=\frac{(a+nb-b)a+\{a+(n-1)b\}nb-ab-a^2}{2b}$$

$$=\frac{a^2+(nb)a-ba+\{a+(n-1)b\}nb-ab-a^2}{2b}$$

$$=\frac{a^2-a^2+nba+\{a+(n-1)b\}nb-ba-ab}{2b}=\frac{nab+n\{a+(n-1)b\}b-(ab+ab)}{2b}$$

$$=\frac{n[a+\{a+(n-1)b\}]b-2ab}{2b}=\frac{n\{a+a+(n-1)b\}b-a\times 2b}{2b}$$

$$=\frac{[n\{2a+(n-1)b\}]b}{2b}-\frac{a\times 2b}{2b}=\frac{n\{2a+(n-1)b\}}{2}-a.$$

Wherefore $a+(a+b)+(a+2b)+\cdots+\{a+(n-1)b\}$

$$=a+b+(a+2b)+\cdots+\{a+(n-1)b\}+a=\frac{n\{2a+(n-1)b\}}{2}-a+a$$

$=\frac{n\{2a+(n-1)b\}}{2}$ and $\therefore$ (art. 124) $=\frac{1}{2}n\{2a+(n-1)b\}$ as in art. 166.

It $\therefore$ also still further $=\frac{1}{2}n\times 2a+\frac{1}{2}n(n-1)b=\frac{1}{2}\times 2na+\left\{\frac{1}{2}n(n-1)\right\}b$

$$=\left(\frac{1}{2}\times 2\right)na+\frac{n(n-1)}{2}b \text{ that is } na+\frac{n(n-1)}{2}b.$$

Again when a is 0 and b 1 $1.2.3\ldots(r-2)(r-1)r+2.3.4\ldots(r-1)r(r+1)+\cdots+n(n+1)(n+2)\ldots(n+r-3)(n+r-2)(n+r-1)$

$=\frac{n(n+1)(n+2)\ldots(n+r-2)(n+r-1)(n+r)}{r+1}$ of which particular cases are

$$1+2+3+\cdots+n=\frac{n(n+1)}{2},$$

$$1.2+2.3+3.4+\cdots+n(n+1)=\frac{n(n+1)(n+2)}{3},$$

$$1.2.3+2.3.4+3.4.5+\cdots+n(n+1)(n+2)=\frac{n(n+1)(n+2)(n+3)}{4}.$$

By help of these theorems simple expressions may be found operationally equivalent to the sums of the terms of several series. Thus

$$a+(a+b)+(a+2b)+\cdots+\{a+(n-1)b\}=a+a+b+a+2b+\cdots+a+(n-1)b$$
$$=a+a+a+\cdots+a+b+2b+\cdots+(n-1)b=na+\{b+2b+\cdots+(n-1)b\}$$
$$=na+\{1+2+\cdots+(n-1)\}b \text{ and } \therefore =na+\frac{(n-1)n}{2}b=na+\frac{n(n-1)}{2}b$$

as was above found.

Again $\because i^2=i\{i+(1-1)\}=i(i+1-1)=i(i+1)-i$

$1^2+2^2+3^2+\cdots+n^2=1.2-1+(2.3-2)+(3.4-3)+\cdots+\{n(n+1)-n\}$ and $\therefore$

$$=1.2-1+2.3-2+3.4-3+\cdots+n(n+1)-n$$
$$=1.2+2.3+3.4+\cdots+n(n+1)-n-\cdots-3-2-1$$
$$=1.2+2.3+3.4+\cdots+n(n+1)-(1+2+3+\cdots+n)=\frac{n(n+1)(n+2)}{3}-\frac{n(n+1)}{2}$$
$$=\frac{\{n(n+1)(n+2)\}\times 2}{3\times 2}-\frac{\{n(n+1)\}\times 3}{2\times 3}=\frac{n(n+1)(n+2)\times 2-n(n+1)\times 3}{6}$$
$$=\frac{n(n+1)(n\times 2+2^2-3)}{6}=\frac{n(n+1)\{n\times 2+(4-3)\}}{6}=\frac{n(n+1)(2n+1)}{6}.$$

Once more $\because i^2=(i+1)i-i=(i+1)(i+2-2)-(i+1-1)$

$$=(i+1)(i+2)-(i+1)\times 2+1-(i+1)=(i+1)(i+2)+1-(i+1)-(i+1)\times 2$$
$$=(i+1)(i+2)+1-\{(i+1)\times 2+(i+1)\}=(i+1)(i+2)+1-(i+1)(2+1)$$

$$i^3 = i\{(i+1)(i+2)+1-(i+1)\times 3\} = i(i+1)(i+2)+i-i(i+1)\times 3 \text{ and } \therefore$$

$$1^3+2^3+3^3+\cdots+n^3 = 1.2.3+1-1.2.3+(2.3.4+2-2.3.3)+(3.4.5+3-3.4.3)+ \cdots +\{n(n+1)(n+2)+n-n(n+1)\times 3\}$$

$$= 1.2.3+1-1.2.3+2.3.4+2-2.3.3+\cdots+n(n+1)(n+2)+n-n(n+1)\times 3$$

$$= 1.2.3+2.3.4+\cdots+n(n+1)(n+2)+1+2+\cdots+n -n(n+1)\times 3-\cdots-2.3.3-1.2.3$$

$$= 1.2.3+2.3.4+\cdots+n(n+1)(n+2)+(1+2+\cdots+n) -[(1.2)\times 3+(2.3)\times 3+\cdots+\{n(n+1)\}\times 3]$$

$$= 1.2.3+2.3.4+\cdots+n(n+1)(n+2)+(1+2+\cdots+n) -\{1.2+2.3+\cdots+n(n+1)\}\times 3$$

$$= \frac{n(n+1)(n+2)(n+3)}{4}+\frac{n(n+1)}{2}-\frac{n(n+1)(n+2)}{3}\times 3$$

$$= \frac{n(n+1)(n+2)(n+3)+\{n(n+1)\}\times 2-\{n(n+1)(n+2)\}\times 4}{4}$$

$$= \frac{n(n+1)\{(n+2)(n+3)+2-(n+2)\times 4\}}{4}$$

$$= \frac{n(n+1)\{n(n+3)+2(n+3)+2-(n\times 4+2\times 4)\}}{4}$$

$$= \frac{n(n+1)\{n^2+n\times 3+(2n+2.3)+2-8-4n\}}{4}$$

$$= \frac{n(n+1)(n^2+3n+2n+6+2-8-4n)}{4} = \frac{n(n+1)(n^2+3n+2n-4n+6+2-8)}{4}$$

$$= \frac{n(n+1)\{n^2+(3n+2n-4n)+(6+2-8)\}}{4} = \frac{n(n+1)\{n^2+(3+2-4)n\}}{4}$$

$$= \frac{\{n(n+1)\}(n^2+n)}{4} = \frac{\{n(n+1)\}^2}{2^2} = \left\{\frac{n(n+1)}{2}\right\}^2.$$

169. If r be any whole number greater than 1 and v_i stand for $a+ib$

$$\frac{1}{v_1}\frac{1}{v_2}\frac{1}{v_3}\cdots\frac{1}{v_{r-2}}\frac{1}{v_{r-1}}\frac{1}{v_r}+\frac{1}{v_2}\frac{1}{v_3}\frac{1}{v_4}\cdots\frac{1}{v_{r-1}}\frac{1}{v_r}\frac{1}{v_{r+1}}+\frac{1}{v_3}\frac{1}{v_4}\frac{1}{v_5}\cdots\frac{1}{v_r}\frac{1}{v_{r+1}}\frac{1}{v_{r+2}}+ \cdots+\frac{1}{v_n}\frac{1}{v_{n+1}}\frac{1}{v_{n+2}}\cdots\frac{1}{v_{n+r-3}}\frac{1}{v_{n+r-2}}\frac{1}{v_{n+r-1}}$$

$$= \frac{\frac{1}{v_1}\frac{1}{v_2}\frac{1}{v_3}\cdots\frac{1}{v_{r-2}}\frac{1}{v_{r-1}}-\frac{1}{v_{n+1}}\frac{1}{v_{n+2}}\cdots\frac{1}{v_{n+r-2}}\frac{1}{v_{n+r-1}}}{(r-1)b}.$$

$$\because v_{i+r-1}-v_i = a+\{i+(r-1)\}b-(a+ib) = a+\{ib+(r-1)b\}-(a+ib)$$
$$= a+ib+(r-1)b-(a+ib) = a+ib-(a+ib)+(r-1)b,$$
$$(r-1)b = v_r-v_1 = v_{r+1}-v_2 = v_{r+2}-v_3 = \cdots = v_{n+r-1}-v_n \text{ and } \therefore$$

$$\left(\begin{array}{l}\frac{1}{v_1}\frac{1}{v_2}\cdots\frac{1}{v_{r-1}}\frac{1}{v_r}+\frac{1}{v_2}\frac{1}{v_3}\cdots\frac{1}{v_r}\frac{1}{v_{r+1}}+\frac{1}{v_3}\frac{1}{v_4}\cdots\frac{1}{v_{r+1}}\frac{1}{v_{r+2}}+\\ \qquad\qquad\cdots+\frac{1}{v_n}\frac{1}{v_{n+1}}\cdots\frac{1}{v_{n+r-2}}\frac{1}{v_{n+r-1}}\end{array}\right)(r-1)b$$

$$=\left(\frac{1}{v_1}\cdots\frac{1}{v_r}\right)(v_r-v_1)+\left(\frac{1}{v_2}\cdots\frac{1}{v_{r+1}}\right)(v_{r+1}-v_2)+\left(\frac{1}{v_3}\cdots\frac{1}{v_{r+2}}\right)(v_{r+2}-v_3)+$$

$$\cdots+\left(\frac{1}{v_n}\cdots\frac{1}{v_{n+r-1}}\right)(v_{n+r-1}-v_n)$$

$$=\left(\frac{1}{v_1}\cdot\cdot\frac{1}{v_r}\right)v_r-\left(\frac{1}{v_1}\cdot\cdot\frac{1}{v_r}\right)v_1+\left\{\left(\frac{1}{v_2}\cdot\cdot\frac{1}{v_{r+1}}\right)v_{r+1}-\left(\frac{1}{v_2}\cdot\cdot\frac{1}{v_{r+1}}\right)v_2\right\}$$

$$+\left\{\left(\frac{1}{v_3}\cdot\cdot\frac{1}{v_{r+2}}\right)v_{r+2}-\left(\frac{1}{v_3}\cdot\cdot\frac{1}{v_{r+2}}\right)v_3\right\}+\cdot\cdot+\left\{\left(\frac{1}{v_n}\cdot\cdot\frac{1}{v_{n+r-1}}\right)v_{n+r-1}-\left(\frac{1}{v_n}\cdot\cdot\frac{1}{v_{n+r-1}}\right)v_n\right\}$$

$$=\frac{1}{v_1}\cdot\cdot\frac{1}{v_{r-1}}\frac{1}{v_r}v_r-\left(v_1\frac{1}{v_1}\right)\frac{1}{v_2}\cdot\cdot\frac{1}{v_r}+\frac{1}{v_2}\cdot\cdot\frac{1}{v_r}\frac{1}{v_{r+1}}v_{r+1}-\left(v_2\frac{1}{v_2}\right)\frac{1}{v_3}\cdot\cdot\frac{1}{v_{r+1}}+$$

$$\cdots\cdot+\frac{1}{v_n}\cdot\cdot\frac{1}{v_{n+r-2}}\frac{1}{v_{n+r-1}}v_{n+r-1}-\left(v_n\frac{1}{v_n}\right)\frac{1}{v_{n+1}}\cdot\cdot\frac{1}{v_{n+r-1}}$$

$$=\frac{1}{v_1}\cdot\cdot\frac{1}{v_{r-1}}-\frac{1}{v_2}\cdot\cdot\frac{1}{v_r}+\frac{1}{v_2}\cdot\cdot\frac{1}{v_r}-\frac{1}{v_3}\cdot\cdot\frac{1}{v_{r+1}}+\frac{1}{v_3}\cdot\cdot\frac{1}{v_{r+1}}-$$

$$\cdots-\frac{1}{v_n}\cdot\cdot\frac{1}{v_{n+r-2}}+\frac{1}{v_n}\cdot\cdot\frac{1}{v_{n+r-2}}-\frac{1}{v_{n+1}}\cdot\cdot\frac{1}{v_{n+r-1}}$$

$$=\frac{1}{v_1}\frac{1}{v_2}\cdots\frac{1}{v_{r-1}}-\frac{1}{v_{n+1}}\cdots\frac{1}{v_{n+r-2}}\frac{1}{v_{n+r-1}}.$$

Whence the proposition at once follows.

If a be 0 and b 1 there is when r is 2

$$\frac{1}{1}\frac{1}{2}+\frac{1}{2}\frac{1}{3}+\frac{1}{3}\frac{1}{4}+\cdots+\frac{1}{n}\frac{1}{n+1}=1-\frac{1}{n+1}$$

and $\therefore=\frac{n+1}{n+1}-\frac{1}{n+1}=\frac{n+1-1}{n+1}=\frac{n+(1-1)}{n+1}$ that is $\frac{n}{n+1}$. When r is 3

$$\frac{1}{1}\frac{1}{2}\frac{1}{3}+\frac{1}{2}\frac{1}{3}\frac{1}{4}+\frac{1}{3}\frac{1}{4}\frac{1}{5}+\cdots+\frac{1}{n}\frac{1}{n+1}\frac{1}{n+2}=\frac{\frac{1}{2}-\frac{1}{n+1}\frac{1}{n+2}}{2}$$

$$\text{and }\therefore=\frac{\frac{(n+1)(n+2)}{2(n+1)(n+2)}-\frac{2}{\{(n+2)(n+1)\}\times 2}}{2}=\frac{\left\{\frac{(n+1)(n+2)-2}{2(n+1)(n+2)}\right\}}{2}$$

$$=\frac{n(n+2)+(n+2)-2}{2\times 2(n+1)(n+2)}=\frac{n^2+n\times 2+n+2-2}{(2\times 2)(n+1)(n+2)}=\frac{n^2+(2n+n)+(2-2)}{4(n+1)(n+2)}$$

$$=\frac{n^2+3n}{4(n+1)(n+2)}=\frac{(n+3)n}{4(n+1)(n+2)}=\frac{n(n+3)}{4(n+1)(n+2)}.$$

170. *Def.* Numbers are said to be in GEOMETRICAL PROGRESSION when the division of every one but the first by the one next before it gives for quotient the same number called the COMMON QUOTIENT.

Let a be the first term and c the common quotient of a series of numbers in geometrical progression.

The second term then by definition is ca, the third term likewise is cca and $\therefore = (cc)a$ that is c^2a, $\therefore$ the 4th term $= cc^2a = (cc^2)a$ that is c^3a, and if generally after the law of the first 4 terms the ith term be operationally equivalent to $c^{i-1}a$ the next following term the $(i+1)$th $= cc^{i-1}a = (cc^{i-1})a$ that is c^ia after the same law. Hence this law which holds for the first 4 terms if it hold for any term whatever holds for the next following term. It holds $\therefore$ for the $(4+1)$th or 5th term $\therefore$ for the $(5+1)$th or 6th and so on throughout. Any term the ith $\therefore = c^{i-1}a$.

Hence if the series be $u_1, u_2, u_3, \ldots u_n$,

$$u_1+u_2+u_3+\cdots+u_n = a+ca+c^2a+\cdots+c^{n-1}a \text{ and } \therefore \text{ also}$$
$$= (1+c+c^2+\cdots+c^{n-1})a.$$

Now $c^{i+1} = c^i - c^i + c^{i+1} = c^i + c^{i+1} - c^i = c^i + (c^{1+i} - c^i)$ if c^{1+i} be not less than c^i, $= c^i + (c^ic - c^i) = c^i + c^i(c-1)$ if c be not less than 1. $\therefore$ if c be not less than 1

$$\begin{aligned}
c^n &= c^{n-1}+c^{n-1}(c-1)\\
&= c^{n-2}+c^{n-2}(c-1)+c^{n-1}(c-1)\\
&= c^{n-3}+c^{n-3}(c-1)+c^{n-2}(c-1)+c^{n-1}(c-1)\\
&= \cdot\quad\cdot\quad\cdot\quad\cdot\quad\cdot\quad\cdot\quad\cdot\quad\cdot\quad\cdot\quad\cdot\quad\cdot\quad\cdot\quad\cdot\\
&= c+c(c-1)+c^2(c-1)+\cdots+c^{n-2}(c-1)+c^{n-1}(c-1)\\
&= 1+(c-1)+c(c-1)+c^2(c-1)+\cdots+c^{n-1}(c-1),
\end{aligned}$$

$$\begin{aligned}
c^n-1 &= 1+(c-1)+c(c-1)+\cdots+c^{n-1}(c-1)-1\\
&= 1-1+(c-1)+c(c-1)+\cdots+c^{n-1}(c-1) = (1+c+c^2+\cdots+c^{n-1})(c-1)
\end{aligned}$$

and if c be greater than 1 $\dfrac{c^n-1}{c-1} = 1+c+c^2+\cdots+c^{n-1}$.

Again $c^i = c^{i+1} - c^{i+1} + c^i = c^{i+1} + c^i - c^{1+i} = c^{i+1} + (c^i - c^ic)$ if c^ic be not greater than c^i, $= c^{i+1} + c^i(1-c)$ if c be not greater than 1. $\therefore$ if c be not greater than 1

$$\begin{aligned}
1 &= c+(1-c)\\
&= c^2+c(1-c)+(1-c)\\
&= c^3+c^2(1-c)+c(1-c)+(1-c)\\
&= \cdot\quad\cdot\quad\cdot\quad\cdot\quad\cdot\quad\cdot\quad\cdot\quad\cdot\quad\cdot\\
&= c^{n-1}+c^{n-2}(1-c)+\cdots+c(1-c)+(1-c)\\
&= c^n+c^{n-1}(1-c)+c^{n-2}(1-c)+\cdots+c(1-c)+(1-c),
\end{aligned}$$

$$1-c^n = c^n+c^{n-1}(1-c)+c^{n-2}(1-c)+\cdots+(1-c)-c^n$$
$$= c^n-c^n+c^{n-1}(1-c)+\cdots+(1-c) = (c^{n-1}+c^{n-2}+\cdots+1)(1-c)$$

and if c be less than 1 $\dfrac{1-c^n}{1-c} = c^{n-1}+c^{n-2}+\cdots+c+1 = 1+c+c^2+\cdots+c^{n-1}$.

If c be 1 $1+c+c^2+\cdots+c^{n-1}$ is n. Hence on the whole

$$u_1+u_2+u_3+\cdots+u_{n-1}+u_n = \begin{cases} \dfrac{c^n-1}{c-1}a & \text{if } c \text{ be greater than } 1 \\ na & \text{if } c \text{ be } 1 \\ \dfrac{1-c^n}{1-c}a & \text{if } c \text{ be less than } 1. \end{cases}$$

171. The conclusions of art. 170 may be put to such uses as the following:—

(1). $1+2+2^2+\cdots+2^{n-1} = \dfrac{2^n-1}{2-1}$ which is 2^n-1, that is the sum of any number of consecutive terms of the series 1, 2, 2^2, ... beginning with the first term is less by 1 than the term next following these.

Thus a horse if sold at 1 farthing for the 1st of the 32 nails fastening his shoes, 2 farthings for the 2nd, 4 for the 3rd, and so on doubling each new time, would bring $1+2+2^2+\cdots+2^{32-1}$ farthings and ∴ $\dfrac{2^{32}-1}{4}$ pence which $= \dfrac{2^{30}\times2^2}{4}-\dfrac{1}{4} = 2^{30}-\dfrac{1}{4}$. Since 2^{10} is 1024 and $2^{3\times10} = (2^{10})^3$, $2^{30} = (1024)^3$ and ∴ is 1073741824. All the farthings amount to 4473924*l.* 5*s.* $3\frac{3}{4}$*d.*

To give 1 grain of corn for the 1st of the 64 squares of a chess board, 2 grains for the 2nd, 4 for the 3rd, and generally 2^{i-1} grains for the ith, there would be needed in all $2^{64}-1$ grains. Since $2^{6\times10+4} = 2^4(2^{10})^{3\times2} = 2^4\{(2^{10})^2\}^3$ and 2^{10} is greater than 1000, this is more than 16 trillions of grains which reckoning at the rate of 5000000 grains to the quarter of corn comes to more than 3 billions of quarters and taking the whole population of the world to be 2000000000 would be more than enough to afford every individual 1500 quarters.

(2). $\dfrac{1}{2}+\left(\dfrac{1}{2}\right)^2+\left(\dfrac{1}{2}\right)^3+\cdots+\left(\dfrac{1}{2}\right)^n = \dfrac{1}{2}\left\{1+\dfrac{1}{2}+\left(\dfrac{1}{2}\right)^2+\cdots+\left(\dfrac{1}{2}\right)^{n-1}\right\}$

$= \dfrac{1}{2}\dfrac{1-\left(\dfrac{1}{2}\right)^n}{1-\dfrac{1}{2}}$ and ∴ (art. 126) $= \dfrac{1-\left(\dfrac{1}{2}\right)^n}{\left(1-\dfrac{1}{2}\right)\times2}$ that is $1-\left(\dfrac{1}{2}\right)^n$. This result may be reached more directly. For if from a magnitude its $\frac{1}{2}$ be cut off its other $\frac{1}{2}$ is left. Hence if from the $\frac{1}{2}$ left its $\frac{1}{2}$ be cut off the new $\frac{1}{2}$ left is $\left(\frac{1}{2}\right)^2$ of the whole magnitude. Hence again if from the $\left(\frac{1}{2}\right)^2$ so left its $\frac{1}{2}$ be cut off there is then left $\left(\frac{1}{2}\right)^3$ of the whole. And so on. Therefore after the nth halving $\left(\frac{1}{2}\right)^n$ of the whole magnitude is left and the parts cut off are together therefore both $\frac{1}{2}+\left(\frac{1}{2}\right)^2+\cdots+\left(\frac{1}{2}\right)^n$ and $1-\left(\frac{1}{2}\right)^n$ of the magnitude.

Likewise more generally

$$\frac{u}{u+v}+\frac{u}{u+v}\,\frac{v}{u+v}+\frac{u}{u+v}\left(\frac{v}{u+v}\right)^2+\cdots+\frac{u}{u+v}\left(\frac{v}{u+v}\right)^{n-1}=1-\left(\frac{v}{u+v}\right)^n.$$

(3). An annuity of a pounds to be entered on m years hence and to last n years thenceforth is (art. 153) now worth in pounds

$$\frac{a}{(1+r)^{m+1}}+\frac{a}{(1+r)^{m+2}}+\frac{a}{(1+r)^{m+3}}+\cdots+\frac{a}{(1+r)^{m+n}},$$

reckoning money at compound interest at the yearly rate of r pounds a pound, and since $\frac{a}{(1+r)^t}=a\frac{1}{(1+r)^t}=a\left(\frac{1}{1+r}\right)^t$ is therefore operationally equivalent to

$$a\left\{\left(\frac{1}{1+r}\right)^{m+1}+\left(\frac{1}{1+r}\right)^{m+2}+\cdots+\left(\frac{1}{1+r}\right)^{m+n}\right\}\ \therefore$$

$$=a\left\{\left(\frac{1}{1+r}\right)^{m+1}+\frac{1}{1+r}\left(\frac{1}{1+r}\right)^{m+1}+\left(\frac{1}{1+r}\right)^2\left(\frac{1}{1+r}\right)^{m+1}+\cdots+\left(\frac{1}{1+r}\right)^{n-1}\left(\frac{1}{1+r}\right)^{m+1}\right\}$$

$$=a\frac{1-\left(\frac{1}{1+r}\right)^n}{1-\frac{1}{1+r}}\left(\frac{1}{1+r}\right)^{m+1}=a\left\{\frac{1-\left(\frac{1}{1+r}\right)^n}{1-\frac{1}{1+r}}\,\frac{1}{1+r}\right\}\left(\frac{1}{1+r}\right)^m$$

$$=\left\{a\frac{1-\left(\frac{1}{1+r}\right)^n}{(1+r)\left(1-\frac{1}{1+r}\right)}\right\}\left(\frac{1}{1+r}\right)^m=\left[\frac{a}{1+r-(1+r)\frac{1}{1+r}}\left\{1-\left(\frac{1}{1+r}\right)^n\right\}\right]\left(\frac{1}{1+r}\right)^m$$

$$=\frac{a}{1+r-1}\left\{1-\left(\frac{1}{1+r}\right)^n\right\}\left(\frac{1}{1+r}\right)^m=\frac{a}{1-1+r}\left\{\left(\frac{1}{1+r}\right)^m-\left(\frac{1}{1+r}\right)^n\left(\frac{1}{1+r}\right)^m\right\}$$

$$=\frac{a}{r}\left\{\left(\frac{1}{1+r}\right)^m-\left(\frac{1}{1+r}\right)^{m+n}\right\}=\frac{a}{r}\,\frac{1}{(1+r)^m}-\frac{a}{r}\,\frac{1}{(1+r)^{m+n}}=\frac{\left(\frac{a}{r}\right)}{(1+r)^m}-\frac{\left(\frac{a}{r}\right)}{(1+r)^{m+n}}.$$

The same result may be got thus:—It is $\frac{a}{r}$ pounds that (art. 152) yields a pounds yearly interest at the rate of r pounds a pound and therefore the present worth of a perpetual annuity of a pounds to be entered on m years hence is $\frac{\left(\frac{a}{r}\right)}{(1+r)^m}$ pounds. But the worth of this annuity is made up of two parts the one the worth of the first n payments and the other the worth of all the after payments. The latter part is $\frac{\left(\frac{a}{r}\right)}{(1+r)^{m+n}}$ pounds the worth of a perpetual annuity of a pounds to be entered on $m+n$ years hence and therefore the former part is $\frac{\left(\frac{a}{r}\right)}{(1+r)^m}-\frac{\left(\frac{a}{r}\right)}{(1+r)^{m+n}}$ pounds.

(4). A number denoted decimally by a knot of digits having its last digit in the αth decimal place and which as a separate row with its last digit the unit's digit would denote a whole number a followed by n knots all alike of β digits each and each of which as a separate row with its last digit the unit's digit would denote a whole number b is (art. 136) operationally equivalent to

$$a\left(\frac{1}{t}\right)^{\alpha}+b\left(\frac{1}{t}\right)^{\alpha+\beta}+b\left(\frac{1}{t}\right)^{\alpha+2\beta}+\cdots+b\left(\frac{1}{t}\right)^{\alpha+n\beta}$$

t standing for 10 and $\therefore$

$$=a\left(\frac{1}{t}\right)^{\alpha}+\left\{b\left(\frac{1}{t}\right)^{\alpha+\beta}+b\left(\frac{1}{t}\right)^{\beta}\left(\frac{1}{t}\right)^{\alpha+\beta}+b\left(\frac{1}{t}\right)^{2\beta}\left(\frac{1}{t}\right)^{\alpha+\beta}+\cdots+b\left(\frac{1}{t}\right)^{(n-1)\beta}\left(\frac{1}{t}\right)^{\alpha+\beta}\right\}$$

$$=a\left(\frac{1}{t}\right)^{\alpha}+b\left[1+\left(\frac{1}{t}\right)^{\beta}+\left\{\left(\frac{1}{t}\right)^{\beta}\right\}^{2}+\cdots+\left\{\left(\frac{1}{t}\right)^{\beta}\right\}^{n-1}\right]\left(\frac{1}{t}\right)^{\alpha+\beta}$$

$$=a\left(\frac{1}{t}\right)^{\alpha}+b\frac{1-\left\{\left(\frac{1}{t}\right)^{\beta}\right\}^{n}}{1-\left(\frac{1}{t}\right)^{\beta}}\left(\frac{1}{t}\right)^{\beta}\left(\frac{1}{t}\right)^{\alpha}=a\frac{1}{t^{\alpha}}+b\left[\frac{1-\left\{\left(\frac{1}{t}\right)^{\beta}\right\}^{n}}{1-\left(\frac{1}{t}\right)^{\beta}}\frac{1}{t^{\beta}}\right]\left(\frac{1}{t}\right)^{\alpha}$$

$$=\frac{a}{t^{\alpha}}+b\frac{1-\left\{\left(\frac{1}{t}\right)^{\beta}\right\}^{n}}{t^{\beta}\left\{1-\left(\frac{1}{t}\right)^{\beta}\right\}}\frac{1}{t^{\alpha}}=\frac{a}{t^{\alpha}}+b\frac{1-\left\{\left(\frac{1}{t}\right)^{\beta}\right\}^{n}}{t^{\alpha}\left\{t^{\beta}-t^{\beta}\left(\frac{1}{t}\right)^{\beta}\right\}}=\frac{a(t^{\beta}-1)}{t^{\alpha}(t^{\beta}-1)}+\frac{b\left[1-\left\{\left(\frac{1}{t}\right)^{\beta}\right\}^{n}\right]}{t^{\alpha}(t^{\beta}-1)}$$

$$=\frac{at^{\beta}-a+\left[b-b\left\{\left(\frac{1}{t}\right)^{\beta}\right\}^{n}\right]}{t^{\alpha}(t^{\beta}-1)}=\frac{at^{\beta}-a+b-b\left\{\left(\frac{1}{t}\right)^{\beta}\right\}^{n}}{t^{\alpha}(t^{\beta}-1)}=\frac{at^{\beta}+b-a-b\left\{\left(\frac{1}{t}\right)^{\beta}\right\}^{n}}{(t^{\beta}-1)t^{\alpha}}$$

$$=\frac{at^{\beta}+b-a}{(t^{\beta}-1)t^{\alpha}}-\frac{\left\{\left(\frac{1}{t}\right)^{\beta}\right\}^{n}b}{(t^{\beta}-1)t^{\alpha}}=\frac{at^{\beta}+b-a}{(t^{\beta}-1)t^{\alpha}}-\left\{\left(\frac{1}{t}\right)^{\beta}\right\}^{n}\frac{b}{(t^{\beta}-1)t^{\alpha}}.$$

As (art. 170) $t^\beta-1=t-1+t(t-1)+t^2(t-1)+\cdots+t^{\beta-1}(t-1)$ and ∴ $=(t-1)t^{\beta-1}+(t-1)t^{\beta-2}+\cdots+(t-1)t+(t-1)$ $(t^\beta-1)t^\alpha$ is denoted by a row made up of β 9s and α 0s. Also $at^\beta+b=$ the whole number denoted by the row made up of the knot of digits which as a separate row would denote a and the knot which as a separate row would denote b.

Thus the number denoted by 42·73681681...681 with the knot of digits 681 written n times

$$=\frac{4273681-4273}{(1000-1)\times 100}-(0\cdot 001)^n\times\frac{681}{(1000-1)\times 100}$$

$$=\frac{4269408}{99900}-(0\cdot 001)^n\times\frac{681}{99900}.$$

The number 0·05959...59 with the digits 59 written n times

$$=\frac{59}{990}-(0\cdot 01)^n\times\frac{59}{990}.$$

And 0·999...9 with 9 written n times $=1-(0\cdot 1)^n$.

172. PROP. *A number may be taken so great that its reciprocal is less than any given number however small. And a number may be taken so small that its reciprocal is greater than any given number however great.*

Let κ be any given number however small. Of the magnitude expressed numerically by κ in reference to the unit magnitude a multiple may (art. 133) be taken greater than the unit magnitude. Let m be the whole number expressing the multiple so taken and let $\frac{\mu}{\nu}$ be any simple fraction greater than m. Since then m times the magnitude expressed by κ is greater than the unit the magnitude expressed by κ (art. 91) is greater than $\frac{1}{m}$ of the unit, in other words $\frac{1}{m}$ is less than κ. Moreover ∵ $\frac{\mu}{\nu}$ is greater than m which (art. 98) is $\frac{m\nu}{\nu}$ μ (art. 100) is greater than $m\nu$ ∴ (art. 91) $\frac{1}{m}\mu$ is greater than $\frac{1}{m}m\nu$ which (art. 108) is $\left(\frac{1}{m}m\right)\nu$ that is ν ∴ (art. 115) $\frac{\frac{1}{m}\mu}{\mu}$ is greater than $\frac{\nu}{\mu}$ that is (arts. 114, 111) $\frac{1}{m}$ is greater than $\frac{1}{\left(\frac{\mu}{\nu}\right)}$.

Much more ∴ (art. 9) is κ greater than $\frac{1}{\left(\frac{\mu}{\nu}\right)}$.

Again let a be any given number however great. Of the unit magnitude a multiple may be taken (art. 133) greater than the magnitude expressed in reference to the unit magnitude by a. Let this multiple be expressed by the whole number n. Then n is greater than a and $\because \dfrac{n}{\frac{1}{n}n}$ (art. 116) $= \dfrac{1}{\left(\frac{1}{n}\right)}$ the reciprocal of $\frac{1}{n}$ is $\therefore$ greater than a. Also if $\frac{\mu}{\nu}$ be any simple fraction less than $\frac{1}{n}$ or its equivalent $\frac{\mu}{n\mu}$, (art. 100) ν is greater than $n\mu$ $\therefore$ (art. 115) $\frac{\nu}{\mu}$ or (art. 111) $\dfrac{1}{\left(\frac{\mu}{\nu}\right)}$ is greater than n, and much greater $\therefore$ is $\dfrac{1}{\left(\frac{\mu}{\nu}\right)}$ than a.

173. PROP. *Of any given number greater by however little than 1 a power may be taken greater than any given number however great. And of any given number less by however little than 1 a power may be taken less than any given number however small.*

If $1+\kappa$ be any given number greater than 1 (art. 170)

$\because 1+\kappa-1 = 1-1+\kappa \quad (1+\kappa)^n - 1 = \{1+(1+\kappa)+(1+\kappa)^2+\cdots+(1+\kappa)^{n-1}\}\kappa.$

Any power of $1+\kappa$ is $\therefore$ greater than 1. $\therefore$ (arts. 11, 12) $1+(1+\kappa)+(1+\kappa)^2+\cdots+(1+\kappa)^{n-1}$ is greater than n. And $\therefore$ (art. 113) $(1+\kappa)^n-1$ is greater than $n\kappa$. But (art. 133) n may be taken so great a whole number that $n\kappa$ is greater than any given number. Therefore n may be taken so great that $(1+\kappa)^n-1$ and much more $(1+\kappa)^n$ is greater than any given number.

Again if $1-\kappa$ be any given number less than 1

$$\because 1-\kappa = \frac{1-\kappa}{\frac{1-\kappa+\kappa}{1-\kappa}(1-\kappa)} = \frac{1}{1+\frac{\kappa}{1-\kappa}} \qquad (1-\kappa)^n = \left(\frac{1}{1+\frac{\kappa}{1-\kappa}}\right)^n = \frac{1}{\left(1+\frac{\kappa}{1-\kappa}\right)^n}.$$

But by what has just been shown n may be taken so great as to make $\left(1+\frac{\kappa}{1-\kappa}\right)^n$ greater than any given number and $\therefore$ (art. 172) n may be taken so great as to make $\dfrac{1}{\left(1+\frac{\kappa}{1-\kappa}\right)^n}$ or the operational equivalent $(1-\kappa)^n$ less than any given number.

174. From arts. 172, 173, and the several operational equivalents found for the sums of the terms of some of the series in arts. 169,

171, it follows that the sum of the terms of a series may become ever nearer and nearer to some fixed number when ever more and more terms are taken and may at length when a great enough number of terms are taken become nearer to that fixed number than by any given number whatever. Since (art. 169) $\frac{1}{1}\frac{1}{2}+\frac{1}{2}\frac{1}{3}+\frac{1}{3}\frac{1}{4}+\cdots+\frac{1}{n}\frac{1}{n+1}=1-\frac{1}{n+1}$ and (art. 172) n may be taken so great that $\frac{1}{n+1}$ is less than any given number however small so many terms may be taken of the series $\frac{1}{1}\frac{1}{2}, \frac{1}{2}\frac{1}{3}, \frac{1}{3}\frac{1}{4}, \ldots.$ as to make their sum although always less than 1 yet less than 1 only by a number less than any given number however small. That the sum of the terms of this series becomes thus endlessly near to 1 as the number of the terms becomes endlessly great is written shortly $\frac{1}{1}\frac{1}{2}+\frac{1}{2}\frac{1}{3}+\frac{1}{3}\frac{1}{4}+\cdots\cdots=1.$

In like manner that (art. 171) there is no end to the nearness to which $\frac{1}{2}+\left(\frac{1}{2}\right)^2+\left(\frac{1}{2}\right)^3+\cdots+\left(\frac{1}{2}\right)^n$ may be made equal to 1 by taking n great enough is written $\frac{1}{2}+\left(\frac{1}{2}\right)^2+\left(\frac{1}{2}\right)^3+\cdots=1$. With the like meaning too (art. 170) when c is less than 1, $\because \frac{1-c^n}{1-c}=\frac{1}{1-c}-\frac{c^n}{1-c}$ and $\frac{c^n}{1-c}$ may by taking n great enough be made less than any given number κ to wit by making c^n less than $\kappa(1-c)$, $1+c+c^2+c^3+\cdots=\frac{1}{1-c}$; $0{\cdot}999\ldots.$ with endless 9s (art. 171) $=1$; $42{\cdot}7368168168 1\ldots.$ the knot 681 being repeated without end (art. 171) $=\frac{4269408}{99900}$.

NOMIC PROCESSES

175. *Def.* A symbolized product of which one factor is some power x^n of a number x and no other factor has x in it at all is called a MONONOMIC EXPRESSION or MONONOME in respect of, in reference to, or in, x. This product being, either itself, or by the commutation and colligation laws of multiplications operationally equivalent to, the product made by multiplying x^n either by or into the product of all the other factors, the product of all the factors into

which x does not enter is called the COEFFICIENT of x^n. Also the degree of the power x^n is called the DEGREE of the mononome.

Although an expression a into which x does not enter is not by itself a mononome in x yet in relation to the mononomes ax, ax^2, ... xa, x^2a, ... it must be held to be so; it may also as in arts. 57, 135, 139, be taken to be ax^0 or x^0a in relation to x.

Since x^n is at once $x^n \times 1$, $x^n \times 1 \times 1$, ... $1 \times x^n$, $1 \times 1 \times x^n$, ... a simple power of x comes straight under the definition.

Def. The symbolized result of an addition or additions, of a subtraction or subtractions, or of one or more additions and one or more subtractions any how mixed, when the numbers operated with in their written order are mononomes in x of ever higher and higher degrees or of ever lower and lower is called a POLYNOMIC EXPRESSION or POLYNOME in x. The mononomes in order operated with are called the TERMS in order of the polynome. A polynome's DEGREE is the degree of its term of highest degree. And the RANGE OF THE POWERS, the POWER RANGE, or the RANGE, of a polynome in x is the number of terms in that series of successive powers of x of which the first term is the lowest power of x in the polynome and the last the highest.

If p_0, p_1, p_2, ... p_n, be the coefficients of successive powers of x in the several terms and $\pm$ stand for "$+$ or $-$" a polynome of the nth degree in ascending powers of x either is or is operationally equivalent to

$$p_0 \pm p_1 x \pm p_2 x^2 \pm \cdots \pm p_n x^n \quad \text{or} \quad p_0 \pm x p_1 \pm x^2 p_2 \pm \cdots \pm x^n p_n,$$

and a polynome of the nth degree in descending powers of x either is or is operationally equivalent to

$$p_0 x^n \pm p_1 x^{n-1} \pm p_2 x^{n-2} \pm \cdots \pm p_{n-1} x \pm p_n \quad \text{or} \quad x^n p_0 \pm x^{n-1} p_1 \pm x^{n-2} p_2 \pm \cdots \pm x p_{n-1} \pm p_n.$$

If the lowest power of x in a polynome be x^m and the highest x^n the range of the polynome is the number of terms in the series x^m, x^{m+1}, x^{m+2}, ... x^n, and if this number be i

$$m+(i-1)=n \quad \text{and} \therefore i=n-m+1.$$

176. *Def.* A mononome in each of any symbolized numbers x_1, x_2, ... x_r, with $x_1^{n_1}$, $x_2^{n_2}$, ... $x_r^{n_r}$, the factors into which alone x_1, x_2, ... x_r, severally enter is called a MONONOME of the $(n_1+n_2+\cdots+n_r)$th DEGREE in x_1, x_2, ... x_r,

and the product of all the other factors is called the COEFFICIENT of $x_r^{n_r} \ldots x_2^{n_2} x_1^{n_1}$. Mononomes in the same symbolized numbers are called HOMOGENEOUS when of the same degree in those numbers jointly, and LIKE when of the same degree in each of those numbers separately.

Thus $4x^3$, $2(7-2a)x^2y$, $3bxy^2$, $5(a+\frac{1}{2}b-6c)a^2y^3$, are homogeneous mononomes of the 3rd degree in x, y, and no two of them are like mononomes in x, y. Again $9a^2bc^3$, $7x^2a^2bc^3$, $(2x-5x^1)a^2bc^3$, being all of the 2nd degree in a of the 1st in b and of the 3rd in c, are like mononomes in a, b, c. Like mononomes are the same either in everything or in everything but the coefficients.

Def. The symbolized result of one or more operations of addition, subtraction, or addition and subtraction, with mononomes, every two of them unlike, in several numbers $x_1, x_2, \ldots x_r$, in the order of ascent or descent of their degrees in each and all of those numbers, is called a POLYNOME in $x_1, x_2, \ldots x_r$. The mononomes are called the TERMS, and the degree of the term or terms of highest degree is called the DEGREE, of the polynome. A polynome is said to be HOMOGENEOUS of which all the terms are homogeneous. That term of a polynome in $x_1, x_2, \ldots x_r$, in which the powers of $x_1, x_2, \ldots x_r$, severally are $x_1^{n_1}, x_2^{n_2}, \ldots x_r^{n_r}$, is distinguished from every other term by calling it the term IN $x_r^{n_r} \ldots x_2^{n_2} x_1^{n_1}$.

Since a polynome in descending powers of a with all terms possible has terms in order in a 1, that is in a^1 a^0, if of the 1st degree, in a^2 a 1 if of the 2nd, in a^3 a^2 a 1 if of the 3rd, and so on a polynome in b, a, with all terms possible arranged in descending powers of b primarily and in descending powers of a secondarily has terms in order in b a 1 if of the 1st degree, in b^2 ba b a^2 a 1 if of the 2nd, in b^3 b^2a b^2 ba^2 ba b a^3 a^2 a 1 if of the 3rd, and so on. Hence a polynome in c, b, a, with all possible terms written in the order of descending powers of c as a 1st principle of arrangement descending powers of b as a 2nd and descending powers of a as a 3rd has terms in order if of the 1st degree in c b a 1, if of the 2nd in c^2 cb ca c b^2 ba b a^2 a 1, if of the 3rd in c^3 c^2b c^2a c^2 cb^2 cba cb ca^2 ca c b^3 b^2a b^2 ba^2 ba b a^3 a^2 a 1, and so on. Hence too in like manner may the order be found of the terms arranged after the like laws of polynomes in 4 or more symbolized numbers.

If the polynomes be in ascending instead of descending powers of their several symbolized numbers of reference while these numbers bear the same rank to one another in the arrangement the order of the terms is exactly reversed. Thus a polynome of the 4th degree in ascending powers of a, b, c, with a ranking before b and b before c has all the terms in order that it can have in 1 c c^2 c^3 c^4 b bc bc^2, bc^3 b^2 b^2c b^2c^2 b^3 b^3c b^4 a ac ac^2 ac^3 ab abc abc^2 ab^2 ab^2c ab^3 a^2 a^2c a^2c^2 a^2b a^2bc a^2b^2 a^3 a^3c a^3b a^4.

Polynomes may be in ascending powers of some of the numbers of reference and descending powers of the rest. Thus a polynome of the 3rd degree in x, y, a, b, in descending powers of x and subordinately of y and ascending powers of a and subordinately of b has all its possible terms in order in x^3 x^2y x^2 bx^2 ax^2 xy^2 xy bxy axy x bx b^2x ax abx a^2x y^3 y^2 by^2 ay^2 y by b^2y ay aby a^2y 1 b b^2 b^3 a ab ab^2 a^2 a^2b a^3.

Polynomes may further be in ascending or descending degrees of their reference numbers jointly and then in each set of homogeneous terms the order of the terms may be either as before in ascending or descending powers of each and all of the reference numbers or after some law of symmetry by which when the reference numbers are taken in a certain order in an endless round with the first following the last and every other the one before it those terms follow one another in order that are in the same powers of such of the reference numbers as hold severally the same relative situations to one another in the round. Thus of a polynome of the 2d degree in descending powers of a, b, c, d, all the possible terms in order are in a^2 ab ac ad b^2 bc bd c^2 cd d^2 a b c d 1 if in each homogeneous set the terms are in descending powers first of all in a next in b then in c and last in d, but by a symmetric arrangement in each homogeneous set the terms in order are in a^2 b^2 c^2 d^2 ab bc cd da ac bd a b c d 1. Again all the terms in order that there can be in a polynome of the 3rd degree in ascending powers of x, y, taking each homogeneous set of terms in ascending powers of y as the primary number of reference and in ascending powers of x as the secondary are either unsymmetrically in 1 x y x^2 xy y^2 x^3 x^2y xy^2 y^3 or symmetrically in 1 x y x^2 y^2 xy x^3 y^3 x^2y xy^2.

177. Since a polynome of the rth degree in a single symbolized number x may have terms in 1 x $x^2 \ldots x^r$ and can have no more it has at most $r+1$ terms.

Hence a polynome of the rth degree in two symbolized numbers one of them x has at most 1 term in x^r, 2 terms in x^{r-1}, 3 in

$x^{r-2}, \ldots r$ in x, and $r+1$ free from x or in x^0. The greatest number of terms it can have is therefore $1+2+3+\cdots+(r+1)$ which (art. 168) $= \frac{(r+1)(r+2)}{2}$.

Hence the greatest possible number of terms of a polynome of the rth degree in 3 symbolized numbers of which one is x is $\frac{1.2}{2}$ in x^r, $\frac{2.3}{2}$ in $x^{r-1}, \ldots \frac{r(r+1)}{2}$ in x, $\frac{(r+1)(r+2)}{2}$ in x^0 or 1, and ∴ in all

$$\frac{1.2}{2}+\frac{2.3}{2}+\frac{3.4}{2}+\cdots+\frac{(r+1)(r+2)}{2} = \frac{1.2+2.3+\cdots+(r+1)(r+2)}{2}$$

$$= \frac{\left\{\frac{(r+1)(r+2)(r+3)}{3}\right\}}{2} = \text{(art. 118) } \frac{(r+1)(r+2)(r+3)}{2.3}.$$

And if generally after the law of polynomes in 1, 2, or 3, symbolized numbers the greatest number of terms that a polynome of the rth degree in any number n of symbolized numbers can have be

$$\frac{(r+1)(r+2)(r+3)\ldots(r+n-2)(r+n-1)(r+n)}{1.2.3\ldots(n-2)(n-1)n}$$

the greatest number that one in $n+1$ symbolized numbers can have must then in like manner be

$$\frac{1.2..(n-1)n}{1.2..(n-1)n}+\frac{2.3..n(n+1)}{1.2..(n-1)n}+\cdots+\frac{(r+1)(r+2)..(r+n-1)(r+n)}{1.2..(n-1)n}$$

$$= \frac{1.2..(n-1)n+2.3..n(n+1)+\cdots+(r+1)(r+2)..(r+n-1)(r+n)}{1.2..(n-1)n}$$

$$= \frac{\left\{\frac{(r+1)(r+2)\ldots(r+n-1)(r+n)(r+n+1)}{n+1}\right\}}{1.2\ldots(n-1)n}$$

$$= \frac{(r+1)(r+2)\ldots(r+n-1)(r+n)(r+n+1)}{\{1.2\ldots(n-1)n\}(n+1)}$$

$$= \frac{(r+1)(r+2)(r+3)\ldots(r+n-2)(r+n-1)(r+n)(r+n+1)}{1.2.3\ldots(n-2)(n-1)n(n+1)}$$

which is after the same law. The law is ∴ true for 4 symbolized numbers, ∴ for 5, and so on for ever.

The greatest number of terms of a polynome of the rth degree in n symbolized numbers being thus

$$\frac{(r+1)(r+2)\ldots(r+n-1)(r+n)}{1.2\ldots(n-1)n} \text{ is ∴ also}$$

$$\frac{\{(r+1)(r+2)\ldots(r+n-1)(r+n)\}1.2\ldots(r-2)(r-1)r}{\{1.2\ldots(n-1)n\}1.2\ldots(r-2)(r-1)r}$$

$$=\frac{\{1.2\ldots(r-1)r\}r(+1)\ldots(r+n-1)(r+n)}{\{1.2\ldots(r-1)r\}1.2\ldots(n-1)n}$$

$$=\frac{1.2.3\ldots r(r+1)\ldots(r+n-2)(r+n-1)(r+n)}{\{1.2.3\ldots(r-2)(r-1)r\}1.2.3\ldots(n-2)(n-1)n}$$

$$=\frac{1.2.3\ldots(n+r-2)(n+r-1)(n+r)}{\{1.2.3\ldots(n-2)(n-1)n\}1.2.3\ldots(r-2)(r-1)r}$$

and ∴ besides in the same way

$$=\frac{(n+1)(n+2)(n+3)\ldots(n+r-2)(n+r-1)(n+r)}{1.2.3\ldots(r-2)(r-1)r}.$$

It hence comes out that the greatest number possible of the terms is the same for a polynome of the rth degree in n symbolized numbers and for a polynome of the nth degree in r symbolized numbers.

Moreover since a mononome of the rth degree in n symbolized numbers if of the ith degree in one of these numbers must be of the $(r-i)$th degree in the other $n-1$ the number of all the terms that any homogeneous polynome of the rth degree in n symbolized numbers can have, which is just the number of all the unlike mononomes there can be of the rth degree in n symbolized numbers, is the same as the number of all the terms that any polynome of the rth degree in $n-1$ symbolized numbers can have and therefore is

$$\frac{n(n+1)(n+2)\ldots(n+r-3)(n+r-2)(n+r-1)}{1.2.3\ldots(r-2)(r-1)r}.$$

178. *Def.* When n is a whole number the symbol $\underline{|n}$ stands for $1.2.3\ldots(n-2)(n-1)n$ and is read "n FACTORIAL".

The greatest possible number of terms then of a polynome of the rth degree in n symbolized numbers is any of the operational equivalents $\dfrac{(r+1)(r+2)\ldots(r+n-1)(r+n)}{\underline{|n}}$, $\dfrac{(n+1)(n+2)\ldots(n+r-1)(n+r)}{\underline{|r}}$, $\dfrac{\underline{|r+n}}{\underline{|r}\,\underline{|n}}$, $\dfrac{\underline{|n+r}}{\underline{|n}\,\underline{|r}}$. And the greatest possible number of unlike mononomes of the rth degree in n symbolized numbers is any of the operational equivalents $\dfrac{n(n+1)\ldots(n+r-2)(n+r-1)}{\underline{|r}}$, $\dfrac{(r+1)(r+2)\ldots(r+n-2)(r+n-1)}{\underline{|n-1}}$, $\dfrac{\underline{|n-1+r}}{\underline{|n-1}\,\underline{|r}}$, $\dfrac{\underline{|r+(n-1)}}{\underline{|r}\,\underline{|n-1}}$.

179. PROP. *However small but not* 0 *may be the given number* p_0 *and however great each of the given numbers* p_1, $p_2, \ldots p_n$, *there are yet endless great enough numbers such that if* x *be any one of them the polynomic expression* $x^n p_0 - x^{n-1} p_1 - x^{n-2} p_2 - \cdots - x p_{n-1} - p_n$ *has meaning and endless small enough numbers such that if* x *be any one of them the polynomic expression* $p_0 - x p_1 - x^2 p_2 - \cdots - x^{n-1} p_{n-1} - x^n p_n$ *has meaning.*

In the first place if a number a be not less than either of two numbers b, b', (arts. 59, 103)

$$(a-b,\ b) = (a-b',\ b') \because \text{each} = a.$$

$\therefore$ (art. 15) if b be greater than b' $a-b$ is less than $a-b'$. Hence and by arts. 91, 89, if p be either that or any one of the coefficients $p_1, p_2, \ldots p_n$, which is not less than any other the polynome $x^n p_0 - x^{n-1} p_1 - x^{n-2} p_2 - \cdots - x p_{n-1} - p_n$ has meaning if $x^n p_0 - x^{n-1} p - x^{n-2} p - \cdots - xp - p$ has and the polynome $p_0 - x p_1 - x^2 p_2 - \cdots - x^{n-1} p_{n-1} - x^n p_n$ if $p_0 - xp - x^2 p - \cdots - x^{n-1} p - x^n p$ has.

In the next place $a-(b+c)$ (arts. 59, 103) has meaning precisely when a is not less than $b+c$ or (arts. 7, 6, 9) its equal (b, c) and this is precisely when a being not less than c $(a-c, c)$ the then equal to a is further not less than (b, c) which again (arts. 13, 14) is precisely when $a-c$ has meaning and what it means is not less than b, that is precisely when $a-c-b$ has meaning. So that $a-(b+c)$ has meaning just when $a-c-b$ has and $\therefore$ generally $x-(a+b+c+\cdots+f+g+h)$ has meaning just when any of the following expressions has, to wit $x-h-(a+b+\cdots+g)$, $x-h-g-(a+b+\cdots+f)$, - - - $x-h-g-f-\cdots-c-b-a$. Hence the polynome $x^n p_0 - x^{n-1} p - x^{n-2} p - \cdots - xp - p$ has meaning just when $x^n p_0 - (p + xp + x^2 p + \cdots + x^{n-1} p)$ has and the polynome $p_0 - xp - x^2 p - \cdots - x^n p$ just when $p_0 - (x^n p + x^{n-1} p + \cdots + x^2 p + xp)$ has.

But in the third place $x^n p_0 - (p + xp + x^2 p + \cdots + x^{n-1} p)$ (art. 170) $= x^n p_0 - \dfrac{x^n - 1}{x-1} p$ if x be greater than 1 which by what has been here first proved and art. 113 has meaning if $x^n p_0 - \dfrac{x^n}{x-1} p$ has and $\therefore$ if any of the following operational equivalents of this last has, to wit

$$\left\{\frac{x^n}{x-1}(x-1)\right\} p_0 - \frac{x^n}{x-1} p,\quad \frac{x^n}{x-1}(x-1) p_0 - \frac{x^n}{x-1} p,\quad \frac{x^n}{x-1}(x p_0 - p_0 - p),$$

$$\frac{x^n}{x-1}\left(x p_0 - \frac{p+p_0}{p_0} p_0\right),\quad \frac{x^n}{x-1}\left(x - \frac{p+p_0}{p_0}\right) p_0.$$ And $\because$ $x^r p = x^{r-1+1} p = (x x^{r-1}) p$ $= x x^{r-1} p$, $\quad p_0 - (x^n p + x^{n-1} p + \cdots + x^2 p + xp) = p - (x x^{n-1} p + x x^{n-2} p + \cdots + xp)$

$=p_0-x(x^{n-1}+x^{n-2}+\cdots+x+1)p=$ (art. 170) $p_0-x\dfrac{1-x^n}{1-x}p$ if x be less than 1 which as before has meaning if either $p_0-x\dfrac{1}{1-x}p$ or any of its following operational equivalents has,

$$\left\{\frac{1}{1-x}(1-x)\right\}p_0-\frac{1}{1-x}xp,\ \frac{1}{1-x}(1-x)p_0-\frac{1}{1-x}xp,\ \frac{1}{1-x}(p_0-xp_0-xp),$$

$$\frac{1}{1-x}\{p_0-(xp+xp_0)\},\ \frac{1}{1-x}\left\{\frac{p_0}{p+p_0}(p+p_0)-x(p+p_0)\right\},\ \frac{1}{1-x}\left(\frac{p_0}{p+p_0}-x\right)(p+p_0).$$

On the whole $\therefore$ $x^np_0-x^{n-1}p_1-x^{n-2}p_2-\cdots-xp_{n-1}-p_n$ has meaning if x be any number not less than $\dfrac{p+p_0}{p_0}$ and $p_0-xp_1-x^2p_2-\cdots-x^{n-1}p_{n-1}-x^np_n$ has meaning if x be any number not greater than $\dfrac{p_0}{p+p_0}$.

Hence much more will any polynome got from either of these polynomes by putting a + instead of a − before any term or each of several terms have meaning under the like conditions severally.

180. PROP. *To find the mononomic or polynomic expression that is operationally equivalent to the result of one or more operations of addition, subtraction, or addition and subtraction, with expressions mononomic, polynomic, or mononomic and polynomic.*

This is done as in the instances following.

(1). $7ax^2-2ax^2+23ax^2+ax^2-10ax^2-4ax^2$ by distribution of the multiplication of ax^2 over the subtractions and additions
$=(7-2+23+1-10-4)ax^2$ which is $15ax^2$.

(2). $4a^2+5a^2-7b^2-8ab+6ab+11b^2-3a^2$ by commutation of the additions and subtractions $=4a^2+5a^2-3a^2+6ab-8ab+11b^2-7b^2$ $\therefore$ by distribution of subtraction and of addition over subtraction
$=4a^2+5a^2-3a^2-(8ab-6ab)+(11b^2-7b^2)$ and $\therefore$ by distribution of multiplication of over addition and subtraction
$=(4+5-3)a^2-(8-6)ab+(11-7)b^2$ which is $6a^2-2ab+4b^2$.

(3). $a\times2+b\times3+c+(a\times4-c\times20)+c\times6-(a\times6+b\times5-c\times7)$
$+(a\times10-b\times4+c\times9)-b\times8+(a\times11+b\times16)+(a-b\times2-c\times3)$
by distribution of the additions and the subtraction over the several additions and subtractions
$=a\times2+b\times3+c+a\times4-c\times20+c\times6+c\times7-b\times5-a\times6+a\times10-b\times4+c\times9$
$-b\times8+a\times11+b\times16+a-b\times2-c\times3$ $\therefore$ by commutation of the additions and subtractions $=a\times2+a\times4-a\times6+a\times10+a\times11+a$
$+b\times16+b\times3-b\times5-b\times4-b\times8-b\times2+c+c\times6+c\times7+c\times9-c\times3-c\times20$

$\therefore$ again by distribution of subtraction and of addition over additions and subtractions $=a\times 2+a\times 4-a\times 6+a\times 10+a\times 11+a$
$-(b\times 2+b\times 8+b\times 4+b\times 5-b\times 3-b\times 16)+(c+c\times 6+c\times 7+c\times 9-c\times 3-c\times 20)$
and $\therefore$ by distribution of multiplication by over additions and subtractions
$=a\times(2+4-6+10+11+1)-b\times(2+8+4+5-3-16)+c\times(1+6+7+9-3-20)$
which is $a\times 22$.

(4). $1-2x-3x^2+4x^3-\left(\frac{1}{2}+\frac{2}{3}x-\frac{3}{4}x^2+\frac{4}{5}x^3\right)-\left(\frac{1}{4}-x^4\right)$

$$+(3-2x^2-x^3)+\left(\frac{3}{5}x+7x^3-\frac{5}{3}x^4\right)$$

by distribution of the subtractions and the additions over the additions and subtractions

$$=1-2x-3x^2+4x^3-\frac{4}{5}x^3+\frac{3}{4}x^2-\frac{2}{3}x-\frac{1}{2}+x^4-\frac{1}{4}+3-2x^2-x^3+\frac{3}{5}x+7x^3-\frac{5}{3}x^4$$

$\therefore$ by commutation of these additions and subtractions

$$=1-\frac{1}{2}-\frac{1}{4}+3+\frac{3}{5}x-2x-\frac{2}{3}x+\frac{3}{4}x^2-3x^2-2x^2+4x^3-\frac{4}{5}x^3-x^3+7x^3+x^4-\frac{5}{3}x^4$$

$\therefore$ by distribution of addition and of subtraction over additions and subtractions

$$=1-\frac{1}{2}-\frac{1}{4}+3-\left(\frac{2}{3}x+2x-\frac{3}{5}x\right)-\left(2x^2+3x^2-\frac{3}{4}x^2\right)$$
$$+\left(4x^3-\frac{4}{5}x^3-x^3+7x^3\right)-\left(\frac{5}{3}x^4-x^4\right)$$

and $\therefore$ by distribution of multiplication of over additions and subtractions

$$=1-\frac{1}{2}-\frac{1}{4}+3-\left(\frac{2}{3}+2-\frac{3}{5}\right)x-\left(2+3-\frac{3}{4}\right)x^2+\left(4-\frac{4}{5}-1+7\right)x^3-\left(\frac{5}{3}-1\right)x^4,$$

that is $\frac{13}{4}-\frac{31}{15}x-\frac{17}{4}x^2+\frac{46}{5}x^3-\frac{2}{3}x^4$.

(5). $4x^3-3xy^2+2y^3-z^3-(7x^2y-6xyz+5y^2z)$
$+(11x^3-9x^2z+7xyz-5y^3+3yz^2)-(2x^2z-4xz^2+6yz^2)$
$+(8x^2y-10xz^2+12z^3)-(3x^2y-4xy^2+4y^2z-3yz^2)+6x^2z$

by distributing commutating and writing like mononomes in the same column

$$\begin{array}{llllllllll}
=4x^3 & & & -3xy^2 & & & +2y^3 & & & -z^3 \\
& -7x^2y & & & +6xyz & & & -5y^2z & & \\
+11x^3 & & -9x^2z & & +7xyz & & -5y^3 & & +3yz^2 & \\
& & -2x^2z & & & +4xz^2 & & & -6yz^2 & \\
& +8x^2y & & & & -10xz^2 & & & & +12z^3 \\
& -3x^2y & & +4xy^2 & & & & -4y^2z & +3yz^2 & \\
& & +6x^2z & & & & & & &
\end{array}$$

∴ by commutation of these additions and subtractions and by distribution of addition subtraction and multiplication of

$$=(4+11)x^3-(3+7-8)x^2y-(2+9-6)x^2z+(4-3)xy^2+(6+7)xyz$$
$$-(10-4)xz^2-(5-2)y^3-(4+5)y^2z+(3+3-6)yz^2+(12-1)z^3$$

that is $15x^3-2x^2y-5x^2z+xy^2+13xyz-6xz^2-3y^3-9y^2z+11z^3$

and this is equivalent operationally to the symmetric polynome

$$15x^3-3y^3+11z^3-2x^2y-9y^2z-6z^2x+xy^2-5zx^2+13xyz.$$

181. PROP. *To find the mononomic or polynomic expression that is operationally equivalent to the product made by multiplying by one mononomic or polynomic expression another.*

The several cases that can arise are to be dealt with as follows.

(1). $(agw^mx^ny^p)a'g'w^{m'}x^{n'}y^{p'}$ by the colligation of multiplications $=agw^mx^ny^pa'g'w^{m'}x^{n'}y^{p'}$ ∴ by commutation of multiplications $=aa'gg'w^mw^{m'}x^nx^{n'}y^py^{p'}$ ∴ again by colligation of multiplications $=(aa')(gg')(w^mw^{m'})(x^nx^{n'})y^py^{p'}$ and ∴ $=(aa')(gg')w^{m'+m}x^{n'+n}y^{p'+p}$.

(2). $\left(2x^5-8x^4-11x^3+\frac{2}{3}x^2+\frac{3}{2}x-5\right)\times18x^3$ by the distribution of the multiplication of $18x^3$

$$=(2x^5)\times18x^3-(8x^4)\times18x^3-(11x^3)\times18x^3+\left(\frac{2}{3}x^2\right)\times18x^3$$
$$+\left(\frac{3}{2}x\right)\times18x^3-5\times18x^3$$

and ∴ by (1) above

$$=(2\times18)x^{3+5}-(8\times18)x^{3+4}-(11\times18)x^{3+3}+\left(\frac{2}{3}\times18\right)x^{3+2}$$
$$+\left(\frac{3}{2}\times18\right)x^{3+1}-(5\times18)x^3$$

which is $36x^8-144x^7-198x^6+12x^5+27x^4-90x^3$.

Again $\left(\frac{7}{2}a^6x\right)(4a^3-3a^2x+5ax^2-12x^3)$ by the distribution of the multiplication by $\frac{7}{2}a^6x$

$$=\left(\frac{7}{2}a^6x\right)\times4a^3-\left(\frac{7}{2}a^6x\right)\times3a^2x+\left(\frac{7}{2}a^6x\right)\times5ax^2-\left(\frac{7}{2}a^6x\right)\times12x^3$$

and ∴ by the case (1) above

$$=\left(\frac{7}{2}\times4\right)a^{3+6}x-\left(\frac{7}{2}\times3\right)a^{2+6}x^{1+1}+\left(\frac{7}{2}\times5\right)a^{1+6}x^{2+1}-\left(\frac{7}{2}\times12\right)a^6x^{3+1}$$

which is $14a^9x-\frac{21}{2}a^8x^2+\frac{35}{2}a^7x^3-42a^6x^4$.

(3). $(3a^4+2a^3c-a^2c^2-2ac^3+3c^4)(4a^3-5a^2c+6ac^2-7c^3)$ by distribution of multiplication of

$$\begin{aligned}&=(3a^4)(4a^3-5a^2c+6ac^2-7c^3)+(2a^3c)(4a^3-5a^2c+6ac^2-7c^3)\\&\quad-(a^2c^2)(4a^3-5a^2c+6ac^2-7c^3)-(2ac^3)(4a^3-5a^2c+6ac^2-7c^3)\\&\qquad\qquad+(3c^4)(4a^3-5a^2c+6ac^2-7c^3)\end{aligned}$$

$\therefore$ by the above case (2)

$$\begin{aligned}&=12a^7-15a^6c+18a^5c^2-21a^4c^3+(8a^6c-10a^5c^2+12a^4c^3-14a^3c^4)\\&\quad-(4a^5c^2-5a^4c^3+6a^3c^4-7a^2c^5)-(8a^4c^3-10a^3c^4+12a^2c^5-14ac^6)\\&\qquad\qquad+(12a^3c^4-15a^2c^5+18ac^6-21c^7)\end{aligned}$$

and $\therefore$ (art. 180)

$$\begin{array}{llllllll}=12a^7 & -15a^6c & +18a^5c^2 & -21a^4c^3 & & & & \\ & +8\cdot\cdot & -10\cdot\cdot & +12\cdot\cdot & -14a^3c^4 & & & \\ & & -4\cdot\cdot & +5\cdot\cdot & -6\cdot\cdot & +7a^2c^5 & & \\ & & & -8\cdot\cdot & +10\cdot\cdot & -12\cdot\cdot & +14ac^6 & \\ & & & & +12\cdot\cdot & -15\cdot\cdot & +18\cdot\cdot & -21c^7.\end{array}$$

$$=12a^7-7a^6c+4a^5c^2-12a^4c^3+2a^3c^4-20a^2c^5+32ac^6-21c^7.$$

This polynomic equivalent may be got otherwise by first distributing the multiplication by. The product then

$$\begin{aligned}&=(3a^4+2a^3c-a^2c^2-2ac^3+3c^4)\times4a^3-(3a^4+2a^3c-a^2c^2-2ac^3+3c^4)\times5a^2c\\&\quad+(3a^4+2a^3c-a^2c^2-2ac^3+3c^4)\times6ac^2-(3a^4+2a^3c-a^2c^2-2ac^3+3c^4)\times7c^3\end{aligned}$$

and $\therefore$

$$\begin{array}{llllllll}=12a^7 & +8a^6c & -4a^5c^2 & -8a^4c^3 & +12a^3c^4 & & & \\ & -15\cdot\cdot & -10\cdot\cdot & +5\cdot\cdot & +10\cdot\cdot & -15a^2c^5 & & \\ & & +18\cdot\cdot & +12\cdot\cdot & -6\cdot\cdot & -12\cdot\cdot & +18ac^6 & \\ & & & -21\cdot\cdot & -14\cdot\cdot & +7\cdot\cdot & +14\cdot\cdot & -21c^7\end{array}$$

$$=12a^7-7a^6c+4a^5c^2-12a^4c^3+2a^3c^4-20a^2c^5+32ac^6-21c^7.$$

When the terms of each or either polynomic factor are arranged symmetrically or when the law of succession of the terms in ascending or descending powers of the reference numbers ranked in some order is not the same for both factors the process can be gone through in a regular and orderly manner only after commutating the operations with the terms so as to have that law alike in the factors. Thus

$$\begin{aligned}&(2y^2+3x^2-4xy^3-5x^3y)(5-4y+3x-2x^2y+y^3)\\&\qquad\qquad=(2y^2-4xy^3+3x^2-5x^3y)(5-4y+y^3+3x-2x^2y)\end{aligned}$$

and $\therefore$ distributing the multiplication of, polynomizing by the (2) case the products so got, distributing the additions and subtractions, commutating the operations equivalent by distribution to the subtractions,

$$
\begin{aligned}
= 10y^2-8y^3+2y^5+6xy^2 &\qquad\qquad\qquad\qquad -4x^2y^3 \\
-20xy^3+16xy^4-4xy^6 &\qquad\qquad\qquad\qquad -12x^2y^3 \\
&+15x^2-12x^2y+3x^2y^3 \\
&\qquad\qquad +8x^3y^4 \\
+9x^3 &\qquad\qquad\qquad\qquad -6x^4y \\
&-25x^3y+20x^3y^2-5x^3y^4-15x^4y+10x^5y^2 \\
= 10y^2-8y^3+2y^5+6xy^2-20xy^3&+16xy^4-4xy^6+15x^2-12x^2y-13x^2y^3 \\
&+9x^3-25x^3y+20x^3y^2+3x^3y^4-21x^4y+10x^5y^2.
\end{aligned}
$$

182. The last written polynome is operationally equivalent to

$$
\begin{aligned}
&10y^2-8y^3+2y^5+(6y^2-20y^3+16y^4-4y^6)x+(15-12y-13y^3)x^2 \\
&+(9-25y+20y^2+3y^4)x^3-21yx^4+10y^2x^5 = 2(5-4y+y^3)y^2 \\
&+2(3-10y+8y^2-2y^4)y^2x+(15-12y-13y^3)x^2+(9-25y+20y^2+3y^4)x^3 \\
&-21yx^4+10y^2x^5 = 10y^2x^5-21yx^4+(9-25y+20y^2+3y^4)x^3 \\
&+(15-12y-13y^3)x^2+2(3-10y+8y^2-2y^4)y^2x+2(5-4y+y^3)y^2.
\end{aligned}
$$

It is also operationally equivalent to

$$
\begin{aligned}
&2(y^3-4y+5)y^2-2(2y^4-8y^2+10y-3)y^2x-(13y^3+12y-15)x^2 \\
&\qquad\qquad\qquad +(3y^4+20y^2-25y+9)x^3-21yx^4+10y^2x^5 \\
&= 10x^5y^2-21x^4y+x^3(3y^4+20y^2-25y+9)-x^2(13y^3+12y-15) \\
&\qquad\qquad\qquad -2xy^2(2y^4-8y^2+10y-3)+2y^2(y^3-4y+5).
\end{aligned}
$$

And in the same way any polynome in one symbolized number having as a factor of any term a polynome in one other symbolized number is equivalent operationally to a polynome in those two symbolized numbers. So again if there be still a polynomic factor in some third symbolized number of any term of this polynome in two symbolized numbers both this polynome and the original polynome in the first symbolized number is equivalent operationally to a polynome in the three symbolized numbers. And so on.

183. To find the polynomic equivalent in x of $(1-kx)(1-jx)(1-ix)\ldots(1-cx)(1-bx)(1-ax)$ if $a, b, c, \ldots i, j, k$, be any n numbers. By the polynomic process of art. 181

$$
\begin{aligned}
(1-bx)(1-ax) &= 1-ax \\
&\qquad -bx+(ba)x^2 = 1-(a+b)x+(ba)x^2.
\end{aligned}
$$

$$
\begin{aligned}
\therefore\ (1-cx)(1-bx)(1-ax) &= (1-cx)\{1-(a+b)x+(ba)x^2\} \\
&= 1-(a+b)x \quad +(ba)x^2 \\
&\qquad\qquad -cx+\{c(a+b)\}x^2-(cba)x^3 \\
&= 1-(a+b+c)x+\{ba+c(a+b)\}x^2-(cba)x^3 \\
&= 1-(a+b+c)x+(ba+ca+cb)x^2-(cba)x^3.
\end{aligned}
$$

So $\therefore$ $(1-dx)(1-cx)(1-bx)(1-ax)$

$$
\begin{aligned}
&= 1-(a+b+c)x+(ba+ca+cb)x^2 \qquad -(cba)x^3 \\
&\qquad -dx+\{d(a+b+c)\}x^2-\{d(ba+ca+cb)\}x^3+(dcba)x^4 \\
&= 1-(a+b+c+d)x+(ba+ca+cb+da+db+dc)x^2 \\
&\qquad\qquad -(cba+dba+dca+dcb)x^3+(dcba)x^4.
\end{aligned}
$$

And if the law of these three polynomes hold for any n numbers $a, b, \ldots j, h$, $(1-kx)\ldots(1-bx)(1-ax)$

$$= 1-s_1x+s_2x^2-s_3x^3+\cdots\pm s_rx^r\mp\cdots\begin{cases}-s_{n-1}x^{n-1}+s_nx^n & \text{if } n \text{ be even}\\ +s_{n-1}x^{n-1}-s_nx^n & \text{if } n \text{ be odd,}\end{cases}$$

where s_1 stands for the sum of the numbers $a, b, c, \ldots k$, and s_r for the sum of the products made by multiplying together (arts. 56, 108) every r of them, and the upper or the lower of the marks $+-$ written before the term in x^r is to be taken according as r is even or odd. By an EVEN number is meant a whole number that is exactly divisible by 2 and by an ODD number one that is not. If then this be the law for any n numbers $a, b, \ldots k$, it follows that for any $n+1$ numbers $a, b, \ldots k, l$,

$$(1-lx)(1-kx)(1-jx)(1-ix)\ldots\ldots(1-cx)(1-bx)(1-ax)$$

$$= 1-s_1x \quad +s_2x^2 \quad -s_3x^3+\cdots\cdots\pm s_rx^r\mp\cdots\cdots\begin{cases}+s_nx^n\\ -s_nx^n\end{cases}$$

$$-lx+(ls_1)x^2-(ls_2)x^3+\cdots\pm(ls_{r-1})x^r\mp\cdots\begin{cases}+(ls_{n-1})x^n-(ls_n)x^{n+1}\\ -(ls_{n-1})x^n+(ls_n)x^{n+1}\end{cases}$$

$$= 1-(s_1+l)x+(s_2+ls_1)x^2-\cdots\pm(s_r+ls_{r-1})x^r\mp\cdots$$

$$\cdots\begin{cases}+(s_n+ls_{n-1})x^n-(ls_n)x^{n+1} & \text{if } n+1 \text{ be odd}\\ -(s_n+ls_{n-1})x^n+(ls_n)x^{n+1} & \text{if } n+1 \text{ be even.}\end{cases}$$

Now s_1+l is the sum and ls_n is the product of the $n+1$ numbers $a, b, \ldots k, l$. Moreover if when r is any whole number between 1 and $n+1$ $p, p', p'', \ldots$ be the products of every r of the n numbers $a, b, \ldots k$, and $q, q', q'', \ldots$ the products of every $r-1$ of them so that s_r is $p+p'+p''+\cdots$ and s_{r-1} $q+q'+q''+\cdots$ the products of every r of the $n+1$ numbers $a, b, \ldots k, l$, are made up of two sets those into which l does not enter as a factor and those into which l does and therefore are $p, p', p'', \ldots lq, lq', lq'', \ldots$ of which the sum

$$p+p'+p''+\cdots+lq+lq'+lq''+\cdots=p+p'+\cdots+(lq+lq'+\cdots)=p+p'+\cdots+l(q+q'+\cdots)$$

that is s_r+ls_{r-1}.

Hence the law that holds for 2 for 3 and for 4 numbers if it hold for any number of numbers holds for the next greater number. It therefore holds universally.

If for the sake of symmetry the symbol s_0 be given such a meaning as makes the sum of the products of every r of the $n+1$ numbers $a, b, \ldots k, l$, equal to s_r+ls_{r-1} even when r is 1 $a+b+\cdots+k+l = s_1+ls_0$ and s_0 can only be 1. The polynome in x then which is operationally equivalent to $(1-kx)(1-jx)(1-ix)\ldots(1-cx)(1-bx)(1-ax)$ is $s_0-s_1x+s_2x^2-s_3x^3+\cdots\pm s_nx^n$ according as n is an even number or an odd.

The coefficients $s_1, s_2, \ldots s_n$, may be found in an orderly way when $a, b, \ldots k$, are given by means of the principle proved that the sum of the products of every r of more than r numbers is equivalent to the sum got by adding to the sum of the products of every r of all but one of those numbers the product made by multiplying by that one the sum of the products of every $r-1$ of all the numbers but that same one. After this principle if the n numbers be taken in the order $a, b, c, d, \dots h, i, j, k$, and k_r stand for the sum of the products of every r of the numbers before k, j_r for the sum of the products of every r of them before j, and so on,

s_1 is $a+b+c+\cdots+h+i+j+k$
k_1 – $a+b+c+\cdots+h+i+j$
j_1 – $a+b+c+\cdots+h+i$
i_1 – $a+b+c+\cdots+h$
\- \- \- \-
d_1 – $a+b+c$
c_1 – $a+b$
b_1 – a

$s_2 = bb_1+cc_1+\cdots+hh_1+ii_1+jj_1+kk_1$
$k_2 = bb_1+cc_1+\cdots+hh_1+ii_1+jj_1$
$j_2 = bb_1+cc_1+\cdots+hh_1+ii_1$
$i_2 = bb_1+cc_1+\cdots+hh_1$
\- \- \- \-
$d_2 = bb_1+cc_1$
c_2 is bb_1

$s_3 = cc_2+\cdots+hh_2+ii_2+jj_2+kk_2$
$k_3 = cc_2+\cdots+hh_2+ii_2+jj_2$
\- \- \- \-
d_3 is cc_2
\- \- \- \-
\- \- \-

$s_{n-2} = ii_{n-3}+jj_{n-3}+kk_{n-3}$
$k_{n-2} = ii_{n-3}+jj_{n-3}$
j_{n-2} is ii_{n-3}

$s_{n-1} = jj_{n-2}+kk_{n-2}$
k_{n-1} is jj_{n-2}

s_n is kk_{n-1}.

And the several results may be found, or CALCULATED, in the following way, or ALGORITHM as it is called.

a	b	c	d	-	-	-	i	j	k	
	b_1	c_1	d_1	-	-	-	i_1	j_1	k_1	s_1
	bb_1	cc_1	dd_1	-	-	-	ii_1	jj_1	kk_1	
		c_2	d_2	-	-	-	i_2	j_2	k_2	s_2
		cc_2	dd_2	-	-	-	ii_2	jj_2	kk_2	
			d_3	-	-	-	i_3	j_3	k_3	s_3
			-	-	-	-	-			
					-		-	-		-
								jj_{n-2}	kk_{n-2}	
									k_{n-1}	s_{n-1}
									kk_{n-1}	
										s_n

where the first row is the numbers written in order, the 2nd and every other even row has its first number the same as the first number of the row next before but written one place to the right and every other the sum got by adding to the number next before it in the row the number right above that number in the row next before always written one place to the right, and each number in every odd row but the first is as its symbol shows the product made by multiplying the number next above it by the number in the first row at the top of the column. Taking for example 1, 2, 3, ... 9, 10,

1	2	3	4	5	6	7	8	9	10	
	1	3	6	10	15	21	28	36	45	55
	2	9	24	50	90	147	224	324	450	
		2	11	35	85	175	322	546	870	1320
		6	44	175	510	1225	2576	4914	8700	
			6	50	225	735	1960	4536	9450	18150
			24	250	1350	5145	15680	40824	94500	
				24	274	1624	6769	22449	63273	157773
				120	1644	11368	54152	202041	632730	
					120	1764	13132	67284	269325	902055
					720	12348	105056	605556	2693250	
						720	13068	118124	723680	3416930
						5040	104544	1063116	7236800	
							5040	109584	1172700	8409500
							40320	986256	11727000	
								40320	1026576	12753576
								362880	10265760	
									362880	10628640
									3628800	
										3628800

so that

$$(1-10x)(1-9x)(1-8x)(1-7x)(1-6x)(1-5x)(1-4x)(1-3x)(1-2x)(1-x)$$
$$=1-55x+1320x^2-18150x^3+157773x^4-902055x^5+3416930x^6$$
$$-8409500x^7+12753576x^8-10628640x^9+3628800x^{10}.$$

By the commutation laws of additions and of multiplications the result is the same whatever be the order of the numbers. The rightness of the ciphering may hence be tried by going through the process with a changed order of the numbers.

When each of the numbers $a, b, \dots k$, is 1 every odd row but the first being the same as the row next before it may be left out and then the process is that of the ARITHMETICAL TRIANGLE of PASCAL (*Œuvres. tome 5me. Paris.* 1819). Hence

1	1	1	1	1	1	1	- - -
	1	2	3	4	5	6	- - -
		1	3	6	10	15	- - -
			1	4	10	20	- - -
				1	5	15	- - -
					1	6	- - -
						1	- - -
							- - -

$$(1-x)^2 = 1-2x+x^2,$$
$$(1-x)^3 = 1-3x+3x^2-x^3,$$
$$(1-x)^4 = 1-4x+6x^2-4x^3+x^4, \text{ and so on.}$$

184. If in art. 183 the n numbers $a, b, c, \dots k$, be all equal to one another each of the products whose sum is s_r is a^r and hence if the symbol $n_{|r}$ stand for the number of these products

$$(1-ax)^n = 1-(n_{|1}a)x+(n_{|2}a^2)x^2-(n_{|3}a^3)x^3+\cdots\pm(n_{|n}a^n)x^n$$
$$= 1-n_{|1}ax+n_{|2}a^2x^2-\cdots\pm n_{|n}a^nx^n.$$

Here $n_{|1}$ being the number of the numbers themselves is n and $n_{|n}$ being the number of the products of all the numbers is 1. Further since as before the $n_{|r+1}$ products of every $r+1$ of the numbers may be made up of those into which one of the numbers does not enter of which the number is $(n-1)_{|r+1}$ and those into which that same one does enter of which the number is $(n-1)_{|r}$

$$n_{|r+1} = (n-1)_{|r+1}+(n-1)_{|r}$$

$\therefore$ doing the like with $n-1$, $n-2$, $n-3$, ... $r+1$, in succession

$$n_{|r+1} = (n-2)_{|r+1}+(n-2)_{|r}+(n-1)_{|r}$$
$$= (n-3)_{|r+1}+(n-3)_{|r}+(n-2)_{|r}+(n-1)_{|r}$$
$$= \cdot \quad \cdot \quad \cdot \quad \cdot \quad \cdot \quad \cdot$$
$$= (r+2)_{|r+1}+(r+2)_{|r}+(r+3)_{|r}+\cdots+(n-2)_{|r}+(n-1)_{|r}$$
$$= (r+1)_{|r+1}+(r+1)_{|r}+(r+2)_{|r}+\cdots+(n-1)_{|r}$$

and $\therefore = r_{|r}+(r+1)_{|r}+(r+2)_{|r}+\cdots+(n-1)_{|r} \quad \because (r+1)_{|r+1} = 1 = r_{|r}$

When r is 1 then, $n_{|2} = 1_{|1}+2_{|1}+3_{|1}+\cdots+(n-1)_{|1}$ that is $1+2+3+\cdots+(n-1)$ and $\therefore$ (art. 168) $= \dfrac{(n-1)n}{2}$. $\therefore$ when r is 2, $n_{|3} = 2_{|2}+3_{|2}+4_{|2}+\cdots+(n-1)_{|2} = \dfrac{1.2}{2}+\dfrac{2.3}{2}+\cdots+\dfrac{(n-2)(n-1)}{2} = \dfrac{1.2+2.3+\cdots+(n-2)(n-1)}{2}$

$$= \frac{\left\{\dfrac{(n-2)(n-1)n}{3}\right\}}{2} = \frac{(n-2)(n-1)n}{2.3}.$$

And if generally after the same law

$$n_{|r} = \frac{(n-r+1)(n-r+2)\dots(n-2)(n-1)n}{\underline{|r}},$$

$$n_{|r+1}=\frac{1.2\ldots(r-1)r}{\underline{|r}}+\frac{2.3\ldots r(r+1)}{\underline{|r}}+\cdots+\frac{(n-r)(n-r+1)\ldots(n-2)(n-1)}{\underline{|r}}$$

$$=\frac{1.2..(r-1)r+2.3..r(r+1)+\cdots+(n-r)(n-r+1)..(n-2)(n-1)}{\underline{|r}}$$

$$=\frac{\left\{\dfrac{(n-r)(n-r+1)\ldots(n-2)(n-1)n}{r+1}\right\}}{\underline{|r}}$$

$$=\frac{(n-r)\ldots(n-2)(n-1)n}{\{1.2\ldots(r-1)r\}(r+1)}=\frac{(n-r)(n-r+1)\ldots(n-2)(n-1)n}{1.2.3\ldots(r-1)r(r+1)},$$

after the same law. $n_{|r}$ is $\therefore$ always the same number as $\dfrac{(n-r+1)(n-r+2)\ldots(n-2)(n-1)n}{\underline{|r}}$ and may $\therefore$ be taken henceforth to symbolize either this number itself or any of this's operational equivalents such as $\dfrac{n-(r-1)}{r}\cdots\dfrac{n-2}{3}\,\dfrac{n-1}{2}\,\dfrac{n}{1}$ (art. 126), $\dfrac{\underline{|n}}{\underline{|r}\,\underline{|n-r}}$ (art. 177), $\dfrac{\underline{|n}}{\underline{|n-r}\,\underline{|r}}$, $\dfrac{(r+1)(r+2)\ldots(n-2)(n-1)n}{\underline{|n-r}}$, $\dfrac{n(n-1)(n-2)\ldots(n-r+2)(n-r+1)}{\underline{|r}}$.

Hence $n_{|r}=n_{|n-r}$, $\because$ each $=$ either $\dfrac{\underline{|n}}{\underline{|r}\,\underline{|n-r}}$ or $\dfrac{\underline{|n}}{\underline{|n-r}\,\underline{|r}}$. The same too follows straight from the fact that to the product of any r out of n numbers there corresponds the product of the other $n-r$.

If meaning be given to $n_{|0}$ by making the law $(n+1)_{|r+1}=n_{|r+1}+n_{|r}$ still hold when r is 0 $(n+1)_{|1}=n_{|1}+n_{|0}$ and $n_{|0}$ must then symbolize 1. The same result is got if taking $n_{|r}$ to stand for $\dfrac{n-(r-1)}{r}\cdots\dfrac{n-2}{3}\,\dfrac{n-1}{2}\,\dfrac{n}{1}$ the consequence $n_{|r+1}=\dfrac{n-r}{r+1}n_{|r}$ is still made to hold when r is 0. With this meaning of $n_{|0}$ the polynomic equivalent of $(1-ax)^n$ may be written without lopsidedness

$$(1-ax)^n=n_{|0}-n_{|1}ax+n_{|2}a^2x^2-n_{|3}a^3x^3+\cdots\begin{cases}+n_{|n}a^nx^n & \text{if } n \text{ be even}\\ -n_{|n}a^nx^n & \text{if } n \text{ be odd}\end{cases}$$

$$=n_{|n}-n_{|n-1}ax+n_{|n-2}a^2x^2-n_{|n-3}a^3x^3+\cdots\pm n_{|0}a^nx^n,$$

where also for greater symmetry the first terms may be severally written $n_{|0}a^0x^0$ and $n_{|n}a^0x^0$.

185. In the same way as in art. 183 it may be shown that if $a, b, c, \ldots i, j, k$, be any n numbers and s_r be the sum of the products of every r of them

$$(1+kx)(1+jx)\ldots(1+cx)(1+bx)(1+ax)=s_0+s_1x+s_2x^2+s_3x^3+\cdots+s_nx^n,$$

$$(x-k)(x-j)\ldots(x-c)(x-b)(x-a)$$
$$=s_0x^n-s_1x^{n-1}+s_2x^{n-2}-s_3x^{n-3}+\cdots\begin{cases}+s_n & \text{if } n \text{ be even}\\ -s_n & \text{if } n \text{ be odd,}\end{cases}$$

$$(x+k)(x+j)\ldots(x+c)(x+b)(x+a) = s_0 x^n + s_1 x^{n-1} + s_2 x^{n-2} + \cdots + s_{n-1}x + s_n,$$

$$(kx-1)(jx-1)\ldots(cx-1)(bx-1)(ax-1)$$

$$= s_n x^n - s_{n-1}x^{n-1} + s_{n-2}x^{n-2} - s_{n-3}x^{n-3} + \cdots \begin{cases} +s_0 & \text{if } n \text{ be even} \\ -s_0 & \text{if } n \text{ be odd,} \end{cases}$$

$$(kx+1)(jx+1)\ldots(cx+1)(bx+1)(ax+1) = s_n x^n + s_{n-1}x^{n-1} + s_{n-2}x^{n-2} + \cdots + s_1 x + s_0,$$

$$(k-x)(j-x)\ldots(c-x)(b-x)(a-x)$$

$$= s_n - s_{n-1}x + s_{n-2}x^2 - s_{n-3}x^3 + \cdots \begin{cases} -s_1 x^{n-1} + s_0 x^n & \text{if } n \text{ be even} \\ +s_1 x^{n-1} - s_0 x^n & \text{if } n \text{ be odd,} \end{cases}$$

$$(k+x)(j+x)\ldots(c+x)(b+x)(a+x) = s_n + s_{n-1}x + s_{n-2}x^2 + \cdots + s_1 x^{n-1} + s_0 x^n.$$

The polynomic expression operationally equivalent to the n factored product

$$(k_,-kx)(j_,-jx)(i_,-ix)\ldots(c_,-cx)(b_,-bx)(a_,-ax)$$

may be found either in the same way or from the polynome got in art. 183 as follows. Since

$$a_,-ax = a_,-\left(a_,\frac{1}{a_,}\right)ax = a_,-a_,\frac{1}{a_,}ax = a_,\left\{1-\left(\frac{1}{a_,}a\right)x\right\}$$

and the like for each of the other factors $(k_,-kx)\ldots(b_,-bx)(a_,-ax)$

$$= \left[k_,\left\{1-\left(\frac{1}{k_,}k\right)x\right\}\right]\ldots\left[b_,\left\{1-\left(\frac{1}{b_,}b\right)x\right\}\right]a_,\left\{1-\left(\frac{1}{a_,}a\right)x\right\}$$

$$= k_,\left\{1-\left(\frac{1}{k_,}k\right)x\right\}\ldots b_,\left\{1-\left(\frac{1}{b_,}b\right)x\right\}a_,\left\{1-\left(\frac{1}{a_,}a\right)x\right\}$$

$$= k_,\ldots b_,a_,\left\{1-\left(\frac{1}{k_,}k\right)x\right\}\ldots\left\{1-\left(\frac{1}{b_,}b\right)x\right\}\left\{1-\left(\frac{1}{a_,}a\right)x\right\}$$

$$= k_,\ldots b_,a_,(s'_0 - s'_1 x + s'_2 x^2 - s'_3 x^3 + \cdots \pm s'_n x^n)$$

if s'_r stand for the sum of the products of every r of the n numbers $\frac{1}{a_,}a, \frac{1}{b_,}b,\ldots\frac{1}{k_,}k,$

$$= k_,\ldots b_,a_,s'_0 - (k_,\ldots b_,a_,s'_1)x + (k_,\ldots b_,a_,s'_2)x^2 - (k_,\ldots b_,a_,s'_3)x^3 + \cdots \pm (k_,\ldots b_,a_,s'_n)x^n.$$

But $k_,\ldots b_,a_,s'_0$ is $k_,\ldots b_,a_,$, $k_,\ldots b_,a_,s'_n$ is $k_,\ldots b_,a_,\left(\frac{1}{k_,}k\right)\ldots\left(\frac{1}{b_,}b\right)\frac{1}{a_,}a$ and $\therefore$

$$= k_,\ldots b_,a_,\frac{1}{k_,}k\ldots\frac{1}{b_,}b\frac{1}{a_,}a = k_,\frac{1}{k_,}k\ldots b_,\frac{1}{b_,}ba_,\frac{1}{a_,}a = \left(k_,\frac{1}{k_,}\right)k\ldots\left(b_,\frac{1}{b_,}\right)b\left(a_,\frac{1}{a_,}\right)a$$

$= k\ldots ba$, and $k_,\ldots b_,a_,s'_r$ when r is one of the whole numbers 1, 2, 3,...$n-1$, is by distributing the successive multiplications by $a_,$, by $b_,,\ldots$ by $k_,$, over the additions that give the sum s'_r operationally equivalent to the sum of the products that have every r of the n numbers $\frac{1}{a_,}a, \frac{1}{b_,}b, \ldots \frac{1}{k_,}k,$ for their several first r factors in order

of operation and $a_{,}$, $b_{,}$, ...$k_{,}$, for their common other n and $\therefore$,—
$\because k_{,}\ldots b_{,}a_{,}\left(\frac{1}{j_{,}}j\right)\ldots\left(\frac{1}{c_{,}}c\right)\frac{1}{a_{,}}a$ taken for example of one of these products
$= k_{,}j_{,}\left(\frac{1}{j_{,}}j\right)i_{,}\ldots c_{,}\left(\frac{1}{c_{,}}c\right)b_{,}a_{,}\frac{1}{a_{,}}a = k_{,}j_{,}\frac{1}{j_{,}}ji_{,}\ldots c_{,}\frac{1}{c_{,}}cb_{,}a_{,}\frac{1}{a_{,}}a$
$= k_{,}\left(j_{,}\frac{1}{j_{,}}\right)ji_{,}\ldots\left(c_{,}\frac{1}{c_{,}}\right)cb_{,}\left(a_{,}\frac{1}{a_{,}}\right)a = k_{,}ji_{,}\ldots cb_{,}a$ and any other may be dealt with in a like way—, also operationally equivalent to the sum of the products made by multiplying together the upper numbers of every r of the n pairs $\left.\begin{matrix}a\\a_{,}\end{matrix}\right\}$, $\left.\begin{matrix}b\\b_{,}\end{matrix}\right\}$, $\left.\begin{matrix}c\\c_{,}\end{matrix}\right\}$, ... $\left.\begin{matrix}i\\i_{,}\end{matrix}\right\}$, $\left.\begin{matrix}j\\j_{,}\end{matrix}\right\}$, $\left.\begin{matrix}k\\k_{,}\end{matrix}\right\}$, and the lower numbers of the other $n-r$. If then S_r symbolize the sum last spoken of

$$(k_{,}-kx)(j_{,}-jx)\ldots(c_{,}-cx)(b_{,}-bx)(a_{,}-ax)$$
$$= S_0-S_1x+S_2x^2-S_3x^3+\cdots\begin{cases}+S_nx^n & \text{if } n \text{ be even}\\ -S_nx^n & \text{if } n \text{ be odd.}\end{cases}$$

So likewise

$$(k_{,}+kx)(j_{,}+jx)\ldots(c_{,}+cx)(b_{,}+bx)(a_{,}+ax) = S_0+S_1x+S_2x^2+\cdots+S_nx^n,$$
$$(kx-k_{,})(jx-j_{,})\ldots(cx-c_{,})(bx-b_{,})(ax-a_{,})$$
$$= S_nx^n-S_{n-1}x^{n-1}+S_{n-2}x^{n-2}-S_{n-3}x^{n-3}+\cdots\begin{cases}+S_0 & \text{if } n \text{ be even}\\ -S_0 & \text{if } n \text{ be odd,}\end{cases}$$
$$(kx+k_{,})(jx+j_{,})\ldots(cx+c_{,})(bx+b_{,})(ax+a_{,})$$
$$= S_nx^n+S_{n-1}x^{n-1}+S_{n-2}x^{n-2}+\cdots+S_1x+S_0.$$

186. When in art. 185 $a=b=c= - - - =k$ and $a_{,}=b_{,}=c_{,}= - - - =k_{,}$, (art. 184)

$$(a_{,}-ax)^n = n|_0a_{,}^n-(n|_1a_{,}^{n-1}a)x+(n|_2a_{,}^{n-2}a^2)x^2-(n|_3a_{,}^{n-3}a^3)x^3+\cdots$$
$$\cdots\begin{cases}+(n|_na^n)x^n & \text{if } n \text{ be even}\\ -(n|_na^n)x^n & - - - \text{ odd}\end{cases}$$
$$= n|_0a_{,}^n-n|_1a_{,}^{n-1}ax+n|_2a_{,}^{n-2}a^2x^2-n|_3a_{,}^{n-3}a^3x^3+\cdots\pm n|_na^nx^n$$
$$= n|_na_{,}^n-n|_{n-1}a_{,}^{n-1}ax+n|_{n-2}a_{,}^{n-2}a^2x^2-n|_{n-3}a_{,}^{n-3}a^3x^3+\cdots\pm n|_0a^nx^n,$$
$$(a_{,}+ax)^n = n|_0a_{,}^n+n|_1a_{,}^{n-1}ax+n|_2a_{,}^{n-2}a^2x^2+\cdots+n|_{n-1}a_{,}a^{n-1}x^{n-1}+n|_na^nx^n,$$
$$(ax-a_{,})^n = n|_0a^nx^n-n|_1a^{n-1}a_{,}x^{n-1}+n|_2a^{n-2}a_{,}^2x^{n-2}-n|_3a^{n-3}a_{,}^3x^{n-3}+\cdots$$
$$\cdots\begin{cases}+n|_na_{,}^n & \text{if } n \text{ be even}\\ -n|_na_{,}^n & - - - \text{ odd,}\end{cases}$$
$$(ax+a_{,})^n = n|_0a^nx^n+n|_1a^{n-1}a_{,}x^{n-1}+n|_2a^{n-2}a_{,}^2x^{n-2}+\cdots+n|_{n-1}aa_{,}^{n-1}x+n|_na_{,}^n.$$

In particular $(u+v)^n = \frac{\lfloor n}{\lfloor n}u^n+\frac{\lfloor n}{\lfloor 1\,\lfloor n-1}u^{n-1}v+\frac{\lfloor n}{\lfloor 2\,\lfloor n-2}u^{n-2}v^2+$

$\cdots+\frac{\lfloor\underline{n}}{\lfloor\underline{n-1}\,\lfloor\underline{1}}uv^{n-1}+\frac{\lfloor\underline{n}}{\lfloor\underline{n}}v^n$. And $\because\ \frac{\lfloor\underline{n}}{\lfloor\underline{r}\,\lfloor\underline{n-r}}u^{n-r}v^r=\left(\lfloor\underline{n}\frac{1}{\lfloor\underline{r}\,\lfloor\underline{n-r}}\right)u^{n-r}v^r$ $=\lfloor\underline{n}\frac{1}{\lfloor\underline{r}\,\lfloor\underline{n-r}}u^{n-r}v^r$ and meaning may be given to $\lfloor\underline{0}$ by making the law $\lfloor\underline{r+1}=\lfloor\underline{r}(r+1)$, got from $1.2\ldots(r-1)r(r+1)=\{1.2\ldots(r-1)r\}(r+1)$, hold even when r is 0 which gives $\lfloor\underline{0}$ a symbol for 1 $(u+v)^n$ $=\lfloor\underline{n}\frac{1}{\lfloor\underline{n}}u^n+\lfloor\underline{n}\frac{1}{\lfloor\underline{1}\,\lfloor\underline{n-1}}u^{n-1}v+\cdots+\lfloor\underline{n}\frac{1}{\lfloor\underline{n}}v^n=\lfloor\underline{n}\left(\frac{1}{\lfloor\underline{n}}u^n+\frac{1}{\lfloor\underline{1}\,\lfloor\underline{n-1}}u^{n-1}v+\cdots+\frac{1}{\lfloor\underline{n}}v^n\right)$ and $\therefore\ \frac{1}{\lfloor\underline{n}}(u+v)^n=\frac{1}{\lfloor\underline{0}\,\lfloor\underline{n}}u^nv^0+\frac{1}{\lfloor\underline{1}\,\lfloor\underline{n-1}}u^{n-1}v+\frac{1}{\lfloor\underline{2}\,\lfloor\underline{n-2}}u^{n-2}v^2+\cdots+\frac{1}{\lfloor\underline{n-1}\,\lfloor\underline{1}}uv^{n-1}$ $+\frac{1}{\lfloor\underline{n}\,\lfloor\underline{0}}u^0v^n$. Still further $\because\ \frac{1}{\lfloor\underline{r}\,\lfloor\underline{n-r}}u^{n-r}v^r=\left(\frac{1}{\lfloor\underline{n-r}}\frac{1}{\lfloor\underline{r}}\right)u^{n-r}v^r$ $=\frac{1}{\lfloor\underline{n-r}}\frac{1}{\lfloor\underline{r}}u^{n-r}v^r=\frac{1}{\lfloor\underline{n-r}}u^{n-r}\frac{1}{\lfloor\underline{r}}v^r=\left(\frac{1}{\lfloor\underline{n-r}}u^{n-r}\right)\frac{1}{\lfloor\underline{r}}v^r=\frac{u^{n-r}}{\lfloor\underline{n-r}}\frac{v^r}{\lfloor\underline{r}}$, $\frac{1}{\lfloor\underline{n}}(u+v)^n$ $=\frac{1}{\lfloor\underline{n}}\frac{1}{\lfloor\underline{0}}u^nv^0+\frac{1}{\lfloor\underline{n-1}}\frac{1}{\lfloor\underline{1}}u^{n-1}v+\frac{1}{\lfloor\underline{n-2}}\frac{1}{\lfloor\underline{2}}u^{n-2}v^2+\cdots\cdot+\frac{1}{\lfloor\underline{0}}\frac{1}{\lfloor\underline{n}}u^0v^n$ and

$$\frac{(u+v)^n}{\lfloor\underline{n}}=\frac{u^n}{\lfloor\underline{n}}\frac{v^0}{\lfloor\underline{0}}+\frac{u^{n-1}}{\lfloor\underline{n-1}}\frac{v}{\lfloor\underline{1}}+\frac{u^{n-2}}{\lfloor\underline{n-2}}\frac{v^2}{\lfloor\underline{2}}+\cdots+\frac{u}{\lfloor\underline{1}}\frac{v^{n-1}}{\lfloor\underline{n-1}}+\frac{u^0}{\lfloor\underline{0}}\frac{v^n}{\lfloor\underline{n}}.$$

Whenever n is given $n_{|2}$, $n_{|3}$, ... may be readily found one after another as in the instances following:—$7_{|1}$ being 7, $7_{|2}=\frac{7-1}{2}\times 7=3\times 7$ or 21, $\therefore\ 7_{|3}=\frac{7-2}{3}\times 21=35$, $\therefore\ 7_{|4}=\frac{7-3}{4}\times 35=35$, which $=7_{|7-4}$ as it ought, and so on.

$\therefore\ (1-ax)^7=1-7ax+21a^2x^2-35a^3x^3+35a^4x^4-21a^5x^5+7a^6x^6-a^7x^7.$

Likewise $8_{|2}=\frac{7}{2}\times 8$ which is 28, $\therefore\ 8_{|3}=\frac{6}{3}\times 28$ which is 56, $\therefore\ 8_{|4}=\frac{5}{4}\times 56$ which is 70, $\therefore\ 8_{|5}=\frac{4}{5}\times 70$ which is $56=8_{|8-5}$, and so on.

$\therefore\ (x-a)^8=x^8-8ax^7+28a^2x^6-56a^3x^5+70a^4x^4-56a^5x^3+28a^6x^2-8a^7x+a^8.$

187. PROP. *To find the mononomic or polynomic expression, when there is one, operationally equivalent to the quotient got by dividing one mononomic or polynomic expression by another.*

The ways of dealing with the several cases are as follows.

(1). $$\frac{afw^m x^p z^r}{bgw^n x^q z^s} = \frac{\left(\frac{a}{b}b\right)\left(\frac{f}{g}g\right)\left(\frac{w^m}{w^n}w^n\right)\left(\frac{x^p}{x^q}x^q\right)\frac{z^r}{z^s}z^s}{bgw^n x^q z^s} = \frac{\frac{a}{b}b\frac{f}{g}g\frac{w^m}{w^n}w^n\frac{x^p}{x^q}x^q\frac{z^r}{z^s}z^s}{bgw^n x^q z^s}$$

$$= \frac{\frac{a}{b}\frac{f}{g}\frac{w^m}{w^n}\frac{x^p}{x^q}\frac{z^r}{z^s}bgw^n x^q z^s}{bgw^n x^q z^s} = \frac{\left(\frac{a}{b}\frac{f}{g}\frac{w^m}{w^n}\frac{x^p}{x^q}\frac{z^r}{z^s}\right)bgw^n x^q z^s}{bgw^n x^q z^s} = \frac{a}{b}\frac{f}{g}\frac{w^m}{w^n}\frac{x^p}{x^q}\frac{z^r}{z^s}$$

and ∴ if m be not less than n, p than q, and r than s, $=\frac{a}{b}\frac{f}{g}w^{m-n}x^{p-q}z^{r-s}$.

(2). $$\frac{6a^5b^3-27a^4b^4+15a^3b^5-42a^2b^6}{3a^2b^3} = \frac{6a^5b^3}{3a^2b^3}-\frac{27a^4b^4}{3a^2b^3}+\frac{15a^3b^5}{3a^2b^3}-\frac{42a^2b^6}{3a^2b^3}$$

$$=\frac{6}{3}a^{5-2}b^{3-3}-\frac{27}{3}a^{4-2}b^{4-3}+\frac{15}{3}a^{3-2}b^{5-3}-\frac{42}{3}a^{2-2}b^{6-3} = 2a^3-9a^2b+5ab^2-14b^3.$$

(3). For finding the polynomic equivalent of

$$\frac{20-27x-23x^2+36x^3+3x^4-9x^5}{4-7x+3x^2} \text{ if there be any,}$$

$20-27x-23x^2+36x^3+3x^4-9x^5$

$$=\frac{20}{4}(4-7x+3x^2)-5(4-7x+3x^2)+20-27x-23x^2+36x^3+3x^4-9x^5$$
$$=5(4-7x+3x^2)-(20-35x+15x^2)+20-27x-23x^2+36x^3+3x^4-9x^5$$
$$=5(4-7x+3x^2)-15x^2+35x-20+20-27x-23x^2+36x^3+3x^4-9x^5$$
$$=5(4-7x+3x^2)+35x-27x-23x^2-15x^2+36x^3+3x^4-9x^5$$
$$=5(4-7x+3x^2)+(35x-27x)-(15x^2+23x^2)+36x^3+3x^4-9x^5$$
$$=5(4-7x+3x^2)+(35-27)x-(15+23)x^2+36x^3+3x^4-9x^5$$
$$=5(4-7x+3x^2)+\left\{\frac{8x}{4}(4-7x+3x^2)-\left(\frac{8}{4}x\right)(4-7x+3x^2)\right\}$$
$$+8x-38x^2+36x^3+3x^4-9x^5$$
$$=5(4-7x+3x^2)+(2x)(4-7x+3x^2)-(8x-14x^2+6x^3)$$
$$+8x-38x^2+36x^3+3x^4-9x^5$$
$$=(5+2x)(4-7x+3x^2)-6x^3+14x^2-8x+8x-38x^2+36x^3+3x^4-9x^5$$
$$=(5+2x)(4-7x+3x^2)-(38-14)x^2+(36-6)x^3+3x^4-9x^5 \text{ as before}$$
$$=(5+2x)(4-7x+3x^2)-\frac{24x^2}{4}(4-7x+3x^2)+\left(\frac{24}{4}x^2\right)(4-7x+3x^2)$$
$$-24x^2+30x^3+3x^4-9x^5$$
$$=(5+2x-6x^2)(4-7x+3x^2)+(24x^2-42x^3+18x^4)-24x^2+30x^3+3x^4-9x^5$$
$$=(5+2x-6x^2)(4-7x+3x^2)+24x^2-42x^3+18x^4-24x^2+30x^3+3x^4-9x^5$$
$$=(5+2x-6x^2)(4-7x+3x^2)+24x^2-24x^2+30x^3-42x^3+18x^4+3x^4-9x^5$$

$$= (5+2x-6x^2)(4-7x+3x^2)+(24x^2-24x^2)-(42x^3-30x^3)$$
$$+(18x^4+3x^4)-9x^5$$
$$= (5+2x-6x^2)(4-7x+3x^2)-(42-30)x^3+(18+3)x^4-9x^5$$
$$= (5+2x-6x^2)(4-7x+3x^2)-\frac{12x^3}{4}(4-7x+3x^2)+\left(\frac{12}{4}x^3\right)(4-7x+3x^2)$$
$$-12x^3+21x^4-9x^5$$
$$= (5+2x-6x^2-3x^3)(4-7x+3x^2)+(12x^3-21x^4+9x^5)-12x^3+21x^4-9x^5$$
$$= (5+2x-6x^2-3x^3)(4-7x+3x^2)+(12x^3-12x^3)-21x^4+21x^4+(9x^5-9x^5)$$

as before

$$= (5+2x-6x^2-3x^3)(4-7x+3x^2)$$
$$\therefore \frac{20-27x-23x^2+36x^3+3x^4-9x^5}{4-7x+3x^2} = 5+2x-6x^2-3x^3.$$

The result of this process may be got in an orderly way without showing the steps if keeping like mononomes always in the same column there be written in order the divisor, the quotient's polynomic equivalent as the successive terms are got, and the dividend with a stroke drawn close above it along its whole length; then in alternate rows with a stroke drawn close under and stretching just as far as each those sets of successive operations with mononomes which in polynomizing the product of the divisor by the quotient's polynomic equivalent come from the several terms of that equivalent; and lastly close under each set and on the other side of the stroke those other operations with mononomes which enter into the process at the same time as the set. So that the operations of all such of the said sets as are above any stroke the operations close under that stroke and the operations with the terms of the dividend beyond the right-hand end of the same stroke give together a result always operationally equivalent to the dividend.

$$\begin{array}{llllll}
4 & -7x & +3x^2 & & & \\
5 & +2x & -6x^2 & -3x^3 & & \\
\hline
20 & -27x & -23x^2 & +36x^3 & +3x^4 & -9x^5 \\
20 & -35x & +15x^2 & & & \\
\cline{1-3}
 & +8x & -38x^2 & & & \\
 & +8x & -14x^2 & +6x^3 & & \\
\cline{2-4}
 & & -24x^2 & +30x^3 & & \\
 & & -24x^2 & +42x^3 & -18x^4 & \\
\cline{3-5}
 & & & -12x^3 & +21x^4 & \\
 & & & -12x^3 & +21x^4 & -9x^5 \\
\cline{4-6}
\end{array}$$

In like manner if a polynomic equivalent be sought for

$$\frac{3x^6-10x^5+\frac{31}{2}x^4-27x^3+\frac{109}{6}x^2+9x-4}{6x^3-12x^2-3x+2}$$ the first term being operationally equivalent to $\frac{3x^6}{6x^3}$ is $\frac{3}{6}x^{6-3}$ or $\frac{1}{2}x^3$ and the dividend

$=\left(\frac{1}{2}x^3\right)(6x^3-12x^2-3x+2)$

$-4x^5+17x^4-28x^3$

$+\frac{109}{6}x^2+9x-4,$

the next term $=\frac{4x^5}{6x^3}$ is $\therefore$

$\frac{4}{6}x^{5-3}$ or $\frac{2}{3}x^2$ and the dividend

$=\left(\frac{1}{2}x^3-\frac{2}{3}x^2\right)(6x^3-12x^2-3x+2)$

$+9x^4-30x^3+\frac{39}{2}x^2+9x-4,$

the third term $=\frac{9x^4}{6x^3}$ is $\therefore$

$\frac{9}{6}x^{4-3}$ that is $\frac{3}{2}x$ and the dividend

$$\begin{array}{r} 6x^3\ -12x^2-3x+2 \\ \frac{1}{2}x^3\ \ -\frac{2}{3}x^2+\frac{3}{2}x-2 \\ \hline 3x^6-10x^5+\frac{31}{2}x^4-27x^3+\frac{109}{6}x^2+9x-4 \end{array}$$

$$\begin{array}{l} 3.\ \ -6.\ \ -\frac{3}{2}.\ \ +1. \\ \hline \quad -4.+17.-28. \\ \quad -4.\ \ +8.\ \ +2.\ \ -\frac{4}{3}. \\ \hline \qquad +9.-30.\ \ +\frac{39}{2}. \\ \qquad +9.-18.\ \ -\frac{9}{2}.+3. \\ \hline \qquad\quad -12.\ \ +24.+6. \\ \qquad\quad -12.\ \ +24.+6.-4 \\ \hline \end{array}$$

$$=\left(\frac{1}{2}x^3-\frac{2}{3}x^2+\frac{3}{2}x\right)(6x^3-12x^2-3x+2)-12x^3+24x^2+6x-4,$$

the last term $=\frac{12x^3}{6x^3}$ is $\therefore$ $\frac{12}{6}$ or 2 and

$$3x^6-10x^5+\frac{31}{2}x^4-27x^3+\frac{109}{6}x^2+9x-4$$

$$=\left(\frac{1}{2}x^3-\frac{2}{3}x^2+\frac{3}{2}x-2\right)(6x^3-12x^2-3x+2).$$

The polynome sought is $\therefore$ $\frac{1}{2}x^3-\frac{2}{3}x^2+\frac{3}{2}x-2.$

For carrying on the process regularly when the law of succession of the terms in ascending or descending powers of the reference numbers is not the same in the dividend and in the divisor the operations with the terms must be commutated so as to get the law the same in both. Thus

$$\frac{a^3+b^3+c^3-3abc}{a^2+b^2+c^2-bc-ca-ab}=\frac{a^3-3abc+b^3+c^3}{a^2-ab-ac+b^2-bc+c^2}=\frac{a^3-3bca+(b^3+c^3)}{a^2-(b+c)a+(b^2-bc+c^2)}$$

and the algorithm for finding this quotient's polynomic equivalent, $a+b+c$ or $a+(b+c)$, takes either of the following shapes:—

$$
\begin{array}{llllll}
a^2 & -ab & -ac & +b^2 & -bc & +c^2 \\
 & & a & & +b & +c \\
\hline
a^3 & -3abc & & +b^3 & & +c^3 \\
a^3-a^2b-a^2c & +ab^2 \; -abc & +ac^2 & & & \\
\hline
+a^2b+a^2c & -ab^2-2abc & -ac^2 & & & \\
+a^2b & -ab^2 \; -abc & & +b^3-b^2c & +bc^2 & \\
\hline
+a^2c & -abc & -ac^2 & +b^2c & -bc^2 & \\
+a^2c & -abc & -ac^2 & +b^2c & -bc^2 & +c^3 \\
\hline
\end{array}
$$

$$
\begin{array}{lll}
a^2 & -(b+c)a & +(b^2-bc+c^2) \\
 & a & +(b+c) \\
\hline
a^3 & -3bca & +(b^3+c^3) \\
a^3-(b+c)a^2+(b^2-bc+c^2)a & & \\
\hline
+(b+c)a^2-(b^2+2bc+c^2)a & & \\
+(b+c)a^2 & -(b+c)^2a+(b^3-b^2c+bc^2 & \\
 & & +b^2c-bc^2+c^3) \\
\hline
\end{array}
$$

188. When there is no polynome or mononome operationally equivalent to the quotient arising from the division of one polynome by another, or in other words none which multiplied into the latter polynome gives a product operationally equivalent to the former, yet by the process of art. 187 a polynome or mononome can be found which multiplied into the latter gives a product differing from the former by some polynome or mononome of a less range of powers than the latter. Thus taking the polynomes

$2x^5+\frac{13}{2}x^4-11x^3-\frac{5}{2}x^2+7x+3$ and $4x^3-3x^2-2x+1$, the first

$$=\left(\frac{1}{2}x^2\right)(4x^3-3x^2-2x+1)+8x^4-10x^3-3x^2+7x+3$$

$$=\left(\frac{1}{2}x^2+2x\right)(4x^3-3x^2-2x+1)-4x^3+x^2+5x+3$$

$$=\left(\frac{1}{2}x^2+2x-1\right)(4x^3-3x^2-2x+1)-2x^2+3x+4$$

$$\text{and } \therefore =\left(\frac{1}{2}x^2+2x-1\right)(4x^3-3x^2-2x+1)-(2x^2-3x-4).$$

Again taking $2+3x-4x^2+19x^3$ and $2-x+4x^2$,

$$
\begin{aligned}
2+3x-4x^2+19x^3 &= (1+2x)(2-x+4x^2)-(6x^2-11x^3) \\
&= (1+2x-3x^2)(2-x+4x^2)+(8x^3+12x^4) \\
&= (1+2x-3x^2+4x^3)(2-x+4x^2)+(16x^4-16x^5) \\
&= (1+2x-3x^2+4x^3+8x^4)(2-x+4x^2)-(8x^5+32x^6)
\end{aligned}
$$

and so on.

Since when $a=qb\pm c$, $\frac{a}{b}=\frac{qb\pm c}{b}=\frac{qb}{b}\pm\frac{c}{b}=q\pm\frac{c}{b}$,

$$\frac{2x^5+\frac{13}{2}x^4-11x^3-\frac{5}{2}x^2+7x+3}{4x^3-3x^2-2x+1}=\frac{1}{2}x^2+2x-1-\frac{2x^2-3x-4}{4x^3-3x^2-2x+1},$$

$$\frac{2+3x-4x^2+19x^3}{2-x+4x^2}=1+2x-\frac{6x^2-11x^3}{2-x+4x^2}=1+2x-3x^2+\frac{8x^3+12x^4}{2-x+4x^2}$$

$$=1+2x-3x^2+4x^3+\frac{16x^4-16x^5}{2-x+4x^2}=1+2x-3x^2+4x^3+8x^4-\frac{8x^5+32x^6}{2-x+4x^2}.$$

And hence the quotient got from the division of a polynome by a polynome of not greater range if not operationally equivalent to some polynome or mononome differs therefrom by the quotient got from the division by the divisor either of a mononome or of some polynome of less range than the divisor.

Also $\because$ $6x^2-11x^3=(6-11x)x^2$, $8x^3+12x^4=(8+12x)x^3$ $=\{4(2+3x)\}x^3$, $16x^4-16x^5=\{16(1-x)\}x^4$, $8x^5+32x^6=\{8(1+4x)\}x^5$, and by the colligation of multiplication of and division $\frac{de}{b}=\frac{d}{b}e$, it follows that

$$\frac{2+3x-4x^2+19x^3}{2-x+4x^2}=1+2x-\frac{6-11x}{2-x+4x^2}x^2=1+2x-3x^2+\frac{4(2+3x)}{2-x+4x^2}x^3$$

$$=1+2x-3x^2+4x^3+\frac{16(1-x)}{2-x+4x^2}x^4=1+2x-3x^2+4x^3+8x^4-\frac{8(1+4x)}{2-x+4x^2}x^5.$$

In like way $\frac{a^n-x^n}{a-x}=a^{n-1}+a^{n-2}x+a^{n-3}x^2+\cdots+ax^{n-2}+x^{n-1}$,

$$\frac{a^n+x^n}{a-x}=a^{n-1}+a^{n-2}x+a^{n-3}x^2+\cdots+ax^{n-2}+x^{n-1}+\frac{2}{a-x}x^n,$$

$$\frac{a^n-x^n}{a+x}=a^{n-1}-a^{n-2}x+a^{n-3}x^2-a^{n-4}x^3+\cdots\begin{cases}-x^{n-1} \text{ if } n \text{ be even}\\ +x^{n-1}-\frac{2}{a+x}x^n \text{ if } n \text{ be odd,}\end{cases}$$

$$\frac{a^n+x^n}{a+x}=a^{n-1}-a^{n-2}x+a^{n-3}x^2-a^{n-4}x^3+\cdots\begin{cases}-x^{n-1}+\frac{2}{a+x}x^n \text{ if } n \text{ be even,}\\ +x^{n-1} \text{ if } n \text{ be odd.}\end{cases}$$

189. If $p_0, p_1, p_2, \ldots p_n$, be given numbers as many numbers q_0, $q_1, q_2, \ldots q_n$, may be found such that $p_0x^n+p_1x^{n-1}+p_2x^{n-2}+\cdots+p_{n-1}x+p_n$ and $(q_0x^{n-1}+q_1x^{n-2}+q_2x^{n-3}+\cdots+q_{n-1})(x-a)+q_n$ are operationally equivalent. For (arts. 181, 180) the latter

$$=q_0x^n \quad +q_1x^{n-1} \quad +q_2x^{n-2}+\cdots \quad +q_{n-1}x$$

$$-(q_0a)x^{n-1}-(q_1a)x^{n-2}-\cdots-(q_{n-2}a)x-q_{n-1}a+q_n$$

$$=q_0x^n+(q_1-q_0a)x^{n-1}+(q_2-q_1a)x^{n-2}+\cdots+(q_{n-1}-q_{n-2}a)x+(q_n-q_{n-1}a).$$

And this is the same as the given polynome if q_0, q_1-q_0a, q_2-q_1a, ... $q_{n-1}-q_{n-2}a$, $q_n-q_{n-1}a$, be severally the same numbers as $p_0, p_1, p_2, \ldots p_{n-1}, p_n$. It is only needful then to make q_0 p_0 and to take $q_1, q_2, q_3, \ldots q_n$, in order so that

$$q_1=p_1+q_0a,\ q_2=p_2+q_1a,\ q_3=p_3+q_2a,\ \ldots\ q_{n-1}=p_{n-1}+q_{n-2}a,\ q_n=p_n+q_{n-1}a.$$

The whole may be handily arranged by writing the ps in order in a row, below each p in another row the product of a by the q next before when got, and below this product with a stroke drawn between and in a third row the new q got by adding the product to the p above it. Thus to find a polynome in x and a number such that the sum got by adding the number to the product by the polynome of $x-3$ is operationally equivalent to $10x^8+4x^7+9x^6+5x^4+8x^3+6x+7$,

p_0	p_1	p_2	p_3	- - -	p_{n-1}	p_n
	q_0a	q_1a	q_2a	- - -	$q_{n-2}a$	$q_{n-1}a$
q_0	q_1	q_2	q_3	- - -	q_{n-1}	q_n

10	4	9	0	5	8	0	6	7
	30	102	333	999	3012	9060	27180	81558
10	34	111	333	1004	3020	9060	27186	81565

and $(10x^7+34x^6+111x^5+333x^4+1004x^3+3020x^2+9060x+27186)(x-3)+81565=$ the polynome last written.

Again for $p_0+p_1x+p_2x^2+\cdots+p_nx^n$ to be operationally equivalent to

$$(q_0+q_1x+q_2x^2+\cdots+q_{n-3}x^{n-3})(1-ax-bx^2-cx^3)+(q_{n-2}+q_{n-1}x+q_nx^2)x^{n-2}$$

when $p_0, p_1, p_2, \ldots p_n$, a, b, c, are given numbers it must

$$\begin{array}{l}
=q_0 \quad +q_1x \quad +q_2x^2 \quad +q_3x^3 \quad +q_4x^4+\cdots \\
\qquad -(q_0a)x-(q_1a)x^2-(q_2a)x^3-(q_3a)x^4-\cdots \\
\qquad\qquad -(q_0b)x^2-(q_1b)x^3-(q_2b)x^4-\cdots \\
\qquad\qquad\qquad -(q_0c)x^3-(q_1c)x^4-\cdots \\
\hline
\cdots\cdots+q_{n-3}x^{n-3} \\
\cdots-(q_{n-4}a)x^{n-3}-(q_{n-3}a)x^{n-2} \\
\cdots-(q_{n-5}b)x^{n-3}-(q_{n-4}b)x^{n-2}-(q_{n-3}b)x^{n-1} \\
\cdots-(q_{n-6}c)x^{n-3}-(q_{n-5}c)x^{n-2}-(q_{n-4}c)x^{n-1}-(q_{n-3}c)x^n \\
\qquad\qquad +q_{n-2}x^{n-2} \quad +q_{n-1}x^{n-1} \quad +q_nx^n
\end{array}$$

$$=q_0+(q_1-q_0a)x+(q_2-q_1a-q_0b)x^2+(q_3-q_2a-q_1b-q_0c)x^3+(q_4-q_3a-q_2b-q_1c)x^4$$
$$+\cdots+(q_{n-2}-q_{n-3}a-q_{n-4}b-q_{n-5}c)x^{n-2}+(q_{n-1}-q_{n-3}b-q_{n-4}c)x^{n-1}+(q_n-q_{n-3}c)x^n.$$

And $\therefore$ inasmuch as $q-r-s-t=p$ just when $q=p+t+s+r=p+r+s+t$

$$
\begin{aligned}
q_0 &= p_0 \\
q_1 &= p_1 + q_0 a \\
q_2 &= p_2 + q_1 a + q_0 b \\
q_3 &= p_3 + q_2 a + q_1 b + q_0 c \\
q_4 &= p_4 + q_3 a + q_2 b + q_1 c \\
&\quad \cdot \quad \cdot \quad \cdot \quad \cdot \\
q_{n-2} &= p_{n-2} + q_{n-3} a + q_{n-4} b + q_{n-5} c \\
q_{n-1} &= p_{n-1} \qquad\qquad + q_{n-3} b + q_{n-4} c \\
q_n &= p_n \qquad\qquad\qquad\quad + q_{n-3} c.
\end{aligned}
$$

Hence the *q*s can be found one by one in order. The process may be carried on by writing in successive columns the sets of numbers of which the *q*s are severally the sums with the *p*s in the first row and the *qa* products the *qb* products and the *qc* products in the other several rows in order.

p_0	p_1	p_2	p_3	p_4	- - -	p_{n-3}	p_{n-2}	p_{n-1}	p_n
	q_0a	q_1a	q_2a	q_3a	- - -	$q_{n-4}a$	$q_{n-3}a$		
		q_0b	q_1b	q_2b	- - -	$q_{n-5}b$	$q_{n-4}b$	$q_{n-3}b$	
			q_0c	q_1c	- - -	$q_{n-6}c$	$q_{n-5}c$	$q_{n-4}c$	$q_{n-3}c$
q_0	q_1	q_2	q_3	q_4	- - -	q_{n-3}	q_{n-2}	q_{n-1}	q_n

If for instance the *p*s in order be 1, 2, 3, 4, 5, 6, 7, 8, 9, 10, and *a*, *b*, *c*, severally 3, 5, 7,

1	2	3	4	5	6	7	8	9	10
	3	15	69	315	1410	6306	28194		
		5	25	115	525	2350	10510	46990	
			7	35	161	735	3290	14714	65786
1	5	23	105	470	2102	9398	42002	61713	65796

So that $\dfrac{1+2x+3x^2+4x^3+5x^4+6x^5+7x^6+8x^7+9x^8+10x^9}{1-3x-5x^2-7x^3}$

$$= 1+5x+23x^2+105x^3+470x^4+2102x^5+9398x^6$$

$$+\frac{42002+61713x+65796x^2}{1-3x-5x^2-7x^3}x^7.$$

190. If P_n stand for a given polynome $p_0x^n+p_1x^{n-1}+p_2x^{n-2}+\cdots+p_{n-1}x+p_n$ of the *n*th degree in descending powers of x a like polynome P_{n-1} of the $(n-1)$th degree and a number Q_n may (art. 189) be found such that when a is any given number

$$P_n = P_{n-1}(x-a)+Q_n.$$

In the same way may a polynome P_{n-2} of the $(n-2)$th degree in descending powers of x be found and a number Q_{n-1} such that $P_{n-1}=P_{n-2}(x-a)+Q_{n-1}$ and then by distribution and colligation

$$P_n=\{P_{n-2}(x-a)+Q_{n-1}\}(x-a)+Q_n=\{P_{n-2}(x-a)\}(x-a)+Q_{n-1}(x-a)+Q_n$$
$$=P_{n-2}(x-a)^2+Q_{n-1}(x-a)+Q_n.$$

Again in the same way $P_{n-2}=P_{n-3}(x-a)+Q_{n-2}$, P_{n-3} being a polynome of the $(n-3)$th degree and Q_{n-2} a number, and ∵ as before

$$\{P_{n-3}(x-a)+Q_{n-2}\}(x-a)^2=\{P_{n-3}(x-a)\}(x-a)^2+Q_{n-2}(x-a)^2$$
$$=P_{n-3}(x-a)^3+Q_{n-2}(x-a)^2$$

it follows that $P_n=P_{n-3}(x-a)^3+Q_{n-2}(x-a)^2+Q_{n-1}(x-a)+Q_n$. By going on in the same way getting from P_{n-3} P_{n-4} and Q_{n-3}, from P_{n-4} P_{n-5} and Q_{n-4},... a polynome P_1 of the 1st degree is at length come to which $=P_0(x-a)+Q_1$ where P_0 is a number independent of x, or what by a stretch of language may be called a polynome of the 0th degree in x, and may ∴ in the system of symbols used, ∵ $P_0=0(x-a)+Q_0$, be symbolized by Q_0. Hence at last a polynome

$$Q_0(x-a)^n+Q_1(x-a)^{n-1}+Q_2(x-a)^{n-2}+\cdots+Q_{n-1}(x-a)+Q_n$$

of the nth degree in descending powers of $x-a$ is thus got operationally equivalent to P_n.

The same then that is done with the coefficients of P_n for finding Q_n has again to be done with the coefficients of P_{n-1} for finding Q_{n-1}, next with the coefficients of P_{n-2} for finding Q_{n-2}, and so on to the end. To find for example the polynome in $x-2$ that is operationally equivalent to $9x^6+7x^5+5x^4+3x^3+4x^2+6x+8$,—leaving out needless repetitions of the first coefficient 9—,

9	7	5	3	4	6	8
	18	50	110	226	460	932
	25	55	113	230	466	940
	18	86	282	790	2040	
	43	141	395	1020	2506	
	18	122	526	1842		
	61	263	921	2862		
	18	158	842			
	79	421	1763			
	18	194				
	97	615				
	18					
	115					

and $9(x-2)^6+115(x-2)^5+615(x-2)^4+1763(x-2)^3+2862(x-2)^2$
$+2506(x-2)+940$

is the polynome sought.

191. PROP. *To find the mononomic or polynomic equivalent if any there be of the product made by multiplying one mononome or polynome by the reciprocal of another.*

The following are the ways of dealing with the several cases.

(1). $\frac{1}{acx^m y^p z^r} bdx^n y^q z^s = \left(\frac{1}{z^r}\frac{1}{y^p}\frac{1}{x^m}\frac{1}{c}\frac{1}{a}\right) bdx^n y^q z^s = \frac{1}{z^r}\frac{1}{y^p}\frac{1}{x^m}\frac{1}{c}\frac{1}{a} bdx^n y^q z^s$

$$= \frac{1}{a}b\frac{1}{c}d\frac{1}{x^m}x^n\frac{1}{y^p}y^q\frac{1}{z^r}z^s = \left(\frac{1}{a}b\right)\left(\frac{1}{c}d\right)\left(\frac{1}{x^m}x^n\right)\left(\frac{1}{y^p}y^q\right)\frac{1}{z^r}z^s$$

$$= \left(\frac{1}{a}b\right)\left(\frac{1}{c}d\right)x^{n-m}y^{q-p}z^{s-r}$$

if n be not less than m, q than p, and s than r.

(2). $\frac{1}{4ac^2x^4}(8ac^5x^4-12ac^4x^5+28ac^3x^6-24ac^2x^7)$

$$= \frac{1}{4ac^2x^4}\times 8ac^5x^4-\frac{1}{4ac^2x^4}\times 12ac^4x^5+\frac{1}{4ac^2x^4}\times 28ac^3x^6-\frac{1}{4ac^2x^4}\times 24ac^2x^7$$

$$= \left(\frac{1}{4}\times 8\right)a^{1-1}c^{5-2}x^{4-4}-\left(\frac{1}{4}\times 12\right)a^{1-1}c^{4-2}x^{5-4}+\left(\frac{1}{4}\times 28\right)a^{1-1}c^{3-2}x^{6-4}$$

$$-\left(\frac{1}{4}\times 24\right)a^{1-1}c^{2-2}x^{7-4}$$

that is $\quad 2c^3-3c^2x+7cx^2-6x^3$.

(3). For polynomizing $\frac{1}{5x^2+6x-7}(20x^5+9x^4-36x^3+28x^2-20x+7)$ if it can be done,

$$20x^5+9x^4-36x^3+28x^2-20x+7$$

$$= (5x^2+6x-7)\times\frac{1}{5x^2}\times 20x^5-(5x^2+6x-7)\left(\frac{1}{5}\times 20\right)x^{5-2}$$
$$+20x^5+9x^4-36x^3+28x^2-20x+7$$

$$=(5x^2+6x-7)\times 4x^3-(20x^5+24x^4-28x^3)+20x^5+9x^4-36x^3+28x^2-20x+7$$

$$=(5x^2+6x-7)\times 4x^3+28x^3-24x^4-20x^5+20x^5+9x^4-36x^3+28x^2-20x+7$$

$$=(5x^2+6x-7)\times 4x^3+9x^4-24x^4+28x^3-36x^3+28x^2-20x+7$$

$$=(5x^2+6x-7)\times 4x^3-(24x^4-9x^4)-(36x^3-28x^3)+28x^2-20x+7$$

$$=(5x^2+6x-7)\times 4x^3-(24-9)x^4-(36-28)x^3+28x^2-20x+7$$

$$=(5x^2+6x-7)\times 4x^3-(5x^2+6x-7)\times\frac{1}{5x^2}\times 15x^4$$
$$+(5x^2+6x-7)\left(\frac{1}{5}\times 15\right)x^{4-2}-15x^4-8x^3+28x^2-20x+7$$

$$=(5x^2+6x-7)(4x^3-3x^2)+(15x^4+18x^3-21x^2)-15x^4-8x^3+28x^2-20x+7$$

$$=(5x^2+6x-7)(4x^3-3x^2)+15x^4+18x^3-21x^2-15x^4-8x^3+28x^2-20x+7$$

$$=(5x^2+6x-7)(4x^3-3x^2)+15x^4-15x^4+18x^3-8x^3+28x^2-21x^2-20x+7$$

$$=(5x^2+6x-7)(4x^3-3x^2)+(15x^4-15x^4)+(18x^3-8x^3)$$
$$+(28x^2-21x^2)-20x+7$$
$$=(5x^2+6x-7)(4x^3-3x^2)+(18-8)x^3+(28-21)x^2-20x+7$$
$$=(5x^2+6x-7)(4x^3-3x^2)$$
$$+\{(5x^2+6x-7)\frac{1}{5x^2}\times 10x^3-(5x^2+6x-7)\left(\frac{1}{5}\times 10\right)x^{3-2}\}+10x^3+7x^2-20x+7$$
$$=(5x^2+6x-7)(4x^3-3x^2)+(5x^2+6x-7)\times 2x-(10x^3+12x^2-14x)$$
$$+10x^3+7x^2-20x+7$$

$=(5x^2+6x-7)(4x^3-3x^2+2x)-(12-7)x^2-(20-14)x+7$ in the same way as before, $=(5x^2+6x-7)(4x^3-3x^2+2x)-(5x^2+6x-7)\frac{1}{5x^2}\times 5x^2$

$$+(5x^2+6x-7)-5x^2-6x+7$$
$$=(5x^2+6x-7)(4x^3-3x^2+2x-1)+(5x^2-5x^2)+(6x-6x)-7+7$$

as before.

Wherefore because the inverse operation of multiplication of is reciprocal multiplication of

$$\frac{1}{5x^2+6x-7}(20x^5+9x^4-36x^3+28x^2-20x+7)=4x^3-3x^2+2x-1.$$

$$\begin{array}{rrrrrr}
 & & 4x^3 & -3x^2 & +2x & -1 \\
 & & & 5x^2 & +6x & -7 \\
\hline
20x^5 & +9x^4 & -36x^3 & +28x^2 & -20x & +7 \\
20. & +24. & -28. & & & \\
\cline{1-3}
 & -15. & -8. & & & \\
 & -15. & -18. & +21. & & \\
\cline{2-4}
 & & +10. & +7. & & \\
 & & +10. & +12. & -14. & \\
\cline{3-5}
 & & & -5. & -6. & \\
 & & & -5. & -6. & +7 \\
\cline{4-6}
\end{array}$$

The result may be got without showing the steps by writing with like mononomes everywhere in the same column first the polynome sought operationally equivalent to the product as it is got term by term, then the polynome whose reciprocal is the multiplier, thirdly the multiplicand with a stroke close above over its whole length, after these in alternate rows and with a stroke close below each stretching throughout its precise length the sets of successive operations with mononomes that in polynomizing the product of the sought polynome by the polynome in the second row arise from the several terms of the former, and lastly next to each set close under the stroke those other operations with mononomes that come into the process with the set and are such that all the sets up to and ending with the set they and the operations with those of the multiplicand's terms which stretch away to the right of the stroke give jointly a result operationally equivalent to the multiplicand.

In a like way for finding the polynomic expression operationally equivalent to

$$\frac{1}{2-3x+4x^2-x^3}(8-6x-3x^2+23x^3-23x^4+5x^5),$$

$$\begin{aligned}&8-6x-3x^2+23x^3-23x^4+5x^5\\&=(2-3x+4x^2-x^3)\times 4+6x-19x^2\\&\qquad+27x^3-23x^4+5x^5\\&=(2-3x+4x^2-x^3)(4+3x)-10x^2\\&\qquad+15x^3-20x^4+5x^5\\&=(2-3x+4x^2-x^3)(4+3x-5x^2)\end{aligned}$$

$$\begin{array}{l}4\ +3x\ -5x^2\\ 2\ -3x\ +4x^2\ \ -x^3\\ \hline 8\ -6x\ -3x^2+23x^3-23x^4+5x^5\\ 8-12.+16.\ -4.\\ \hline \quad +6.-19.+27.\\ \quad +6.\ -9.+12.\ -3.\\ \hline \qquad -10.+15.-20.\\ \qquad -10.+15.-20.+5.\\ \hline\end{array}$$

and the polynomic expression sought is $\therefore$

$$4+3x-5x^2.$$

192. When art. 191's process shows that there is no mononome or polynome operationally equivalent to the product of one given polynome by the reciprocal of another, or which comes to the same none which multiplied by the latter given polynome gives a product operationally equivalent to the former, a mononome or polynome may still be found by means of the process the product of the multiplication of which by the latter differs from the former by some polynome of less range than the latter. Then too if a be the former and b the latter of the given polynomes v the mononome or polynome and c the polynome of less range than b so found $a=bv\pm c$ and $\therefore$

$$\frac{1}{b}a=\frac{1}{b}(bv\pm c)=\frac{1}{b}bv\pm\frac{1}{b}c=\left(\frac{1}{b}b\right)v\pm\frac{1}{b}c=v\pm\frac{1}{b}c.$$

NUMERICAL AND SYMBOLIC WHOLES MEASURES MULTIPLES AND FRACTIONS

193. *Def.* By a WHOLE EXPRESSION is meant either a whole number or a symbolic expression in which there is no fractional symbol.

Any letter in a whole symbolic expression may stand for a fractional number; the wholeness has to do only with the appearance, look, form, or shape, of the expression. In the definition the word *quotient* may (art. 114) be used instead of *fractional;* it is only to contrast with *whole* that the latter word is chosen.

194. *Def.* When a whole expression is equal to the product made by multiplying another whole expression by some

third whole expression, or which is the same thing when the quotient of the division of the first whole expression by the second is equal to some whole expression, the first is called a MULTIPLE of the second and the second a SUBMULTIPLE or MEASURE of the first.

As to whole numbers this definition agrees with the definition of art. 87 but as to whole symbolics it may or may not. As to whole symbolic expressions FORMAL or SYMBOLIC multiples and measures are all that is to be understood.

195. *Def.* A whole expression is said to be PRIME which has no measure but itself and unity.

196. *Def.* Two whole expressions are said to be PRIME TO ONE ANOTHER which have no common measure but unity.

197. PROP. *Any measure of a whole expression is a measure of any multiple of it.*

For if m be any measure of a whole expression a there is (art. 194) some whole expression k such that

$$a = km$$

and therefore any multiple of a

$$pa = pkm = (pk)m.$$

But k, p, being whole expressions pk is a whole expression. Wherefore m is a measure of pa.

198. PROP. *If two whole expressions be prime to one another any measure of the one is prime to any measure of the other.*

For any common measure of the measures (art. 197) is a measure of any multiples of them and ∴ of the whole expressions. There can ∴ be none but 1.

199. PROP. *All the common measures of a whole expression and any multiple of it are precisely all the measures of the whole expression itself.*

For any measure of a whole expression a is a measure of pa any multiple of it and ∴ is a common measure of a and pa. Conversely any common measure of a and pa is of course a measure of a. All the common measures of a and pa are ∴ precisely all the measures of a.

200. PROP. *Any common measure of two whole expressions is a measure both of their sum and of their difference.*

For if m be any common measure of two whole expressions a a'

$$a = km \qquad a' = k'm$$

k k' being whole expressions. ∴

$$a+a' = km+k'm = (k+k')m \qquad a-a' = km-k'm = (k-k')m.$$

$k+k'$ $k-k'$ then being whole expressions m measures each of the expressions $a+a'$ $a-a'$.

201. PROP. *If a whole expression differ from a multiple of another whole expression by a third whole expression all the common measures of the first and second whole expressions are precisely all the common measures of the second and third.*

Let a b c be three whole expressions and kb a multiple of the second such that

$$a = \text{either } kb+c \text{ or } kb-c$$

and ∴ also such that $c = \text{either } a-kb \text{ or } kb-a$.

Any measure of b is a measure of kb. All common measures of a and b ∴ are common measures of a and kb ∴ (art. 200) are measures of either $a-kb$ or $kb-a$ ∴ (art. 194) are measures of the equal thereto c and ∴ are common measures of b and c. Likewise too all common measures of b and c are common measures of kb and c ∴ are measures of either $kb+c$ or $kb-c$ ∴ are measures of a the equal to this and ∴ are common measures of a and b. Since then all common measures of a and b are common measures of b and c and all common measures of b and c common measures of a and b all the common measures of a and b are the very same as all the common measures of b and c.

202. PROP. *To find the greatest common measure of two given whole numbers.*

Let a b be any two given whole numbers of which a is not less than b.

Divide a by b without using fractions (arts. 81, 83). If the division be exact a is a multiple of b, all the common measures of a and b (art. 199) are then precisely all the measures of b, and as a whole number is clearly (art. 194) its own greatest measure b is the greatest common measure of a and b. But if the division be inexact let q_1 be the quotient and r_1 the remainder, then since a differs from q_1b by r_1 all the common measures of a and b (art. 201) are precisely all the common measures of b

a	b
$q_1)\ \underline{q_1 b}$	$\underline{q_2 r_1}\ (q_2$
r_1	r_2
$q_3)\ \underline{q_3 r_2}$	$\underline{q_4 r_3}\ (q_4$
r_3	r_4

and r_1. Divide unfractionally b by r_1 and if the division be exact all the common measures of b and r_1 and ∴ also all those of a and b are precisely all the measures of r_1, as before then r_1 is the greatest common measure at once of b and r_1 and of a and b. But if the division be inexact let q_2 be the quotient and r_2 the remainder, then all the common measures of b and r_1 are precisely all the common measures of r_1 and r_2, ∴ all the common measures of a and b are precisely all the common measures of r_1 and r_2. Again divide unfractionally r_1 by r_2. If the division be exact all the common measures of r_1 and r_2 and ∴ all those both of a and b and of b and r_1 are precisely all the measures of r_2 and the greatest of them then is r_2. But if the division yield an inexact quotient q_3 and a remainder r_3 all the common measures of r_1 and r_2 are the same as all those of r_2 and r_3 and all the common measures are the same of a and b of b and r_1 of r_1 and r_2 and of r_2 and r_3. Now in going on with the process of dividing unfractionally by each new remainder that arises the remainder of the inexact division next before the remainders become ever less and less and by not less than 1 at each step, ∴ in not more than b unfractional divisions from starting a remainder 0 must at length be come to. Let it be the ith remainder r_i that divides exactly the remainder r_{i-1} next before. All the common measures then of r_{i-1} and r_i are precisely all the measures of r_i. But all the common measures of r_{i-1} and r_i are precisely all the common measures of any two consecutive terms of the series

$$a,\ b,\ r_1,\ r_2,\ r_3,\ \ldots\ r_{i-1},\ r_i.$$

Therefore all the common measures of a and b are precisely all the measures of r_i. And ∴ r_i being itself the greatest of all its own measures is the greatest common measure of a and b.

Since if r' be less than r $r_{,}=qr+r'$ just when $r_{,}=qr+(r-r)+r'$ $=qr+r-r+r'=(q+1)r+r'-r=(q+1)r-(r-r')$ and ∴ all the common measures of $r_{,}$ and r are precisely all the common measures of r and $r-r'$, in the above process $r-r'$ may be anywhere used instead of r'.

203. COROLLARY. *All the common measures of two whole numbers are precisely all the measures of their greatest common measure.*

In art. 202 it is shown that all the common measures of any two whole numbers a b are precisely all the measures of their greatest common measure r_i.

204. PROP. *A common measure of two whole numbers is their greatest common measure precisely when the whole numbers are prime to one another that express what multiples of it they are.*

First let m be the greatest common measure of two whole numbers a b so that

$$a = hm \qquad b = km$$

h k being whole numbers. If μ be any common measure that h k can have so that

$$h = \eta\mu \qquad k = \kappa\mu$$

η κ being whole numbers

$$a = (\eta\mu)m = \eta\mu m \qquad b = (\kappa\mu)m = \kappa\mu m$$

and $\therefore$ μm is a common measure of a and b. But $\because$ m is the greatest common measure μ cannot be else than 1. The only common measure $\therefore$ that h k can have is 1, that is h k are prime to one another.

Again let m be a common measure of two whole numbers a b such that

$$a = hm \qquad b = km$$

where h k are whole numbers that are prime to one another. $\because$ (art. 203) all common measures of two whole numbers are measures of their greatest common measure the greatest common measure of a and b is gm, g being some whole number, and $\therefore$

$$a = h'gm \qquad b = k'gm$$

h' k' being whole numbers. $\therefore$

$$hm = h'gm = (h'g)m \qquad km = k'gm = (k'g)m$$

and $\therefore$ $$h = h'g \qquad k = k'g.$$

But h k being prime to one another have no common measure but 1. $\therefore$ g is 1 and $\therefore$ gm the greatest common measure of a and b is m.

205. PROP. *To find the greatest common measure of more than two given whole numbers.*

First if there be three given whole numbers a b c find (art. 202) the greatest common measure m of any two a b and then the greatest common measure m' of m and the third c. Since all the common measures of a b (art. 203) are precisely all the measures of m all the common measures of a b c are precisely all the common measures of m and c. But all the common measures of m and c are precisely all the measures of m'. Therefore all the common measures of $a\ b\ c$ are precisely all the measures of m'. Hence since of all m''s measures m' itself is the greatest m' is the greatest common measure of $a\ b$ and c.

Next if there be a fourth given whole number d find the greatest common measure m'' of m' and d. All the common measures then of $a\ b\ c$ being precisely all the measures of m' all the common measures of $a\ b\ c\ d$ are precisely all the common measures of m' and d. But these again are precisely all the measures of m''. All the common measures therefore of $a\ b\ c\ d$ are precisely all the measures of m'' and therefore the greatest of them is m''.

In like manner all the measures of the greatest common measure of m'' and a fifth number e are precisely all the common measures of the five numbers $a\ b\ c\ d\ e$ and that greatest common measure is therefore the greatest common measure of the five. And so on for six or more whole numbers.

The greatest common measure of four given whole numbers may also be got by finding the greatest common measure m of any two of them the greatest common measure $m_{,}$ of the other two and then the greatest common measure $m_{,,}$ of m and $m_{,}$. For all the common measures of the first two being precisely all the measures of m and all the common measures of the other two precisely all the measures of $m_{,}$ all the common measures of the four are precisely all the common measures of $m\ m_{,}$, hence are precisely all the measures of $m_{,,}$, and hence have $m_{,,}$ the greatest of them. In a like way may the greatest common measure of more than four given whole numbers be found.

206. COR. *All the common measures of any whole numbers are precisely all the measures of their greatest common measure.*

207. PROP. *A whole number prime to each of two whole numbers is prime to the product made by multiplying by either of them the other.*

Let a whole number a be prime to each of two whole numbers $b\ c$.

If a be 1 1 is the only measure of a and ∴ the only common measure of a and bc. Again if b be 1 bc is c if c be 1 bc is b and to both c and b a is prime.

But if neither $a\ b$ nor c be 1 let the unfractional division of b by

a give the quotient q_1 and the remainder r_1; should b be less than a q_1 is 0 and r_1 is b. Because a is not a measure of b r_1 is not 0. Because b differs from q_1a by r_1 all the common measures of b and a are precisely all the common measures of a and r_1 and ∴ as b is prime to a so also is a to r_1. Wherefore and because when a is a multiple of r_1 all the common measures of a and r_1 are precisely all the measures of r_1 a can be a multiple of r_1 only when r_1 is 1. When then r_1 is not 1 let the unfractional division of a by r_1 give the quotient q_2 and the remainder r_2 else than 0. As before because a is prime to r_1 r_1 is prime to r_2 and r_1 can be a multiple of r_2 only when r_2 is 1. Again when r_2 is not 1 let the unfractional division of r_1 by r_2 give the quotient q_3 and the remainder r_3 else than 0, then r_2 is prime to r_3 and r_2 can be a multiple of r_3 only when r_3 is 1. And so on. Now in carrying on these unfractional divisions by each new remainder of the last one the remainders go on lessening by 1 or more at each step and must at length ∴ be always brought down to 0. Let it be the unfractional division by the ith remainder r_i that gives this remainder 0, then every two consecutive terms of the series

	b
a	q_1a (q_1
q_2) q_2r_1	r_1
r_2	q_3r_2 (q_3
q_4) q_4r_3	r_3
r_4	

$$b \quad a \quad r_1 \quad r_2 \quad r_3 \quad . \; . \; . \quad r_{i-1} \quad r_i$$

are prime to one another and ∴ and because r_{i-1} is a multiple of r_i r_i is 1.

Because $r_1 = b - q_1a$ $r_1c = (b-q_1a)c = bc-(q_1a)c = bc-q_1ac$ and ∴ any common measure of bc and q_1ac is a measure of r_1c. But because any measure of a is a measure of the multiple thereof ca, ∴ of this multiple's equal ac, and ∴ of the multiple of this last q_1ac, any common measure of a and bc is a common measure of q_1ac and bc. Therefore any common measure that a and bc can have is a measure both of ac and of r_1c. In like manner because $r_2 = a - q_2r_1$ $r_2c = ac - q_2r_1c$ ∴ any common measure of ac and r_1c is a measure of r_2c and ∴ any common measure that a and bc can have is a measure at once of ac of r_1c and of r_2c. In the same way it follows step by step that any common measure of a and bc is a measure of every term of the series

$$ac \quad r_1c \quad r_2c \quad r_3c \quad . \; . \; . \quad r_{i-1}c \quad r_ic.$$

Because then r_ic is c any common measure that a and bc can have is a measure of c and ∴ a common measure of a and c. But a is prime to c. Therefore a is prime to bc.

208. PROP. *If each of any whole numbers be prime to each of any other whole numbers the product of all the first whole numbers is prime to the product of all the others.*

Let each of the whole numbers $a\ b\ c \ldots g\ h$ be prime to each of the whole numbers $a'\ b'\ c' \ldots m'\ n'$.

Because a' is prime to a and to b it is (art. 207) prime to ba, because then prime to ba and also prime to c it is prime to cba, and so on. ∴ a' is prime to $hg \ldots dcba$, that is a whole number prime to each of any whole numbers is prime to their product. So each of the other whole numbers $b'\ c' \ldots m'\ n'$ is prime to the same product $hg \ldots cba$. Hence $hg \ldots cba$ because prime to each of the whole numbers $a'\ b'\ c' \ldots m'\ n'$ is prime to their product $n'm' \ldots c'b'a'$.

209. COR. *If one whole number be prime to another any power of the one is prime to any power of the other.*

This is only the particular case of art. 208's prop. when $a = b = c = \cdots = h$ and $a' = b' = c' = \cdots = n'$.

210. PROP. *All the common measures of two whole numbers are precisely all the common measures of one of them and the product made by multiplying the other by any whole number prime to the first.*

Let $a\ b$ be any two whole numbers and p any whole number prime to a.

If m be the greatest common measure of $a\ b$

$$a = hm \qquad b = km$$

$h\ k$ being whole numbers that (art. 204) are prime to one another, and

$$pb = pkm = (pk)m.$$

But ∵ $a = hm$ and ∴ $= mh$ h is a measure of a and (art. 198) ∴ is prime to p. h then is prime to both k and p and ∴ (art. 207) is prime to pk, that is the whole numbers are prime to one another that express what multiples a and pb are of m. ∴ (art. 204) m is the greatest common measure of a and pb. All the common measures ∴ of a and b (art. 203) are precisely all the common measures of a and pb.

211. PROP. *If of two equal simple fractions the terms of one be prime to one another the terms of the other are those multiples of some whole number that are expressed by the corresponding terms of the first.*

Let $\frac{a}{b} = \frac{a'}{b'}$, $a\ b\ a'\ b'$ being whole numbers of which $a\ b$ are prime to one another.

Then (art. 113) $a' = \frac{a}{b}b'$ and ∴ (art. 124) $= \frac{ab'}{b}$. ∴ (art. 194) ab' is a multiple of b and ∴ (art. 199) all the common measures of ab' and b are precisely all the measures of b. But ∵ a is prime to b all the common measures of b' and b (art. 210) are precisely all the common measures of ab' and b. ∴ all the common measures of b' and b are precisely all the measures of b. ∴ b is a measure of b', that is there is some whole number k such that

$$b' = kb \text{ and } \therefore \text{ also } = bk.$$

$$\therefore\ a' = \frac{akb}{b} = \frac{(ak)b}{b} = ak.$$

212. PROP. *To find the least common multiple of two given whole numbers.*

Let $a\ b$ be two given whole numbers.

Any whole number that is at once a multiple of a and a multiple of $b =$

$$pa = qb,$$

$p\ q$ being whole numbers. If then m be the greatest common measure of a and b and

$$a = hm \qquad\qquad b = km$$

$h\ k$ being whole numbers that are prime to one another

$$phm = qkm.$$

This happens precisely when singly $(ph)m = (qk)m$, $ph = qk$, $ph = kq$,

$$\frac{ph}{qh} = \frac{kq}{hq},\ \frac{p}{q} = \frac{k}{h}.$$

∴ (art. 211) $p = kr$ $q = hr$, r being some whole number, and ∴ any common multiple of a and b

$$= (kr)hm = (hr)km = krhm = hrkm = rkhm = rhkm.$$

Any of these equals is least when r is 1 and ∴ of all the common multiples of a and b the least

$$= khm = hkm = \frac{b}{m}a = \frac{a}{m}b = \frac{ba}{m} = \frac{ab}{m}.$$

213. COR. *All the common multiples of two whole numbers are precisely all the multiples of their least common multiple.*

In art. 212 any common multiple of two whole numbers $a\ b$ is shown $= rhkm$ where r is a whole number and $hkm =$ the least common multiple.

214. PROP. *To find the least common multiple of more than two given whole numbers.*

If there be three given whole numbers a b c find (art. 212) first the least common multiple v of any two of them a b and then the least common multiple v' of v and the third c. ∵ (art. 213) all the common multiples of a and b are precisely all the multiples of v all the common multiples of a b and c are precisely all the common multiples of v and c. But all the common multiples of v and c are precisely all the multiples of their least common multiple v'. ∴ all the common multiples of a b and c are precisely all the multiples of v'. Now the least of all the multiples of any whole number is clearly just that whole number itself. ∴ v' is its own least multiple and ∴ is the least common multiple of a b and c.

Again if there be a fourth given whole number d find the least common multiple v'' of v' and d. All the common multiples of a b and c being precisely all the multiples of v' all the common multiples of a b c and d are precisely all the common multiples of v' and d. But these last are precisely all the multiples of v''. ∴ all the common multiples of a b c and d are precisely all the multiples of v'' and ∴ the least common multiple of a b c and d is the least of v'''s multiples, that is v'' itself.

So all the common multiples of five given whole numbers a b c d e are precisely all the common multiples of v'' and e ∴ are precisely all the multiples of the least common multiple of v'' and e and ∴ the least common multiple of a b c d e is the least common multiple of v'' e. And so on for more than five given whole numbers.

If u be the least common multiple of two or more whole numbers u' the least common multiple of other two or more whole numbers and u'' the least common multiple of u and u', all the common multiples of the first two or more whole numbers being precisely all the multiples of u and all the common multiples of the other two or more precisely all the multiples of u' all the common multiples of all those whole numbers are precisely all the common multiples of u and u'. As then all the common multiples of u and u' are precisely all the multiples of u'' so too are all the common multiples of all the whole numbers and the least of them is therefore u''. This may be made use of for finding the least common multiple of more than three given whole numbers.

215. COR. *All the common multiples of any whole numbers are precisely all the multiples of their least common multiple.*

216. It is by means of the foregoing propositions about the measures and multiples of whole numbers that the fractional operations of arts. 101, 103, 106, 109, and the process of art. 99 on which these operations all hang may be most readily gone through.

First and foremost a fraction is expressed as a simple fraction in terms the least possible only when the terms are whole numbers that are prime to one another. For (art. 211) any whole number terms in which it can be expressed are multiples by these of some whole number and therefore (arts. 88, 89) neither of them ever can be less than the corresponding term. If then in the handling of fractions the least whole numbers are to be used that need be the greatest common measure of the terms of every given simple fraction and the numbers expressing what multiples the terms are of the greatest common measure must be found. Taking for example the fraction $\frac{2231}{3104}$ the greatest common measure of 2231 and 3104 is 97 $\frac{2231}{97} = 23$ $\frac{3104}{97} = 32$ $\frac{2231}{3104} = \frac{23 \times 97}{32 \times 97} = \frac{23}{32}$ and hence $\frac{23}{32}$ is the simple fraction with least terms that $= \frac{2231}{3104}$.

2231	3104
3) 2619	2231 (1
388	873
4) 388	776 (2
	97

	3104
2231	2231 (1
2) 1746	873
485	485 (1
1) 388	388
97	388 (4
23) 194	291 (32
29	19
291	194

When given simple fractions are all expressed in the least whole terms those of them that have particular terms different must next be expressed so as to have the corresponding terms the same and the least whole numbers through which this can be done must be found by making the terms that are to be the same all equal to the least common multiple of the corresponding terms that differ. For example to find the simple fraction with least terms that $= \frac{45}{713} - \frac{33}{527} + \frac{19}{391}$ each of the fractions $\frac{45}{713}$, $\frac{33}{527}$, $\frac{19}{391}$, is already in its least whole terms the greatest common measure of 713 and 527 is 31 the least common multiple of these is hence $\frac{527}{31} \times 713$, $\frac{713}{31} \times 527$, or 12121, which being equal to 31×391 is a multiple of 391 and the fraction sought

$$=\frac{45\times17}{713\times17}-\frac{33\times23}{527\times23}+\frac{19\times31}{391\times31}=\frac{765-759+589}{12121}=\frac{595}{12121}=\frac{35\times17}{713\times17}$$

and because 17 is the greatest common measure of 595 and 12121 therefore is $\frac{35}{713}$.

In like manner

$$\frac{53563}{28913}\times\frac{82751}{154463}=\frac{1847\times29}{997\times29}\times\frac{997\times83}{1861\times83}=\frac{1847}{997}\times\frac{997}{1861}=\frac{1847}{1861},$$

$$\frac{63}{1729}\times\frac{2090}{60}=\frac{9\times7}{247\times7}\times\frac{209\times10}{6\times10}=\frac{9}{247}\times\frac{209}{6}=\frac{9\times11}{247\times11}\times\frac{209\times13}{6\times13}$$

$$=\frac{11\times3\times3}{2717}\times\frac{2717}{13\times2\times3}=\frac{(11\times3)\times3}{(13\times2)\times3}=\frac{33}{26},$$

$$\frac{\left(\frac{264}{4235}\right)}{\left(\frac{424}{1848}\right)}=\frac{\left(\frac{24\times11}{385\times11}\right)}{\left(\frac{53\times8}{231\times8}\right)}=\frac{\left(\frac{24}{385}\right)}{\left(\frac{53}{231}\right)}=\frac{\left(\frac{24\times3}{385\times3}\right)}{\left(\frac{53\times5}{231\times5}\right)}=\frac{\left(\frac{72}{1155}\right)}{\left(\frac{265}{1155}\right)}=\frac{72}{265}.$$

217. Any given whole number is either prime or equal to the product of two whole numbers each else than 1, if the latter each of the factors is either prime or equal to the product of two whole numbers each else than 1 and then the given whole number either immediately or by colligation and commutation of the multiplications is equal to the product of three or four whole numbers each else than 1, again each of these factors is either prime or equal to the product of two whole numbers each else than 1 and the given whole number is then as before equal to the product of more than three but not more than eight whole numbers each else than 1, and so on. But because any whole number that is not prime is equal to the product of two whole numbers each of which is less than itself and by never less than 2 the process cannot go on for ever. Therefore any given whole number either is itself a prime number or is equal to the product of whole numbers of which every one is a prime number. If then α of the factors of this product be the prime number a β of them the prime number b γ of them the prime number c - - - - ν of them the prime number n the given whole number

$$=nn..nnmm..mmll..ll......cc..ccbb..bbaa..aa$$

$$=(n..nn)(m..mm)(l..ll)....(c..cc)(b..bb)a..aa$$

that is $\quad n^{\nu}m^{\mu}l^{\lambda}...c^{\gamma}b^{\beta}a^{\alpha}.$

Thus 1, 2, 3, 5, 7, 11, 13, 17, 19, 23, 29, 31, 37, 41, 43,... are prime numbers. $4=2^2$, $6=2\times 3$, $8=2^3$, $9=3^2$, $10=2\times 5$, $12=3\times 4$ $=3.2^2$, $14=2\times 7$, $15=3\times 5$, $16=4\times 4=2^2\times 2^2=2^4$, $360=6\times 60$ $=(2\times 3).2.30=2.3.2.2.15=2.3.2.2.3.5=2.2.2.3.3.5=(2.2.2)(3.3).5$ $=2^3.3^2.5$,

$$70\,000=7(5\times 2)^4=7.5^4.2^4=2^4.5^4.7=7.2^4.5^4.$$

Any number equal to $n^{\nu_{,}}m^{\mu_{,}}l^{\lambda_{,}}\ldots c^{\gamma_{,}}b^{\beta_{,}}a^{\alpha_{,}}$, where $\alpha_{,}\ \beta_{,}\ \gamma_{,}\ldots\nu_{,}$ are whole numbers not greater severally than $\alpha\ \beta\ \gamma\ldots\nu$ and of which any may be 0, is a measure of the number equal to $n^\nu m^\mu l^\lambda\ldots c^\gamma b^\beta a^\alpha$. For

$$n^\nu\ldots b^\beta a^\alpha=n^{\nu-\nu_{,}+\nu_{,}}\ldots b^{\beta-\beta_{,}+\beta_{,}}a^{\alpha-\alpha_{,}+\alpha_{,}}=(n^{\nu_{,}}n^{\nu-\nu_{,}})\ldots(b^{\beta_{,}}b^{\beta-\beta_{,}})a^{\alpha_{,}}a^{\alpha-\alpha_{,}}$$
$$=n^{\nu_{,}}n^{\nu-\nu_{,}}\ldots b^{\beta_{,}}b^{\beta-\beta_{,}}a^{\alpha_{,}}a^{\alpha-\alpha_{,}}=n^{\nu-\nu_{,}}\ldots b^{\beta-\beta_{,}}a^{\alpha-\alpha_{,}}n^{\nu_{,}}\ldots b^{\beta_{,}}a^{\alpha_{,}}$$
$$=(n^{\nu-\nu_{,}}m^{\mu-\mu_{,}}l^{\lambda-\lambda_{,}}\ldots c^{\gamma-\gamma_{,}}b^{\beta-\beta_{,}}a^{\alpha-\alpha_{,}})n^{\nu_{,}}m^{\mu_{,}}l^{\lambda_{,}}\ldots c^{\gamma_{,}}b^{\beta_{,}}a^{\alpha_{,}}.$$

But no other whole number can be a measure whether equal to $vu\ldots rqpn^{\nu_{,}}l^{\lambda_{,}}\ldots b^{\beta_{,}}$ where $\beta_{,}\ldots\lambda_{,}\ \nu_{,}$ are as before and $p\ q\ r\ldots u\ v$ are powers of other primes than $a\ b\ c\ldots l\ m\ n$ of which one at least is not 1, to $n^{\nu_{,}}m^{\mu'}l^{\lambda_{,}}\ldots c^{\gamma'}b^{\beta_{,}}a^{\alpha'}$ where $\alpha'\ \gamma'\ldots\mu'$ are whole numbers severally greater than $\alpha\ \gamma\ldots\mu$, or to $vu\ldots rqpn^{\nu_{,}}m^{\mu'}l^{\lambda_{,}}\ldots c^{\gamma'}b^{\beta_{,}}a^{\alpha'}$,—taking these as instances of the equal product having severally as factors powers of other primes than $a\ b\ldots n$ and no higher powers of $a\ b\ldots n$ than the product $n^\nu\ldots b^\beta a^\alpha$, powers of no other primes but higher powers of some of these, and powers of other primes and higher powers of some of the primes $a\ b\ldots n$. For

$$\frac{n^\nu m^\mu l^\lambda\ldots c^\gamma b^\beta a^\alpha}{vu\ldots rqpn^{\nu_{,}}m^{\mu'}l^{\lambda_{,}}\ldots c^{\gamma'}b^{\beta_{,}}a^{\alpha'}}=\frac{(n^{\nu-\nu_{,}}l^{\lambda-\lambda_{,}}\ldots b^{\beta-\beta_{,}})n^{\nu_{,}}m^\mu l^{\lambda_{,}}\ldots c^\gamma b^{\beta_{,}}a^\alpha}{(vu\ldots rqpm^{\mu'-\mu}\ldots c^{\gamma'-\gamma}a^{\alpha'-\alpha})n^{\nu_{,}}m^\mu l^{\lambda_{,}}\ldots c^\gamma b^{\beta_{,}}a^\alpha}$$

as above and $\therefore$ $$=\frac{n^{\nu-\nu_{,}}l^{\lambda-\lambda_{,}}\ldots b^{\beta-\beta_{,}}}{vu\ldots rqpm^{\mu'-\mu}\ldots c^{\gamma'-\gamma}a^{\alpha'-\alpha}}.$$

Likewise $$\frac{n^\nu m^\mu l^\lambda\ldots c^\gamma b^\beta a^\alpha}{vu\ldots rqpn^{\nu_{,}}l^{\lambda_{,}}\ldots b^{\beta_{,}}}=\frac{n^{\nu-\nu_{,}}m^\mu l^{\lambda-\lambda_{,}}\ldots c^\gamma b^{\beta-\beta_{,}}a^\alpha}{vu\ldots rqp},$$

$$\frac{n^\nu m^\mu l^\lambda\ldots c^\gamma b^\beta a^\alpha}{n^{\nu_{,}}m^{\mu'}l^{\lambda_{,}}\ldots c^{\gamma'}b^{\beta_{,}}a^{\alpha'}}=\frac{n^{\nu-\nu_{,}}l^{\lambda-\lambda_{,}}\ldots b^{\beta-\beta_{,}}}{m^{\mu'-\mu}\ldots c^{\gamma'-\gamma}a^{\alpha'-\alpha}}.$$

And in the latter fraction of each of these pairs of equals every factor of the numerator product is prime to every factor of the denominator product, therefore the numerator is prime to the denominator. Besides none of the denominators is 1.

Hence all the measures of the number equal to $n^\nu m^\mu l^\lambda\ldots c^\gamma b^\beta a^\alpha$ are the same as all the products that can be made by multiplying together numbers taken one from each of the sets

$$\left.\begin{matrix}1\\a\\a^2\\ \cdot\\ \cdot\\ a^\alpha\end{matrix}\right\}\quad\left.\begin{matrix}1\\b\\b^2\\ \cdot\\ \cdot\\ b^\beta\end{matrix}\right\}\quad\left.\begin{matrix}1\\c\\c^2\\ \cdot\\ \cdot\\ c^\gamma\end{matrix}\right\}\ \cdots\ \left.\begin{matrix}1\\l\\l^2\\ \cdot\\ \cdot\\ l^\lambda\end{matrix}\right\}\quad\left.\begin{matrix}1\\m\\m^2\\ \cdot\\ \cdot\\ m^\mu\end{matrix}\right\}\quad\left.\begin{matrix}1\\n\\n^2\\ \cdot\\ \cdot\\ n^\nu\end{matrix}\right\}.$$

It follows from this that the same whole number can be equal to the product of none but the very same powers of the very same primes. Else all the number's measures would not be precisely all the products aforesaid.

When the products of powers of primes are found to which given whole numbers are severally equal it hence further follows that the greatest common measure of the numbers is equal to the product of the least powers in the products of such primes as are common to them all and that the least common multiple is equal to the product of the greatest powers in them of all the several primes. Thus of the numbers equal severally to $2^{11}.7^7.(13)^5$, $5^4.7^9.(13)^6$, $7^{10}.(13)^4.59$, the greatest common measure $=7^7.(13)^4$ and the least common multiple $=2^{11}.5^4.7^{10}.(13)^6.59$.

218. PROP. *Any fraction that can be denoted decimally must when expressed as a simple fraction with least terms be equal to* $\frac{m}{2^u.5^v}$ *where m u v are whole numbers.*

For if $\frac{m}{n}$ be a simple fraction with least terms that is denoted decimally by a row of digits with its last digit in the αth decimal place and which were it a row without a decimal dot with its last digit the unit's digit would denote a whole number a

$$\frac{m}{n}=a\left(\frac{1}{10}\right)^\alpha$$

by the principle of digit knitting and $\therefore = a\times\frac{1}{(10)^\alpha}=\frac{a}{(10)^\alpha}=\frac{a}{(2\times5)^\alpha}$ $=\frac{a}{2^\alpha.5^\alpha}$. Also (art. 216) m is prime to n. Hence (art. 211) $2^\alpha.5^\alpha=nk$, k being some whole number, and $\therefore$

$$2^\alpha.5^\alpha=kn,$$

that is (art. 194) n is a measure of $2^\alpha.5^\alpha$. Wherefore (art. 217)

$$n=2^u.5^v$$

where u v are whole numbers of which neither is greater than α and either may be 0.

219. *Def.* The FULLEST COMMON MEASURE of whole symbolic expressions is that common measure of them which has in it the greatest number or numbers and the most letters, any power of a number symbolized by a letter counting of course for as many letters as the index of the power expresses.

The fullest measure of any whole expression is clearly just the whole expression itself.

220. PROP. *To find the fullest common measure of whole symbolic expressions.*

PROP. *A whole symbolic expression prime to each of two whole symbolic expressions is prime to the mononomic or polynomic equivalent of their product.*

These propositions have their several cases so closely bound up with one another that they must go together.

Since (arts. 175, 176, 180, 181, 193) whole symbolic expressions either are or are operationally equivalent to whole mononomic or polynomic expressions these last alone need be dealt with.

Let then $a\ b\ c \ldots$ be whole mononomes or polynomes whose fullest common measure is sought. Let $f\ h\ k$ be three whole mononomes or polynomes such that f is prime to h and to k, and let $\overline{hk}$ symbolize the mononomic or polynomic equivalent of hk.

(1). If $a\ b\ c \ldots$ be absolute mononomes,—that is expressions mononomic relative to every number in them symbolized by a letter—, their fullest common measure is the mononome having as factors the greatest common measure of all the numerical coefficients and the lowest powers that anywhere enter of such letter symbolized numbers as are common to them all. For (art. 181 or 187) the only measures that a whole absolute mononome can have are either the measures of its several factors or the products of some or all of these measures.

Hence all the common measures of absolute mononomes are precisely all the measures of their fullest common measure.

(i). If $f\ h\ k$ be absolute mononomes the numerical coefficient of f being prime to the numerical coefficient of h and to the numerical coefficient of k is (art. 207) prime to the numerical coefficient of $\overline{hk}$, and f having no letter symbolizing a number in common with either h or k can have none in common with $\overline{hk}$. Therefore f is prime to $\overline{hk}$.

(2). If some of the expressions $a\ b\ c \ldots$ be absolute mononomes and the rest polynomes having absolute mononomes as terms, since (art. 181 or 187) the only mononomic measures of a polynome are the common measures of its terms all the common measures of *a b c*... are precisely all the common measures of all those of them that are mononomes and all the terms of all those of them that are polynomes and therefore the fullest common measure of the one is the fullest common measure of the other.

Hence all the common measures of whole expressions some of them absolute mononomes and the rest absolutely mononomic termed polynomes are precisely all the measures of their fullest common measure.

(ii). If f h be absolute mononomes and k be a polynome whose terms are absolute mononomes let κ be the fullest common measure of all k's terms. Since all the absolutely mononomic measures of k are precisely all the measures of κ all the absolutely mononomic measures of $\overline{hk}$ are precisely all the measures of $\overline{h\kappa}$ and therefore all the common measures of f and $\overline{hk}$ are precisely all the common measures of f and $\overline{h\kappa}$. But f being prime to h and to k is (art. 198) prime to h and to k's measure κ and therefore by the (i) case above to $\overline{h\kappa}$. Hence 1 being the only common measure of f and $\overline{h\kappa}$ 1 is the only common measure of f and $\overline{hk}$.

(iii). If f be an absolute mononome and h k be polynomes whereof each's terms are absolute mononomes any common measure of f and $\overline{hk}$ can only be an absolute mononome. Now any whole absolute mononome is either prime or equal to the product of two whole absolute mononomes each else than 1, each of these is again either prime or equal to the product of two whole absolute mononomes each else than 1, of each of these the same holds, and so on. But since a whole absolute mononome is the product of a whole number and powers of numbers each symbolized by a letter this process cannot be endless. Hence and by colligation and commutation of the multiplications any whole absolute mononome is either prime or equal to the product of whole absolute mononomes each prime and else than 1. Let then μ be either any common measure that f and $\overline{hk}$ can have if prime or any factor else than 1 of the prime factored product equal thereto if not prime and so (art. 197) still a prime common measure. Because f is prime to h f's measure μ is prime to h and therefore h must have one or more terms prime to μ; h may have one or more other terms not prime to μ of which then μ must because a prime be a measure. Hence by the processes of arts. 180, 181, 187, if u v be the several sums of the nonsubtracted and of the subtracted terms of h that are prime to μ and u' v' the several mononomic or polynomic equivalents of the quotients got from dividing by μ the sums of the nonsubtracted and of the subtracted terms of h that are not prime to μ $h=u+u'\mu-v-v'\mu$ and therefore is operationally equivalent to one or other of

$$\overline{u-v}+\overline{u'-v'}\mu \qquad \overline{u-v}-\overline{v'-u'}\mu \qquad \overline{u'-v'}\mu-\overline{v-u}$$

which for shortness may be severally symbolized by $h_{,}+h'\mu$ $h_{,}-h'\mu$ $h'\mu-h_{,}$ where $h_{,}$ is a whole mononome or polynome with each of its terms prime to μ and never can be 0 and h' is either a whole mononome or polynome or 0. In the same way k is operationally equivalent to one or other of

$$k_{,}+k'\mu \qquad k_{,}-k'\mu \qquad k'\mu-k_{,}$$

where $k_{,}$ is a whole mononome or polynome with each of its terms prime to μ and can never be 0 and k' is either a whole mononome or polynome or 0. Should h be equal to the second of the former three expressions and k to the first of the latter three $\overline{hk}$ = either $\overline{h_{,}k_{,}}+\{h_{,}k'-h'(k_{,}+k'\mu)\}\mu$ or $\overline{h_{,}k_{,}}-\{h'(k_{,}+k'\mu)-h_{,}k'\}\mu$, again should h be equal to the first of the former three and k to the third of the latter $\overline{hk}=\{h_{,}k'+h'(k'\mu-k_{,})\}\mu-\overline{h_{,}k_{,}}$, and each of the nine cases that can so arise may be treated in the like way. Hence $\overline{hk}$ is operationally equivalent to one or other of

$$\overline{h_{,}k_{,}}+w\mu \qquad \overline{h_{,}k_{,}}-w\mu \qquad w\mu-\overline{h_{,}k_{,}}$$

where w is some whole expression or 0 and therefore $\overline{h_{,}k_{,}}$ is equal to one or other of

$$\overline{\frac{hk}{\mu}-w\mu} \qquad \overline{\frac{hk}{\mu}+w\mu} \qquad \overline{w-\frac{hk}{\mu}\mu},$$

that is (art. 194) μ is a measure of $\overline{h_{,}k_{,}}$. Therefore μ is a measure of every term of $\overline{h_{,}k_{,}}$. But if t t' be severally the terms of highest degree in $h_{,}$ and in $k_{,}$ (t being $h_{,}$ when $h_{,}$ is a mononome and t' $k_{,}$ when $k_{,}$ is a mononome) the term of highest degree in $\overline{h_{,}k_{,}}$ is (art. 181) $\overline{tt'}$ and since μ is prime to t and to t' μ is prime to $\overline{tt'}$. The terms of lowest degree would do just as well. Hence μ cannot be aught but 1 and hence 1 is the only common measure of f and $\overline{hk}$.

(iv). If f be a polynome whose terms are absolute mononomes and h k be absolute mononomes since $\overline{hk}$ is then an absolute mononome any common measure that f and $\overline{hk}$ can have must be an absolute mononome. But because f is prime to h and to k so likewise is any measure of f and therefore any mononomic measure. Therefore any common measure of f and $\overline{hk}$ is an absolute mononome prime to h and to k and therefore also to $\overline{hk}$. It can therefore be no other than 1.

(3). If a b be absolutely mononomic termed polynomes in the same reference number let α be the fullest common measure of a's terms β the fullest common measure of b's and μ the fullest common measure of α and β all found by the (1) case above. All the absolutely mononomic measures of a are precisely all the common measures of a's terms and therefore precisely all the measures of these's fullest common measure α. Likewise all the absolutely mononomic measures of b are precisely all the measures of β. Therefore all the absolutely mononomic common measures of a and b are precisely all the common measures of α and β therefore again are precisely all the measures of μ and therefore the fullest of them is μ.

Because α is a's fullest mononomic measure $a=a'\alpha$ where a' is not only a whole polynome but also an absolutely pure polynome, calling an absolutely mononomic termed whole polynome PURE which has besides 1 none but polynomic measures or in other words which is prime to every absolute mononome. For if a' had any absolutely mononomic measure α' $a'=a''\alpha'$, a'' being a whole expression, $a=(a''\alpha')\alpha=a''\alpha'\alpha$ and if α' were else than 1 a would have a fuller mononomic measure $\alpha'\alpha$ than the fullest α. So $b=b'\beta$ where b' is an absolutely pure polynome.

All the measures of a that are pure polynomes are precisely all the measures but 1 of a'. For since $a=a'\alpha=\alpha a'$ a is a multiple of a' and therefore any measure of a' is a measure of a. Also if p be a purely polynomic measure of a $a=p'p$, p' being a whole mononome or polynome, $a'\alpha=p'p$ and α is a measure of $p'p$. Hence and because α is prime to p if α be a prime else than 1 α cannot be prime to p', since by either the (ii) or the (iii) case above α would then be prime to $p'p$, therefore α is a measure of p' and therefore $a'=\frac{p'p}{\alpha}=\left|\frac{p'}{\alpha}\right. p$ where $\left|\frac{p'}{\alpha}\right.$ is whole. But if α be not a prime α as shown above in (iii) $=\psi\ldots\tau\sigma\rho$ where ρ σ ... ψ are primes each else than 1, $p'p$ $=a'\psi\ldots\tau\sigma\rho=(a'\psi\ldots\tau\sigma)\rho$ and hence as before p is a measure of $a'\psi\ldots\tau\sigma$, hence in the same way p is a measure of $a'\psi\ldots\tau$, and so on until at last it follows that p is a measure of a'. Since then any measure of a' is a measure of a and any purely polynomic measure of a is a measure of a' all the purely polynomic measures are the same of a and of a'. So too all the purely polynomic measures of

b are precisely all the measures but 1 of b'. Therefore all the purely polynomic common measures of a and b are precisely all the common measures but 1 of a' and b'.

Of a' b' let it be a' that has not the less power range, then (arts. 187, 188) there is a mononomic or polynomic expression the product of the multiplication by which of b' either is equal to a' or differs from a' by a mononomic or polynomic expression of less range than b'. These expressions however need not be whole. But if the first term of the former of them were $\frac{\delta'}{\delta}x^i$ where x is the reference number and δ' δ are whole mononomes of which δ is not a measure of δ' the first step in the process for finding them would give

$$a' = \left(\frac{\delta'}{\delta}x^i\right)b' \pm l \pm m \pm \ldots \pm s$$

where l m ... s are unlike mononomes that can be other than whole only by reason of the fractionality of $\frac{\delta'}{\delta}$ and hence it would follow that

$$\overline{a'\delta} = \left\{\left(\frac{\delta'}{\delta}x^i\right)b' \pm l \pm \ldots \pm s\right\}\delta = (\delta' x^i)b' \pm \overline{l\delta} \pm \overline{m\delta} \pm \ldots \pm \overline{s\delta}$$

where $\overline{l\delta}$ $\overline{m\delta}$ ... $\overline{s\delta}$ are all wholes. The next step would then give

$$\overline{a'\delta} = (\delta' x^i \pm \frac{\zeta'}{\zeta}x^{i\pm})b' \pm m' \pm n' \pm \ldots \pm s'$$

ζ' ζ being whole mononomes of which ζ might not be a measure of ζ' and m' n' ... s' unlike mononomes that can be other than whole only by reason of $\frac{\zeta'}{\zeta}$'s fractionality, hence as before $\overline{(a'\delta)\zeta}$ and hence the operational equivalent of this

$$\overline{a'\delta\zeta} = \{(\delta'\zeta)x^i \pm \zeta' x^{i\pm j}\}b' \pm \overline{m'\zeta} \pm \overline{n'\zeta} \pm \ldots \pm \overline{s'\zeta}$$

$\overline{m'\zeta}$ $\overline{n'\zeta}$ ... $\overline{s'\zeta}$ being wholes. And so on. Therefore when the mononome or polynome is not whole which multiplied into b' gives a product either equal to a' or differing from a' by a mononome or polynome of less range than b' a whole mononome α' may be found such that $\overline{a'\alpha'}$ either is equal to, or differs by a whole mononome or polynome of less range than b' from, the product of b' by some whole mononome or polynome.

In the former alternative $\overline{a'\alpha'}$ is a multiple of b' and because b' is a pure polynome a' is by what is above shown also a multiple of b'. All the common measures of a' and b' are then (art. 199) precisely all the measures of b' and the fullest common measure of a' and b' is therefore b'.

In the latter alternative let r_1 be the whole mononome or polynome of less range than b' by which $\overline{a'\alpha'}$ differs from a multiple q_1b' of b'. All the common measures of $\overline{a'\alpha'}$ and b' are then (art. 201) precisely all the common measures of b' and r_1. But because b' is a pure polynome all the common measures of $\overline{a'\alpha'}$ and b' are by what is above shown precisely all the common measures of a' and b'. Therefore all the common measures of a' and b' are precisely all the common measures of b' and r_1. If r_1 be a mononome b' is prime to r_1 and therefore a' is prime to b'. If r_1 be a polynome let ρ_1 be the fullest common measure of all r_1's terms and therefore r_1's fullest mononomic measure, then $r_1 = r'_1\rho_1$ r'_1 being a pure polynome of the same range as r_1 and as before all the common measures of b' and r_1, and therefore all those of a' and b', are precisely all the common measures of b' and r'_1.

Again in like manner either b' is a multiple of r'_1 and then all the common measures at once of b' and r'_1 and of a' and b' are precisely all the measures of r'_1 and the fullest of them is therefore r'_1 or a whole mononome β' may be got such that $\overline{b'\beta'}$ differs from a multiple $q_2r'_1$ of r'_1 by a whole mononome or polynome r_2 of less range than r'_1 and then all the common measures at once of b' and r'_1 and of a' and b' are precisely all the common measures of r'_1 and r_2. If r_2 be a mononome the pure polynome r'_1 is prime to r_2 and therefore so is b' to r'_1 and a' to b' but if r_2 be a polynome let $r_2 = r'_2\rho_2$ where ρ_2 is r_2's fullest mononomic measure and therefore r'_2 a pure polynome of the same range as r_2 then all the common measures are the same of a' and b' of b' and r'_1 and of r'_1 and r'_2.

The like process may now be gone through with r'_1 and r'_2. And so on. But since the range of the successive rs goes on lessening at every step by at least 1 there must sooner or later always arise one of them r_ι which either is a mononome or is such that the purely polynomic equivalent r'_ι of the quotient got from dividing it by its fullest mononomic measure ρ_ι is a measure of the purely polynomic equivalent $r'_{\iota-1}$ of the like quotient got from $r_{\iota-1}$. In the first hap the pure polynome $r'_{\iota-1}$ is prime to the mononome r_ι in the other $r'_{\iota-1}$ being a multiple of r'_ι all the common measures of $r'_{\iota-1}$ and r'_ι are precisely all the measures of r'_ι. But all the common

$a'\alpha = a$	$b = b'\beta$
$\overline{a'\alpha'}$ $q_1)$ q_1b' $r'_1\rho_1 = r_1$	b'
r'_1	$\overline{b'\beta'}$ $q_2r'_1$ $(q_2$ $r_2 = r'_2\rho_2$
$\overline{r'_1\rho'_1}$ $q_3)$ $q_3r'_2$ $r'_3\rho_3 = r_3$	r'_2

measures are the same of every two consecutive terms of the series

$$a' \quad b' \quad r'_1 \quad r'_2 \quad r'_3 \ldots r'_{i-1} \begin{cases} \text{either } r_i \\ \text{or } r'_i. \end{cases}$$

Therefore if r_i be a mononome the only common measure of a' and b' is 1 and if r_i be a polynome all the common measures of a' and b' are precisely all the measures of r'_i.

Now any common measure of a and b is either a mononome or a polynome and if a polynome either a pure polynome or operationally equivalent to the product of a pure polynome and a mononome each of them a common measure of a and b. But all the mononomic common measures of a and b are precisely all the measures of μ and all the purely polynomic common measures of a and b are precisely all the common measures but 1 of a' and b'. Therefore if r_i be a mononome all the common measures of a and b are precisely all the measures of μ and the fullest common measure of a and b is then μ. And if r_i be a polynome since of all the whole products equal to common measures of a and b all the mononomic factors are precisely all the measures of μ and all the purely polynomic factors precisely all the measures but 1 of r'_i all the common measures of a and b are precisely all the measures of $r'_i\mu$ and the fullest common measure of a and b is therefore $r'_i\mu$.

Hence all the common measures of two absolutely mononomic termed polynomes in the same reference number are precisely all the measures of their fullest common measure.

(4). If a b c... be three or more absolutely mononomic termed polynomes all in the same reference number find by the foregoing (3) case the fullest common measure m of a and b and the fullest common measure m' of m and c. Because all the common measures of a and b are precisely all the measures of m all the common measures of a b and c are precisely all the common measures of m and c. But all the common measures of m and c are precisely all the measures of m'. Therefore all the common measures of a b and c are precisely all the measures of m' and therefore the fullest common measure of a b and c is m'. Again find the fullest common measure m'' of m' and d then all the common measures of a b c being precisely all the measures of m' all the common measures of a b c d are precisely all the common measures of m' and d therefore are precisely all the measures of m'' and therefore the fullest of them is m''. In the like way all the common measures of a b c d e are precisely all the common measures of m'' and e precisely all the measures of these's fullest common measure m''' and m''' is the fullest common measure of $a\ b\ c\ d\ e$. And so on.

Hence all the common measures of whole polynomes in the same reference number of each of which the terms are all absolute mononomes are precisely all the measures of the fullest common measure of those polynomes.

(v). If $f\ h$ be polynomes in a common reference number with absolutely mononomic terms and k be either an absolute mononome or a polynome in that common reference number with absolutely mononomic terms let $f=f'\phi$ $h=h'\eta$ where $\phi\ \eta$ are the fullest mononomic measures severally of $f\ h$ and $f'\ h'$ are therefore pure polynomes. Because f is prime to h and to k so is f's measure ϕ and therefore by either the (ii) or the (iii) case above ϕ is prime to $\overline{hk}$. Also all the mononomic measures of f are precisely all the measures of ϕ and as $\overline{hk}=(h'\eta)k=\overline{h'k}\eta$ all the purely polynomic measures are the same of $\overline{hk}$ and of $\overline{h'k}$. Therefore f and $\overline{hk}$ can have besides 1 none but purely polynomic common measures and all the common measures of f and $\overline{hk}$ are precisely all the common measures of f' and $\overline{h'k}$.

Since $f'\phi=\phi f'$ and $h'\eta=\eta h'$ f' is a measure of f and h' of h therefore f' is prime to h' and to k. The polynome h' then being prime to, and therefore no multiple of, the polynome f' there is some whole absolute mononome η' such that $\overline{h'\eta'}$ differs from a multiple q_1f' of f' by a whole mononome or polynome r_1 of less range than f'; should f''s range be greater than h''s η' is 1 q_1 is 0 r_1 is h' and everything still holds good. Because $r_1=$ either $\overline{h'\eta'}-q_1f'$ or $q_1f'-\overline{h'\eta'}$ r_1k either $=(\overline{h'\eta'}-q_1f')k=\eta'\overline{h'k}-q_1f'k$ or $=q_1f'k-\eta'\overline{h'k}$ and any common measure of f' and $\overline{h'k}$ being a measure at once of f''s multiple kf' of this multiple's equal $f'k$ of this equal's multiple $q_1f'k$ and of $\overline{h'k}$'s multiple $\eta'\overline{h'k}$ is hence a measure of r_1k.

If now r_1 be a mononome all the purely polynomic measures are the same of r_1k as of k so that any common measure but 1 of f' and $\overline{h'k}$ is then a purely polynomic common measure of f' and k and therefore is none whatever. But if r_1 be a polynome $r_1=r'_1\rho_1$ ρ_1 being r_1's fullest mononomic measure and r'_1 a pure polynome of the same range as r_1 and then since $(r'_1\rho_1)k=(r'_1k)\rho_1$ all the purely polynomic measures of r_1k are precisely all those of r'_1k and among them therefore any common measures but 1 of f' and $\overline{h'k}$. Then too as shown in the (3) case above all the common measures of h' and f' are precisely all the common measures of f' and r'_1 and therefore f' is prime to r'_1. Hence again there is some whole absolute mononome ϕ' such that $\overline{f'\phi'}$ differs from $q_2r'_1$ a multiple of r'_1 by a whole mononome or polynome r_2 of less range than r'_1 and since as before $r_2k =$ either $\phi'f'k-q_2r'_1k$ or $q_2r'_1k-\phi'f'k$ it follows as before that any common measure of f' and $\overline{h'k}$ because a measure of $f'k$ and of r'_1k is further a measure of r_2k.

Here again if r_2 be a mononome all the purely polynomic measures are the same of r_2k as of k and therefore because f' is prime to k so is f' to $\overline{h'k}$. But if r_2 be a polynome $r_2 = r'_2\rho_2$ ρ_2 being r_2's fullest mononomic measure and r'_2 a pure polynome of r_2's range such that r'_2k has the very same purely polynomic measures as r_2k then again in the like way r'_1 is prime to r'_2 some whole absolute mononome ρ'_1 makes $\overline{r'_1\rho'_1}$ differ from a multiple $q_3r'_2$ of r'_2 by a whole mononome or polynome r_3 of less range than r'_2 and any common measure of f' and $\overline{h'k}$ is a measure not only of $f'k$ of r'_1k and of r'_2k but also of r_3k. And so on.

But since the rs keep on lessening in range by 1 or more at every step and each r' is prime to the r' next following an r must at last be reached,—call it r_ι—, which is an absolute mononome. Then any common measure of f' and $\overline{h'k}$ is a measure at once of $f'k$ of r'_1k of r'_2k - - - of $r'_{\iota-1}k$ and of $r_\iota k$ all the purely polynomic measures of $r_\iota k$ are precisely all the purely polynomic measures of k all common measures of f' and $\overline{h'k}$ are common measures of f' and k and therefore f' is prime to $\overline{h'k}$ and f to $\overline{hk}$.

(5). If $a\ b\ c$... be absolutely mononomic termed polynomes but not all in the same reference number since they could only be polynomes in the reference number of a polynomic common measure all their common measures are absolute mononomes are therefore precisely all the common measures of all their terms and are therefore precisely all the measures of the fullest common measure at once of themselves and of all their terms.

(vi). If $f\ h$ be absolutely mononomic termed polynomes in different reference numbers and k be either an absolute mononome or an absolutely mononomic termed polynome f and $\overline{hk}$ can have a polynomic common measure only when they are polynomes in a common reference number and therefore only when k is a polynome in f's reference number. If then k be either a mononome or a polynome in another reference number than f's f and $\overline{hk}$ have no polynomic common measure. If k be a polynome in f's reference number h must be a measure of every coefficient of a power of that number in $\overline{hk}$ and therefore any measure of $\overline{hk}$ which is a polynome in that number must have a purely polynomic measure in common with k so that still f and $\overline{hk}$ have no polynomic common measure because f and k have none. Moreover by the (ii) and (iii) cases above any mononomic measure of f is always prime to $\overline{hk}$. Therefore f is prime to $\overline{hk}$.

(6) and (vii). Next in order of complexity to a polynome whose terms are all absolute mononomes is a polynome of which some of the coefficients of the several powers of the symbolized reference number are polynomes of that kind and the others if there be any are absolute mononomes. Of a whole polynome of this kind the terms though not absolute mononomes are yet mononomes relative to the reference number and any measure is either a relatively pure polynome,—calling a whole polynome RELATIVELY PURE when it is prime to every relative mononome and therefore has besides 1 only relatively polynomic measures—, a relative mononome which is a common measure of all the terms or an operational equivalent of the product of a relatively pure polynome and a relative mononome each of which is itself a measure. Hence as after the cases (1) and (i) above of absolute mononomes the cases (2) (ii) (iii) (iv) (3) (4) (v) (5) and (vi) of absolutely mononomic termed polynomes follow so now in like manner by putting all the foregoing cases viewed as the cases of relative mononomes in the place of the two first and using throughout the nine others relative instead of absolute mononomes and relatively pure instead of absolutely pure polynomes do the corresponding cases follow of relatively mononomic termed polynomes.

Since by taking polynomes of any complexity for some of the coefficients of the several powers of a new reference number a polynome is got of a greater complexity the orders of polynomic complexity are endless. But as from the cases of absolute mononomes the cases are passed on to of polynomes with absolute mononomes for terms then from all these the cases are passed on to of polynomes with absolutely mononomic termed polynomes for coefficients of powers of their reference numbers so generally in the same way from the cases of mononomes and polynomes up to any given order of complexity may the cases be passed on to of polynomes of the next higher order of complexity by always holding as relatively mononomic all the terms of these last and all polynomes of lower complexity in other reference numbers.

221. COR. *All the common measures of whole symbolic expressions are precisely all the measures of their fullest common measure.*

Hence any common factor of whole products is a factor of their fullest common measure and in finding the fullest common measure the other factors then need only be sought.

222. PROP. *A common measure of two whole symbolic expressions is their fullest common measure precisely when the whole expressions equal to the quotients of their divisions by it are prime to one another.*

Let a b be two whole symbolic expressions. And first let m be their fullest common measure so that

$$a = hm \qquad b = km$$

h k being whole expressions. Then if μ be any common measure of h and k $h = \eta\mu$ $k = \kappa\mu$ η κ being whole expressions and

$$a = (\eta\mu)m = \eta\mu m \qquad b = (\kappa\mu)m = \kappa\mu m.$$

Therefore μm is a common measure of a and b. And if μ were else than 1 μm would be a fuller common measure than the fullest m. Wherefore h is prime to k. Conversely let m be such a common measure of a and b that

$$a = hm \qquad b = km$$

where h k are whole expressions prime to one another. Then because m is a common measure of a and b and all the common measures of a and b are (art. 221) precisely all the measures of the fullest common measure of a and b this fullest common measure $= gm$ g being some whole expression and $a = h'gm$ $b = k'gm$ h' k' being whole expressions. Therefore

$$hm = h'gm = (h'g)m \qquad km = k'gm = (k'g)m$$

$$h = h'g \qquad k = k'g.$$

Therefore g is a common measure of h and k. But h and k have no common measure but 1. Hence g is 1 and gm the fullest common measure of a and b is m.

223. PROP. *All the common measures of two whole symbolic expressions are precisely all the common measures of one of them and the mononomic or polynomic equivalent of the product made by multiplying the other by any whole expression prime to the first.*

Let a b be any two whole symbolic expressions and p any whole expression prime to a.

If m be the fullest common measure of a and b

$$a = hm \qquad b = km$$

where h k are whole expressions that (art. 222) are prime to one another and

$$\overline{pb} = pkm = \overline{pkm}.$$

Now since $hm = mh$ h is a measure of a and therefore is prime to p. Hence h being prime to both p and k is (art. 220) prime to $\overline{pk}$. Since then h $\overline{pk}$ the whole expressions severally equal to $\frac{a}{m}$ $\frac{\overline{pb}}{m}$ are prime to one another m is (art. 222) the fullest common measure of a and $\overline{pb}$. Hence the fullest common measure being the same of a and b as of a and $\overline{pb}$ and all the common measures of whole symbolic expressions being (art. 221) the same as all the measures of their fullest common measure all the common measures are the same of a and b as of a and $\overline{pb}$.

224. PROP. *If two symbolically whole termed fractional expressions be equal and the one's terms be prime to each other the other's terms are the multiples by the first's corresponding terms of some whole expression.*

Let $\frac{a}{b} = \frac{a'}{b'}$ where a b a' b' are symbolically whole expressions and a is prime to b.

Then $a' = \frac{a}{b}b' = \frac{\overline{ab'}}{b}$ therefore $\overline{ab'}$ is a multiple of b and therefore all the common measures of $\overline{ab'}$ and b are precisely all the measures of b. But (art. 223) all the common measures of b' and b are precisely all the common measures of $\overline{ab'}$ and b because a is prime to b. Therefore all the common measures of b' and b are precisely all the measures of b. Therefore b is a measure of b', that is there is some whole expression k such that

$$b' = kb \text{ and } \therefore \text{ also} = bk.$$

$$\therefore \text{ too } a' = \frac{akb}{b} = \frac{(ak)b}{b} = ak.$$

225. *Def.* The SIMPLEST COMMON MULTIPLE of whole symbolic expressions is that common multiple of them which has in it the least number or numbers and the fewest letters, any power of a number symbolized by a letter counting of course for as many letters as the index expresses.

The simplest multiple of a whole expression is clearly just the whole expression itself.

226. PROP. *To find the simplest common multiple of whole symbolic expressions.*

First let a b be two whole symbolic expressions. Any whole expression that is both a multiple of a and a multiple of $b =$

$$pa = qb$$

p q being whole expressions. And if m be the fullest common measure of a and b

$$a = hm \qquad\qquad b = km$$

h k being whole expressions that (art. 222) are prime to one another.

$$\therefore\ phm = qkm.$$

But this happens precisely when severally $(ph)m = (qk)m$ $ph = qk$ $= kq$ $\frac{ph}{qh} = \frac{kq}{hq}$

$$\frac{p}{q} = \frac{k}{h}.$$

Hence (art. 224) p q are the multiples by k h severally of some whole expression r and hence any common multiple of a and b

$$= (kr)hm = (hr)km = krhm = hrkm = rkhm = rhkm.$$

each of these equals is simplest when r is 1 and therefore the simplest common multiple of a and b

$$= khm = hkm = \left\lceil\frac{b}{m}\right. a = \left\lceil\frac{a}{m}\right. b = \left\lceil\frac{ba}{m}\right. = \left\lceil\frac{ab}{m}\right. .$$

Since any common multiple of a and $b = rhkm$ all the common multiples of two whole symbolic expressions are precisely all the multiples of their simplest common multiple.

Next for three whole symbolic expressions a b c find the simplest common multiple v of any two of them a b and the simplest common multiple v' of v and the third one c. Then since all the common multiples of a and b are precisely all the multiples of v all the common multiples of a b and c are precisely all the common multiples of v and c. But all the common multiples of v and c are precisely all the multiples of v'. Therefore all the common multiples of a b and c are precisely all the multiples of v' and therefore v' its own simplest multiple is the simplest common multiple of a b and c.

In like manner for a b c and a fourth whole symbolic expression d find the simplest common multiple v'' of v' and d. Then all the common multiples of a b and c being precisely all the multiples of v' all the common multiples of a b c and d are precisely all the common multiples of v' and d therefore are precisely all the multiples of v'' and therefore v'' is itself the simplest of them all. And so on for five or more whole symbolics.

227. COR. *All the common multiples of whole symbolic expressions are precisely all the multiples of their simplest common multiple.*

228. What has been now shown about symbolic measures and multiples gives the means of simplifying to the utmost the processes of arts. 126, 129, 132, for finding single quotient expressions operationally equivalent to the results of operations with quotient expressions.

First and as the ground whereon rests everything else a fraction is expressed in the simplest whole terms precisely when these terms are prime to one another. For (art. 224) any other whole terms of an equal fraction are the multiples by them of some whole expression and therefore must be severally less simple than they.

Hence any fraction with whole symbolic terms is expressed most simply in terms of the whole expressions equivalent operationally to the quotients got from dividing the terms by their fullest common measure. Thus for the simplification of $\frac{30+140x+70x^2-360x^3-320x^4}{75x^2+140x^3-115x^4+120x^5}$ the algorithm may handily be what is on the next following page whence

$$30+140x+70x^2-360x^3-320x^4=(3+14x+7x^2-36x^3-32x^4)\times 10$$

$$75x^2+140x^3-115x^4+120x^5=(15+28x-23x^2+24x^3)\times 5x^2,$$

10 and $5x^2$ being the fullest mononomic measures and their fullest common measure is 5,

$$(3+14x+7x^2-36x^3-32x^4)\times 5^2=(1\times 5+14x)(15+28x-23x^2+24x^3)$$
$$-(51+349x+568x^2)\times 2x^2$$

$$(15+28x-23x^2+24x^3)(17)^2=(5\times 17-423x)(51+349x+568x^2)$$
$$+(3+8x)\times 30900x^2$$

$$51+349x+568x^2=(17+71x)(3+8x)$$

so that of every consecutive two of the polynomes
$3+14x+7x^2-36x^3-32x^4$ $15+28x-23x^2+24x^3$ $51+349x+568x^2$
$3+8x$ all the common measures are precisely all the measures of $3+8x$ and this is the fullest purely polynomic common measure

$$\frac{30+140x+70x^2-360x^3-320x^4}{75x^2+140x^3-115x^4+120x^5}=\frac{\{(1+2x-3x^2-4x^3)\times 2\}(3+8x)\times 5}{\{(5-4x+3x^2)x^2\}(3+8x)\times 5}$$

$$=\frac{(1+2x-3x^2-4x^3)\times 2}{(5-4x+3x^2)x^2}=\frac{2+4x-6x^2-8x^3}{5x^2-4x^3+3x^4}.$$

	10	30	$+140ax$	$+70ax^2$	$-360ax^3$	$-320ax^4$
		3	+14	+7	−36	−32
	5	15	+70	+35	−180	−160
1)		15	+28	−23	+24	
			+42	+58	−204	
	5		+210	+290	−1020	−800
$+14x$)			+210	+392	−322	+336
	$2x^2$			−102	−698	−1136
		51	+349	+568		
$17+71x$)		51	+136			
			+213			
			+213	+568		
$1+2x-3x^2-4x^3$)		3	+8			
			+6			
			+6	+16		
				−9		
				−9	−24	
					−12	
					−12	−32

		$75x^2$	$+140ax^3-115x^4+120ax^5$	$5x^2$	
15	$+28x$	$-23x^2$	$+24x^3$		
255	+476	−391	+408	17	(5
255	+1745	+2840			
	−1269	−3231			
	−21573	−54927	+6936	17	
	−21573	−147627	−240264		$(-423x$
		+92700	+247200	$30900ax^2$	
3	+8				
15	+40				$(5-4x+3x^2$
	−12				
	−12	−32			
		+9			
		+9	+24		

Likewise $\dfrac{(3-5c+2c^2)x^2+(4+4c-15c^2+7c^3)x-(4-4c-7c^2+10c^3-3c^4)}{(3+c-2c^2)x^2-(8-8c+9c^2-5c^3)x+(4-8c+11c^2-10c^3+3c^4)}$

$$=\frac{[\{x+(2+3c)\}(1-c)][\{(3-2c)x-(2-3c+c^2)\}]}{\{(1+c)x-(2-c+3c^2)\}\{(3-2c)x-(2-3c+c^2)\}}=\frac{\{x+(2+3c)\}(1-c)}{(1+c)x-(2-c+3c^2)}$$

$$=\frac{(1-c)x+(2+c-3c^2)}{(1+c)x-(2-c+3c^2)}.$$

Again simplest whole termed fractional expressions with certain of their terms unlike are most simply expressed so as to have the corresponding terms alike precisely when the latter terms are made all equal to the simplest common multiple of the former. Thus

$$\frac{2x-1}{x^3+1}\quad\frac{2x^3-x^2+x+1}{x-1}=\frac{2x-1}{(x+1)(x^2-x+1)}\quad\frac{(2x+1)(x^2-x+1)}{x-1}$$

$$=\frac{(2x-1)(2x+1)}{(x^3+1)(2x+1)}\quad\frac{(2x^3-x^2+x+1)(x+1)}{(x-1)(x+1)}$$

$$=\frac{(2x)^2-1}{2x^4+x^3+2x+1}\quad\frac{2x^4+x^3+2x+1}{x^2-1}=\frac{4x^2-1}{x^2-1},$$

$$\frac{2a^3+5a^2b+5ab^2+3b^3}{3a^3+5a^2b+5ab^2+2b^3}\quad\frac{6a^3+a^2b+7ab^2+6b^3}{4a^3-8a^2b+9ab^2-9b^3}$$

$$=\frac{(2a+3b)(a^2+ab+b^2)}{(3a+2b)(a^2+ab+b^2)}\quad\frac{(3a+2b)(2a^2-ab+3b^2)}{(2a-3b)(2a^2-ab+3b^2)}=\frac{2a+3b}{3a+2b}\quad\frac{3a+2b}{2a-3b}=\frac{2a+3b}{2a-3b},$$

$$\frac{\left(\dfrac{2+a-3a^2}{4+4a+a^2+6a^3}\right)}{\left(\dfrac{3+a-2a^2}{6-7a+8a^2-4a^3}\right)}=\frac{\left\{\dfrac{(1-a)(2+3a)}{(2-a+2a^2)(2+3a)}\right\}}{\left\{\dfrac{(1+a)(3-2a)}{(2-a+2a^2)(3-2a)}\right\}}=\frac{\left(\dfrac{1-a}{2-a+2a^2}\right)}{\left(\dfrac{1+a}{2-a+2a^2}\right)}=\frac{1-a}{1+a},$$

$$\frac{\left(\dfrac{2x^2-5x+2}{2x^2+x-1}\right)}{\left(\dfrac{2x^2+5x+2}{2x^2-x-1}\right)}=\frac{\left\{\dfrac{(x-2)(2x-1)}{(x+1)(2x-1)}\right\}}{\left\{\dfrac{(x+2)(2x+1)}{(x-1)(2x+1)}\right\}}=\frac{\left(\dfrac{x-2}{x+1}\right)}{\left(\dfrac{x+2}{x-1}\right)}=\frac{\left\{\dfrac{(x-2)(x-1)}{(x+1)(x-1)}\right\}}{\left\{\dfrac{(x+2)(x+1)}{(x-1)(x+1)}\right\}}=\frac{x^2-3x+2}{x^2+3x+2},$$

$$\frac{119ax^2}{272a^2xy^4}-\frac{65a^3x}{156a^3xy^3}+\frac{209a^2xy}{76a^3x^2y^3}-\frac{143a^2xy}{88ax^3y^2}$$

$$=\frac{(7x)\times17ax}{(16ay^4)\times17ax}-\frac{5\times13a^3x}{(12y^3)\times13a^3x}+\frac{11\times19a^2xy}{(4axy^2)\times19a^2xy}-\frac{(13a)\times11axy}{(8x^2y)\times11axy}$$

$$=\frac{7x}{16ay^4}-\frac{5}{12y^3}+\frac{11}{4axy^2}-\frac{13a}{8x^2y}=\frac{(7x)\times3x^2}{(16ay^4)\times3x^2}-\frac{5\times4ax^2y}{(12y^3)\times4ax^2y}$$

$$+\frac{11\times12xy^2}{(4axy^2)\times12xy^2}-\frac{(13a)\times6ay^3}{(8x^2y)\times6ay^3}=\frac{21x^3-20ax^2y+132xy^2-78a^2y^3}{48ax^2y^4},$$

$$\frac{5c+2}{10c^2+21c-10}+\frac{5c-2}{10c^2+29c+10}-\frac{2c+5}{25c^2-4}-\frac{17}{20c}$$

$$=\frac{5c+2}{(5c-2)(2c+5)}+\frac{5c-2}{(5c+2)(2c+5)}-\frac{2c+5}{(5c-2)(5c+2)}-\frac{17}{20c}$$

$$=\frac{(5c+2)^2}{(10c^2+21c-10)(5c+2)}+\frac{(5c-2)^2}{(10c^2+29c+10)(5c-2)}-\frac{(2c+5)^2}{(25c^2-4)(2c+5)}-\frac{17}{20c}$$

$$=\frac{25c^2+20c+4+(25c^2-20c+4)-(4c^2+20c+25)}{50c^3+125c^2-8c-20}-\frac{17}{20c}$$

$$=\frac{(46c^2-20c-17)\times 20c}{(50c^3+125c^2-8c-20)\times 20c}-\frac{17(50c^3+125c^2-8c-20)}{(20c)(50c^3+125c^2-8c-20)}$$

$$=\frac{70c^3-2525c^2-204c+340}{20(50c^3+125c^2-8c-20)c}.$$

CHAPTER III

MAGNITUDE IN RELATION TO NUMBER

229. *Def.* When capital letters are taken to stand for magnitudes and small letters for whole numbers the symbol $n(A)$ stands for n magnitudes taken together each equal to the magnitude A, $\frac{1}{n}(A)$ for a magnitude such that n magnitudes each equal to it are together equal to A, and $\frac{m}{n}(A)$ for m magnitudes together each equal to a magnitude such that n magnitudes each equal to the same are together equal to A.

These symbols for the multiple the submultiple and the fraction of A expressed severally by n $\frac{1}{n}$ and $\frac{m}{n}$ serve to mark that A is the unit magnitude referred to.

230. *Def.* The symbols $>$, not$>$, $<$, not$<$, severally stand for *is greater than, is not greater than, is less than, is not less than,* or for such other parts of the several verbs *to be greater than, to be not greater than, to be less than, to be not less than,* as their manner of use may need.

231. *Any given magnitude may be wholly cut into so many equal parts that each of them is less than any given magnitude of the same kind however small.*

For if A be any given magnitude and G any given magnitude of the same kind however small a multiple $n(G)$ of G may be taken (art. 133) greater than A and (art. 97) A may be cut into as many equal parts n as there are parts each equal to G in the multiple so taken. Then (art. 91)

$$\because n(G) > A \qquad G > \frac{1}{n}(A).$$

232. *If from any given magnitude there be taken not less than its half and from the remainder left not less than its half and from the remainder then left not less than its half and so on there is at length left a remainder less than any given magnitude of the same kind however small.* (EUCLID Bk. x Prop. I with *not less* put for *greater*).

Let a straight line AB represent any given magnitude and another straight line C any given magnitude of the same kind however small; straight lines may fitly represent magnitudes of any kind if those only of their qualities be heeded which they have in common with all magnitudes. Of the magnitude represented by C a multiple may (art. 133) be taken greater than the magnitude represented by AB, let then a straight line DE represent a multiple of C so taken greater than AB and let EH HG GF FD represent the parts each equal to C of which this multiple is wholly made up. From AB let a part AK be taken not less than $\frac{1}{2}(AB)$ and from the remainder KB let a part KL be taken not less than $\frac{1}{2}(KB)$ and from the new remainder LB let LM be taken not less than $\frac{1}{2}(LB)$ and so on until at last there are as many parts AK KL LM MB in AB as there are parts DF FG GH HE in DE.

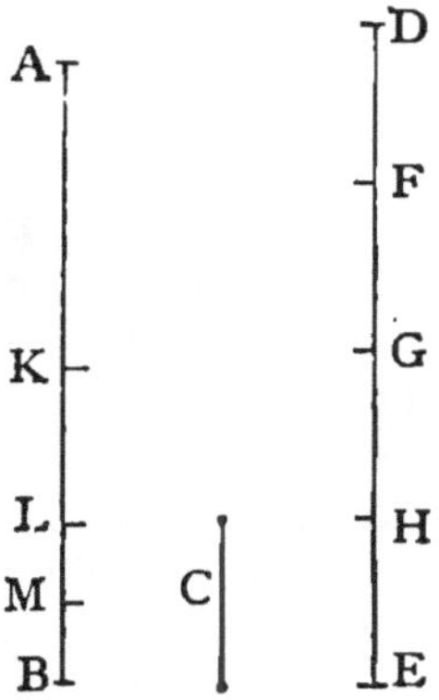

Because $DE > AB$ and the part DF of the greater DE not> $\frac{1}{2}(DE)$ and the part AK of the less AB not< $\frac{1}{2}(AB)$ it follows (arts. 91, 9, 16) that the remaining part $FE >$ the remaining part KB. In the like way $\because$ $FE > KB$ and FG not> $\frac{1}{2}(FE)$ and KL not< $\frac{1}{2}(KB)$ it follows that the remainder $GE >$ the remainder LB. And at last $\because$ $GE > LB$ and GH not> $\frac{1}{2}(GE)$ and LM not< $\frac{1}{2}(LB)$ it follows that the remainder $HE >$ the remainder MB. But $C = HE$. Therefore (art. 9) $C > MB$.

233. *Any magnitude greater than the least and less than the greatest of a set of magnitudes in ascending order of*

greatness is either equal to some one of these magnitudes or is greater than the one and less than the other of some consecutive two.

For if A_1 A_2 A_3...A_{n-1} A_n be any magnitudes in ascending order of greatness a magnitude X greater than A_1 and less than A_n either not> A_2 or > A_2 and if not greater than A_2 either > A_1 and < A_2 or $=A_2$. But if greater than A_2 X either not> or > A_3 and if not greater either > A_2 and < A_3 or $=A_3$. Again X if greater than A_3 either not> A_4 or > A_4 and if the former either > A_3 and < A_4 or $=A_4$. And so on until at length X if greater than A_{n-1} is by hypothesis less than A_n.

234. *If magnitudes each equal to the less of two unequal magnitudes be taken away in succession from the greater and from the remainders successively left there is at length left either no remainder or a remainder less than the less of the two unequals.*

Since (art. 133) of the less of two unequals a multiple may be taken greater than the greater (art. 233) of the successive multiples of the less got by taking it once twice thrice - - - the greater either is equal to some one or is greater than the one and less than the other of some consecutive two. Magnitudes then each equal to the less taken severally from the greater and the successive remainders left are in the one case together equal to the greater and therefore can leave no remainder but are in the other together less than the greater by a less than the less and therefore must leave this less than the less as a remainder.

235. PROP. *To find whether or not two given magnitudes of the same kind have any common measure and if they have to find the greatest one.*

Let two straight lines AB and C represent any two given magnitudes of the same kind and let AB represent the one that is not less than the other.

If AB be a multiple of C any measure of C is a measure of each of the one or more parts each equal to C that make up AB therefore is a measure of the whole AB and therefore is a common measure of AB and C. Any common measure too of AB and C is of course a measure of C. Therefore all the common measures of AB and C are precisely all the measures of C and since (art. 8) a magnitude is greater than any measure but itself of itself the greatest common measure of AB and C is therefore C.

All the common measures then of a magnitude and any multiple of it are precisely all the measures of the magnitude itself.

But if AB be not a multiple of C let AD DE EF be the parts each equal to C taken away in succession from AB and the successive remainders DB EB until at length a remainder FB is left less than C. If AB and C have any common measure each of the parts AD DE EF is wholly made up of parts each equal to that common measure and therefore so likewise is AF the whole made up of AD DE EF. But AF is a part of AB and AB is wholly made up of parts each equal to the common measure. Therefore FB the remaining part of AB is wholly made up of parts each equal to that same common measure, that is any common measure of AB and C is a measure of FB and therefore is a common measure of C and FB. Again if C and FB have any common measure each of the parts AD DE EF and also the part FB is wholly made up of parts each equal thereto and hence so likewise is the whole AB made up of AD DE EF FB, that is any common measure of C and FB is a measure of AB and therefore is a common measure of AB and C. Since then any common measure of AB and C is a common measure of C and FB and any common measure of C and FB is a common measure of AB and C all the common measures of AB and C are precisely all the common measures of C and FB.

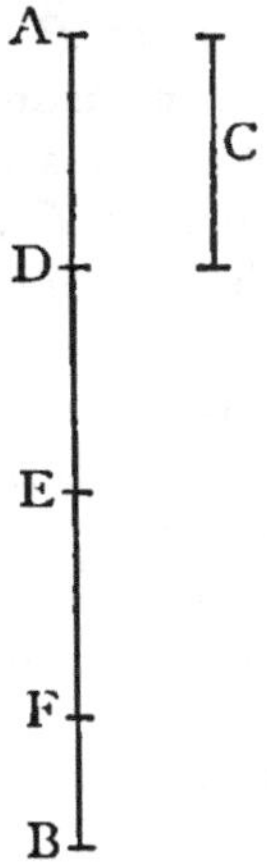

Now let P Q R_1 be the magnitudes represented severally by AB C FB. Because $R_1 < Q$ magnitudes each equal to R_1 may be taken in succession from Q and the remainders successively left until at length (art. 234) either no remainder is left or a remainder less than R_1. If no remainder be left Q is a multiple of R_1 and then as shown above all the common measures of Q and R_1 are precisely all the measures of R_1. But all the common measures of P and Q are precisely all the common measures of Q and R_1. Therefore all the common measures of P and Q are precisely all the measures of R_1 and therefore R_1 its own greatest measure is the greatest common measure of P and Q.

But if a remainder be left call it R_2 then as shown above all the common measures of Q and R_1 and therefore also all the common measures of P and Q are precisely all the common measures of R_1 and R_2. As before $\because$ $R_2 < R_1$ one or more magnitudes each equal to

R_2 may be taken successively from R_1 and the successive remainders got until there is left either no remainder or a remainder less than R_2. If R_1 be a multiple of R_2 all the common measures at once of P and Q of Q and R_1 and of R_1 and R_2 are precisely all the measures of R_2 and the greatest of all is therefore R_2. But if there be a remainder R_3 less than R_2 all the common measures at once of P and Q of Q and R_1 and of R_1 and R_2 are precisely all the common measures of R_2 and R_3 one or more magnitudes each equal to R_3 may then be taken in succession from R_2 and the remainders successively got as before and so on.

The series of constantly decreasing magnitudes R_1 R_2 R_3... can come to an end only when there arises some one R_i that is a measure of R_{i-1} the one next before. Then since all the common measures are the same of every consecutive two of the magnitudes P Q R_1 R_2...R_{i-1} R_i and all the common measures of the last two R_{i-1} and R_i are precisely all the measures of R_i all the common measures of P and Q are precisely all the measures of R_i and therefore the greatest common measure of P and Q is R_i.

But if the series is endless P and Q have no common measure. For Q being not a measure of P is either greater or less than $\frac{1}{2}(P)$. If greater R_1 is what is left of P after taking from it not less than its half and if less $\because$ $R_1 < Q$ R_1 is still what is left of P after taking from P not less than $\frac{1}{2}(P)$. In the same way R_2 is what is left of Q after taking from Q not less than $\frac{1}{2}(Q)$, R_3 what is left of R_1 after taking from R_1 not less than $\frac{1}{2}(R_1)$, and so on. If then P and Q could have any common measure M since from P there is taken not less than $\frac{1}{2}(P)$ and from the remainder R_1 left not less than $\frac{1}{2}(R_1)$ and from the new remainder R_3 left not less than $\frac{1}{2}(R_3)$ and so on there would (art. 232) be at length left some one remainder of the series R_1 R_3 R_5... and much more every after one less than M. But every common measure of P and Q is a measure of every one of the magnitudes R_1 R_2 R_3 R_4.... Therefore M would be a measure of every one of the magnitudes R_1 R_3 R_5... and therefore would be a measure of each of endless magnitudes each less than M itself.

236. COR. *All the common measures of two magnitudes are precisely all the measures of their greatest common measure.*

Hence a fraction is expressed in the least whole terms by the numbers expressing what multiples the fraction and the unit magnitude severally are of their greatest common measure and hence what was shown in art. 216 comes out afresh but by a straighter road that a fraction is expressed in terms of the least whole numbers precisely when these are prime to one another.

237. PROP. *The base of a right angled isosceles triangle has no measure in common with either of the equal sides.*

Let ABC be a right angled isosceles triangle of which AB AC are the equal sides. Because AB is equal to AC the angle ABC is equal to the angle ACB and therefore these two angles are together twice either of them.

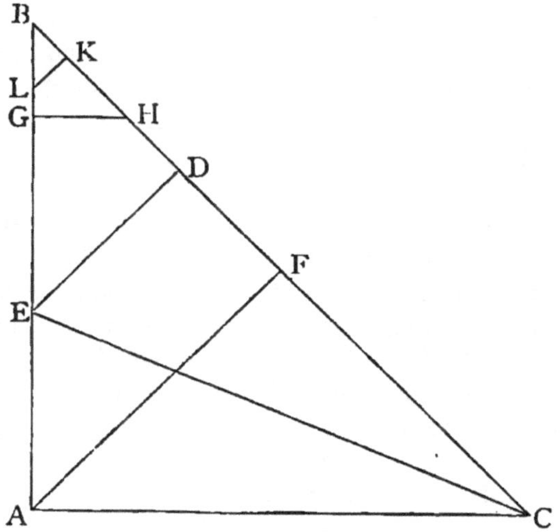

But these two angles of the triangle ABC are together less than twice a right angle and so therefore is twice either of them. Therefore each of the angles ABC ACB is less than a right angle. Wherefore it is the angle BAC that is the right angle.

Because the angles of the triangle ABC are altogether twice a right angle and one of them BAC is a right angle the other two ABC ACB are together equal to a right angle. Hence twice each of the angles ABC ACB is equal to a right angle, that is each of them is half a right angle. Since then the right angle BAC is greater than either of the half right angles ABC ACB the side BC is greater than either of the sides AC AB. But the two sides AB AC of the triangle are together greater than the third side BC and therefore so is twice AB or twice AC the equal to AB AC together. From BC which thus $> AB$ or AC but

$< 2(AB)$ or $2(AC)$ cut off close to the end C a part DC equal to AB or AC and the remaining part $BD < AB$ or AC.

From the point D in BC draw on the same side as the triangle a straight line DE at right angles to BC. Bisect BC in F and join FA. Because ABC is a triangle the point A is not in the straight line BC and therefore BFA CFA are triangles. In these triangles because the two sides BF FA are equal severally to the two sides CF FA and the base BA is equal to the base CA the angle BFA is equal to the angle CFA, that is FA is at right angles to BC. Because then of the triangle CFA the angle CFA is a right angle the other two angles FCA FAC are together equal to a right angle, but one of these FCA is half a right angle therefore the other FAC is half a right angle and therefore equal to the first FCA therefore again the sides FC FA over against these equal angles are equal so that the triangle FAC is a right angled isosceles one. In the like way the triangle FAB is a right angled isosceles one. Because of the triangle FAC the angle $AFC >$ the angle FAC the side $AC >$ the side FC and therefore also $DC > FC$. Since then DC FC are unequal straight lines having a common end C and lying in the same straight line upon the same side of their common end the less FC is a part of the greater DC and therefore the point F is on the same side of the point D as the point C. But because BD DC are the parts of BC the point B is on the opposite side of D to that on which C is. Therefore B and F are on opposite sides of D. Again the straight lines DE FA make with the straight line BC cut by them (if produced) at the different points D F interior angles FDE DFA upon one side that are severally right angles and that are together therefore twice a right angle therefore DE is parallel to FA. And because in the straight line BC the points B F are on opposite sides of the point D and through D F parallel straight lines DE FA are drawn the point B and the straight line FA are on opposite sides of the straight line DE. But the point A is on the same side of DE as FA therefore B and A are on opposite sides of DE. The line BA then ends at two points B A on opposite sides of the endless line in which DE is and therefore is cut somewhere between B A by that endless line. Let E be the point where BA is so cut. Join EC.

Because E is in the straight line AB but not at either A or B the only points that AB has in common with the straight lines AC BC severally EAC EDC are triangles. And because the angles EAC EDC of these triangles are right angles (square of EA, square of AC) = square of EC

(square of ED, square of DC) = square of EC

$\therefore$ (sq. of EA, sq. of AC) = (sq. of ED, sq. of DC).

But the square of AC is equal to the square of DC because the straight line AC is equal to the straight line DC. Therefore the square of EA is equal to the square of ED and therefore the straight line EA is equal to the straight line ED. Moreover since E is not in BC EBD is a triangle and since the angle BDE of this triangle is a right angle and the angle DBE is half a right angle it follows as before that DBE is a right angled isosceles triangle and therefore AE ED DB are all equal to one another.

Hence as before since DBE is a right angled isosceles triangle $BE > BD$ and $< 2(BD)$. From BE therefore cut off close to E a part GE equal to BD and a part BG remains less than BD. Then because AE is equal to ED or DB AB is wholly made up of three parts AE EG GB of which each of the two AE $EG = BD$ and the third $GB < BD$ so that $AB > 2(BD)$ but $< 3(BD)$. Also to each of the equals AE DE putting EB AB is equal to DE EB together and BD is one of the equal sides of the right angled isosceles triangle DBE of which the other two sides DE EB are together equal to AB.

Again in the same way from G in BE draw on the same side as the triangle DBE a straight line GH at right angles to BE and BD is cut between B and D by GH. Let the point where BD is so cut be H and DH HG GB are all equal GBH is a right angled isosceles triangle $BH > BG$ but $< 2(BG)$ from BH may be cut close to H KH equal to BG leaving BK less than BG $BD > 2(BG)$ but $< 3(BG)$ and BG is one of the equal sides of the right angled isosceles triangle GBH of which the other two sides BH HG are together equal to BD.

From K in like manner KL may now be drawn at right angles to BH cutting BG in L between B and G and making a right angled isosceles triangle KBL of which one of the equal sides is BK and the other two sides BL LK are together equal to BG, and so on. In the series of ever lessening straight lines BC BA BD BG BK - - - thus got from the first two BC BA by each after one being that part less than the one next before of the one next but one before by which the latter exceeds a multiple of the former it hence comes out that every one but the first two is one of the equal sides of a right angled isosceles triangle of which the other two sides are together equal to the one next before so that the latter is greater than twice and less than thrice the former. The series therefore goes on for ever and therefore (art. 235) BC has no measure in common with BA.

238. *Def.* Two magnitudes of the same kind that have no common measure are said to be INCOMMENSURABLE with one another.

Since (arts. 92, 112) one of two magnitudes can be a fraction of the other only when they are multiples of some one magnitude a magnitude may be of the same kind as another magnitude and yet be no fraction of it. One magnitude is or is not a fraction of another according as it is or is not commensurable therewith.

239. PROP. *Of any two magnitudes of the same kind the first either is a fraction of the second or is greater than the one and less than the other of two fractions of the second differing from one another by less than any given fraction however small.*

Let A B be any two magnitudes of the same kind. Cut B into any number n of equal parts so that each of them (art. 231) is less than A and take magnitudes each equal to $\frac{1}{n}(B)$ from A and from the successive remainders left until at last (art. 234) either no remainder is left or a remainder less than $\frac{1}{n}(B)$. Let m be the number of the magnitudes each equal to $\frac{1}{n}(B)$ so taken. If no remainder be left A is $\frac{m}{n}(B)$ but if a remainder be left $A > \frac{m}{n}(B)$ and $< \frac{m+1}{n}(B)$ and the fractions $\frac{m}{n}$ $\frac{m+1}{n}$ differ from one another by $\frac{1}{n}$ which (art. 172) may be made less than any given fraction however small by taking n a great enough whole number.

If A and B have no common measure A is greater than the less and less than the greater of two successive multiples of $\frac{1}{n}(B)$ whatever whole number n be and then A can be known numerically relative to B only through one or other pair of fractions of B intercepting as it were A between them. If A and B have a common measure let u v be the whole numbers expressing what multiples they severally are of their greatest common measure then since all the common measures of A and B are (art. 236) precisely all the measures of their greatest common measure A can be expressed as a simple fraction of B only by $\frac{ni}{vi}$, where i is a whole number and hence it is only when $n = vi$ that A is a multiple of $\frac{1}{n}(B)$ or in other words only when n is such that $\frac{1}{v}n$ is equal to a whole number. Wherefore A is a multiple of $\frac{1}{n}(B)$ whatever whole number n be only when v is 1, that is only when A is a multiple of B. But even when A is a multiple of B and is therefore $\frac{m}{n}(B)$ whatever whole number n be still $A > \frac{m-c}{n}(B)$ and $< \frac{m+d}{n}(B)$ c d being any whole numbers of which c not$>$ m and $\frac{m+d}{n} - \frac{m-c}{n} = \frac{m+d-(m-c)}{n}$ $= \frac{m+d+c-m}{n} = \frac{m-m+d+c}{n}$, that is $\frac{d+c}{n}$ which because equal to $\frac{1}{\left(\frac{n}{d+c}\right)}$ may (arts. 133, 172) by keeping $d+c$ fixed and taking n great enough be made less than any given fraction however small.

240. *Def.* Two fractions of a magnitude such that another magnitude is greater than the less of them but less than the greater are called a PAIR OF FRACTIONS INTERCEPTING THIS OTHER MAGNITUDE RELATIVE TO THE FIRST.

241. PROP. *If any pair of fractions be given intercepting one magnitude relative to another each of the fractions of any other pair intercepting the first magnitude relative to the other if only these fractions differ from one another by less than a certain fraction is greater than the less and less than the greater of the fractions of the given pair.*

Let two straight lines OA and B represent any two magnitudes of the same kind and let a straight line OC of which OA is a part represent the greater and a straight line OD which is a part of OA the less of any two given fractions intercepting the magnitude represented by OA relative to the magnitude represented by B so that AC and DA represent the differences between the magnitude represented by OA and the greater and the less of the given fractions severally. Cut B into any such number n of equal parts (art. 231) that each of them is less than either AC or DA and take magnitudes each equal to $\frac{1}{n}(B)$ from OA and the remainders successively left until (art. 234) there be left either no remainder or a remainder less than $\frac{1}{n}(B)$. Let there be m magnitudes so taken. If no remainder be left OA is $\frac{m}{n}(B)$ and if a straight line OE a part of OA represent the next less multiple of $\frac{1}{n}(B)$ to wit $\frac{m-1}{n}(B)$ and a straight line OF of which OA is a part the next greater multiple to wit $\frac{m+1}{n}(B)$ so that each of the differences AE AF represents $\frac{1}{n}(B)$ since each of the magnitudes AD $AC > \frac{1}{n}(B)$ $AD > AE$ and $AC > AF$. Hence from OA taking the former unequals $OD < OE$ and to OA putting the latter $OC > OF$, that is $\frac{m-1}{n}(B) >$ the less of the two given fractions intercepting OA relative to B and $\frac{m+1}{n}(B) <$ the greater.

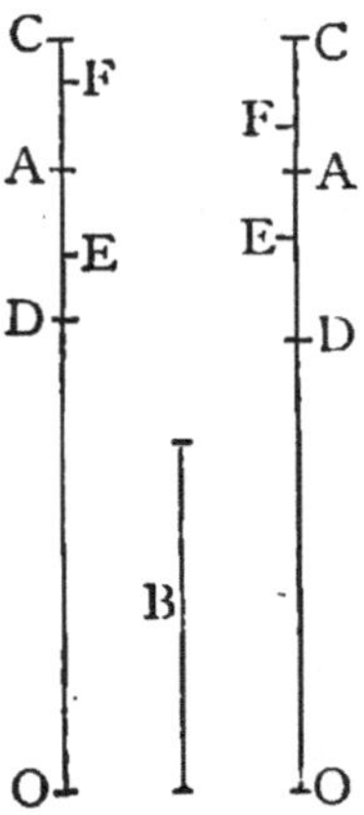

But if a remainder be left $OA > \frac{m}{n}(B)$ and $< \frac{m+1}{n}(B)$ and if OE a part of OA now represent $\frac{m}{n}(B)$ and OF a whole of which OA is a part $\frac{m+1}{n}(B)$ much more $AE < AD$ and $AF < AC$. Therefore $OE > OD$ and $OF < OC$, that is $\frac{m}{n}(B) >$ the less and $\frac{m+1}{n}(B) <$ the greater of the given fractions intercepting OA relative to B.

Much more if n were taken a still greater whole number would each of the fractions of the pair got in the same way intercepting OA relative to B be greater than the less and less than the greater of the fractions of the given pair.

242. PROP. *Each of the fractions of any pair intercepting the greater of two unequal magnitudes relative to an-*

other magnitude is greater than each of the fractions of any pair intercepting the less relative to the same if only the fractions of each pair differ from one another by less than a certain fraction.

Let a whole straight line OA represent the greater and a part OB of OA the less of any two unequal magnitudes so that the other part BA represents their difference and let a straight line C represent any other magnitude of the same kind. Cut BA (art. 97) into two equal parts BD DA so that OD represents a magnitude greater than the less of the two unequals and less than the greater by the very same amount that to wit represented by BD or DA. Cut C into so many equal parts n (art. 231) that each $\frac{1}{n}(C)$ is less than BD or DA and also less than OB and let (art. 234)

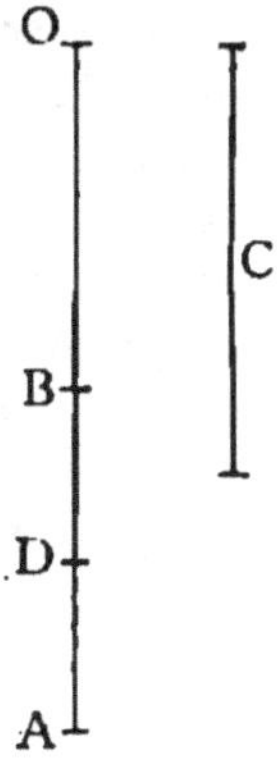

$$OA \text{ not} < \frac{m}{n}(C) \text{ but } < \frac{m+1}{n}(C)$$

$$OB \text{ not} < \frac{m'}{n}(C) \text{ but } < \frac{m'+1}{n}(C)$$

where m m' are whole numbers.

Because $BA > \frac{2}{n}(C)$ $OA > \frac{m'+2}{n}(C)$. If then OA be $\frac{m}{n}(C)$ $m > m'+2$ and $\therefore m-1 > m'+1$, each of the fractions $\frac{m-1}{n}$ $\frac{m+1}{n}$ therefore is greater than $\frac{m'+1}{n}$ and much more therefore than either $\frac{m'}{n}$ or $\frac{m'-1}{n}$. But if $OA > \frac{m}{n}(C)$ $\frac{m+1}{n}(C) > \frac{m'+2}{n}(C)$ $\therefore m > m'+1$ and therefore each of the fractions $\frac{m}{n}$ $\frac{m+1}{n}$ is greater both than each of the fractions $\frac{m'}{n}$ $\frac{m'+1}{n}$ and than each of the fractions $\frac{m'-1}{n}$ $\frac{m'+1}{n}$.

If the whole number n were taken still greater much more would each of the fractions so got intercepting OA relative to C be greater than each of the fractions so got intercepting OB relative to C.

243. Since (arts. 239, 241) all the pairs of fractions that intercept however closely equal magnitudes relative to any one and the same magnitude are the very same and (art. 242) all those pairs of fractions that intercept closely enough the greater of two unequal mag-

nitudes relative to any third magnitude have each of their fractions greater than each of the fractions of all those pairs of fractions that intercept closely enough the less relative to the same third magnitude, so that the tracks of greater and less intercepters of equal magnitudes are everywhere the same however close together the tracks go and the tracks of greater and less intercepters of the greater of two unequal magnitudes run apart from the tracks of greater and less intercepters of the less wherever the tracks of each pair go close enough together, not only is the order of greatness of magnitudes that are fractions of a magnitude settled (art. 100) by the fractions which they severally are thereof but generally the order of greatness of any magnitudes of the same kind whether fractions or not of some one magnitude is settled by the pairs of fractions which intercept them severally closer than by any given degree of closeness relative to that magnitude.

Def. That definite numerical relation of any magnitude to any magnitude of the same kind in virtue of which the former either is a fraction of the latter or is greater than the one and less than the other of two fractions of the latter differing from one another by less than any given fraction however small is called the RATIO of the former magnitude to the latter.

244. *Def.* The symbol $A:B$ stands for the ratio of the magnitude A to the magnitude B, A B are called the TERMS of the ratio, A is called the FIRST TERM the LEADING MAGNITUDE or the ANTECEDENT of the ratio and B the SECOND TERM the FOLLOWING MAGNITUDE or the CONSEQUENT.

245. *Def.* The first of four magnitudes is said to have to the second the SAME RATIO as the third has to the fourth when if the first be a fraction of the second the third is the same fraction of the fourth or if the first be greater than the one and less than the other of two fractions of the second by less than however small a given fraction they may differ from one another the third is also greater than the former and less than the latter of the same two fractions of the fourth.

Def. That A has to B the same ratio as C has to D is symbolized by $A:B=C:D$ and this is read "A is to B

as C to D" or "as A is to B so is C to D". Likewise $A_1 : B_1 = A_2 : B_2 = A_3 : B_3 = - - - = A_n : B_n$ symbolizes that the ratios of A_1 to B_1 of A_2 to B_2 of A_3 to B_3 - - - of A_n to B_n are all the same. The sameness of ratios is called a PROPORTIONALITY. The terms of the ratios in a proportionality are called the TERMS OF THE PROPORTIONALITY and are said to be PROPORTIONAL MAGNITUDES PROPORTIONALS or PROPORTIONATE. Magnitudes of the same kind $A\ B\ C\ D \ldots G\ H$ such that $A : B = B : C = C : D = - - - = G : H$ are said to be CONTINUEDLY PROPORTIONATE or IN A CONTINUED PROPORTIONALITY.

246. *Def.* The first of four magnitudes is said to have to the second a GREATER RATIO than the third has to the fourth when the first is greater than some fraction of the second but the third is not greater than the same fraction of the fourth. The third is then also said to have to the fourth a LESS RATIO than the first has to the second.

Def. That A has to B a greater ratio than C to D is symbolized by $A : B > C : D$ and that C has to D a less ratio than A to B by $C : D < A : B$.

247. PROP. *Magnitudes of the same kind that are like multiples severally of as many other magnitudes are together the like multiple of those other magnitudes together.* (EUCLID V. 1).

Let $A\ B\ C \ldots F\ G\ H$ be any magnitudes of one kind and let n be any whole number. Because $n(A)$ is wholly made up of n parts each equal to A and $n(B)$ is wholly made up of n parts each equal to B (arts. 7, 10) $\{n(A),\ n(B)\}$ is wholly made up of n parts each equal to (A, B), that is

$$\{n(A),\ n(B)\} = n(A, B).$$

Hence $\{n(A),\ n(B),\ n(C)\} = \{n(A, B),\ n(C)\}$ and therefore as before $= n(A, B, C)$. In like manner therefore $\{n(A),\ n(B),\ n(C),\ n(D)\} = \{n(A, B, C),\ n(D)\} = n(A, B, C, D)$. And so on till at last

$$\{n(A),\ n(B),\ n(C), \ldots n(G),\ n(H)\} = n(A, B, C, \ldots G, H).$$

248. COR. *If a magnitude be a multiple of another magnitude that like multiple is any multiple of the first magnitude of the same multiple of the other.*

This is only the particular case of the general proposition in art. 247 when $A=B=C=\text{- - - -}=G=H$. If there be m of these magnitudes the proposition then is that $m\{n(A)\}=n\{m(A)\}$.

249. *Def.* The symbol $\substack{>\\=\\<}$ stands for *is greater than equal to or less than* and $\substack{<\\=\\>}$ for *is less than equal to or greater than.*

250. Prop. *A magnitude and any fraction of a magnitude of the same kind are in the same order of greatness as the multiples of those magnitudes expressed severally by the denominator and by the numerator of the fraction.*

Let A B be any two magnitudes of one kind and m n any two whole numbers. Since (art. 229) $\frac{m}{n}(B)$ is $m\left\{\frac{1}{n}(B)\right\}$ (art. 248) $n\left\{\frac{m}{n}(B)\right\}$ $=m\left[n\left\{\frac{1}{n}(B)\right\}\right]$ which is $m(B)$. Hence (arts. 88, 89)

$$\text{according as } A \substack{>\\=\\<} \frac{m}{n}(B) \text{ so is } n(A) \substack{>\\=\\<} m(B)$$

and (arts. 90, 91) according as $n(A) \substack{>\\=\\<} m(B)$ so is $A \substack{>\\=\\<} \frac{m}{n}(B)$.

251. Prop. *The first of four magnitudes has to the second the same ratio as the third has to the fourth precisely when any multiples whatever of the first and second are in the same order of greatness as the like multiples severally of the third and fourth.* (This is The Multiple Test of the sameness of ratios and is made the *Definition* in Euclid's Fifth Book).

First let A B C D be four magnitudes such that $A:B=C:D$ and let m n be any two whole numbers. Cut B into n equal parts and let it be μ magnitudes each equal to $\frac{1}{n}(B)$ that can be taken from A and the successive remainders left so as to leave at last either no remainder or a remainder less than $\frac{1}{n}(B)$.

If there be no remainder $A=\frac{\mu}{n}(B)$ and then (art. 245) $C=\frac{\mu}{n}(D)$. Hence (art. 250)

$$n(A)=\mu(B) \qquad n(C)=\mu(D).$$

If then $n(A)=m(B)$ m can only be μ and $\therefore$ $n(C)=m(D)$. If

$n(A) > m(D)$ m must $< \mu$ and $\therefore$ $n(C) > m(D)$. And if $n(A) < m(B)$ m must $> \mu$ and $\therefore$ $n(C) < m(D)$.

But if there be a remainder $A > \frac{\mu}{n}(B)$ but $< \frac{\mu+1}{n}(B)$ and then (art. 245) $C > \frac{\mu}{n}(D)$ but $< \frac{\mu+1}{n}(D)$. Hence (art. 250)

$$n(A) > \mu(B) \text{ but} < (\mu+1)(B) \qquad n(C) > \mu(D) \text{ but} < (\mu+1)(D).$$

$n(A)$ cannot now $= m(B)$ for then (art. 250) would $A = \frac{m}{n}(B)$ or A would be a multiple of $\frac{1}{n}(B)$. But if $n(A) > m(B)$ m not$> \mu$ $\therefore$ $n(C) > m(D)$. And if $n(A) < m(B)$ m not$< \mu+1$ $\therefore$ $n(C) < m(D)$.

Hence according as $n(A) \gtreqless m(B)$ so is $n(C) \gtreqless m(D)$.

Conversely let A B C D be four magnitudes such that whatever whole numbers m n are according as $n(A) \gtreqless m(B)$ so is $n(C) \gtreqless m(D)$. As before cut B into n equal parts and let μ be the number of magnitudes each equal to $\frac{1}{n}(B)$ taken from A and the remainders successively left when at length either no remainder or a remainder less than $\frac{1}{n}(B)$ is left.

If no remainder be left $A = \frac{\mu}{n}(B)$ and $\therefore$ (art. 250) $n(A) = \mu(B)$. Now m being any whole number whatever may be taken μ so that $n(A) = m(B)$. Then by hypothesis $n(C) = m(D)$, that is $\mu(D)$. And $\therefore$ (art. 250) $C = \frac{\mu}{n}(D)$.

If a remainder be left $A > \frac{\mu}{n}(B)$ and $< \frac{\mu+1}{n}(B)$ and $\therefore$ $n(A) > \mu(B)$ and $< (\mu+1)(B)$. But m being any whole number may be taken μ so that $n(A) > m(B)$ and then by the hypothesis $n(C) > m(D)$, that is $> \mu(D)$ $\therefore$ $C > \frac{\mu}{n}D$. Also m may be taken $\mu+1$ so that $n(A) < m(B)$ then $n(C) < m(D)$, that is $< (\mu+1)(D)$ and $\therefore$ $C < \frac{\mu+1}{n}(D)$. Moreover n being any whole number may (art. 172) be taken so great that $\frac{1}{n}$ by which $\frac{\mu}{n}$ $\frac{\mu+1}{n}$ differ is made less than any given fraction however small.

Hence if A be a fraction of B C is the same fraction of D or if A be greater than the one and less than the other of two fractions of B by less than however small a given fraction they may differ from one another C is also greater than the former and less than the latter of the same two fractions of D, that is (art. 245) $A : B = C : D$.

On the whole then since four magnitudes that are proportionate have any multiples whatever of the first and second of them in the same order of greatness as the like multiples severally of the third and fourth and four magnitudes that have any multiples whatever of the first and second of them in the same order of greatness as the like multiples severally of the third and fourth are proportionate it follows that four magnitudes are proportionate precisely when any multiples whatever of the first and second of them are in the same order of greatness as the like multiples severally of the third and fourth.

252. PROP. *The first of four magnitudes has to the second a greater ratio than the third has to the fourth precisely when some multiples of the first and second are in descending order of greatness but the like multiples severally of the third and fourth are not.* (This THE MULTIPLE TEST of the greaterness and lessness of ratio is made the *Definition* in EUCLID'S Fifth Book.)

First let A B C D be four magnitudes such that $A : B > C : D$. There are (art. 246) whole numbers m n such that $A > \frac{m}{n}(B)$ but C not$> \frac{m}{n}(D)$ and $\therefore$ (art. 250) $n(A) > m(B)$ but $n(C)$ not$> m(D)$.

Again let A B C D be four magnitudes such that for some whole numbers m n $n(A) > m(B)$ but $n(C)$ not$> m(D)$. Then (art. 250) $A > \frac{m}{n}(B)$ but C not$> \frac{m}{n}(D)$, that is (art. 246) $A : B > C : D$.

253. PROP. *If when the first of four magnitudes is greater than the one and less than the other of two fractions of the second the third is greater than the former and less than the latter of the same two fractions of the fourth by less than however small a given fraction the two fractions may differ from one another then if the first be a fraction of the second the third is the same fraction of the fourth.*

Let A B C D be four magnitudes such that

$$A > \frac{m}{n}(B) \text{ and } < \frac{m+c}{n}(B) \qquad C > \frac{m}{n}(D) \text{ and } < \frac{m+c}{n}(D)$$

m n c being whole numbers such that $\frac{c}{n}$ may by taking n great enough be made less than any given fraction however small and let $A = \frac{u}{v}(B)$. Then $\frac{u}{v} > \frac{m}{n}$ and $< \frac{m+c}{n}$ and hence

$$\frac{u}{v}(D) > \frac{m}{n}(D) \text{ and } < \frac{m+c}{n}(D).$$

Now since if $C > \frac{u}{v}(D)$ n could (art. 242) be taken so great as to make each of the fractions $\frac{m}{n}(D)$ $\frac{m+c}{n}(D)$ greater than each of two fractions intercepting $\frac{u}{v}(D)$ relative to D and therefore so that $\frac{m}{n} > \frac{u}{v}$ and if $C < \frac{u}{v}(D)$ n could be taken so great as to make each of the fractions $\frac{m}{n}(D)$ $\frac{m+c}{n}(D)$ less than each of two fractions intercepting $\frac{u}{v}(D)$ relative to D and therefore so that $\frac{m+c}{n} < \frac{u}{v}$ it follows that $C = \frac{u}{v}(D)$.

254. PROP. *Ratios the same as the same ratio are the same as one another.* (EUCLID V. 11).

Let $A : B = P : Q$ and $C : D = P : Q$.

If n be any whole number there is (art. 239) a whole number m such that A not$< \frac{m}{n}(B)$ but $< \frac{m+1}{n}(B)$. ∴ and ∵ $A : B = P : Q$ (art. 245) P not$< \frac{m}{n}(Q)$ but $< \frac{m+1}{n}(Q)$. ∴ again and ∵ $C : D = P : Q$ (art. 245) C not$< \frac{m}{n}(D)$ but $< \frac{m+1}{n}(D)$. If then A be a fraction of B,—to wit when "not<" here means "="—, C is the same fraction of D and if A be greater than the one and less than the other of two fractions of B by less than however small a given fraction they may differ from one another,—to wit when "not<" means ">"—, C is greater than the former and less than the latter of the same two fractions of D, that is (art. 245) $A : B = C : D$.

255. PROP. *Equal magnitudes have to the same magnitude, and the same has to equals, the same ratio.* (EUCL. v. 7).

For if A B be equal magnitudes and P a magnitude of the same kind A not$<\frac{m}{n}(P)$ but $<\frac{m+1}{n}(P)$, m n being whole numbers n any whatever, and $\therefore$ (arts. 6, 9) B not$<\frac{m}{n}(P)$ but $<\frac{m+1}{n}(P)$, that is $A:P=B:P$. Also P not$<\frac{r}{s}(A)$ but $<\frac{r+1}{s}(A)$, r s being whole numbers s any whatever, and $\therefore$ (arts. 90, 88, 6, 9) P not$<\frac{r}{s}(B)$ but $<\frac{r+1}{s}(B)$, that is $P:A=P:B$.

256. PROP. *If two magnitudes have to a single magnitude the same ratios severally as other two magnitudes have to another single magnitude the first has to the second of the first two magnitudes the same ratio as the first has to the second of the other two.*

Let two magnitudes A B have to a magnitude H the same ratios severally as two magnitudes C D have to a magnitude K, that is let $A:H=C:K$ and $B:H=D:K$.

If n s be any whole numbers (art. 239) there are whole numbers m r such that

$$A \text{ not}<\frac{m}{n}(H) \text{ and } <\frac{m+1}{n}(H) \qquad B \text{ not}<\frac{r}{s}(H) \text{ and } <\frac{r+1}{s}(H).$$

Therefore and by reason of the given proportionalities (art. 245)

$$C \text{ not}<\frac{m}{n}(K) \text{ and } <\frac{m+1}{n}(K) \qquad D \text{ not}<\frac{r}{s}(K) \text{ and } <\frac{r+1}{s}(K).$$

If then "not<" mean "=" throughout (arts. 112, 114) $A=\frac{\left(\frac{m}{n}\right)}{\left(\frac{r}{s}\right)}(B)$ $C=\frac{\left(\frac{m}{n}\right)}{\left(\frac{r}{s}\right)}(D)$. But if "not<" do not mean "=" throughout

A not$<\frac{\left(\frac{m}{n}\right)}{\left(\frac{r+1}{s}\right)}\left\{\frac{r+1}{s}(H)\right\}$ and $<\frac{\left(\frac{m+1}{n}\right)}{\left(\frac{r}{s}\right)}\left\{\frac{r}{s}(H)\right\}$ and therefore (art. 112 or 129)

$$A \text{ not} < \frac{ms}{(r+1)n}\left\{\frac{r+1}{s}(H)\right\} \text{ and } < \frac{(m+1)s}{rn}\left\{\frac{r}{s}(H)\right\}.$$

But $\because \frac{r+1}{s}(H) > B$ (art. 91) $\frac{1}{(r+1)n}\left\{\frac{r+1}{s}(H)\right\} > \frac{1}{(r+1)n}(B)$ and $\therefore$ (art. 89) $\frac{ms}{(r+1)n}\left\{\frac{r+1}{s}(H)\right\} > \frac{ms}{(r+1)n}(B)$. Likewise and by arts. 90, 88, $\because \frac{r}{s}(H)$ not$> B$ $\frac{(m+1)s}{rn}\left\{\frac{r}{s}(H)\right\}$ not$> \frac{(m+1)s}{rn}(B)$. Hence (art. 9)

$$A > \frac{ms}{(r+1)n}(B) \text{ and } < \frac{(m+1)s}{rn}(B).$$

In the same way

$$C > \frac{ms}{(r+1)n}(D) \text{ and } < \frac{(m+1)s}{rn}(D).$$

Now $\frac{(m+1)s}{rn} - \frac{ms}{(r+1)n} = \frac{1}{n}\left(\frac{m+1}{r} - \frac{m}{r+1}\right)s = \frac{1}{n}\left\{\frac{m}{r(r+1)} + \frac{1}{r}\right\}s$ $= \frac{1}{s}\frac{m}{n}\frac{s}{r+1}\frac{s}{r} + \frac{1}{n}\frac{s}{r}$. And if $\frac{\mu}{\nu}$ $\frac{\mu+\theta}{\nu}$ be any given pair of fractions intercepting A relative to H (art. 241) n may be taken so great that each of the fractions $\frac{m}{n}$ $\frac{m+1}{n}$ is greater than $\frac{\mu}{\nu}$ but less than $\frac{\mu+\theta}{\nu}$ and if $\frac{\rho}{\sigma}$ $\frac{\rho+\kappa}{\sigma}$ be any given pair of fractions intercepting B relative to H s may be taken so great that each of the fractions $\frac{r}{s}$ $\frac{r+1}{s}$ is greater than $\frac{\rho}{\sigma}$ but less than $\frac{\rho+\kappa}{\sigma}$. Let then n s be taken thus great and $\because \frac{r}{s} > \frac{\rho}{\sigma}$ $\frac{r\rho}{s\rho} > \frac{\rho r}{\sigma r}$ $\therefore$ (art. 100) $s\rho < \sigma r$ $\therefore$ (art. 115) $\frac{s\rho}{r\rho} < \frac{\sigma r}{\rho r}$ and $\therefore \frac{s}{r} < \frac{\sigma}{\rho}$. Much more $\therefore$ (arts. 100, 9) $\frac{s}{r+1} < \frac{\sigma}{\rho}$. Since then $\frac{s}{r} < \frac{\sigma}{\rho}$ (arts. 91, 89) $\frac{s}{r+1}\frac{s}{r} < \frac{s}{r+1}\frac{\sigma}{\rho}$ and $\therefore$ much more (arts. 113, 9) $< \left(\frac{\sigma}{\rho}\right)^2$. In the same way $\because \frac{m}{n} < \frac{\mu+\theta}{\nu}$ $\frac{m}{n}\frac{s}{r+1}\frac{s}{r} < \frac{\mu+\theta}{\nu}\left(\frac{\sigma}{\rho}\right)^2$ and $\therefore$ (art. 91) $\frac{1}{s}\frac{m}{n}\frac{s}{r+1}\frac{s}{r} < \frac{1}{s}\frac{\mu+\theta}{\nu}\left(\frac{\sigma}{\rho}\right)^2$. Also $\frac{1}{n}\frac{s}{r} < \frac{1}{n}\frac{\sigma}{\rho}$. Hence (art. 12) $\frac{1}{s}\frac{m}{n}\frac{s}{r+1}\frac{s}{r} + \frac{1}{n}\frac{s}{r} < \frac{1}{s}\frac{\mu+\theta}{\nu}\left(\frac{\sigma}{\rho}\right)^2 + \frac{1}{n}\frac{\sigma}{\rho}$. But since if k be any given fraction however small s n may (art. 231) be severally taken so great that $\frac{1}{s}\frac{\mu+\theta}{\nu}\left(\frac{\sigma}{\rho}\right)^2 < \frac{1}{2}k$ and $\frac{1}{n}\frac{\sigma}{\rho} < \frac{1}{2}k$ s n may be taken so great that

$\frac{1}{s}\frac{\mu+\theta}{\nu}\left(\frac{\sigma}{\rho}\right)^2+\frac{1}{n}\frac{\sigma}{\rho}<\frac{1}{2}k+\frac{1}{2}k$ that is k. Much more therefore (art. 9) may s n be taken so great that $\frac{1}{s}\frac{m}{n}\frac{s}{r+1}\frac{s}{r}+\frac{1}{n}\frac{s}{r}$ and therefore also its equal $\frac{(m+1)s}{m}-\frac{ms}{(r+1)n}$ is less than any given fraction however small. Hence too (art. 253) should even in this case A be a fraction of B C is the same fraction of D.

Wherefore in both cases if A be a fraction of B C is the same fraction of D or if A be greater than the one and less than the other of two fractions of B by less than however small a given fraction they may differ from one another C is also greater than the former and less than the latter of the same two fractions of D, that is (art. 245) $A:B=C:D$.

257. COR. *The reciprocal of the greater of two unequal fractions is less than the reciprocal of the less.*

For if $\frac{r}{s}>\frac{\rho}{\sigma}$ it is shown in art. 256 that the reciprocal of $\frac{r}{s}$ $\frac{s}{r}<\frac{\sigma}{\rho}$ the reciprocal of $\frac{\rho}{\sigma}$.

258. COR. *A product of which the factors are as many as and severally greater than the factors of another product is greater than that other.*

The proof in art. 256 that if $\frac{\mu+\theta}{\nu}$ $\frac{\sigma}{\rho}$ $\frac{\sigma}{\rho}$ be severally greater than $\frac{m}{n}$ $\frac{s}{r+1}$ $\frac{s}{r}$ the product $\frac{\mu+\theta}{\nu}\left(\frac{\sigma}{\rho}\right)^2$ is greater than the product $\frac{m}{n}\frac{s}{r+1}\frac{s}{r}$ is quite general.

259. PROP. *If the first of four magnitudes be to the second in the same ratio as the third is to the fourth the second is to the first in the same ratio as the fourth is to the third.*

Let $A:B=C:D$. Then $\because$ (art. 245) $B:B=D:D$ B A have to B the same ratios severally as D C have to D and $\therefore$ (art. 256) $B:A=D:C$.

260. *Def.* The ratio of the consequent to the antecedent of a ratio is called the INVERSE RATIO.

The proposition of art. 259 then is that of ratios the same as one another the inverses are the same as one another. Since the magnitudes A B C D are in the inverse order D C B A and to say that $D:C=B:A$ is just (art. 245) to say that $B:A=D:C$ the proposition may also be stated—Magnitudes proportionate are inversely proportionate.

261. PROP. *Magnitudes are equal to one another either which have to the same magnitude or to which the same magnitude has the same ratio.* (EUCL. V. 9).

First let $A:Q=B:Q$. Then if A B could be unequal and A were taken as the greater each of two fractions $\frac{m}{n}$ $\frac{m+c}{n}$ intercepting A relative to Q would (art. 242) be greater than each of two fractions intercepting B relative to Q if only each two fractions differed from one another by less than some fraction and therefore (art. 9) $B<\frac{m}{n}(Q)$. But $\because$ $A:Q=B:Q$ and $A>\frac{m}{n}(Q)$ (art. 245) $B>\frac{m}{n}(Q)$. Therefore A cannot be unequal to B, that is $A=B$.

Again let $Q:A=Q:B$. Then (art. 259) $A:Q=B:Q$ and therefore by what has just now been shown $A=B$.

262. PROP. *A ratio the same as the greater of two ratios is greater than the less.* (EUCL. V. 13). *And a ratio the same as the less of two ratios is less than the greater.*

Let $A:B=C:D$ and $C:D>E:F$. Because $C:D>E:F$ there is (art. 246) a fraction $\frac{m}{n}$ such that $C>\frac{m}{n}(D)$ but E not$>\frac{m}{n}(F)$. And $\because$ $A:B=C:D$ and $C>\frac{m}{n}(D)$ (art. 245) $A>\frac{m}{n}(B)$. Hence $A>\frac{m}{n}(B)$ but E not$>\frac{m}{n}(F)$, that is (art. 246) $A:B>E:F$.

Again let $A:B=C:D$ and $C:D<E:F$. $\because$ $C:D<E:F$ there is a fraction $\frac{m}{n}$ such that C not$>\frac{m}{n}(D)$ but $E>\frac{m}{n}(F)$. And because $A:B=C:D$ and C not$>\frac{m}{n}(D)$ A not$>\frac{m}{n}(B)$. A then not$>\frac{m}{n}(B)$ but $E>\frac{m}{n}(F)$ and that is to say that $A:B<E:F$.

263. PROP. *The greater of two unequal magnitudes has to the same magnitude, and the same has to the less of two unequals, the greater ratio.* (EUCL. V. 8).

Let a magnitude A be greater than a magnitude B and let K be any other magnitude of the same kind.

Because $A > B$ there are (art. 242) fractions $\frac{m}{n}\ \frac{m+c}{n}$ intercepting A relative to K each greater than each of two fractions intercepting B relative to K and $\therefore A > \frac{m}{n}(K)$ but $B <$ and $\therefore$ not$> \frac{m}{n}(K)$, that is (art. 246) $A : K > B : K$. Again $\because \frac{m}{n}(K) > B$ and $< A$

$K > \frac{n}{m}(B)$ and $<$ and $\therefore$ not$> \frac{n}{m}(A)$, that is $K : B > K : A$.

264. PROP. *A ratio greater than the greater of two ratios is greater than the less.*

Let $A : B > C : D$ and $C : D > E : F$. Then (art. 246) there are fractions $\frac{m}{n}\ \frac{r}{s}$ such that $A > \frac{m}{n}(B)$ but C not$> \frac{m}{n}(D)$ $C > \frac{r}{s}(D)$ but E not$> \frac{r}{s}(F)$. Hence (art. 9) $\frac{m}{n} > \frac{r}{s}$ and much more $A > \frac{r}{s}(B)$. But to say that $A > \frac{r}{s}(B)$ but E not$> \frac{r}{s}(F)$ is (art. 246) to say that $A : B > E : F$.

265. PROP. *Of two magnitudes that is the greater which has to the same magnitude, and that is the less to which the same magnitude has, the greater ratio.* (EUCL. v. 10).

For if $A : Q > B : Q$ (art. 246) $A > \frac{m}{n}(Q)$ but B not$> \frac{m}{n}(Q)$ $m\ n$ being whole numbers and $\therefore$ (art. 9) $A > B$.

Again if $Q : B > Q : A$ $Q > \frac{r}{s}(B)$ but not$> \frac{r}{s}(A)$ $r\ s$ being some whole numbers $\therefore \frac{r}{s}(B) < \frac{r}{s}(A)$ $\frac{1}{s}(B) < \frac{1}{s}(A)$ and $B < A$.

266. PROP. *The inverse of the greater of two ratios is less than the inverse of the less.*

Let $A : B > C : D$. A fraction $\frac{m}{n}$ is such that $A > \frac{m}{n}(B)$ but C not$> \frac{m}{n}(D)$ and $\therefore B < \frac{n}{m}(A)$ but D not$< \frac{n}{m}(C)$.

If $D > \frac{n}{m}(C)$ then $\because B < \frac{n}{m}(A)$ B not$> \frac{n}{m}(A)$ and $D : C > B : A$.

But if $D = \frac{n}{m}(C)$ $\because \frac{n}{m}(A) > B$ there are fractions $\frac{r}{s}\ \frac{r+k}{s}$ intercepting $\frac{n}{m}(A)$ relative to A each greater than each of two fractions intercepting B relative to A and therefore so that $\frac{n}{m}(A) > \frac{r}{s}(A)$ and $B < \frac{r}{s}(A)$. $\therefore \frac{n}{m} > \frac{r}{s}$ $\therefore \frac{n}{m}(C) > \frac{r}{s}(C)$ and $\therefore D > \frac{r}{s}(C)$. $\therefore D > \frac{r}{s}(C)$ but $B <$ and $\therefore$ not$> \frac{r}{s}(A)$, that is $D:C > B:A$.

267. PROP. *If there be more than two magnitudes of one kind and as many other magnitudes of one kind which taken two and two in order have severally the same ratios the first has to the last of the former magnitudes the same ratio as the first has to the last of the latter.* (EUCL. v. 22).

First let there be three magnitudes $A\ B\ C$ of one kind and other three $A'\ B'\ C'$ of one kind which taken two and two in order have severally the same ratios to wit $A:B = A':B'$ $B:C = B':C'$. $\because B:C = B':C'$ inversely (art. 259) $C:B = C':B'$ so that $A\ C$ have to B the same ratios severally as $A'\ C'$ have to B'. $\therefore$ (art. 256) $A:C = A':C'$.

A	B	C
A'	B'	C'

Next let there be four magnitudes $A\ B\ C\ D$ and other four magnitudes $A'\ B'\ C'\ D'$ such that $A:B = A':B'$ $B:C = B':C'$ and $C:D = C':D'$. From the first two proportionalities by what has just been shown $A:C = A':C'$ and hence from this and the third given proportionality in the same way $A:D = A':D'$.

A	B	C	D
A'	B'	C'	D'

In like manner if a fourth proportionality $D:E = D':E'$ be given since there are three magnitudes of one kind $A\ D\ E$ and other three $A'\ D'\ E'$ of one kind which taken two and two in order have severally the same ratios $A:E = A':E'$. And so on.

A	B	C	D	E
A'	B'	C'	D'	E'

268. *Def.* When from the ratios $A:B$ $B:C$ - - - $G:H$ of the first to the second of all the consecutive twos in order of magnitudes $A\ B\ C \ldots G\ H$ being severally the same as the ratios $A':B'$ $B':C'$ - - - $G':H'$ of the first to the second of all the consecutive twos in order of as many magnitudes $A'\ B'\ C' \ldots G'\ H'$ the ratio $A:H$ of the first to the last of the former magnitudes is inferred (art. 267) to be the same as the ratio $A':H'$ of the first to the last of the latter

this proportionality is said to arise BY EQUALITY OF RATIO IN ORDERLY PROPORTIONALITIES.

269. PROP. *If there be more than two magnitudes of one kind and as many other magnitudes of one kind which taken two and two in cross order have severally the same ratios,—that is the first two of the former magnitudes and the last two of the latter the two next after the first of the former and the two next before the last of the latter and so on—, the first has to the last of the first magnitudes the same ratio as the first has to the last of the other.* (EUCL. V. 23).

A	B	C
A'	B'	C'

First let there be three magnitudes of one kind $A\ B\ C$ and other three magnitudes of one kind $A'\ B'\ C'$ such that $A : B = B' : C'$ and $B : C = A' : B'$.

If $n\ s$ be any whole numbers there are (art. 239) whole numbers $m\ r$ such that

$$A \text{ not} < \frac{m}{n}(B) \text{ but} < \frac{m+1}{n}(B) \qquad B \text{ not} < \frac{r}{s}(C) \text{ but} < \frac{r+1}{s}(C)$$

and then (art. 245)

$$B' \text{ not} < \frac{m}{n}(C') \text{ but} < \frac{m+1}{n}(C') \qquad A' \text{ not} < \frac{r}{s}(B') \text{ but} < \frac{r+1}{s}(B').$$

If "not<" mean "=" in all these statements (art. 106)

$$A = \frac{m}{n}\frac{r}{s}(C) \qquad A' = \frac{r}{s}\frac{m}{n}(C') \text{ and } \therefore \text{ (art. 108)} = \frac{m}{n}\frac{r}{s}(C').$$

Otherwise $\because$ (arts. 90, 91, 88, 89)

$$\frac{m}{n}(B) \text{ not} < \frac{m}{n}\left\{\frac{r}{s}(C)\right\} \text{ and } \frac{m+1}{n}(B) < \frac{m+1}{n}\left\{\frac{r+1}{s}(C)\right\}$$

$$\text{(arts. 9, 106) } A > \frac{m}{n}\frac{r}{s}(C) \text{ but} < \frac{m+1}{n}\frac{r+1}{s}(C).$$

And in like manner $A' > \frac{r}{s}\frac{m}{n}(C')$ but $< \frac{r+1}{s}\frac{m+1}{n}(C')$ and $\therefore$ (arts. 108, 9) $> \frac{m}{n}\frac{r}{s}(C')$ but $< \frac{m+1}{n}\frac{r+1}{s}(C')$. Moreover $\frac{m+1}{n}\frac{r+1}{s} - \frac{m}{n}\frac{r}{s} = \frac{m}{n}\left(\frac{r+1}{s}-\frac{r}{s}\right)+\frac{1}{n}\frac{r+1}{s} = \frac{1}{s}\frac{m}{n}+\frac{1}{n}\frac{r+1}{s}$. But (art. 241) if $\frac{\mu}{\nu}\ \frac{\mu+\theta}{\nu}$ be any given pair of fractions intercepting A relative to B n may be taken so great that both $\frac{m}{n}$ and $\frac{m+1}{n}$ is greater than $\frac{\mu}{\nu}$ and less than $\frac{\mu+\theta}{\nu}$ and if $\frac{\rho}{\sigma}\ \frac{\rho+\kappa}{\sigma}$ be any given pair of fractions intercepting B relative to C s may be taken so great that both $\frac{r}{s}$ and $\frac{r+1}{s} > \frac{\rho}{\sigma}$ and $< \frac{\rho+\kappa}{\sigma}$.

When then n s are taken thus great (art. 91) $\frac{1}{s}\frac{m}{n} < \frac{1}{s}\frac{\mu+\theta}{\nu}$ $\frac{1}{n}\frac{r+1}{s} < \frac{1}{n}\frac{\rho+\kappa}{\sigma}$ and $\therefore$ (art. 12) $\frac{1}{s}\frac{m}{n}+\frac{1}{n}\frac{r+1}{s} < \frac{1}{s}\frac{\mu+\theta}{\nu}+\frac{1}{n}\frac{\rho+\kappa}{\sigma}$. Now by taking n s great enough each of the fractions $\frac{1}{n}\frac{\rho+\kappa}{\sigma}$ $\frac{1}{s}\frac{\mu+\theta}{\nu}$ may (art. 231) be made less than $\frac{1}{2}k$ k being any given number and therefore (art. 12) $\frac{1}{s}\frac{\mu+\theta}{\nu}+\frac{1}{n}\frac{\rho+\kappa}{\sigma}$ may be made less than k. Much more therefore (art. 9) may $\frac{1}{s}\frac{m}{n}+\frac{1}{n}\frac{r+1}{s}$ or its equal $\frac{m+1}{n}\frac{r+1}{s}-\frac{mr}{ns}$ be made less than k. Hence also (art. 253) if A be still a fraction of C although A is not a fraction of B and B not a fraction of C A' is the same fraction of C'.

Wherefore universally if A be a fraction of C A' is the same fraction of C' and if A be greater than the one and less than the other of two fractions of C by less than however small a given fraction they may differ from one another A' is greater than the former and less than the latter of the same two fractions of C', that is (art. 245) $A : C = A' : C'$.

Next let there be four magnitudes A B C D and other four magnitudes A' B' C' D' such that $A : B = C' : D'$ $B : C = B' : C'$ and $C : D = A' : B'$. From the first two of these proportionalities there are three magnitudes A B C and other three magnitudes B' C' D' which taken two and two in cross order have severally the same ratios therefore by what has just been proved $A : C = B' : D'$. Then from this and the third given proportionality there are three magnitudes A C D and the three A' B' D' which taken two and two in cross order have severally the same ratios and therefore again $A : D = A' : D'$.

A	B	C	D
A'	B'	C'	D'

A	B	C
B'	C'	D'

A	C	D
A'	B'	D'

Again let there be five magnitudes A B C D E and other five A' B' C' D' E' such that $A : B = D' : E'$ $B : C = C' : D'$ $C : D = B' : C'$ and $D : E = A' : B'$. Then because there are four magnitudes A B C D and other four B' C' D' E' which taken two and two in cross order have severally the same ratios $A : D = B' : E'$ and because there are now three magnitudes

A	B	C	D	E
A'	B'	C'	D'	E'

A	B	C	D
B'	C'	D'	E'

A	D	E
A'	B'	E'

$A\ D\ E$ and other three $A'\ B'\ E'$ which taken two and two in cross order have severally the same ratios $A : E = A' : E'$. And so on for six or more magnitudes in each of the sets.

270. *Def.* When from the ratios $A:B\ \ B:C\ \ldots\ G:H$ of the first to the second of all the consecutive twos in order of magnitudes $A\ B\ C \ldots G\ H$ being severally the same as the ratios $G':H'\ \ F':G'\ \ldots\ A':B'$ of the first to the second of all the consecutive twos in backward order of as many magnitudes $A'\ B'\ C' \ldots G'\ H'$ the ratio $A:H$ of the first to the last of the former magnitudes is inferred (art. 269) to be the same as the ratio $A':H'$ of the first to the last of the latter this proportionality is said to arise BY EQUALITY OF RATIO IN CROSSORDERLY PROPORTIONALITIES.

The proportionalities of the extreme magnitudes by equality of ratio in orderly and in crossorderly proportionalities are got by striking out the mean magnitudes in the following contrasted ways:—

$A : B = A' : B'$	$A : B = G' : H'$
$B : C = B' : C'$	$B : C = F' : G'$
$C : D = C' : D'$	$C : D = E' : F'$
- - - -	- - - -
- - - -	- - - -
$G : H = G' : H'.$	$G : H = A' : B'.$

271. Since (art. 267) the first has to the last of more than two magnitudes of one kind the same ratio as the first to the last of as many other magnitudes of one kind which taken two and two in order have severally the same ratios and (art. 254) ratios are the same as one another which are the same as the same ratio there arises the following

Def. The ratio of the first to the last of more than two magnitudes of the same kind is said to be COMPOUNDED of ratios severally the same as the ratios of the first to the second of all the consecutive twos in order of the magnitudes. And the ratio compounded of the ratios $P_1 : Q_1$ $P_2 : Q_2 \ldots P_n : Q_n$ is symbolized by $\begin{bmatrix} P_1 : Q_1 \\ P_2 : Q_2 \\ - - - \\ P_n : Q_n \end{bmatrix}$.

Thus if $X_1\ X_2\ X_3 \ldots X_n\ X_{n+1}$ be more than two magnitudes of one kind such that $X_1 : X_2 = P_1 : Q_1$ $X_2 : X_3 = P_2 : Q_2$ - - - $X_n : X_{n+1} = P_n : Q_n$ and Y_1 Y_2 Y_3 ... Y_n Y_{n+1} be as many other magnitudes of one kind such that $Y_1 : Y_2 = P_1 : Q_1$ $Y_2 : Y_3 = P_2 : Q_2$ - - - $Y_n : Y_{n+1} = P_n : Q_n$ then $X_1 : X_2 = Y_1 : Y_2$ $X_2 : X_3 = Y_2 : Y_3$ - - - $X_n : X_{n+1} = Y_n : Y_{n+1}$ and $\begin{bmatrix} P_1 : Q_1 \\ P_2 : Q_2 \\ - - - \\ P_n : Q_n \end{bmatrix}$ stands indifferently for $X_1 : X_{n+1}$ and for $Y_1 : Y_{n+1}$. If too $P_1 : Q_1 = R_1 : S_1$ $P_2 : Q_2 = R_2 : S_2$ - - - $P_n : Q_n = R_n : S_n$ then

$$X_1 : X_{n+1} = \begin{bmatrix} X_1 : X_2 \\ X_2 : X_3 \\ - - - \\ X_n : X_{n+1} \end{bmatrix} = \begin{bmatrix} P_1 : Q_1 \\ P_2 : Q_2 \\ - - - \\ P_n : Q_n \end{bmatrix} = \begin{bmatrix} R_1 : S_1 \\ R_2 : S_2 \\ - - - \\ R_n : S_n \end{bmatrix} = \begin{bmatrix} Y_1 : Y_2 \\ Y_2 : Y_3 \\ - - - \\ Y_n : Y_{n+1} \end{bmatrix} = Y_1 : Y_{n+1}.$$

Again if $A : B = P_1 : Q_1 = B' : C'$ and $B : C = P_2 : Q_2 = A' : B'$ (art. 269) $A : C = A' : C'$, that is

$$\begin{pmatrix} P_1 : Q_1 \\ P_2 : Q_2 \end{pmatrix} = \begin{pmatrix} P_2 : Q_2 \\ P_1 : Q_1 \end{pmatrix}.$$

And $\because$ $X_1 : X_{n+1} = \begin{bmatrix} X_1 : X_2 \\ X_2 : X_3 \\ - - - \\ X_r : X_{r+1} \\ X_{r+1} : X_{n+1} \end{bmatrix} = \begin{bmatrix} X_1 : X_{r+1} \\ X_{r+1} : X_{r+2} \\ X_{r+2} : X_{r+3} \\ - - - \\ X_n : X_{n+1} \end{bmatrix} = \begin{pmatrix} X_1 : X_{r+1} \\ X_{r+1} : X_{n+1} \end{pmatrix}$

$$= \begin{bmatrix} \begin{bmatrix} X_1 : X_2 \\ X_2 : X_3 \\ - - - \\ X_r : X_{r+1} \end{bmatrix} \\ X_{r+1} : X_{n+1} \end{bmatrix} = \begin{bmatrix} X_1 : X_{r+1} \\ \begin{bmatrix} X_{r+1} : X_{r+2} \\ X_{r+2} : X_{r+3} \\ - - - \\ X_n : X_{n+1} \end{bmatrix} \end{bmatrix}$$ it follows that

$$\begin{bmatrix} P_1 : Q_1 \\ P_2 : Q_2 \\ - - - \\ P_n : Q_n \end{bmatrix} = \begin{bmatrix} P_1 : Q_1 \\ - - - \\ P_r : Q_r \\ \begin{bmatrix} P_{r+1} : Q_{r+1} \\ - - - \\ P_n : Q_n \end{bmatrix} \end{bmatrix} = \begin{bmatrix} \begin{bmatrix} P_1 : Q_1 \\ - - - \\ P_r : Q_r \end{bmatrix} \\ P_{r+1} : Q_{r+1} \\ - - - \\ P_n : Q_n \end{bmatrix} = \begin{bmatrix} \begin{bmatrix} P_1 : Q_1 \\ - - - \\ P_r : Q_r \end{bmatrix} \\ \begin{bmatrix} P_{r+1} : Q_{r+1} \\ - - - \\ P_n : Q_n \end{bmatrix} \end{bmatrix}.$$

Hence $\begin{bmatrix} P_1 : Q_1 \\ P_2 : Q_2 \\ P_3 : Q_3 \end{bmatrix} = \begin{bmatrix} P_1 : Q_1 \\ \begin{bmatrix} P_2 : Q_2 \\ P_3 : Q_3 \end{bmatrix} \end{bmatrix} = \begin{bmatrix} P_1 : Q_1 \\ \begin{bmatrix} P_3 : Q_3 \\ P_2 : Q_2 \end{bmatrix} \end{bmatrix} = \begin{bmatrix} P_1 : Q_1 \\ P_3 : Q_3 \\ P_2 : Q_2 \end{bmatrix}$ and so too

$$\begin{bmatrix} P_2 : Q_2 \\ P_1 : Q_1 \\ P_3 : Q_3 \end{bmatrix} = \begin{bmatrix} P_2 : Q_2 \\ P_3 : Q_3 \\ P_1 : Q_1 \end{bmatrix} \quad \begin{bmatrix} P_3 : Q_3 \\ P_1 : Q_1 \\ P_2 : Q_2 \end{bmatrix} = \begin{bmatrix} P_3 : Q_3 \\ P_2 : Q_2 \\ P_1 : Q_1 \end{bmatrix}.$$

But $\left\{\begin{matrix} P_1 : Q_1 \\ \left\{\begin{matrix} P_2 : Q_2 \\ P_3 : Q_3 \end{matrix}\right\} \end{matrix}\right\} = \left\{\begin{matrix} \left\{\begin{matrix} P_2 : Q_2 \\ P_3 : Q_3 \end{matrix}\right\} \\ P_1 : Q_1 \end{matrix}\right\} = \left\{\begin{matrix} P_2 : Q_2 \\ P_3 : Q_3 \\ P_1 : Q_1 \end{matrix}\right\}$ and $\left\{\begin{matrix} P_1 : Q_1 \\ \left\{\begin{matrix} P_3 : Q_3 \\ P_2 : Q_2 \end{matrix}\right\} \end{matrix}\right\}$ likewise $= \left\{\begin{matrix} P_3 : Q_3 \\ P_2 : Q_2 \\ P_1 : Q_1 \end{matrix}\right\}$. Therefore (art. 254) ratios compounded of the same three ratios are all the same as one another in whichever of their six orders of succession the three are taken. Next for ratios compounded of the same four ratios it hence follows by taking the four ratios in any order and compounding the last three that the compound ratios are the same as one another in each set of six which have the same component ratio first in order then by interchanging the first component and the compound of the other three with each of the three made first in turn that four compound ratios taken one from each set of six are the same as one another and that therefore all the twenty-four compound ratios differing in the orders of their component ratios are the same as one another. And so on for ratios compounded of the same five or more ratios. Therefore generally ratios compounded of the same ratios are the same as one another in whatever order these ratios are taken.

If further the terms of the component ratios be all magnitudes of the same kind

$$\left\{\begin{matrix} P_1 : Q_1 \\ P_2 : Q_2 \\ P_3 : Q_3 \\ P_4 : Q_4 \end{matrix}\right\} = \left\{\begin{matrix} \left\{\begin{matrix} P_1 : Q_4 \\ Q_4 : Q_1 \end{matrix}\right\} \\ \left\{\begin{matrix} P_2 : Q_3 \\ Q_3 : Q_2 \end{matrix}\right\} \\ \left\{\begin{matrix} P_3 : Q_1 \\ Q_1 : Q_3 \end{matrix}\right\} \\ \left\{\begin{matrix} P_4 : Q_2 \\ Q_2 : Q_4 \end{matrix}\right\} \end{matrix}\right\} = \left\{\begin{matrix} P_1 : Q_4 \\ Q_4 : Q_1 \\ P_2 : Q_3 \\ Q_3 : Q_2 \\ P_3 : Q_1 \\ Q_1 : Q_3 \\ P_4 : Q_2 \\ Q_2 : Q_4 \end{matrix}\right\} = \left\{\begin{matrix} P_1 : Q_4 \\ P_2 : Q_3 \\ P_3 : Q_1 \\ P_4 : Q_2 \\ Q_4 : Q_1 \\ Q_1 : Q_3 \\ Q_3 : Q_2 \\ Q_2 : Q_4 \end{matrix}\right\} = \left\{\begin{matrix} P_1 : Q_4 \\ P_2 : Q_3 \\ P_3 : Q_1 \\ P_4 : Q_2 \\ \left\{\begin{matrix} Q_4 : Q_1 \\ Q_1 : Q_3 \\ Q_3 : Q_2 \\ Q_2 : Q_4 \end{matrix}\right\} \end{matrix}\right\} = \left\{\begin{matrix} P_1 : Q_4 \\ P_2 : Q_3 \\ P_3 : Q_1 \\ P_4 : Q_2 \end{matrix}\right\}.$$

And in this way may it be shown that a ratio compounded of any ratios having terms all of the same kind is the same as a ratio compounded of ratios having the same antecedents in the same order and the same consequents in any order whatever. Hence and from what is shown above a ratio compounded of ratios with all their terms of one kind is the same as a ratio compounded of other ratios having the same antecedents and the same consequents anyhow taken.

272. *Def.* The ratio compounded of ratios that are all the same as one another is called the MULTIPLICATE COM-

POUND RATIO of any one of these ratios of the DEGREE OF COMPOUNDNESS marked by the number of the ratios. Particularly ratios compounded of two of three of four . . . ratios all the same as $P:Q$ are called severally the DUPLICATE the TRIPLICATE the QUADRUPLICATE . . . compound ratio of $P:Q$. The ratio compounded of n ratios each the same as $P:Q$ is symbolized by $P \overset{n}{:} Q$.

Thus of magnitudes enough in a continued proportionality the first has to the third of any consecutive three the duplicate compound ratio of any one to the next following the first has to the fourth of any consecutive four the triplicate compound ratio of the same and so on.

273. PROP. *If two magnitudes have to a single magnitude the same ratios severally as another single magnitude has to other two magnitudes the first has to the second of the first two magnitudes the same ratio as the second has to the first of the other two.*

Let two magnitudes A B have to a magnitude H the same ratios severally as a magnitude K has to two magnitudes C D, that is let $A:H=K:C$ $B:H=K:D$.

Because $B:H=K:D$ inversely (art. 259) $H:B=D:K$. There are then three magnitudes A H B and other three D K C which taken two and two in cross order have severally the same ratios therefore by equality of ratio in crossorderly proportionalities (art. 269) $A:B=D:C$.

274. PROP. *If of four magnitudes of the same kind the first be to the second as the third is to the fourth the first is to the third as the second to the fourth.* (EUCL. V. 16).

Let A B C D be four magnitudes of the same kind such that $A:B=C:D$. Then since (art. 245) $B:C=B:C$ the three magnitudes A B C and the three B C D taken two and two in cross order have severally the same ratios and therefore (art. 269) $A:C=B:D$.

275. *Def.* Four magnitudes are said to be taken ALTERNATELY when the second and third change places.

Thus art. 274's proposition is that four proportionate magnitudes of the same kind are proportionate alternately.

If A B C D be four such magnitudes of one kind that $A:B=C:D \because A:C=B:D$

$$A:D = \begin{pmatrix} A:B \\ B:D \end{pmatrix} = \begin{pmatrix} A:B \\ A:C \end{pmatrix} \text{ and also} = \begin{pmatrix} C:D \\ B:D \end{pmatrix} = \begin{pmatrix} B:D \\ C:D \end{pmatrix},$$

that is the first of four proportionates of the same kind has to the fourth the ratio compounded of the ratios of the first to the second and to the third which is also the same as the ratio compounded of the ratios of the second and third to the fourth.

276. PROP. *The first and third of four proportionate magnitudes of the same kind are in the same order of greatness as the second and fourth.* (EUCL. V. 14).

For if $A:B=C:D$ where $A\ B\ C\ D$ are magnitudes of one kind alternately (art. 274) $A:C=B:D$. Therefore (art. 251) any multiples whatever of A and C are in the same order of greatness as the like multiples severally of B and D and in particular A and C are in the same order of greatness as B and D.

277. PROP. *If magnitudes have to a single magnitude the same ratios severally as as many other magnitudes have to another single magnitude all the first magnitudes together have to the first single magnitude the same ratio as all the other magnitudes together have to the other single magnitude.* (EUCL. V. 24).

First let there be two magnitudes A_1 A_2 and a magnitude A and other two magnitudes B_1 B_2 and another magnitude B such that $A_1:A=B_1:B$ and $A_2:A=B_2:B$.

If n be any whole number there are (art. 239) whole numbers m_1 m_2 such that

A_1 not$<\frac{m_1}{n}(A)$ and $<\frac{m_1+1}{n}(A)$ $\qquad$ A_2 not$<\frac{m_2}{n}(A)$ and $<\frac{m_2+1}{n}(A)$

$\therefore$ (art. 245)

B_1 not$<\frac{m_1}{n}(B)$ and $<\frac{m_1+1}{n}(B)$ $\qquad$ B_2 not$<\frac{m_2}{n}(B)$ and $<\frac{m_2+1}{n}(B)$.

$$\therefore (A_1, A_2) \text{ not}< \frac{m_1+m_2}{n}(A) \text{ and } < \frac{m_1+m_2+2}{n}(A)$$

$$(B_1, B_2) \text{ not}< \frac{m_1+m_2}{n}(B) \text{ and } < \frac{m_1+m_2+2}{n}(B).$$

And $\frac{m_1+m_2+2}{n}-\frac{m_1+m_2}{n}=\frac{2}{n}$ which may be made less than any given fraction k by taking (art. 231) n so great that $\frac{1}{n}<\frac{1}{2}k$. Hence and by art. 253 if (A_1, A_2) be a fraction of A (B_1, B_2) is the same fraction of B or if (A_1, A_2) be greater than the one and less than the other of two fractions of A by less than however small a given fraction they may differ from one another (B_1, B_2) is greater than the former and less than the latter of the same two fractions of B, that is $(A_1, A_2) : A = (B_1, B_2) : B$.

Next let there be more than two magnitudes $A_1\ A_2\ A_3 \dots A_n$ as many others $B_1\ B_2\ B_3 \dots B_n$ and two magnitudes $A\ B$ such that $A_1 : A = B_1 : B \quad A_2 : A = B_2 : B \quad A_3 : A = B_3 : B \dots A_n : A = B_n : B$. Then since by the first case above $(A_1, A_2) : A = (B_1, B_2) : B$ and by hypothesis $A_3 : A = B_3 : B$ it follows by the first case that $(A_1, A_2, A_3) : A = (B_1, B_2, B_3) : B$. Then in the same way because of this and the fourth given proportionality $(A_1, A_2, A_3, A_4) : A = (B_1, B_2, B_3, B_4) : B$. And so on until at last

$$(A_1, A_2, A_3, \dots A_n) : A = (B_1, B_2, B_3, \dots B_n) : B.$$

278. Prop. *If two magnitudes be proportionate to two magnitudes any multiples whatever of the first two are proportionate to the like multiples severally of the other two.* (Eucl. v. 4).

Let two magnitudes $A\ B$ be proportionate to two $C\ D$, that is $A : B = C : D$, and let $p\ q$ be any two whole numbers. Because p magnitudes each equal to A are severally to B as p magnitudes each equal to C are to D (art. 277) $p(A) : B = p(C) : D$. Likewise because (art. 245) q magnitudes each equal to B are severally to B as q magnitudes each equal to D to D $q(B) : B = q(D) : D$. Therefore (art. 256) $p(A) : q(B) = p(C) : q(D)$.

279. Prop. *If the first of four magnitudes be to the second as the third is to the fourth the first and second together are to the second as the third and fourth together to the fourth.* (Eucl. v. 18).

Let $A : B = C : D$. Then since also (art. 245) $B : B = D : D$ (art. 277) $(A, B) : B = (C, D) : D$.

280. *Def.* Four magnitudes are said to be taken Jointly when the first and second together are put instead of the first and the third and fourth together instead of the third.

Art. 279's proposition then is—Four proportionate magnitudes are proportionate jointly.

281. Prop. *Magnitudes of one kind that have severally to as many other magnitudes the same ratio have together to those other together that same ratio.* (Eucl. v. 12).

Let $A_1 A_2 A_3 \ldots A_n$ be magnitudes of one kind and $B_1 B_2 B_3 \ldots B_n$ as many other such that

$$A_1 : B_1 = A_2 : B_2 = A_3 : B_3 = \text{- - - -} = A_n : B_n.$$

If $A : B$ be any one of these ratios $\because A_1 : B_1 = A : B$ alternately $A_1 : A = B_1 : B$ and in like way $A_2 : A = B_2 : B$ $A_3 : A = B_3 : B$ - - - $A_n : A = B_n : B$. Since then $A_1\ A_2 \ldots A_n$ have to A the same ratios severally as $B_1\ B_2 \ldots B_n$ have to B (art. 277) $(A_1, A_2, \ldots A_n) : A = (B_1, B_2, \ldots B_n) : B$. Therefore alternately

$$(A_1, A_2, A_3, \ldots A_n) : (B_1, B_2, B_3, \ldots B_n) = A : B$$
$$= A_1 : B_1 = A_2 : B_2 = \text{- - -} = A_n : B_n.$$

282. COR. *Two magnitudes of one kind are proportionate to any equimultiples of them.* (EUCL. V. 15).

If in art. 281 $A_1 = A_2 = \text{- - - -} = A_n$ and each $= A$ and $\therefore$ (art. 276) $B_1 = B_2 = \text{- - - -} = B_n = B$ the proposition becomes in particular $n(A) : n(B) = A : B$.

283. *Def.* When from A the greater of two unequal magnitudes $A\ B$ a part equal to B the less is taken the remaining part is symbolized by $A|B$.

284. PROP. *If two unequal magnitudes have to a single magnitude the same ratios severally as other two magnitudes have to another single magnitude the difference of the first two magnitudes has to the first single magnitude the same ratio as the difference of the other two magnitudes has to the other single magnitude.*

Let A be the greater of two unequal magnitudes $A\ B$ which have to a magnitude H the same ratios severally as other two magnitudes $C\ D$ have to a magnitude K, that is so that $A : H = C : K$ and $B : H = D : K$.

Since $A\ B\ H$ are magnitudes of the same kind and A is greater than B (art. 242) a whole number r may be taken so great that there are whole numbers $m\ n$ such that not only

$$A \text{ not} < \frac{m}{r}(H) \text{ and } < \frac{m+1}{r}(H) \qquad B \text{ not} < \frac{n}{r}(H) \text{ and } < \frac{n+1}{r}(H)$$

but also each of the fractions $\frac{m}{r}\ \frac{m+1}{r}$ is greater than each of the fractions $\frac{n}{r}\ \frac{n+1}{r}$. Then (art. 245)

$$C \text{ not} < \frac{m}{r}(K) \text{ and } < \frac{m+1}{r}(K) \qquad D \text{ not} < \frac{n}{r}(K) \text{ and } < \frac{n+1}{r}(K).$$

Should "not<" mean "=" in all these statements $A|B=\frac{m-n}{r}(H)$ and $C|D=\frac{m-n}{r}(K)$.

But whether this be so or not (arts. 14, 15, 16) $\because \frac{m}{r}-\frac{n+1}{r}=\frac{m-n-1}{r}$ and $\frac{m+1}{r}-\frac{n}{r}=\frac{m-n+1}{r}$

$$A|B>\frac{m-n-1}{r}(H) \text{ and } <\frac{m-n+1}{r}(H)$$

$$C|D>\frac{m-n-1}{r}(K) \text{ and } <\frac{m-n+1}{r}(K).$$

And $\frac{m-n+1}{r}-\frac{m-n-1}{r}=\frac{2}{r}$ which can be made less than any given fraction κ however small by simply taking (art. 231) r so great that $\frac{1}{r}<\frac{1}{2}\kappa$. Wherefore and by art. 253 if $A|B$ be a fraction of H $C|D$ is the same fraction of K and if $A|B$ be greater than the one and less than the other of two fractions of H by less than however small a given fraction the two may differ from one another $C|D$ is greater than the former and less than the latter of the same two fractions of K, that is

$$A|B:H=C|D:K.$$

285. PROP. *If the first of four magnitudes be to the second as the third is to the fourth and the first be greater than the second the difference between the first and second is to the second as the difference between the third and fourth to the fourth.* (EUCL. V. 17).

Let $A:B=C:D$ and let $A>B$. Then (art. 251) $C>D$ and $\because$ (art. 245) $B:B=D:D$ the two unequals A B have to B the same ratios severally as C D have to D. Therefore (art. 284)

$$A|B:B=C|D:D.$$

286. *Def.* Four magnitudes of which the first is greater than the second and the third than the fourth are said to be taken DISJOINTLY when the differences of the first and second and of the third and fourth are put severally for the first and the third and REVERSELY DISJOINTLY when those differences are put severally for the second and the fourth.

Art. 285's proposition then is that Four proportionates are proportionate disjointly and art. 287's below that Four proportionates are reversely disjointly proportionate.

287. PROP. *If the first of four magnitudes be to the second as the third is to the fourth and the first be greater than the second the first is to the difference between the first and second as the third to the difference between the third and fourth.*

Let $A:B=C:D$ and $A>B$ so that (art. 251) $C>D$. Since then (art. 285) $A|B:B=C|D:D$ A and $A|B$ have to B the same ratios severally as C and $C|D$ have to D and therefore (art. 256)

$$A:A|B=C:C|D.$$

288. PROP. *Two whole magnitudes proportionate to parts of themselves are proportionate to their remaining parts.* (EUCL. V. 19).

Let A B be two whole magnitudes proportionate to parts A' B' of themselves to wit $A:B=A':B'$. Because $A:B=A':B'$ alternately $A:A'=B:B'$ therefore reversely disjointly (art. 287) $A:A|A'=B:B|B'$ and therefore alternately $A:B=A|A':B|B'$.

289. PROP. *The greatest and least of four proportionates of the same kind are together greater than the other two together.* (EUCL. V. 25).

Let A be the greatest of four magnitudes A B C D of one kind such that $A:B=C:D$. Because $A>B$ (art. 251) $C>D$ and because $A>C$ (art. 276) $B>D$. Wherefore and because by supposition $A>D$ D is the least of the four. Reversely disjointly then $A:A|B=C:C|D$ and hence $\because$ $A>C$ $A|B>C|D$. To each of these unequals put (B, D) and (art. 11) $(A, D)>(B, C)$.

290. *Def.* The ratio of a magnitude to a magnitude of the same kind is determined and may therefore be represented either by the number which expresses what multiple or fraction the first magnitude is of the other or by that which although not a number is yet so closely akin to a number that it can be known only as greater than the less and less than the greater of two numbers differing from one another by less than any given number however small which express what fractions magnitudes severally less and greater than the first magnitude are of the other.

A magnitude estimated numerically in reference to a magnitude of the same kind as unit is called a QUANTITY.

The numerical representative of the ratio of a magnitude to a magnitude of the same kind being the very same as the numerical expression of the former magnitude in reference to the latter as unit is called a NUMERICAL QUANTITY. A numerical quantity then can only be said to be either a number or a something akin to a number from which there are numbers that differ by less than any assignable number. A numerical quantity is called COMMENSURABLE if a number and INCOMMENSURABLE if not.

291. PROP. *If each of several magnitudes can be expressed numerically to any required degree of nearness in reference to a common unit magnitude a magnitude equal to the magnitudes together may be expressed numerically to any required degree of nearness in reference to the same unit magnitude.*

Let a magnitude in reference to a unit magnitude

$$\text{not}< a-\kappa \quad \text{and} \quad \text{not}> a+\kappa$$

and let another magnitude in reference to the same unit magnitude

$$\text{not}< a'-\kappa' \quad \text{and} \quad \text{not}> a'+\kappa'$$

where a a' κ κ' are numbers such that κ κ' may each be taken less than any given number however small. By using "not<" "not>" instead of ">" "<" severally the cases are included of either of the magnitudes being expressible by a number, that is of either of the numbers κ κ' being 0. Then (art. 101) a magnitude equal to the two magnitudes together in reference to the unit magnitude

$$\text{not}< a-\kappa+(a'-\kappa') \quad \text{and} \quad \text{not}> a+\kappa+(a'+\kappa')$$

$$\text{and} \therefore \ \text{not}< a+a'-(\kappa+\kappa') \quad \text{and} \quad \text{not}> a+a'+(\kappa+\kappa').$$

Further by taking κ κ' each less than half of any given number however small $\kappa+\kappa'$ is made less than that number. Hence the magnitude equal to the two magnitudes together may thus be expressed numerically in reference to the unit magnitude to any required degree of nearness.

If a third magnitude be expressible numerically to any required degree of nearness in reference to the unit magnitude then in the same way by putting the magnitude equal to the two magnitudes together instead of the first of the two and the third magnitude instead of the other a magnitude equal to all three magnitudes together may be expressed numerically to any required degree of nearness in reference to the unit magnitude. And so on for a fourth magnitude a fifth or any other.

Hence too if in reference to a common unit magnitude

a 1st	magnitude not<	$a_1-\kappa_1$	and not>	$a_1+\kappa_1$
- 2nd	- - -	$a_2-\kappa_2$	- -	$a_2+\kappa_2$
- 3rd	- - -	$a_3-\kappa_3$	- -	$a_3+\kappa_3$
-	- - -	-	- -	-
- nth	- - -	$a_n-\kappa_n$	- -	$a_n+\kappa_n$

then a magnitude equal to all these n magnitudes together in reference to the same unit magnitude

$$\text{not}< a_1+a_2+\cdots+a_n-(\kappa_1+\kappa_2+\cdots+\kappa_n)$$
$$\text{and not}> a_1+a_2+\cdots+a_n+(\kappa_1+\kappa_2+\cdots+\kappa_n)$$

or as it may be written for shortness not$<\Sigma(a)-\Sigma(\kappa)$ and not$>\Sigma(a)+\Sigma(\kappa)$ taking $\Sigma(a)$ to symbolize the sum of the as and $\Sigma(\kappa)$ the sum of the κs.

292. *Def.* The numerical quantity which (art.291) can be found to any required degree of nearness expressing in reference to a unit magnitude a magnitude equal to several magnitudes together when the numerical quantities severally expressing these in reference to that unit magnitude may each be found to any required degree of nearness is called the SUM of the numerical quantities. The operation by which by adding to one of two numbers the other near enough severally to numerical quantities a b expressing two magnitudes in reference to a common unit magnitude a number is found as near as may be required to the numerical quantity expressing in reference to the same unit magnitude a magnitude equal to the two magnitudes together is called the ADDITION TO THE FIRST NUMERICAL QUANTITY a OF THE OTHER b and the result of the operation which is at once the sum got by this addition and the numerical quantity expressing in reference to the unit magnitude the magnitude equal to the two

magnitudes together is symbolized by $a+b$. The sum got by adding in like manner to $a+b$ a numerical quantity c is symbolized by $a+b+c$ and so on.

293. PROP. *If each of two unequal magnitudes can be expressed numerically to any required degree of nearness in reference to a common unit magnitude their difference may be expressed numerically to any required degree of nearness in reference to that unit magnitude.*

In reference to one unit magnitude let the greater of two unequal magnitudes be not less than $\alpha-\kappa$ and not greater than $\alpha+\kappa$ and the less not less than $\alpha'-\kappa'$ and not greater than $\alpha'+\kappa'$ where $\alpha\ \alpha'\ \kappa\ \kappa'$ are numbers of which both κ and κ' may be less than any given number however small. Then (art. 242) $\kappa\ \kappa'$ may be each taken so small that each of the numbers $\alpha-\kappa\ \ \alpha+\kappa$ is greater than each of the numbers $\alpha'-\kappa'\ \ \alpha'+\kappa'$. Hence (art. 103) taking $\kappa\ \kappa'$ each thus small the difference of the unequal magnitudes in reference to the unit magnitude

$$\text{not}<\alpha-\kappa-(\alpha'+\kappa') \quad \text{and} \quad \text{not}>\alpha+\kappa-(\alpha'-\kappa')$$

$$\text{and}\ \therefore\ \text{not}<\alpha-\alpha'-(\kappa+\kappa') \quad \text{and} \quad \text{not}>\alpha-\alpha'+(\kappa+\kappa').$$

And $\kappa+\kappa'$ may be made less than any given number k however small by taking $\kappa\ \kappa'$ each less than $\frac{1}{2}k$. The difference of the unequal magnitudes therefore may in this way be expressed numerically to any sought degree of nearness in reference to the unit magnitude.

294. *Def.* The operation by which when numerical quantities $a\ b$ expressing two unequal magnitudes in reference to a unit magnitude can be found near enough the numerical quantity may be found to any required degree of nearness (art. 293) expressing the difference of the unequal magnitudes in reference to the unit magnitude is called the SUBTRACTION FROM THE NUMERICAL QUANTITY a EXPRESSING THE GREATER UNEQUAL OF THE NUMERICAL QUANTITY b EXPRESSING THE LESS. The two numerical quantities are called UNEQUAL and the former is called the GREATER and the latter the LESS. The result of the operation is called the REMAINDER got by this subtraction and is symbolized by $a-b$. The remainder

got by subtracting in like manner from $a-b$ a not greater numerical quantity c is symbolized by $a-b-c$ and so on.

295. PROP. *If every one but the last of more than two magnitudes of the same kind can be expressed numerically to any required degree of nearness in reference to the next following one as unit the first of the magnitudes may be expressed numerically to any required degree of nearness in reference to the last of them as unit.*

First let there be three magnitudes A_3 A_2 A_1 of the same kind such that

A_3 in reference to A_2 as unit not$<\alpha-\kappa$ and not$>\alpha+\kappa$

A_2 in reference to A_1 as unit not$<\alpha'-\kappa'$ and not$>\alpha'+\kappa'$

α α' κ κ' being such numbers that κ κ' may be each less than any given number however small. Then (art. 106) A_3 in reference to A_1 as unit not$<(\alpha-\kappa)(\alpha'-\kappa')$ and not$>(\alpha+\kappa)(\alpha'+\kappa')$

and $\therefore$ not$<\alpha\alpha'-\{\kappa(\alpha'-\kappa')+\kappa'\alpha\}$ and not$>\alpha\alpha'+\{\kappa(\alpha'+\kappa')+\kappa'\alpha\}$.

But if β be the greater and γ the less of any given pair of fractions intercepting A_3 relative to A_2 (art. 241) κ may be taken so small that each of the numbers $\alpha-\kappa$ $\alpha+\kappa$ is greater than γ and less than β also if β' be the greater and γ' the less of any given pair of fractions intercepting A_2 relative to A_1 κ' may be taken so small that each of the numbers $\alpha'-\kappa'$ $\alpha'+\kappa'$ is greater than γ' and less than β'. Let then κ κ' be taken each thus small and $\kappa(\alpha'-\kappa')$ $\kappa(\alpha'+\kappa')$ are each less than $\kappa\beta'$ and $\kappa'\alpha$ is less than $\kappa'\beta$ and therefore each of the numbers $\kappa(\alpha'-\kappa')+\kappa'\alpha$ $\kappa(\alpha'+\kappa')+\kappa'\alpha$ is less than $\kappa\beta'+\kappa'\beta$. Now $\kappa\beta'+\kappa'\beta$ may be made less than any given number x however small by taking κ less than $\frac{\frac{1}{2}x}{\beta'}$ and κ' less than $\frac{\frac{1}{2}x}{\beta}$. Much more therefore may $\kappa(\alpha'-\kappa')+\kappa'\alpha$ $\kappa(\alpha'+\kappa')+\kappa'\alpha$ be made each less than any given number. And hence A_3 may thus be expressed numerically to any required degree of nearness in reference to A_1 as unit.

Again if another magnitude A_4 can be expressed numerically to any required degree of nearness in reference to A_3 as unit since A_3 can be expressed numerically to any required degree of nearness in reference to A_1 as unit it follows by what has been shown that A_4 can be expressed numerically to any required degree of nearness in reference to A_1 as unit. And so on for five or more magnitudes.

296. *Def.* The operation by which when numerical quantities b a can be found to any required degree of nearness

expressing severally the first of three magnitudes in reference to the second as unit and the second in reference to the third as unit the numerical quantity may be found to any required degree of nearness (art. 295) expressing the first of the three in reference to the third as unit is called the MULTIPLICATION BY THE FIRST NUMERICAL QUANTITY *b* OF THE OTHER *a*. The result of the operation to wit the numerical quantity expressing the first magnitude in reference to the third as unit is called the PRODUCT of this multiplication and is symbolized by *ba*. The product made by multiplying in the like way *ba* by a numerical quantity *c* is symbolized by *cba* and so on.

The ratio compounded of ratios represented severally by numerical quantities $h\ g\ f \ldots c\ b\ a$ is represented by the product $hgf \ldots cba$.

If a be the numerical expression of any one of many enough continuedly proportionate magnitudes in reference to the next following one as unit a^n is the numerical expression of the first in reference to the last as unit of any consecutive $n+1$. And if the ratio of each to the next following of these continued proportionates be the same as $P:Q$ the numerical representative of $P \overset{n}{:} Q$ is a^n.

297. PROP. *If each of two magnitudes can be expressed numerically to any required degree of nearness in reference to a third magnitude as unit either of the two may be expressed numerically to any required degree of nearness in reference to the other as unit.*

Let one of two magnitudes in reference to a third magnitude as unit not< $\alpha-\kappa$ and not> $\alpha+\kappa$ and the other not< $\alpha'-\kappa'$ and not> $\alpha'+\kappa'$ where $\alpha\ \alpha'\ \kappa\ \kappa'$ are numbers such that $\kappa\ \kappa'$ may be each less than any given number however small. Then the first in reference to the other as unit of the two magnitudes (art. 114) as in art. 256

$$\text{not}< \frac{\alpha-\kappa}{\alpha'+\kappa'} \text{ and not}> \frac{\alpha+\kappa}{\alpha'-\kappa'}$$

$$\text{and } \therefore \text{ not}< \frac{\alpha}{\alpha'} - \left\{\kappa \frac{1}{\alpha'+\kappa'} + \kappa' \frac{\alpha}{\alpha'(\alpha'+\kappa')}\right\}$$

$$\text{and not}> \frac{\alpha}{\alpha'} + \left\{\kappa \frac{1}{\alpha'-\kappa'} + \kappa' \frac{\alpha}{\alpha'(\alpha'-\kappa')}\right\}.$$

If now β be the greater and γ the less of the fractions of any given pair intercepting the first magnitude and β' the greater and γ' the less of the fractions of any given pair intercepting the second each relative to the third magnitude (art. 241) κ may be taken so small that both $\alpha-\kappa$ and $\alpha+\kappa>\gamma$ and $<\beta$ and κ' so small that both $\alpha'-\kappa'$ and $\alpha'+\kappa'>\gamma'$ and $<\beta'$. Let then κ κ' be taken each thus small and $\because$ $\alpha'-\kappa'>\gamma'$ (art. 257) $\frac{1}{\alpha'-\kappa'}<\frac{1}{\gamma'}$ and $\therefore$ (arts. 91, 89) $\kappa\frac{1}{\alpha'-\kappa'}<\kappa\frac{1}{\gamma'}$. Again $\because$ $\alpha'-\kappa'$ and much more $\therefore$ $\alpha'>\gamma'$ (art. 258) $\alpha'(\alpha'-\kappa')>\gamma'^2$ and $\therefore$ as before $\frac{1}{\alpha'(\alpha'-\kappa')}<\frac{1}{\gamma'^2}$. Wherefore $\alpha\frac{1}{\alpha'(\alpha'-\kappa')}<\alpha\frac{1}{\gamma'^2}$ and $\therefore$ much more (arts. 113, 9) $<\beta\frac{1}{\gamma'^2}$ $\because$ $\beta>\alpha+\kappa$ and much more $\therefore$ $>\alpha$. Hence (art. 9) $\frac{\alpha}{\alpha'(\alpha'-\kappa')}<\frac{\beta}{\gamma'^2}$ and $\therefore$ as before $\kappa'\frac{\alpha}{\alpha'(\alpha'-\kappa')}<\kappa'\frac{\beta}{\gamma'^2}$. On the whole $\therefore$ (art. 12) $\kappa\frac{1}{\alpha'-\kappa'}+\kappa'\frac{\alpha}{\alpha'(\alpha'-\kappa')}<\kappa\frac{1}{\gamma'}+\kappa'\frac{\beta}{\gamma'^2}$. In the same way $\kappa\frac{1}{\alpha'+\kappa'}+\kappa'\frac{\alpha}{\alpha'(\alpha'+\kappa')}<\kappa\frac{1}{\gamma'}+\kappa'\frac{\beta}{\gamma'^2}$. But $\kappa\frac{1}{\gamma'}+\kappa'\frac{\beta}{\gamma'^2}$ may be made less than any given number x however small by taking κ less than $\frac{\frac{1}{2}x}{\left(\frac{1}{\gamma'}\right)}$ and κ' less than $\frac{\frac{1}{2}x}{\left(\frac{\beta}{\gamma'^2}\right)}$. Therefore much more may each of the numbers $\kappa\frac{1}{\alpha'+\kappa'}+\kappa'\frac{\alpha}{\alpha'(\alpha'+\kappa')}$ and $\kappa\frac{1}{\alpha'-\kappa'}+\kappa'\frac{\alpha}{\alpha'(\alpha'-\kappa')}$ be made less than any given number. And therefore the first of the two magnitudes may in this way be expressed numerically to any sought degree of nearness in reference to the other as unit.

298. COR. *A number's division by the greater of two numbers gives a less quotient than its division by the less and a still lesser than a greater number's division by the less.*

In art. 297 $\alpha'(\alpha'-\kappa')$ being greater than γ'^2 it is shown that $\alpha\frac{1}{\alpha'(\alpha'-\kappa')}$ and therefore the operational equivalent $\frac{\alpha}{\alpha'(\alpha'-\kappa')}$ is less than $\alpha\frac{1}{\gamma'^2}$ and therefore than the operational equivalent $\frac{\alpha}{\gamma'^2}$ and much more less than $\frac{\beta}{\gamma'^2}$ β being greater than α. This latter also follows from the former by arts. 115, 9.

299. *Def.* The operation by which the numerical quantity expressing one of two magnitudes in reference to the

other as unit may (art. 297) be found to any required degree of nearness when the numerical quantities a b can be found to any required degree of nearness expressing those magnitudes severally in reference to a third magnitude as common unit is called the DIVISION OF THE NUMERICAL QUANTITY a EXPRESSING THE FIRST OF THE TWO MAGNITUDES BY THE NUMERICAL QUANTITY b EXPRESSING THE OTHER. The result of the operation the numerical quantity expressing the one magnitude in reference to the other as unit is called the QUOTIENT of this division and is symbolized by $\frac{a}{b}$.

As a particular case of this operation if one magnitude be expressed in reference to another as unit by the numerical quantity a since this other magnitude is expressed in reference to itself as unit by the numerical quantity 1 it is expressed in reference to the first magnitude as unit by the numerical quantity $\frac{1}{a}$ which is called the RECIPROCAL of a.

If the first of three magnitudes be expressed numerically in reference to the second as unit by a and in reference to the third as unit by b the second is expressed numerically in reference to the first as unit by $\frac{1}{a}$ and therefore (art. 295) in reference to the third as unit by $\frac{1}{a}b$.

According to the manner of its first and every after use (arts. 93, 112, 114) the symbol $\frac{a}{b}$ even now when a b may be incommensurable numerical quantities still stands for and may be read as the numerical expression of one magnitude in reference to another as unit of which the former is expressed numerically by a and the latter by b each in reference to a common magnitude as unit.

300. If in different operations or sets of operations which give the same results with the same numbers be these numbers what they may,—and which for this reason (art. 35) are called EQUIVALENT—, any or all of the numbers operated with be such as approximate to the same incommensurable numerical quantities be these incommensurables what they may the results are the

same which approximate (arts. 291, 293, 295, 297) to the results of the corresponding different operations or sets of operations with the incommensurables. But the results with the incommensurables are known no otherwise than by the results with their commensurable approximates so that it is only the sameness or difference of the latter that settles the sameness or difference of the former. Hence the

PROP. *The Laws of Operational Equivalence are the same for incommensurable as for commensurable numerical quantities.*

Hence also whatever operational equivalences hold good for numbers hold good for numerical quantities generally.

Moreover on the same ground,—that the results of operations with incommensurable numerical quantities are only the results that can be endlessly neared by the results of the like operations with numbers near enough to the incommensurables—, all definitions hitherto made for numbers whole and fractional are henceforth to be understood as made for all numerical quantities commensurable and incommensurable and then all propositions proved for the former are proved for the latter.

301. Not only are there magnitudes which from their very nature (art. 237) are not expressible as numbers but even in all numerical estimation of magnitudes practically by sight feeling hearing and the rest whether alone or helped by instruments such as measuring rods microscopes weighing machines and so on it is no further than some more or less certain degree of nearness that can be reached. What is done throughout arts. 291, 293, 295, 297, may be used for finding the degree of nearness to which the result can be got of operations with numbers thus given only to some more or less certain degree of nearness.

A magnitude is usually expressed by a decimally denoted number of which those digits only are understood to be given aright which refer to multiples or submultiples of the unit magnitude greater than of some given rank. For instance when a French gramme is said to weigh 15·434 English grains this is to be understood only as nearly as can be denoted by using three digits to the right of the decimal dot and the digit in the fourth or any after decimal place is not meant to be 0. It is to be understood that in finding this number there was no way of distinguishing between two weights differing from one another by 0·0001 grain so that for aught said the number of grains equal to a gramme may be indifferently 15·434±0·0001, 15·434±0·0002, or indeed any numerical quantity 15·434±x where x is not greater than 0·0005. But since 15·434+0·0005 is 15·4345 and 15·434−0·0005 is 15·4335 of which the former is just as near to 15·435 and the latter to 15·433 as to 15·434 if x were greater by however little than 0·0005 15·434+x would be nearer to 15·435 and 15·434−x to 15·433 than to 15·434. All the knowledge given of a gramme then is that it is not less than 15·434−0·0005 or 15·4335 grains and is not greater than 15·434+0·0005 or 15·4345.

In like manner if the Sun's distance be 93000000 miles right to a million it can only be gathered that the distance is not less than 92500000 miles and not greater than 93500000.

What can be known about the length of a line wholly made up of three parts severally 34·70 52·693 and 80·1 chains long each number right to the last written digit? Since by what has just been shown the 1st part not< 34·70−0·005 chains and not> 34·70+0·005 the 2nd not< 52·693−0·0005 and not> 52·693+0·0005 and the 3rd not< 80·1−0·05 and not> 80·1+0·05 it follows (art. 291) that the whole line made up of the three parts in reference to a chain as unit not< 34·7+52·693+80·1−(0·005+0·0005+0·05), that is 167·493−0·0555, and not> 167·493+0·0555. But as the digit in the 1st decimal place is not thoroughly known of the sum approximated to nothing whatever is known of the digit in any after place and hence 167·5 the nearest number with a single digit to the right of the decimal dot to 167·493 is to be taken as the approximate sum. And the greatest difference therefrom that can arise, ∵ 167·5−167·493 is 0·007 and ∴ 167·493−0·0555 = 167·5−0·007−0·0555 = 167·5 −(0·0555+0·007) or 167·5−0·0625, is 0·0625. Hence the length sought differs from 167·5 chains by less than 0·07 chain.

```
 34·7      34·7|0
 52·7      52·6|93
 80·1      80·1|
-----     ---------
167·5     167·4|93
```

```
167·493
167·5
-------
  0·007
  0·0555
-------
  0·0625
```

Again if from a line made up of 34·70 and 52·693 chains as parts 80·1 chains be taken each number being given right to the last written digit what can be known of the remainder? As before the line may differ from 87·393 chains by not more than 0·0055 chain and therefore (art. 293) the remainder not< 87·393−80·1−(0·0055+0·05) or 7·293−0·0555 chains and not> 7·293+0·0555 chains. Since then 7·293 = 7·3−0·007 and 0·0555+0·007 is 0·0625

```
34·70|
52·69|3
-------
87·3|9|3
80·1|
-------
 7·2|9 3
```

which < 0·07 the remainder can differ from 7·3 chains only by less than 0·07 chain.

If the distance between the centers of the Earth and Moon be expressed by 59·9643 in reference to the Earth's equatorial radius as unit and this radius by 3962·8 in reference to a mile as unit each right to the last written digit what can be known of the expression in miles of the distance between the Earth's center and the Moon's? Since the former number not< 59·9643–0·00005 and not> 59·9643+0·00005 and the latter not< 3962·8–0·05 and not> 3962·8+0·05 the distance in miles (art. 295) does not differ from 59·9643×3962·8 by more than

0·00005×3962·85+0·05×59·9643,

that is does not differ from 237626·52804 by more than 0·1981425+2·998215 or 3·1963575. And therefore ∵ 237627–237626·52804 is 0·47196 and 3·1963575+0·47196 is 3·6683175 the distance cannot differ from 237627 miles by so much as 3·7 miles.

```
  3962·8              3962·8
     59|9643             59|9643
 ------|------      -------|--------
 179892|9            179892|9
  53967|9             53967|87
   3597|9              3597|858
    119|9               119|9286
     48|0                47|97144.
 ------|------      -------|--------
 237627|             237626|52804
                     237627
                    ----------------
                           0·47196
```

If a cubic inch of water weigh 252·5 grains right to the last digit what can be known about the weight of a cubic foot (1728 cubic inches) of water in reference to an ounce (437·5 grains) as unit? The weight of a cubic foot of water does not differ from 1728×252·5 grs. or 436320 by more than 1728×0·05 or 86·4 grs. and therefore (art. 297) does not differ from

$\frac{436320}{437·5}$ oz. by more than

$86·4 \times \frac{1}{437·5}$ oz. As then the former of the last two numbers differs from 997·30 by less than 0·003 and the latter (equivalent to $\frac{86·4}{437·5}$) from 0·1975 by less than 0·00002 a cubic foot of water weighs more nearly 997·3 oz. than by 0·21 oz.

```
   2 52·5        4 37·5       437·5
    17 28          9 97·30     0·1975
 ---------     -----------    --------
 432|          4363|20         86·4
  43|2         3937|5          43 75
 ---|---       ----|--        -------
4363|20         425|7          42 65
     86·4       393|75         39 375
               ----|---       -------
                 31|95          3 275
                 30|62·5        3 0625
                 --|-----     -------
                  1|32 5          2125
                  1|31 25        21875
                 --|-----
                   | 1 25
```

What can be known about the speed of a railway train which passes over 91·83 fathoms in 7·4 seconds each of these numbers being right to the last written digit? If the first of three magnitudes in reference to the second as unit

not$<\alpha-\kappa$ and not$>\alpha+\kappa$

and in reference to the third as unit

not$<\alpha'-\kappa'$ and not$>\alpha'+\kappa'$

the third in reference to the second as unit after the manner of arts. 295, 297,

12·40946	0·13606	1·6884
7·4	7·35	7·35
91·83	1	12·40946
74	735	7 35
17	265	5 05
14 8	2205	4410
3 0	445	649
2 96	4410	5880
7	40	614
666		5880
34		266
296		
44		

$$\text{not}<\frac{1}{\alpha'+\kappa'}(\alpha-\kappa)\ \text{ which}=\frac{1}{\alpha'}\alpha-\left\{\kappa\frac{1}{\alpha'+\kappa'}+\kappa'\frac{1}{\alpha'+\kappa'}\frac{1}{\alpha'}\alpha\right\}$$

$$\text{and not}>\frac{1}{\alpha'-\kappa'}(\alpha+\kappa)\ \text{ which}=\frac{1}{\alpha'}\alpha+\left\{\kappa\frac{1}{\alpha'-\kappa'}+\kappa'\frac{1}{\alpha'-\kappa'}\frac{1}{\alpha'}\alpha\right\}.$$

Since then $\frac{1}{7{\cdot}4}\times 91{\cdot}83$ falls short of 12·40946 by less than 0·000001 $\frac{1}{7{\cdot}35}$ of 0·13606 by less than 0·000006 and $\frac{1}{7{\cdot}35}\times 12{\cdot}40946$ of 1·6884 by less than 0·00004

$$0{\cdot}005\times\frac{1}{7{\cdot}35}+0{\cdot}05\times\frac{1}{7{\cdot}35}\times\frac{1}{7{\cdot}4}\times 91{\cdot}83<0{\cdot}0006803+0{\cdot}08442 \text{ which is } 0{\cdot}0851003$$

also since 12·40946+0·0851003 = 12·4+(0·00946+0·0851003) or 12·4+0·0945603 and 0·0945603<0·095 the speed is nearer to 12·4 fathoms a second than by 0·095 fathom. Moreover 3600 seconds making an hour and 880 fathoms a mile 3600×12·40946 being 44674·056 and 0·0851003×3600 306·36108 $\frac{44674{\cdot}056}{880}$ falling short of 50·766 by less than 0·00003 and $306{\cdot}36108\times\frac{1}{880}$ of 0·34814 by less than 0·000003 50·8−50·766 being 0·034 and 0·34814+0·034 0·38214 the speed is not different from 50·8 miles an hour by so much as 0·39 mile.

What can be known about the length of a straight line which has to a straight line 22·8 miles long the same ratio as a straight line 3·792 yards long has to a straight line 2·5146 yards long when all the three numbers are given right only as far as the last written digits? Since $\frac{3·792}{2·5146}$ falls short of 1·508 by less than 0·000007, $\frac{1}{2·51455}$ of 0·398 by less than 0·0004 and $\frac{\left(\frac{3·792}{2·5146}\right)}{2·51455}$, operationally equivalent to $\frac{3·792}{2·51455 \times 2·5146}$, of 0·6 by less than 0·0003 the numerical representative of the ratio falls short of 1·508 by less than 0·000007+0·000229 or 0·000236. Hence the length sought differs from 34·3824 miles by less than 0·0807926 mile and therefore is within 0·0983926 (which is 0·0807926+0·0176) mile and much more within 0·099 mile of 34·4 miles.

```
2·5146          2·51455        2·51455
1·508           0·398          0·6
-----------     ----------     ---------
3·792|          1              1·508
2 514|6           754365       1 50873
------            245635
1 277|4           2263095
1 257|3            193255
   20|1            201164
   20|1168
```

0·0005×0·398+0·00005×0·6 is $\left\{\begin{array}{l}0·000199\\+0·00003\end{array}\right\}$ or 0·000229

```
 22·8           22·85
  1·5 08        0·000236
---------       ----------
 30 1|6            457
  3 0|16             6855
  1 2|064            1371
---------       ----------
 34·3|824       0·0053926
 34·4              754
---------          176
 0·0 176        ----------
                0·0983926
```

0·000236×22·85+0·05×1·508 is 0·0807926

How much is there of a certain substance in 77·3 grains of a certain compound in 142·653 grains of which there are 9·87 grains of the substance each of the three numbers being right to the last written digit? In 1 gr. of the compound there is 0·0692 gr. of the substance nearer than by 0·00002+0·000036 or 0·000056 gr. and hence in so much

```
 0·06 92         0·00701005       0·00049
142·65 3         142·6525         142·6525
-----------      -----------      ----------
  9·87|          1                0·0692
  8 55|918         9985675          57061
-----------      -----------      ----------
  1 31|082          14325           12139
  1 28|3877         1426525         12838725
-----------      -----------
     2|6943          5975
     2|85306         7132625
```

0·05×0·00701005+0·0005×0·00049 is 0·00003529525

```
 0·069 2
 77·3
---------
 4·844|
  484|4
   20|76
---------
 5·349|16
```

0·05×0·069256+0·000056×77·3 is 0·0077916

0·0077916+0·00016 is 0·0079516

of the compound as does not differ from 77·3 grs. by more than 0·05 grs. there are more nearly 5·349 grs. of the substance than by 0·008 gr.

If in reference to the weight of a cubic inch of water as unit the weight of a cubic inch of gold be 19·258 of silver 10·474 and of a mixture of gold and silver 13·263 each right to the third decimal place what can be known about the bulks and the weights of gold and silver in one cubic inch of the mixture? Let g s m be the numbers severally approximated to then if there be in 1 cubic inch of the mixture x cubic inches of gold and therefore $1-x$ cubic inches of silver the weight of the cubic inch of mixture in reference to the water inch weight as unit is

$$xg+(1-x)s=m \text{ and } \therefore x=\frac{m-s}{g-s} \quad 1-x=\frac{g-m}{g-s}$$

$m-s$ $g-s$ $g-m$ are not further than by 0·001 from 2·789 8·784 5·995 severally x is within 0·0001512 of 0·31751 and therefore within 0·00017 of 0·3175 $1-x$ is within 0·000202 of 0·6825 and therefore within 0·00021 of 0·6825 there is of gold within 0·0039 water inch weight of 6·115 wr. in. wts. and there is of silver within 0·0033 wr. in. wt. of 7·149 wr. in. wts.

302. PROP. *If the first of continuedly proportionate magnitudes can be expressed numerically to any required degree of nearness in reference to the last of them as unit each of them but the last may be expressed numerically to any required degree of nearness in reference to the next following one as unit.*

Let $A_r\, A_{r-1} \ldots A_2\, A_1\, A_0$ be any continuedly proportionate magnitudes. If a be the numerical expression of A_r in reference to A_0 as unit and x the common numerical expression of $A_r \ldots A_2\, A_1$ in reference to $A_{r-1} \ldots A_1\, A_0$ severally as unit (art. 295) $x^r=a$. Now taking n any whole number any number of terms of the series $\frac{0}{n}\ \frac{1}{n}\ \frac{2}{n}\ \frac{3}{n}$ - - - may be found and therefore too any number of terms of the series $\left(\frac{0}{n}\right)^r \left(\frac{1}{n}\right)^r \left(\frac{2}{n}\right)^r \left(\frac{3}{n}\right)^r$ - - -. But since the former series may be carried on until (art. 100) there are terms in it greater than any given number much more after terms are come to greater than 1 (art. 258) may the latter series be carried on until the corresponding terms are greater than any given number and hence (art. 241) until there are terms greater than any given numerical quantity. If then a can be found to any required degree of nearness since the series

$\left(\frac{0}{n}\right)^r \left(\frac{1}{n}\right)^r \left(\frac{2}{n}\right)^r$ - - - starts from 0 and may be carried on till terms are come to greater than a (art. 233) a either is equal to some one term of this series or is greater than the one and less than the other of some consecutive two terms and either the single term $\left(\frac{m}{n}\right)^r$ may be known to which a is equal or the consecutive two terms $\left(\frac{m}{n}\right)^r \left(\frac{m+1}{n}\right)^r$ between which a lies. If $\left(\frac{m}{n}\right)^r = a$ x is $\frac{m}{n}$ and because (art. 258) any greater number than $\frac{m}{n}$ has a greater rth power and any number less a less rth power x is no other than $\frac{m}{n}$. But if $a > \left(\frac{m}{n}\right)^r$ and $< \left(\frac{m+1}{n}\right)^r$ $x > \frac{m}{n}$ and $< \frac{m+1}{n}$ and $\frac{m+1}{n} - \frac{m}{n} = \frac{1}{n}$ which by taking n great enough may be made less than any given fraction however small. Here too x can only be one numerical quantity for if it could be either of two numerical quantities each of two fractions $\frac{s}{t}$ $\frac{s+1}{t}$ between which the greater lies would (art. 242) be greater than each of two fractions $\frac{u}{v}$ $\frac{u+1}{v}$ between which the less lies so that $\left(\frac{s}{t}\right)^r$ would be greater than $\left(\frac{u+1}{v}\right)^r$ and yet $a > \left(\frac{s}{t}\right)^r$ and $< \left(\frac{u+1}{v}\right)^r$.

303. *Def.* That numerical quantity of which a given power is a given numerical quantity is called the ROOT of the given numerical quantity of the same DEGREE as the power. The root of the rth degree of a numerical quantity a or shortly the rth root of a is symbolized by $\sqrt[r]{a}$ and the whole number r is called the INDEX or EXPONENT of the root. The second root of a is usually written $\sqrt{a}$ without the index.

By definition then $(\sqrt[r]{a})^r$ is a. And since (art. 302) there is only one numerical quantity of which the rth power is a certain numerical quantity $\sqrt[r]{a^r} = a$.

That ratio is represented numerically by $\sqrt[r]{a}$ of which the $r$plicate compound ratio is represented numerically by a.

304. PROP. *A numerical quantity has a commensurable root only when the quantity can be expressed as a simple frac-*

tion with prime terms each a power of a whole number of the root's degree.

For a numerical quantity whose rth root is a prime termed fraction $\frac{m}{n}$ is equal to $\left(\frac{m}{n}\right)^r$ and therefore also equal to $\frac{m^r}{n^r}$. And (art. 209) because m is prime to n m^r is prime to n^r.

305. PROP. $\sqrt[r]{\sqrt[s]{\sqrt[t]{\cdots\sqrt[x]{\sqrt[y]{\sqrt[z]{a}}}}}} = \sqrt[zyx\ldots tsr]{a}$.

First if n be any whole number (art. 302) there is a whole number m such that

$$\sqrt[sr]{a} \text{ not}< \frac{m}{n} \text{ but} < \frac{m+1}{n},$$

that is a not$<\left(\frac{m}{n}\right)^{sr}$ but $<\left(\frac{m+1}{n}\right)^{sr}$. Therefore a not$<\left\{\left(\frac{m}{n}\right)^r\right\}^s$ but $<\left\{\left(\frac{m+1}{n}\right)^r\right\}^s$, that is $\sqrt[s]{a}$ not$<\left(\frac{m}{n}\right)^r$ but $<\left(\frac{m+1}{n}\right)^r$ and that again is

$$\sqrt[r]{\sqrt[s]{a}} \text{ not}< \frac{m}{n} \text{ but} < \frac{m+1}{n}.$$

$\sqrt[sr]{a}$ $\sqrt[r]{\sqrt[s]{a}}$ then (art. 302) either are equal to the same fraction $\frac{m}{n}$ or lie between the same two fractions $\frac{m}{n}$ $\frac{m+1}{n}$ which by taking n great enough may be made to differ from one another by less than any given fraction however small and

$$\therefore \sqrt[r]{\sqrt[s]{a}} = \sqrt[sr]{a}.$$

Hence $\sqrt[r]{\sqrt[s]{\sqrt[t]{\cdots\sqrt[x]{\sqrt[y]{\sqrt[z]{a}}}}}} = \sqrt[sr]{\sqrt[t]{\sqrt[u]{\cdots\sqrt[y]{\sqrt[z]{a}}}}} = \sqrt[tsr]{\sqrt[u]{\cdots\sqrt[y]{\sqrt[z]{a}}}} = \text{- - -} = \sqrt[zyx\ldots tsr]{a}$.

306. In particular since (art. 303) $a = \sqrt[x]{a^x}$.

COR. $\sqrt[r]{a} = \sqrt[xr]{a^x}$.

307. PROP. $(\sqrt[r]{h})(\sqrt[r]{g})(\sqrt[r]{f})\ldots(\sqrt[r]{c})(\sqrt[r]{b})\sqrt[r]{a} = \sqrt[r]{hgf\ldots cba}$.

First if n n' be any whole numbers there are (art. 302) whole numbers m m' such that

$$\sqrt[r]{a} \text{ not}< \frac{m}{n} \text{ and} < \frac{m+1}{n} \qquad \sqrt[r]{b} \text{ not}< \frac{m'}{n'} \text{ and} < \frac{m'+1}{n'}$$

or which is the same thing

$$a \text{ not}< \left(\frac{m}{n}\right)^r \text{ and} < \left(\frac{m+1}{n}\right)^r \qquad b \text{ not}< \left(\frac{m'}{n'}\right)^r \text{ and} < \left(\frac{m'+1}{n'}\right)^r.$$

$$\therefore (\sqrt[r]{a})\sqrt[r]{b} \text{ not}< \frac{m}{n}\frac{m'}{n'} \text{ and} < \frac{m+1}{n}\frac{m'+1}{n'}$$

$$ab \text{ not}< \left(\frac{m}{n}\right)^r\left(\frac{m'}{n'}\right)^r \text{ and} < \left(\frac{m+1}{n}\right)^r\left(\frac{m'+1}{n'}\right)^r$$

$$\text{and } \therefore \text{ not} < \left(\frac{m}{n}\frac{m'}{n'}\right)^r \text{ and} < \left(\frac{m+1}{n}\frac{m'+1}{n'}\right)^r$$

$$\text{that is } \sqrt[r]{ab} \text{ not} < \frac{m}{n}\frac{m'}{n'} \text{ and } < \frac{m+1}{n}\frac{m'+1}{n'}.$$

Moreover $\frac{m+1}{n}\frac{m'+1}{n'} - \frac{m}{n}\frac{m'}{n'} = \frac{1}{n}\frac{m'+1}{n'} + \frac{1}{n'}\frac{m}{n}$ and $\therefore < \frac{1}{n}\beta + \frac{1}{n'}\alpha$ when n n' are taken great enough if α be the greater of any two given fractions intercepting that magnitude relative to another magnitude which in reference to this other as unit is expressed numerically by $\sqrt[r]{a}$ and β the greater of any two given fractions intercepting that magnitude relative to another magnitude which in reference to this other as unit is expressed numerically by $\sqrt[r]{b}$. But both $\frac{1}{n}\beta$ and $\frac{1}{n'}\alpha$ and therefore too $\frac{1}{n}\beta + \frac{1}{n'}\alpha$ may be made less than any given fraction however small by taking n n' great enough. Much more therefore by taking n n' each great enough may $\frac{m+1}{n}\frac{m'+1}{n'} - \frac{m}{n}\frac{m'}{n'}$ be made less than any given fraction however small. So then $(\sqrt[r]{a})\sqrt[r]{b}$ and $\sqrt[r]{ab}$ are either equal to precisely the same fraction or lie between precisely the same two fractions that may be so taken as to differ from one another by less than any given fraction

$$\therefore (\sqrt[r]{a})\sqrt[r]{b} = \sqrt[r]{ab}.$$

$$\begin{aligned} \text{Hence } (\sqrt[r]{h})(\sqrt[r]{g})(\sqrt[r]{f})\dots(\sqrt[r]{c})(\sqrt[r]{b})\sqrt[r]{a} &= (\sqrt[r]{h})(\sqrt[r]{g})(\sqrt[r]{f})\dots(\sqrt[r]{c})\sqrt[r]{ba} \\ &= (\sqrt[r]{h})(\sqrt[r]{g})\dots(\sqrt[r]{d})\sqrt[r]{cba} \\ &= \text{ - - - - } \\ &= \sqrt[r]{hgf\dots cba}. \end{aligned}$$

308. If in art. 307 $a=b=c=---=f=g=h$ and there be x of them the proposition is the

COR. $(\sqrt[r]{a})^x = \sqrt[r]{a^x}$.

In particular when x is r there is anew $a = \sqrt[r]{a^r}$ as in art. 303.

309. PROP. *To find an expression with a single root symbol operationally equivalent to the product of any powers of any roots.*

$(\sqrt[r]{a})\sqrt[s]{b} = (\sqrt[r]{\sqrt[r']{a^{r'}}})\sqrt[s]{\sqrt[s']{b^{s'}}}$ (art. 303) if r' s' be any whole numbers
$= (\sqrt[r'r]{a^{r'}})\sqrt[s's]{b^{s'}}$ (art. 305) and if r' s' be so taken (art. 212) that $r'r$ $s's$ are a common multiple λ of r s
$= \sqrt[\lambda]{a^{\frac{\lambda}{r}}b^{\frac{\lambda}{s}}}$ (art. 307).

In like manner if λ be any common multiple of r s t...x y z

$$(\sqrt[r]{a})(\sqrt[s]{b})(\sqrt[t]{c})\dots(\sqrt[x]{f})(\sqrt[y]{g})\sqrt[z]{h} = \sqrt[\lambda]{a^{\frac{\lambda}{r}}b^{\frac{\lambda}{s}}c^{\frac{\lambda}{t}}\dots f^{\frac{\lambda}{x}}g^{\frac{\lambda}{y}}h^{\frac{\lambda}{z}}}.$$

And $\because (\sqrt[r]{a})^{r'} = \sqrt[r]{a^{r'}}$ and $(a^{r'})^{\frac{\lambda}{r}} = a^{\frac{\lambda}{r}r'}$

$$(\sqrt[r]{a})^{r'}(\sqrt[s]{b})^{s'}(\sqrt[t]{c})^{t'}\dots(\sqrt[x]{f})^{x'}(\sqrt[y]{g})^{y'}(\sqrt[z]{h})^{z'} = \sqrt[\lambda]{a^{\frac{\lambda}{r}r'}b^{\frac{\lambda}{s}s'}c^{\frac{\lambda}{t}t'}\dots f^{\frac{\lambda}{x}x'}g^{\frac{\lambda}{y}y'}h^{\frac{\lambda}{z}z'}}.$$

The λth root is of course of the least possible degree when λ is taken the least common multiple.

310. PROP. $\dfrac{\sqrt[r]{a}}{\sqrt[r]{b}} = \sqrt[r]{\dfrac{a}{b}}$.

If n n' be any whole numbers (art. 302) there are whole numbers m m' such that

$$\sqrt[r]{a} \text{ not} < \frac{m}{n} \text{ but } < \frac{m+1}{n} \qquad \sqrt[r]{b} \text{ not} < \frac{m'}{n'} \text{ but } < \frac{m'+1}{n'}$$

or in other words

$$a \text{ not} < \left(\frac{m}{n}\right)^r \text{ but } < \left(\frac{m+1}{n}\right)^r \qquad b \text{ not} < \left(\frac{m'}{n'}\right)^r \text{ but } < \left(\frac{m'+1}{n'}\right)^r.$$

Then if "not<" mean "=" in all these statements

$$\frac{\sqrt[r]{a}}{\sqrt[r]{b}} = \frac{\left(\frac{m}{n}\right)}{\left(\frac{m'}{n'}\right)} \text{ and } \frac{a}{b} = \frac{\left(\frac{m}{n}\right)^r}{\left(\frac{m'}{n'}\right)^r} = \left\{\frac{\left(\frac{m}{n}\right)}{\left(\frac{m'}{n'}\right)}\right\}^r, \text{ that is } \sqrt[r]{\frac{a}{b}} = \frac{\left(\frac{m}{n}\right)}{\left(\frac{m'}{n'}\right)}.$$

But if "not<" do not mean "=" in all these statements

$$\frac{\sqrt[r]{a}}{\sqrt[r]{b}} > \frac{\left(\frac{m}{n}\right)}{\left(\frac{m'+1}{n'}\right)} \text{ but } < \frac{\left(\frac{m+1}{n}\right)}{\left(\frac{m'}{n'}\right)}$$

$$\frac{a}{b} > \frac{\left(\frac{m}{n}\right)^r}{\left(\frac{m'+1}{n'}\right)^r}, \text{ but } < \frac{\left(\frac{m+1}{n}\right)^r}{\left(\frac{m'}{n'}\right)^r} \text{ and } \therefore > \left\{\frac{\left(\frac{m}{n}\right)}{\left(\frac{m'+1}{n'}\right)}\right\}^r \text{ but } < \left\{\frac{\left(\frac{m+1}{n}\right)}{\left(\frac{m'}{n'}\right)}\right\}^r,$$

$$\text{that is } \sqrt[r]{\frac{a}{b}} > \frac{\left(\frac{m}{n}\right)}{\left(\frac{m'+1}{n'}\right)} \text{ but } < \frac{\left(\frac{m+1}{n}\right)}{\left(\frac{m'}{n'}\right)}.$$

$$\text{And } \frac{\left(\frac{m+1}{n}\right)}{\left(\frac{m'}{n'}\right)} - \frac{\left(\frac{m}{n}\right)}{\left(\frac{m'+1}{n'}\right)} = \frac{\frac{m+1}{n}\frac{m'+1}{n'} - \frac{m}{n}\frac{m'}{n'}}{\frac{m'}{n'}\frac{m'+1}{n'}} = \frac{1}{n}\frac{1}{\left(\frac{m'}{n'}\right)} + \frac{1}{n'}\frac{\left(\frac{m}{n}\right)}{\frac{m'+1}{n'}\frac{m'}{n'}} \text{ and}$$

$\therefore < \dfrac{1}{n}\dfrac{1}{\beta_1} + \dfrac{1}{n'}\dfrac{\alpha}{\beta_1^2}$ when n n' are taken great enough if α be the greater of any two given fractions intercepting that one of two magnitudes relative to the other which in reference to this other as unit is expressed numerically by $\sqrt[r]{a}$ and $\beta_{,}$ the less of any two given fractions intercepting that one of two magnitudes relative to the other which in reference to this other as unit is expressed numerically by $\sqrt[s]{b}$. But $n\ n'$ may be taken so great that each of the fractions $\frac{1}{n}\frac{1}{\beta_{,}}\ \frac{1}{n'}\frac{\alpha}{\beta_{,}^2}$ and therefore also $\frac{1}{n}\frac{1}{\beta_{,}}+\frac{1}{n'}\frac{\alpha}{\beta_{,}^2}$ is less than any given fraction. Much more therefore may $\frac{\left(\frac{m+1}{n}\right)}{\left(\frac{m'}{n'}\right)}-\frac{\left(\frac{m}{n}\right)}{\left(\frac{m'+1}{n'}\right)}$ be made less than any given fraction however small by taking $n\ n'$ each great enough. Thus it comes out that $\frac{\sqrt[r]{a}}{\sqrt[r]{b}}\ \sqrt[r]{\frac{a}{b}}$ either are equal to the very same fraction or lie between the very same two fractions differing from one another by less than any given fraction and $\therefore \frac{\sqrt[r]{a}}{\sqrt[r]{b}}=\sqrt[r]{\frac{a}{b}}$.

311. PROP. *To find a root operationally equivalent to the quotient of a power of a root by a power of a root.*

$\frac{\sqrt[r]{a}}{\sqrt[s]{b}}=\frac{\sqrt[r']{\sqrt[r]{a^{r'}}}}{\sqrt[s']{\sqrt[s]{b^{s'}}}}$ if $r'\ s'$ be any whole numbers

$=\frac{\sqrt[r'r]{a^{r'}}}{\sqrt[s's]{b^{s'}}}$ and if $r'\ s'$ be so taken that $r'r\ s's$ are a common multiple λ of $r\ s$

$$=\sqrt[\lambda]{\frac{a^{\frac{\lambda}{r}}}{b^{\frac{\lambda}{s}}}}.$$

Hence $\frac{(\sqrt[r]{a})^\rho}{(\sqrt[s]{b})^\sigma}=\frac{\sqrt[r]{a^\rho}}{\sqrt[s]{b^\sigma}}=\frac{\sqrt[\lambda]{(a^\rho)^{\frac{\lambda}{r}}}}{\sqrt[\lambda]{(b^\sigma)^{\frac{\lambda}{s}}}}=\sqrt[\lambda]{\frac{a^{\frac{\lambda}{r}\rho}}{b^{\frac{\lambda}{s}\sigma}}}.$

312. PROP. *To find the decimally denoted second root of a given decimally denoted number.*

First of all if $u\ v$ be any two numerical quantities

$$(u+v)^2=u(u+v)+v(u+v)=u^2+uv+v(u+v)=u^2+\{vu+v(u+v)\}$$
$$=u^2+v\{u+(u+v)\}=u^2+v(2u+v).$$

Hence in the next place if $a\ b\ c\ldots f\ g\ h$ be any numerical quantities

$$(a+b+c+\cdots+f+g+h)^2 = (a+b+c+\cdots+f+g)^2+h\begin{Bmatrix}2(a+b+c+\cdots+f+g)\\+h\end{Bmatrix}$$

$$= (a+b+c+\cdots+f)^2+g\begin{Bmatrix}2(a+b+c+\cdots+f)\\+g\end{Bmatrix}+h\begin{Bmatrix}2(a+b+c+\cdots+f+g)\\+h\end{Bmatrix}$$

$$= - \quad - \quad - \quad - \quad - \quad - \quad -$$

$$= a\begin{Bmatrix}2\times 0\\+a\end{Bmatrix}+b\begin{Bmatrix}2a\\+b\end{Bmatrix}+c\begin{Bmatrix}2(a+b)\\+c\end{Bmatrix}+d\begin{Bmatrix}2(a+b+c)\\+d\end{Bmatrix}+$$

$$\cdots\cdots +h\begin{Bmatrix}2(a+b+c+\cdots+f+g)\\+h\end{Bmatrix}.$$

And the numerical quantity $2(u+v)$ added to in the expression within each pair of crow wing brackets after the first $= 2u+(v+v)$ $= 2u+v+v$ the sum got by adding to the expression $2u+v$ within the pair of crow wing brackets next before the numerical quantity v added in this expression.

Thirdly the decimally denoted second power of a given decimally denoted number may thence be found by taking $a\ b\ c \ldots f\ g\ h$ the numbers denoted severally by the digits in order and for the better finding the products a^2 $b(2a+b)$ $c\{2(a+b)+c\}\ldots$ the multiplicands a $2a+b$ $2(a+b)+c\ldots$, got each after the first from the one next before it by the theorem $2(u+v) = 2u+v+v$, may be handily written over against them in an auxiliary column.

	46·30 7
40	16
86	5 16
92·3	27 69
92·607	64 82 49
	21 44·33 82 49

Hence fourthly and lastly for finding the decimally denoted second root of a given decimally denoted number this number has only to be viewed as the sum of the products a^2 $b(2a+b)$ $c\{2(a+b)+c\}\ldots$ made with the numbers $a\ b\ c\ldots$ denoted by the digits in order of the root sought and then by subtracting those products in succession, as the digits become successively known by easy trial, from that sum and the successive remainders got the digits themselves become known. Thus 2144·338249 is at once seen to be

	46·307
	21 44·33 82 49
40	16
	5
86	5 16
	28
92·3	27 69
	64
92·607	64 82 49

greater than $(10)^2$ and less than $(100)^2$. Trying then $(10)^2$ $(20)^2$ $(30)^2\ldots(90)^2$ 2144·338249 is found to be greater than $(40)^2$ and less than $(50)^2$. The greatest number x denoted by a single significant digit has now to be found which makes $x(2\times40+x)$ not greater than 2144·338249–$(40)^2$ or 544·338249. Because 6×86 is 516 and 7×87 is 609 x is 6. Then $\because k-a^2-b(2a+b)=k-\{b(2a+b)+a^2\}$ $=k-\{a^2+b(2a+b)\}=k-(a+b)^2$ 2144·338249–$(46)^2$ = 544·338249–516 or 28·338249. Again 2×46 being equal to 86+6 or 92 and 0·3×92·3 being less and 0·4×92·4 greater than 28·338249 as before 2144·338249–$(46\cdot3)^2$ = 28·338249–27·69 or 0·648249. And so at last 2144·338249–$(46\cdot307)^2$ = 0·648249–0·007(2×46·3+0·007) or 0, that is 2144·338249 is $(46\cdot307)^2$ and that again is $\sqrt{2144\cdot338249}$ is 46·307.

When the root is incommensurable this NOTATIONAL SECOND ROOT PROCESS has only (art. 134) to be carried on far enough to get the root denoted decimally to any required degree of nearness.

313. But after several digits of a decimally denoted second root are got by art. 312's process the rest may be simply approximated to by a divisive process. For let the row of digits denoting a second root be made with a knot of m digits followed by a knot of n, let u be the whole number which a row of the m digits followed by n 0s without a decimal dot would denote v the whole number which a row of the n digits without a decimal dot would denote and ω the multiple or submultiple of the unit to which as unit the last digit of the row denoting the root refers. By the principle of digit knitting then the second root $=(u+v)\omega$ and therefore the number of which this is equal to the second root $=\{(u+v)\omega\}^2=(u+v)^2\omega^2=$ $\{u^2+v(2u+v)\}\omega^2$. Hence if after finding u by the second root notational process instead of going on to find v by the process, which would give v as $\dfrac{v(2u+v)}{2u+v}$, a divisive process be set about to give $\dfrac{v(2u+v)}{2u}=v+\dfrac{v^2}{2u}$ instead of finding the root $(u+v)\omega$ there is found $\left\{u+\left(v+\dfrac{v^2}{2u}\right)\right\}\omega$ which $=\left(u+v+\dfrac{v^2}{2u}\right)\omega=(u+v)\omega+\dfrac{v^2}{2u}\omega$.
But $\because u$ not< $(10)^{n+m-1}$ the least whole number denoted with $m+n$ digits and $v<(10)^n$

$$\frac{v^2}{2u}<\frac{\{(10)^n\}^2}{2(10)^{n+m-1}}\therefore<\frac{(10)^{2n}}{2(10)^{n+m-1}}\text{ and }\therefore<\frac{1}{2}$$

if $2n$ not> $n+m-1$, that is if n not> $m-1$. If then n not> $m-1$ $(u+v)\omega+\dfrac{v^2}{2u}\omega$ is greater than $(u+v)\omega$ only by less than $\dfrac{1}{2}\omega$ so that when several digits of the root are found by the second root process one fewer than as many more may be found aright by the process for finding the quotient of the division of the remainder got at that stage by twice the number denoted by the digits already found. Moreover in the divisive process whereby the other digits of the second root are found it is clearly needless to take account of any digits but those which are more or less nearly the same as the digits which would have been in their places had the root process itself been gone on with. Thus because $2 = (1{\cdot}4142)^2 + 0{\cdot}00003836$ and $\dfrac{0{\cdot}00003836}{2 \times 1{\cdot}4142} = 0{\cdot}00001356$ right to the 8th decimal place $\sqrt{2}$ differs from $1{\cdot}41421356$ by less than $0{\cdot}000000005$.

	1·41421356
	2
1	1
	1
2·4	96
	4
2·81	281
	119
2·824	11296
	604
2·8282	56564
	3836
2·8284	28284
	10076
	8485
	1591
	1414
	177
	170

314. PROP. *To find the mononome or polynome, if there be one, operationally equivalent to the second root of a mononome or polynome.*

Since (art. 181) $(bqx^ry^s)^2 = b^2q^2x^{2r}y^{2s}$ there can be a mononomic equivalent of $\sqrt{apx^my^n}$ only when the whole numbers m n are the doubles of whole numbers and then $\sqrt{apx^my^n} = (\sqrt{a})(\sqrt{p})x^{\frac{1}{2}m}y^{\frac{1}{2}n}$.

For finding polynomic equivalents of second roots of polynomes besides the theorems

$$(u+v)^2 = u^2+v(2u+v) \qquad\qquad 2(u+v) = 2u+v+v$$

already given in art. 312 other two theorems are needed to wit

$$(u-v)^2 = u(u-v)-v(u-v) = u^2-uv-v(u-v) = u^2-\{v(u-v)+vu\}$$
$$= u^2-v(u-v+u) = u^2-v(2u-v)$$
$$\text{and}\quad 2(u-v) = 2u-(v+v) = 2u-v-v.$$

Then the polynomic expression equivalent operationally to the second root of a given polynomic expression is to be found as in the following instance:—

$$16x^6-40x^5+49x^4-46x^3+29x^2-12x+4$$
$$= (\sqrt{16x^6})^2-\{(\sqrt{16})x^{\frac{1}{2}\times 6}\}^2+16x^6-40x^5+\cdots$$
$$= (4x^3)^2-40x^5+49x^4-46x^3+29x^2-12x+4$$
$$= (4x^3)^2-\frac{40x^5}{2\times 4x^3}\left(2\times 4x^3-\frac{40x^5}{8x^3}\right)+\left(\frac{40}{8}x^{5-3}\right)(8x^3-5x^2)-40x^5+49x^4$$
$$-46x^3+\cdots;$$

$$= (4x^3-5x^2)^2+(40x^5-25x^4)-40x^5+49x^4-46x^3+\cdots$$

$$= (4x^3-5x^2)^2+(40x^5-40x^5)+(49-25)x^4-46x^3+29x^2-12x+4$$

$$= (4x^3-5x^2)^2+\left[\frac{24x^4}{2\times4x^3}\left\{2(4x^3-5x^2)+\frac{24x^4}{8x^3}\right\}-\left(\frac{24}{8}x^{4-3}\right)\begin{pmatrix}8x^3-5x^2 & \\ & -5x^2+3x\end{pmatrix}\right]$$
$$+24x^4-46x^3+\cdots$$

$$= (4x^3-5x^2)^2+(3x)\{2(4x^3-5x^2)+3x\}-(3x)(8x^3-10x^2+3x)$$
$$+24x^4-46x^3+29x^2-12x+4$$

$$= (4x^3-5x^2+3x)^2-9x^2+30x^3-24x^4+24x^4-46x^3+29x^2-12x+4$$

$$= (4x^3-5x^2+3x)^2-(46-30)x^3+(29-9)x^2-12x+4$$

$$= (4x^3-5x^2+3x)^2-\frac{16x^3}{2\times4x^3}\left\{2(4x^3-5x^2+3x)-\frac{16x^3}{8x^3}\right\}$$
$$+\frac{16}{8}\begin{pmatrix}8x^3-10x^2+3x & \\ & +3x-2\end{pmatrix}-16x^3+20x^2-12x+4$$

$$= (4x^3-5x^2+3x-2)^2+2(8x^3-10x^2+6x-2)-16x^3+20x^2-12x+4$$

$$= (4x^3-5x^2+3x-2)^2+(16x^3-16x^3)-(20x^2-20x^2)+(12x-12x)-(4-4),$$

that is $(4x^3-5x^2+3x-2)^2$.

$$\therefore \sqrt{(16x^6-40x^5+49x^4-46x^3+29x^2-12x+4)}=4x^3-5x^2+3x-2.$$

The result of this process may be got by the following algorithm without showing the steps.

	$4x^3\ -5x^2\ +3x-2$
	$16x^6-40x^5+49x^4-46x^3+29x^2-12x+4$
$4x^3$	16
$8x^3-5x^2$	$-40\ \ +25$
	$+24$
$8x^3-10x^2+3x$	$+24\ \ -30\ \ +9$
	$-16\ \ +20$
$8x^3-10x^2+6x-2$	$-16\ \ +20\ \ -12\ \ +4$

In the same way $25+70c+9c^2-116c^3-58c^4+62c^5+28c^6-12c^7+c^8$

$$= 5^2+(7c)(10+7c)-(70c+49c^2)+70c+9c^2-116c^3-\,-\,-$$

$$= (5+7c)^2-40c^2-116c^3-\,-\,-$$

$$= (5+7c)^2-(4c^2)(10+14c-4c^2)+(40c^2+56c^3-16c^4)-40c^2-\,-\,-$$

$$= (5+7c-4c^2)^2-60c^3-74c^4+62c^5+-\,-\,-$$

$$=(5+7c-4c^2)^2-(6c^3)(10+14c-8c^2-6c^3)+(60c^3+84c^4-48c^5-36c^6)-60c^3-\,-\,-$$

$$= (5+7c-4c^2-6c^3)^2+10c^4+14c^5-8c^6-12c^7+c^8$$

$$= (5+7c-4c^2-6c^3+c^4)^2.$$

$$\therefore \sqrt{(25+70c+9c^2-116c^3-58c^4+62c^5+28c^6-12c^7+c^8)} = 5+7c-4c^2-6c^3+c^4.$$

315. When there is no polynome operationally equivalent to the second root of a polynome p yet if there be a mononome operationally equivalent to the second root of p's first term art. 314's process may be used for finding a polynome u and a polynome v of less range than u such that p differs from u^2 by v. Then (art. 300) since

either $p=u^2+v$	or $p=u^2-v$
$p-u^2=v$	$u^2-p=v$
$(\sqrt{p}-u)(\sqrt{p}+u)=v$	$(u-\sqrt{p})(u+\sqrt{p})=v$
$\sqrt{p}-u=\dfrac{v}{\sqrt{p}+u}$	$u-\sqrt{p}=\dfrac{v}{u+\sqrt{p}}$
$\sqrt{p}=\dfrac{v}{\sqrt{p}+u}+u=u+\dfrac{v}{\sqrt{p}+u}$	$\sqrt{p}=u-\dfrac{v}{u+\sqrt{p}}=u-\dfrac{v}{\sqrt{p}+u}.$

Also if $v=w\omega$ $\sqrt{p}=u+\dfrac{w}{\sqrt{p}+u}\omega$ in the first case and $u-\dfrac{w}{\sqrt{p}+u}\omega$ in the other.

Thus $\because$ $1+x=1^2+x$

$$=\left(1+\frac{1}{2}x\right)^2-\frac{1}{4}x^2$$

$$=\left(1+\frac{1}{2}x-\frac{1}{8}x^2\right)^2+\left(\frac{1}{8}x^3-\frac{1}{64}x^4\right)$$

$$=\left(1+\frac{1}{2}x-\frac{1}{8}x^2\right)^2+\left\{\frac{1}{8}\left(1-\frac{1}{8}x\right)\right\}x^3$$

$$=\left(1+\frac{1}{2}x-\frac{1}{8}x^2+\frac{1}{16}x^3\right)^2-\left(\frac{5}{64}x^4-\frac{1}{64}x^5+\frac{1}{256}x^6\right)$$

$$=\left(1+\frac{1}{2}x-\frac{1}{8}x^2+\frac{1}{16}x^3\right)^2-\left\{\frac{1}{64}\left(5-x+\frac{1}{4}x^2\right)\right\}x^4$$

$$\begin{array}{l|l}
 & 1+\frac{1}{2}x-\frac{1}{8}x^2+\frac{1}{16}x^3 \\
\hline
 & 1\ +x \\
1 & 1 \\
2+\frac{1}{2}x & +x+\frac{1}{4}x^2 \\
 & -\frac{1}{4} \\
2+x-\frac{1}{8}x^2 & -\frac{1}{4}\quad -\frac{1}{8}x^3+\frac{1}{64}x^4 \\
 & +\frac{1}{8}\quad -\frac{1}{64} \\
2+x-\frac{1}{4}x^2+\frac{1}{16}x^3 & +\frac{1}{8}\quad +\frac{1}{16}\quad -\frac{1}{64}x^5+\frac{1}{256}x^6 \\
\hline
 & -\frac{5}{64}\quad +\frac{1}{64}\quad -\frac{1}{256}
\end{array}$$

and so on $\quad \sqrt{(1+x)}=1+\dfrac{1}{\sqrt{(1+x)}+1}x=1+\dfrac{1}{2}x-\dfrac{\frac{1}{4}}{\sqrt{(1+x)}+\left(1+\frac{1}{2}x\right)}x$

$$= 1+\frac{1}{2}x-\frac{1}{8}x^2+\frac{\frac{1}{8}\left(1-\frac{1}{8}x\right)}{\sqrt{(1+x)}+\left(1+\frac{1}{2}x-\frac{1}{8}x^2\right)}x^3$$

$$= 1+\frac{1}{2}x-\frac{1}{8}x^2+\frac{1}{16}x^3-\frac{\frac{1}{64}\left(5-x+\frac{1}{4}x^2\right)}{\sqrt{(1+x)}+\left(1+\frac{1}{2}x-\frac{1}{8}x^2+\frac{1}{16}x^3\right)}x^4 \quad \text{and so on.}$$

316. PROP. *To find the decimally denoted third root of a given decimally denoted number.*

First if u v be any two numerical quantities $(u+v)^3$
$= (u+v)\{u^2+v(2u+v)\}$ (art. 312) and $\therefore = u\{u^2+v(2u+v)\}+v\{u^2+(v\times 2u+v^2)\}$
$= u^3+uv(2u+v)+v(u^2+v\times 2u+v^2) = u^3+\{vu(2u+v)+v(u^2+v\times 2u+v^2)\}$
$= u^3+v\{u\times 2u+uv+(u^2+v\times 2u+v^2)\} = u^3+v(2u^2+vu+u^2+v\times 2u+v^2)$
$= u^3+v\{2u^2+u^2+(v\times 2u+vu)+v^2\} = u^3+v(3u^2+v\times 3u+v^2)$.

Hence next if a b $c \dots f$ g h be any numerical quantities

$$(a+b+c+\cdots+f+g+h)^3$$

$$= (a+b+c+\cdots+f+g)^3+h\left\{\begin{matrix} 3(a+b+c+\cdots+f+g)^2 \\ +h\times 3(a+b+c+\cdots+f+g) \\ +h^2 \end{matrix}\right\}$$

$$= (a+b+c+\cdots+f)^3+g\left\{\begin{matrix} 3(a+b+c+\cdots+f)^2 \\ +g\times 3(a+b+c+\cdots+f) \\ +g^2 \end{matrix}\right\}$$

$$+h\left\{\begin{matrix} 3(a+b+c+\cdots+f+g)^2 \\ +h\times 3(a+b+c+\cdots+f+g) \\ +h^2 \end{matrix}\right\}$$

$$= - \quad - \quad - \quad - \quad - \quad - \quad - \quad -$$

$$= a\left\{\begin{matrix} 3\times 0^2 \\ +a\times 3\times 0 \\ +a^2 \end{matrix}\right\}+b\left\{\begin{matrix} 3a^2 \\ +b\times 3a \\ +b^2 \end{matrix}\right\}+c\left\{\begin{matrix} 3(a+b)^2 \\ +c\times 3(a+b) \\ +c^2 \end{matrix}\right\}+d\left\{\begin{matrix} 3(a+b+c)^2 \\ +d\times 3(a+b+c) \\ +d^2 \end{matrix}\right\}+$$

$$---+h\left\{\begin{matrix} 3(a+b+c+\cdots+f+g)^2 \\ +h\times 3(a+b+c+\cdots+f+g) \\ +h^2 \end{matrix}\right\}.$$

Moreover $3(u+v)^2$ the first of the three terms summed in each pair of crow wing brackets but the first $=3(u^2+uv+vu+v^2)$ (arts. 162, 300) and $\therefore = 3u^2+3uv+3vu+(2v^2+v^2) = 3u^2+v\times 3u+v\times 3u+2v^2+v^2$ $= 3u^2+v\times 3u+v^2+v\times 3u+2v^2$ the sum of $3u^2+v\times 3u+v^2$ the sum of the three terms summed in the pair of crow wing brackets next before $v\times 3u$ the second of these terms and $2v^2$ twice the third.

From these theorems the decimally denoted third power may be found of a given decimally denoted number by taking $a\ b\ c \ldots$ the numbers denoted by the several digits in order and the more easily to get the products aa^2 $b(3a^2+b\times 3a+b^2)$ $c\left\{\begin{matrix}3(a+b)^2 \\ +c\times 3(a+b)+c^2\end{matrix}\right\}\ldots$ two auxiliary columns may be used the one for finding $3a$ $3(a+b)$ $3(a+b+c)\ldots$ (which are severally equal to $3a$ $3a+3b$ $3a+3b+3c\ldots$) and the other for finding through these and the theorem

		92·085
	8100	729
	162	
	243	
270	54	
	4	
	24844	49 688
	54	
	8	
	25392	
276	22·08	
	64	
	25414·0864	2 033·126 912
	22 08	
	128	
	25436·1792	
276·24	1 3812	
	25	
	25437·560425	127 187 802 125
		780 848·314 714 125

$$3(u+v)^2 = \left\{\begin{matrix}3u^2+v\times 3u\ +v^2 \\ \quad +v\times 3u+2v^2\end{matrix}\right.$$ the several multiplicands a^2 $3a^2+b\times 3a+b^2$ $3(a+b)^2+c\times 3(a+b)+c^2\ldots.$

Hence lastly the decimally denoted third root of a given decimally denoted number k may be found by taking k to be the sum of the products aa^2 $b(3a^2+b\times 3a+b^2)$ $c\{3(a+b)^2+c\times 3(a+b)+c^2\}\ldots$ got from the numbers $a\ b\ c\ldots$ denoted severally by the digits in order of the root sought and so finding these digits by trial in the following way. First the greatest number a denoted by a single significant digit must be found whose third power is not greater than k and then the greatest number b denoted by a single significant digit such that $b(3a^2+b\times 3a+b^2)$ is not greater than $k-a^3$. Since $k-a^3-b(3a^2+b\times 3a+b^2) = k-\{b(3a^2+b\times 3a+b^2)+a^3\}$ $= k-\{a^3+b(3a^2+b\times 3a+b^2)\} = k-(a+b)^3$ the greatest number c denoted by a single significant digit must next be found such that $c\{3(a+b)^2+c\times 3(a+b)+c^2\}$ is not greater than $k-a^3-b(3a^2+b\times 3a+b^2)$ or than its equal $k-(a+b)^3$. And so on. Thus step by step

$$780848{\cdot}314714125-(90)^3 = 51848{\cdot}314714125$$
$$780848{\cdot}314714125-(92)^3 = 2160{\cdot}314714125$$
$$780848{\cdot}314714125-(92{\cdot}08)^3 = 127{\cdot}187802125$$

780848·314714125 $-(92{\cdot}085)^3 = 0$, that is

780848·314714125 $= (92{\cdot}085)^3$

and this again is

$\sqrt[3]{780848{\cdot}314714125} = 92{\cdot}085$.

When the third root is incommensurable a decimally denoted number differing from it by less than any given number may be found by this NOTATIONAL THIRD ROOT PROCESS carried on far enough.

```
                                          92·085
                                   |780|848·314|714|125
              8100                 |729
              162                   |51
             -----
              243
      270      54
                4
             -----                  |49|688
             24844                   |2|160
              54
                8
             -----
             25392
      276      22·08
                  64
             -----------             |2|033 126|912
             25414·0864                |127 187|802
              22 08
                 128
             -----------
             25436·1792
      276·24   1 3812
                   25
             -------------
             25437·560425              |127 187|802·125
```

317. After several digits of a decimally denoted third root are found by art. 316's process one or more others may be found by a divisive process. Let the root be denoted by a row of $m+n$ digits of which the first m followed by n 0s would as a row without a decimal dot denote a whole number u and the other n as a dotless row a whole number v and let ω be the multiple or submultiple of the unit to which as unit the last digit of the row refers so that by the principle of digit knitting the root $= (u+v)\omega$ and therefore the number of which this is equal to the third root $= \{(u+v)\omega\}^3 = \{u^3+v(3u^2+v\times 3u+v^2)\}\omega^3$. If then after finding u by the notational third root process instead of going on

```
                                             0·736 806 30
                                            |0·4
              0·49                          | 343
               98                             |57
             ------
              1·47
      2·1      63
                9
             ------
              1·5339                          |46|017
               63                              |10|983
               18
             ------
              1·5987
      2·19     1314
                 36
             --------
              1·611876                         9|671|256
               1314                             |1|311|744
                 72
             --------
              1·625088
      2·208     17664
                   64
             ----------
              1·62685504                        |1|301|484|032
                17664                             |10|259|9|68
                  128
             ----------
              1·62862272                          9 771 7|4
      2·2104       |                               488 2|3
                   |                               488 5|9
```

further to find v as the equivalent of $\frac{v(3u^2+v\times 3u+v^2)}{3u^2+v\times 3u+v^2}$ there be taken $\frac{v(3u^2+v\times 3u+v^2)}{3u^2}$ which $=\frac{v\times 3u^2}{3u^2}+\frac{v^2\times 3u}{u\times 3u}+\frac{v^3}{3u^2}=v+\frac{v^2}{u}+\frac{v^3}{3u^2}$ there is got not $u+v$ but $u+\left(v+\frac{v^2}{u}+\frac{v^3}{3u^2}\right)$ which $=u+v+\left(\frac{v^2}{u}+\frac{v^3}{3u^2}\right)$. But $\because$ u not$<(10)^{n+m-1}$ and $v<(10)^n$

$$\frac{v^2}{u}+\frac{v^3}{3u^2}<\frac{\{(10)^n\}^2}{(10)^{n+m-1}}+\frac{\{(10)^n\}^3}{3\{(10)^{n+m-1}\}^2} \text{ and } \therefore <\frac{(10)^{2n}}{(10)^{n+m-1}}+\frac{(10)^{3n}}{3(10)^{2(n+m-1)}}.$$

Hence if $n+m-1$ not$<2n+1$, which is just if either m not$<n+2$ or $2(n+m-1)$ not$<3n+m$,

$$\frac{v^2}{u}+\frac{v^3}{3u^2}<\frac{1}{10}+\frac{1}{3(10)^m} \text{ and much more } \therefore <\frac{1}{5}.$$

And then instead of the root $(u+v)\omega$ there is got $(u+v)\omega+\left(\frac{v^2}{u}+\frac{v^3}{3u^2}\right)\omega$ which differs from the root by less than $\frac{1}{5}\omega$. Thus for example

$0{\cdot}4=(0{\cdot}7368)^3+0{\cdot}000010259968 \quad 3(0{\cdot}7368)^2=1{\cdot}62862272$

$\frac{0{\cdot}000010259968}{1{\cdot}62862272}$ is less than $0{\cdot}0000063$ by less than $0{\cdot}000000\,0003$ and $\sqrt[3]{0{\cdot}4}$ is less than $0{\cdot}7368063$ by less than $0{\cdot}000000\,002$. In going through the divisive process those digits only need be heeded that would have been the same or nearly the same had the root process been gone on with.

318. PROP. *To find the mononome or polynome, if there be one, operationally equivalent to the third root of a mononome or polynome.*

Inasmuch as $(bqx^ry^s)^3=b^3q^3x^{3r}y^{3s}$ the third root of apx^my^n can have a mononomic equivalent only when m n are multiples by 3 and then $\sqrt[3]{apx^my^n}=(\sqrt[3]{a})(\sqrt[3]{p})x^{\frac{1}{3}m}y^{\frac{1}{3}n}$.

For finding polynomic equivalents of the third roots of polynomes along with the two theorems

$(u+v)^3=u^3+v(3u^2+v\times 3u+v^2) \quad 3(u+v)^2=3u^2+v\times 3u+v^2+v\times 3u+2v^2$

given in art. 316 these other two are needed

$$\begin{aligned}(u-v)^3 &= (u-v)\{u^2-v(2u-v)\}=u\{u^2-v(2u-v)\}-v\{u^2-(v\times 2u-v^2)\}\\ &= u^3-uv(2u-v)-v(u^2+v^2-v\times 2u)=u^3-\{v(u^2+v^2-v\times 2u)+vu(2u-v)\}\\ &= u^3-v\{u^2+v^2-v\times 2u+(u\times 2u-uv)\}=u^3-v(u^2+v^2-v\times 2u+2u^2-vu)\\ &= u^3-v(2u^2+u^2-vu-v\times 2u+v^2)=u^3-v(3u^2-v\times 3u+v^2)\end{aligned}$$

$$3(u-v)^2 = 3(u^2-uv+v^2-vu) = 3u^2-3uv+(2v^2+v^2)-3vu$$
$$= 3u^2-v\times 3u+2v^2+v^2-v\times 3u = 3u^2-v\times 3u+v^2-v\times 3u+2v^2.$$

Then the polynomic expression operationally equivalent to the third root of a given polynomic expression is to be found as in the instance following:—

$$8+60x+102x^2-31x^3+216x^4+429x^5-610x^6+1071x^7-588x^8+343x^9$$
$$=(\sqrt[3]{8})^3-2^3+8+60x+\cdots$$
$$=2^3+\left[\frac{60x}{3\times 2^2}\left\{3\times 2^2+\frac{60x}{12}\times 3\times 2+\left(\frac{60}{12}x\right)^2\right\}-(5x)(12+30x+25x^2)\right]$$
$$+60x+102x^2-\cdots$$
$$=2^3+(5x)\{3\times 2^2+(5x)\times 3\times 2+(5x)^2\}-125x^3-150x^2-60x+60x$$
$$+102x^2-31x^3+\cdots$$
$$=(2+5x)^3-(150-102)x^2-(125+31)x^3+216x^4+\cdots$$
$$=(2+5x)^3-\frac{48x^2}{3\times 2^2}\left\{3(2+5x)^2-\frac{48x^2}{12}\times 3(2+5x)+\left(\frac{48}{12}x^2\right)^2\right\}$$
$$+(4x^2)\left\{\begin{matrix}12+30x+25x^2\\ +30x+50x^2-(4x^2)(6+15x)+(4x^2)^2\end{matrix}\right\}-48x^2-\cdots$$
$$=(2+5x-4x^2)^3+(4x^2)\begin{pmatrix}12+60x+75x^2\\ -24x^2-60x^3\\ +16x^4\end{pmatrix}$$
$$-48x^2-156x^3+216x^4+\cdots$$
$$=(2+5x-4x^2)^3+(4x^2)(12+60x+51x^2-60x^3+16x^4)-48x^2-156x^3+\cdots$$
$$=(2+5x-4x^2)^3+(48x^2-48x^2)+(240-156)x^3+(204+216)x^4$$
$$+(429-240)x^5-(610-64)x^6+1071x^7-\cdots$$
$$=(2+5x-4x^2)^3$$
$$+\left[\begin{matrix}\frac{84x^3}{3\times 2^2}\left\{3(2+5x-4x^2)^2+\frac{84x^3}{12}\times 3(2+5x-4x^2)+\left(\frac{84}{12}x^3\right)^2\right\}\\ -(7x^3)\left\{\begin{matrix}12+60x+51x^2-60x^3+16x^4\\ -24x^2-60x^3\\ +32x^4+(7x^3)(6+15x-12x^2)+(7x^3)^2\end{matrix}\right\}\end{matrix}\right]$$
$$+84x^3+420x^4+189x^5-546x^6+1071x^7-588x^8+343x^9$$
$$=(2+5x-4x^2)^3+(7x^3)\{3(2+5x-4x^2)^2+(7x^3)\times 3(2+5x-4x^2)+(7x^3)^2\}$$
$$-(7x^3)\begin{pmatrix}12+60x+27x^2-120x^3+48x^4\\ +42x^3+105x^4-84x^5\\ +49x^6\end{pmatrix}+\cdots$$
$$=(2+5x-4x^2+7x^3)^3-(7x^3)(12+60x+27x^2-78x^3+153x^4-84x^5+49x^6)$$
$$+84x^3+\cdots$$

$$= (2+5x-4x^2+7x^3)^3+(546x^6-546x^6)-(1071x^7-1071x^7)$$
$$+(588x^8-588x^8)-(343x^9-343x^9).$$

$$\therefore \sqrt[3]{(8+60x+102x^2-31x^3+216x^4+429x^5-610x^6+1071x^7-588x^8+343x^9)}$$
$$= 2+5x-4x^2+7x^3.$$

The result of the above process may be got without showing the steps by the algorithm on page 236 following.

In like manner $27c^6-108c^5+279c^4-424c^3+465c^2-300c+125$

$$= (3c^2)^3-(4c)\{3(3c^2)^2-(4c)\times 3\times 3c^2+(4c)^2\}+(4c)(27c^4-36c^3+16c^2)$$
$$-108c^5+ - - -$$

$$= (3c^2-4c)^3+108c^5-144c^4+64c^3-108c^5+279c^4-424c^3+ - - -$$

$$= (3c^2-4c)^3+135c^4-360c^3+465c^2-300c+125$$

$$= (3c^2-4c)^3+5\{3(3c^2-4c)^2+5\times 3(3c^2-4c)+5^2\}$$
$$-5\left\{\begin{matrix}27c^4-36c^3+16c^2 \\ -36c^3+32c^2+5(9c^2-12c)+25\end{matrix}\right\}+135c^4- - - -$$

$$= (3c^2-4c+5)^3-5\left(\begin{matrix}27c^4-72c^3+48c^2 \\ +45c^2-60c+25\end{matrix}\right)+135c^4-360c^3+ - - -$$

$$= (3c^2-4c+5)^3-125+300c-465c^2+360c^3-135c^4+135c^4-360c^3+465c^2$$
$$-300c+125$$

$$= (3c^2-4c+5)^3-125+\{300c-465c^2+(360c^3-360c^3)+465c^2-300c\}+125$$

and $\sqrt[3]{(27c^6-108c^5+279c^4-424c^3+465c^2-300c+125)} = 3c^2-4c+5.$

319. If a polynome p be not operationally equivalent to the third power of any polynome yet if p's first term be operationally equivalent to the third power of a mononome art. 318's process may be made use of to find two polynomes u v such that u^3 differs from p by v and v's range is not greater than twice a range less by 1 than u's. Then too

either $p = u^3+v$ and	or $p = u^3-v$ and
$v=p-u^3=(\sqrt[3]{p}-u)\{(\sqrt[3]{p})^2+(\sqrt[3]{p})u+u^2\}$	$v = u^3-p = (u-\sqrt[3]{p})\{u^2+u\sqrt[3]{p}+(\sqrt[3]{p})^2\}$
$\sqrt[3]{p} = u+\dfrac{v}{(\sqrt[3]{p})^2+(\sqrt[3]{p})u+u^2}$	$\sqrt[3]{p} = u-\dfrac{v}{(\sqrt[3]{p})^2+(\sqrt[3]{p})u+u^2}.$

And if $v = w\omega$ $\dfrac{w\omega}{(\sqrt[3]{p})^2+(\sqrt[3]{p})u+u^2} = \dfrac{w}{(\sqrt[3]{p})^2+(\sqrt[3]{p})u+u^2}\omega$. For example

$x^3+3x^5 = (x+x^3)^3-(3x^7+x^9) = (x+x^3)^3-(3+x^2)x^7$

$= (x+x^3-x^5)^3+(5x^9-3x^{13}+x^{15}) = (x+x^3-x^5)^3+(5-3x^4+x^6)x^9$ and so on.

$2 + 5x \quad - 4x^2 \quad + 7x^3$

$8 + 60x + 102x^2 - 31x^3 + 216x^4 + 429x^5 - 610x^6 + 1071x^7 - 588x^8 + 343x^9$

8

6

4
$+8$
12
$+30x$
$+25x^2$

$12 + 30 + 25$
$+30$
$+50$

$+60 \quad +150 \quad +125$

$-48 \quad -156$

$6 + 15x$

$12 + 60 + 75$
$-24 \quad -60x^3$
$+16x^4$

$12 + 60 + 51 \quad -60 \quad +16$
$-24 \quad -60$
$+32$

$-48 \quad -240 \quad -204 \quad +240 \quad -64$

$+84 \quad +420 \quad +189 \quad -546$

$6 + 15x - 12x^2$

$12 + 60 + 27 \quad -120 \quad +48$
$+42 \quad +105 \quad -84x^5$
$+49x^6$

$12 + 60 + 27 \quad -78 \quad +153 \quad -84 \quad +49$

$+84 \quad +420 \quad +189 \quad -546 \quad +1071 \quad -588 \quad +343$

						$x+x^3-x^5$			
						x^3+3x^5			
	x^3					x^3			
	$+2$								
	3								
$3x$		$+3x^4$							
			$+x^6$						
	3	$+3$	$+1$			$+3$	$+3x^7+x^9$		
		$+3$				-3	-1		
			$+2$						
	3	$+6$	$+3$						
$3x+3x^3$			-3	$-3x^8$					
					$+x^{10}$				
	3	$+6$		-3	$+1$	-3	-6	$+3x^{13}-x^{15}$	
							$+5$	-3	$+1$

Therefore $\sqrt[3]{(x^3+3x^5)}$

$$=x+x^3-\frac{3+x^2}{\{\sqrt[3]{(x^3+3x^5)}\}^2+\{\sqrt[3]{(x^3+3x^5)}\}(x+x^3)+(x+x^3)^2}x^7$$

$$=x+x^3-x^5+\frac{5-3x^4+x^6}{\{\sqrt[3]{(x^3+3x^5)}\}^2+\{\sqrt[3]{(x^3+3x^5)}\}(x+x^3-x^5)+(x+x^3-x^5)^2}x^9$$

and so on.

320. The notational and polynomic second and third root processses of arts. 312, 314, 316, 318, may be generalized into corresponding processes for roots of any degree. For the theorems whereon those processes rest are only particular cases of the following. Since (art. 186)

$$(u+v)^n = n|_0u^n+n|_1u^{n-1}v+n|_2u^{n-2}v^2+\cdots+n|_nv^n$$
$$(u-v)^n = n|_0u^n-n|_1u^{n-1}v+n|_2u^{n-2}v^2-\cdots\pm n|_nv^n$$

it follows that

$$(u+v)^n=u^n+v(nu^{n-1}+vn|_2u^{n-2}+v^2n|_3u^{n-3}+\cdots+v^{n-2}n|_{n-1}u+v^{n-1})$$
$$\text{and }(u-v)^n=u^n-v(nu^{n-1}-vn|_2u^{n-2}+v^2n|_3u^{n-3}-\cdots\cdots\pm v^{n-1}).$$

And $\because\ n(n-1)_{|r-1}=\frac{\{n-1-(r-1)+1\}\ldots(n-2)(n-1)}{\underline{|r-1}}n$

$$=\left(r\frac{1}{r}\right)\frac{(n-r+1)\ldots(n-1)n}{\underline{|r-1}}=r\frac{(n-r+1)\ldots(n-1)n}{\underline{|r}}\text{, that is } rn|_r,$$

$$n(u+v)^{n-1}=nu^{n-1}+vn|_2u^{n-2}+v^2n|_3u^{n-3}+\cdots\qquad+v^{r-1}n|_ru^{n-r}+\cdots\qquad+v^{n-1}$$
$$+vn|_2u^{n-2}+2v^2n|_3u^{n-3}+\cdots+(r-1)v^{r-1}n|_ru^{n-r}+\cdots+(n-1)v^{n-1}$$

$$n(u-v)^{n-1}=nu^{n-1}-vn|_2u^{n-2}+v^2n|_3u^{n-3}-v^3n|_4u^{n-4}+\cdots\cdots\qquad\pm v^{n-1}$$
$$-vn|_2u^{n-2}+2v^2n|_3u^{n-3}-3v^3n|_4u^{n-4}+\cdots\pm(n-1)v^{n-1}.$$

321. From the matter of art. 190 a general process may be drawn for finding the decimally denoted nth root of a given decimally denoted number a. Let $c_1\ c_2\ c_3 \ldots c_i$ be the numbers denoted by the digits in order of the root and let $\Pi(x-c)$ symbolize the polynome in $x-c$ operationally equivalent to x^n viewed as the polynome $x^n+0x^{n-1}+0x^{n-2}+\cdots+0x+0$. If $c_1\ c_2\ c_3 \ldots c_i$ were known there could be found in turn the polynomes $\Pi(x-c_1)$ $\Pi(x-c_1-c_2)$ $\Pi(x-c_1-c_2-c_3)$ - - - $\Pi(x-c_1-c_2-\cdots-c_i)$ and hence,—$\because$ $x-c_1-c_2-\cdots-c_i$ $=x-(c_i+\cdots+c_2+c_1)=x-(c_1+c_2+\cdots+c_i)$—, the polynomes $\Pi(x-c_1)$ $\Pi\{x-(c_1+c_2)\}$ $\Pi\{x-(c_1+c_2+c_3)\}$ - - - $\Pi\{x-(c_1+c_2+\cdots+c_i)\}$ of which the last terms,—because every term but the last of $\Pi(x-c)$ is 0 when x is c—, are severally c_1^n $(c_1+c_2)^n$ $(c_1+c_2+c_3)^n$ - - - $(c_1+c_2+\cdots+c_i)^n$. If then e_1 e_1+e_2 $e_1+e_2+e_3$ - - - $e_1+e_2+\cdots+e_i$ be those several last terms $a=e_1+e_2+\cdots e_i=e_2+e_3+\cdots+e_i+e_1$ $\therefore$ $a-e_1=e_2+e_3+\cdots+e_i$ and likewise $a-e_1-e_2=e_3+e_4+\cdots+e_i$ $a-e_1-e_2-e_3=e_4+e_5+\cdots+e_i$ - - - $a-e_1-e_2-\cdots-e_i=0$. Hence c_1 is the greatest number denoted by a single significant digit that makes e_1 not greater than a c_2 the greatest number denoted by a single significant digit that makes e_2 not greater than $a-e_1$ c_3 the greatest number denoted by a single significant digit that makes e_3 not greater than $a-e_1-e_2$ and so on and hence $c_1\ c_2\ c_3 \ldots$ may be found by trial. The process then of art. 190 has to be gone through with $c_1\ c_2\ c_3 \ldots$ in turn when so found but in the last column a $a-e_1$ $a-e_1-e_2$ $a-e_1-e_2-e_3$ - - - are to be written instead of 0 e_1 e_1+e_2 $e_1+e_2+e_3$ - - - severally. If the root can only be approximately denoted it is clear that after finding several digits of the denoting row one or more others may be found by taking heed of those digits alone which serve to fix the digits sought and which are either quite or almost unchanged throughout the whole after process. Thus from the work on the page next following

$$100=2^5+68=(2{\cdot}5)^5+2{\cdot}34375 \text{ and nearly } =(2{\cdot}51)^5+0{\cdot}3749374$$
$$(2{\cdot}511)^5+0{\cdot}1763229 \quad (2{\cdot}5118)^5+0{\cdot}0172037 \quad (2{\cdot}51188)^5+0{\cdot}0012805$$
$$(2{\cdot}511886)^5+0{\cdot}0000862 \quad (2{\cdot}5118864)^5+0{\cdot}0000066$$

and $\sqrt[5]{100}=2{\cdot}5118864$ right as far as the seventh place of decimals.

322. The root laws of arts. 305, 307, 310, may be put to such uses as the following.

(1). The order of greatness may be found of $a\sqrt[r]{b}$ $c\sqrt[s]{d}$ without finding them. For

$$a\sqrt[r]{b}=(\sqrt[r]{a^r})\sqrt[r]{b}=\sqrt[r]{a^r b}=\sqrt[r]{\sqrt[\rho]{(a^r b)^\rho}}=\sqrt[\rho r]{(a^r)\,b^\rho}=\sqrt[\rho r]{a^{\rho r} b^\rho}$$
$$\text{and } c\sqrt[s]{d}=\sqrt[\sigma s]{c^{\sigma s} d^\sigma}.$$

				2·5118864
0	0	0	0	100
2	4	8	16	32
2	4	8	16	68
2	8	24	64	
4	12	32	80	
2	12	48		
6	24	80		
2	16			
8	40			
2				
10				
0·5	5·25	22·625	51·3125	65·65625
10·5	45·25	102·625	131·3125	2·34375
5	5 5	25 375	64	
11	50 75	128	195 3125	
5	5 75	28 25		
11 5	56 5	156 25		
5	6			
12	62 5			
5				
12 5				
1	13	626	1 56876	1 9688126
13	62·6	156·876	196·88126	0·3749374
	13	628	1 57504	
	62·8	157 504	198 45630	
	13	629		
	62 9	158 133		
	13			
	63 0			
		63	15820	1986145
		158·20	198·6145	0·1763229
		63	15826	
		158 26	198 7728	
		63		
		158 32		
			1267	1591192
			198·899	0·0172037
			1267	
			199 026	
			127	159232
			199·04	0·0012805
			1	
			199 05	
				11943
				0·0000862
				796
				66

Hence if ρ σ be so taken that $\rho r = \sigma s =$ any common multiple v of r and s $a\sqrt[r]{b}$ $c\sqrt[s]{d}$ are in the same order of greatness as $a^v b^\rho$ $c^v d^\sigma$. Taking for example $3\sqrt{2}$ and $2\sqrt[5]{41}$ $3\sqrt{2} = \sqrt[5\times2]{3^{5\times2}.2^5}$ $2\sqrt[5]{41} = \sqrt[2\times5]{2^{2\times5}.(41)^2}$ and $3^{10}.2^5 = 2^5\times59049$ $2^{10}.(41)^2 = 2^5\times53792$ $\therefore 3\sqrt{2} > 2\sqrt[5]{41}$.

Instead of c d s putting b a r severally $a\sqrt[r]{b} = \sqrt[r]{a^r b} = \sqrt[r]{(ba)a^{r-1}}$ $b\sqrt[r]{a} = \sqrt[r]{b^r a} = \sqrt[r]{(ab)b^{r-1}}$ and therefore $a\sqrt[r]{b}$ $b\sqrt[r]{a}$ are in the same order of greatness as a^{r-1} b^{r-1} and therefore also as a b.

Making a and d both r and b and c both s

$$r\sqrt[r]{s} = \sqrt[sr]{r^{sr}s^s} = \sqrt[sr]{(s^s r^r)r^{(s-1)r}} \qquad s\sqrt[s]{r} = \sqrt[rs]{(r^r s^s)s^{(r-1)s}}$$

and therefore $r\sqrt[r]{s}$ $s\sqrt[s]{r}$ are in the same order of greatness as $r^{(s-1)r}$ $s^{(r-1)s}$. Thus $3\sqrt[3]{5} < 5\sqrt[5]{3}$ because $3^{(5-1)\times3} < 5^{(3-1)\times5}$.

And generally $r^{(s-1)r}$ $s^{(r-1)s}$ or their equivalents $(r^r)^{s-1}$ $(s^s)^{r-1}$ are in the same order of greatness as r s. For $(n+1)^n$ $= n^n + nn^{n-1} + n_{|2}n^{n-2} + \cdots + n_{|n-1}n + 1$ and $\therefore$ if $n > 1$ $(n+1)^n > n^n + nn^{n-1}$ or $2n^n$ $\therefore (n+1)^{n+1} > (n+1)\times 2n^n$ $\therefore \{(n+1)^{n+1}\}^{n-1} > \{2^{n-1}(n+1)^{n-1}\}(n^n)^{n-1}$. But $\because 2^{n-1}$ not $< n$ and $(n+1)^{n-1} > n^{n-1}$ $2^{n-1}(n+1)^{n-1} > nn^{n-1}$ or n^n. $\therefore \{(n+1)^{n+1}\}^{n-1} > n^n(n^n)^{n-1}$ or $(n^n)^n$. In like way

$$(n+i)^{n+i} > 2(n+i)(n+i-1)^{n+i-1} > 2(n+i)\times2(n+i-1)(n+i-2)^{n+i-2} > \text{- - -}$$
$$> 2(n+i)\ldots 2(n+2)\times 2(n+1)n^n$$

$\therefore \{(n+i)^{n+i}\}^{n-1} > \{2^{n-1}(n+i)^{n-1}\}\ldots\{2^{n-1}(n+2)^{n-1}\}\{2^{n-1}(n+1)^{n-1}\}(n^n)^{n-1}$ and $\therefore$ much more $> (n^n)^{n+i-1}$. Wherefore $r\sqrt[r]{s}$ $s\sqrt[s]{r}$ are in the same order of greatness as r s.

Again if a d be both s and b c both r $s\sqrt[r]{r} = \sqrt[sr]{s^{sr}r^s} = \sqrt[sr]{(r^s s^r)s^{sr-r}}$ $r\sqrt[s]{s} = \sqrt[rs]{(s^r r^s)r^{rs-s}}$ and therefore the order of greatness is the same of $s\sqrt[r]{r}$ $r\sqrt[s]{s}$ as of s^{sr-r} r^{rs-s} and therefore as of s r.

Also $\sqrt[r]{r} = \sqrt[sr]{r^s}$ $\sqrt[s]{s} = \sqrt[rs]{s^r}$ and therefore $\sqrt[r]{r}$ $\sqrt{s}$ are in the order of greatness of r^s s^r.

(2). Other expressions may be dealt with after the same manner. Since $\sqrt[r]{a}\sqrt[s]{b} = \sqrt[r]{\sqrt[s]{a^s b}} = \sqrt[sr]{(ba)a^{s-1}}$ and $\sqrt[r]{b}\sqrt[s]{a} = \sqrt[sr]{(ab)b^{s-1}}$ the order of greatness is the same of $\sqrt[r]{a}\sqrt[s]{b}$ $\sqrt[r]{b}\sqrt[s]{a}$ and of a b. Since $\sqrt[s]{a}\sqrt[r]{b} = \sqrt[rs]{(ba)a^{r-1}}$ $\sqrt[r]{a}\sqrt[s]{b}$ $\sqrt[s]{a}\sqrt[r]{b}$ are in the same order of greatness either as s r if $a > 1$ or as r s if $a < 1$. And since $(\sqrt[r]{a})\sqrt[s]{b} = (\sqrt[r]{\sqrt[s]{a^s}})\sqrt[s]{\sqrt[r]{b^r}} = \sqrt[sr]{a^s b^r} = \sqrt[sr]{(a^s b)b^{r-1}}$ the order of greatness of $\sqrt[r]{a}\sqrt[s]{b}$ $(\sqrt[r]{a})\sqrt[s]{b}$ is that of 1 b^{r-1} and therefore that of 1 b. Thus $\sqrt{3}\sqrt[3]{\frac{1}{2}}$ $(\sqrt{3})\sqrt[3]{\frac{1}{2}}$ $\sqrt[3]{3}\sqrt{\frac{1}{2}}$ $(\sqrt[3]{3})\sqrt{\frac{1}{2}}$ $\sqrt[3]{\frac{1}{2}}\sqrt{3}$ $\sqrt{\frac{1}{2}}\sqrt[3]{3}$ are in descending order of greatness.

(3). It often happens with other root expressions as with the above that there are expressions either equal or operationally equivalent thereto having fewer root symbols. Besides the cases given of this in arts. 309, 311, others are—

$$\sqrt[r]{xa^r}\pm\sqrt[r]{xb^r}\pm\sqrt[r]{xc^r}=(\sqrt[r]{x})a\pm(\sqrt[r]{x})b\pm(\sqrt[r]{x})c=(\sqrt[r]{x})(a\pm b\pm c).$$

$$(\sqrt[r]{a}+\sqrt[r]{b})\{(\sqrt[r]{a})^{r-1}-(\sqrt[r]{a})^{r-2}\sqrt[r]{b}+(\sqrt[r]{a})^{r-3}(\sqrt[r]{b})^2-\cdots-(\sqrt[r]{a})(\sqrt[r]{b})^{r-2}+(\sqrt[r]{b})^{r-1}\}$$
$$=a+b \text{ if } r \text{ be odd.}$$

$$\sqrt[u]{d\sqrt[t]{c\sqrt[s]{b\sqrt[r]{a}}}}=\sqrt[u]{\sqrt[t]{d^tc\sqrt[s]{b\sqrt[r]{a}}}}=\sqrt[tu]{\sqrt[s]{(d^t)^sc^sb\sqrt[r]{a}}}=\sqrt[stu]{d^{st}c^sb\sqrt[r]{a}}$$
$$=\sqrt[stu]{\sqrt[r]{(d^{st})^r(c^s)^rb^ra}}=\sqrt[rstu]{d^{rst}c^{rs}b^ra}.$$

And those expressions equal to the root expressions that have the fewest roots of the lowest degrees may be found as in the following instances.

$$\sqrt[3]{648}+\sqrt[3]{24}-\sqrt[3]{375}=\sqrt[3]{216\times3}+\sqrt[3]{8\times3}-\sqrt[3]{125\times3}=6\sqrt[3]{3}+2\sqrt[3]{3}-5\sqrt[3]{3}$$
$$=(6+2-5)\sqrt[3]{3}=3\sqrt[3]{3}.$$

$$\frac{1}{2}\sqrt{63}-2\sqrt{\frac{7}{9}}+5\sqrt{\frac{175}{36}}=\frac{1}{2}\sqrt{9\times7}-2\sqrt{\frac{1}{9}\times7}+5\sqrt{\frac{25}{36}\times7}$$
$$=\left(\frac{1}{2}\times3\right)\sqrt{7}-\left(2\times\frac{1}{3}\right)\sqrt{7}+\left(5\times\frac{5}{6}\right)\sqrt{7}=\left(\frac{3}{2}-\frac{2}{3}+\frac{25}{6}\right)\sqrt{7}=5\sqrt{7}.$$

$$(\sqrt{3\sqrt[3]{3}})(\sqrt{3})\sqrt[3]{3}=(\sqrt{\sqrt[3]{3^3.3}})(\sqrt{\sqrt[3]{3^3}})\sqrt[3]{\sqrt{3^2}}=\sqrt[6]{3^4.3^3.3^2}$$
$$=\sqrt[6]{3^{2+3+4}}=\sqrt[6]{3^6.3^3}=3\sqrt{\sqrt[3]{3^3}}=3\sqrt{3}.$$

$$\left(\sqrt[4]{\frac{5}{4}}\right)\left(\sqrt[3]{\frac{4}{3}}\right)\left(\sqrt{\frac{3}{2}}\right)\sqrt[4]{\frac{1}{2}\sqrt[3]{\frac{2}{3}\sqrt{\frac{3}{4}}}}$$
$$=\left\{\sqrt[3\times4]{\left(\frac{5}{4}\right)^3\left(\frac{4}{3}\right)^4\left(\frac{3}{2}\right)^6}\right\}\sqrt[2\times3\times4]{\left(\frac{1}{2}\right)^{2\times3}\times\left(\frac{2}{3}\right)^2\times\frac{3}{4}}=\left(\sqrt[12]{\frac{5^3.4^4.3^6}{2^6.3^4.4^3}}\right)\sqrt[24]{\frac{2^2.3}{4.3^2.2^6}}$$
$$=\left\{\sqrt[2\times12]{\left(\frac{5^3.3^2}{2^4}\right)^2}\right\}\sqrt[24]{\frac{1}{3.2^6}}=\sqrt[24]{\frac{5^6.3^4}{2^8}\times\frac{1}{3.2^6}}=\sqrt[24]{\frac{5^6.3^3}{2^{14}}}.$$

$$(5\sqrt{3})(3\sqrt{2}-2\sqrt{3})=(5\sqrt{3})\times3\sqrt{2}-(5\sqrt{3})\times2\sqrt{3}$$
$$=(5\times3)(\sqrt{3})\sqrt{2}-5\times2(\sqrt{3})^2=15\sqrt{3\times2}-30=15(\sqrt{6}-2).$$

$$\left(3\sqrt{\frac{3}{2}}-2\sqrt{2}\right)\left(2\sqrt{\frac{3}{2}}+3\sqrt{2}\right)$$
$$=\left(3\sqrt{\frac{3}{2}}\right)\left(2\sqrt{\frac{3}{2}}+3\sqrt{2}\right)-(2\sqrt{2})\left(2\sqrt{\frac{3}{2}}+3\sqrt{2}\right)$$
$$=3\times2\times\frac{3}{2}+3^2\sqrt{\frac{3}{2}\times2}-2\times3\times2-2^2\sqrt{2\times\frac{3}{2}}=9\sqrt{3}-4\sqrt{3}+9-12$$
$$=(9-4)\sqrt{3}-(12-9)=5\sqrt{3}-3.$$

$$\frac{\frac{3}{2}\sqrt[4]{\frac{5}{7}}}{\frac{5}{7}\sqrt[4]{\frac{3}{2}}}=\frac{\sqrt[4]{\left(\frac{3}{2}\right)^4\times\frac{5}{7}}}{\sqrt[4]{\left(\frac{5}{7}\right)^4\times\frac{3}{2}}}=\sqrt[4]{\frac{\left(\frac{3}{2}\right)^3\times\frac{3}{2}\times\frac{5}{7}}{\left(\frac{5}{7}\right)^3\times\frac{5}{7}\times\frac{3}{2}}}=\sqrt[4]{\left\{\frac{\left(\frac{3}{2}\right)}{\left(\frac{5}{7}\right)}\right\}^3}=\sqrt[4]{\left(\frac{3\times7}{5\times2}\right)^3}.$$

$$\frac{(\sqrt{3\sqrt[3]{3}})\sqrt[3]{3\sqrt{3}}}{(\sqrt{3})\sqrt[3]{3}}=\frac{(\sqrt[3\times2]{3^3.3})\sqrt[2\times3]{3^2.3}}{(\sqrt[3\times2]{3^3})\sqrt[2\times3]{3^2}}=\frac{\sqrt[6]{3^4.3^3}}{\sqrt[6]{3^3.3^2}}=\sqrt[6]{\frac{3^7}{3^5}}=\sqrt[6]{3^2}=\sqrt[3]{\sqrt{3^2}}=\sqrt[3]{3}.$$

$$\frac{(\sqrt{3})\times 2+(\sqrt[3]{4})\times 3}{(\sqrt[4]{5})\times 4}=\frac{(\sqrt{3})\times 2}{\{(\sqrt[4]{5})\times 2\}\times 2}+\frac{3\sqrt[3]{4}}{(\sqrt[4]{5})\times 4}=\frac{1}{2}\frac{\sqrt[2\times 2]{3^2}}{\sqrt[4]{5}}+3\frac{\sqrt[4\times 3]{4^4}}{\sqrt[3\times 4]{5^3.4^{3\times 4}}}$$

$$=\frac{1}{2}\sqrt[4]{\frac{3^2}{5}}+3\sqrt[12]{\frac{4^4}{(5^3.4^8).4^4}}=\frac{1}{2}\sqrt[4]{\frac{3^2}{5}}+3\sqrt[12]{\frac{1}{5^3.4^8}}.$$

(4). Quotients of divisions by divisors free from roots may be found that are equal to quotients of divisions by results of additions and subtractions with roots as in the examples below.

$$\frac{\sqrt{3}-\sqrt{2}}{\sqrt{3}+\sqrt{2}}=\frac{(\sqrt{3}-\sqrt{2})^2}{(\sqrt{3}+\sqrt{2})(\sqrt{3}-\sqrt{2})}=\frac{(\sqrt{3})^2+(\sqrt{2})^2-2(\sqrt{3})\sqrt{2}}{(\sqrt{3})^2-(\sqrt{2})^2}$$

$$=\frac{3+2-2\sqrt{3\times 2}}{3-2}=5-2\sqrt{6}.$$

$$\frac{1}{\sqrt{2}+\sqrt{3}-\sqrt{5}}=\frac{\sqrt{2}+\sqrt{3}+\sqrt{5}}{(\sqrt{2}+\sqrt{3}-\sqrt{5})(\sqrt{2}+\sqrt{3}+\sqrt{5})}=\frac{\sqrt{2}+\sqrt{3}+\sqrt{5}}{(\sqrt{2}+\sqrt{3})^2-5}$$

$$=\frac{(\sqrt{2}+\sqrt{3}+\sqrt{5})(\sqrt{2})\sqrt{3}}{\{2(\sqrt{2})\sqrt{3}\}(\sqrt{2})\sqrt{3}}=\frac{(\sqrt{2})^2\sqrt{3}+(\sqrt{3})^2\sqrt{2}+\sqrt{5\times 2\times 3}}{2(\sqrt{2})^2(\sqrt{3})^2}=\frac{2\sqrt{3}+3\sqrt{2}+\sqrt{30}}{12}.$$

$$\frac{4+\sqrt{3}+\sqrt{2}}{4-\sqrt{3}-\sqrt{2}}=\frac{(4+\sqrt{3}+\sqrt{2})(4-\sqrt{3}+\sqrt{2})}{(4-\sqrt{3}-\sqrt{2})(4-\sqrt{3}+\sqrt{2})}=\frac{(4+\sqrt{2})^2-3}{(4-\sqrt{3})^2-2}$$

$$=\frac{15+(2\times 4)\sqrt{2}}{17-(2\times 4)\sqrt{3}}=\frac{(15+8\sqrt{2})(17+8\sqrt{3})}{(17-8\sqrt{3})(17+8\sqrt{3})}=\frac{(15+8\sqrt{2})(17+8\sqrt{3})}{(17)^2-8^2.3}.$$

$$\frac{a}{\sqrt{x}+\sqrt{(a+x)}}=\frac{a\{\sqrt{(a+x)}-\sqrt{x}\}}{\{\sqrt{(a+x)}+\sqrt{x}\}\{\sqrt{(a+x)}-\sqrt{x}\}}=\frac{\{\sqrt{(a+x)}-\sqrt{x}\}a}{a+(x-x)}$$

$$=\sqrt{(a+x)}-\sqrt{x}.$$

$$\frac{\sqrt{\frac{u}{v}}+\sqrt{\frac{v}{u}}}{\sqrt{\frac{u}{v}}-\sqrt{\frac{v}{u}}}=\frac{\left(\sqrt{\frac{u}{v}}+\sqrt{\frac{v}{u}}\right)(\sqrt{v})\sqrt{u}}{\left(\sqrt{\frac{u}{v}}-\sqrt{\frac{v}{u}}\right)(\sqrt{v})\sqrt{u}}=\frac{\left\{\left(\sqrt{\frac{u}{v}}\right)\sqrt{v}\right\}\sqrt{u}+\left\{\left(\sqrt{\frac{v}{u}}\right)\sqrt{u}\right\}\sqrt{v}}{\left\{\left(\sqrt{\frac{u}{v}}\right)\sqrt{v}\right\}\sqrt{u}-\left\{\left(\sqrt{\frac{v}{u}}\right)\sqrt{u}\right\}\sqrt{v}}$$

$$=\frac{\left(\sqrt{\frac{u}{v}v}\right)\sqrt{u}+\left(\sqrt{\frac{v}{u}u}\right)\sqrt{v}}{\left(\sqrt{\frac{u}{v}v}\right)\sqrt{u}-\left(\sqrt{\frac{v}{u}u}\right)\sqrt{v}}=\frac{u+v}{u-v}.$$

$$\frac{1}{\sqrt[3]{5}-\sqrt[3]{4}}=\frac{(\sqrt[3]{5})^2+(\sqrt[3]{5})\sqrt[3]{4}+(\sqrt[3]{4})^2}{(\sqrt[3]{5}-\sqrt[3]{4})\{(\sqrt[3]{5})^2+(\sqrt[3]{5})\sqrt[3]{4}+(\sqrt[3]{4})^2\}}=(\sqrt[3]{5})^2+(\sqrt[3]{5})\sqrt[3]{4}+(\sqrt[3]{4})^2.$$

$$\frac{a}{b+c\sqrt[3]{d}}=\frac{a\{b^2-bc\sqrt[3]{d}+(c\sqrt[3]{d})^2\}}{(b+c\sqrt[3]{d})\{b^2-bc\sqrt[3]{d}+(c\sqrt[3]{d})^2\}}=\frac{a\{b^2-bc\sqrt[3]{d}+c^2(\sqrt[3]{d})^2\}}{b^3+c^3d}.$$

$$\frac{1}{\sqrt[3]{a}+\sqrt{b}}=$$

$$\frac{(\sqrt[6]{a^2})^5-(\sqrt[6]{a^2})^4\sqrt[6]{b^3}+(\sqrt[6]{a^2})^3(\sqrt[6]{b^3})^2-(\sqrt[6]{a^2})^2(\sqrt[6]{b^3})^3+(\sqrt[6]{a^2})(\sqrt[6]{b^3})^4-(\sqrt[6]{b^3})^5}{(\sqrt[6]{a^2}+\sqrt[6]{b^3})\{(\sqrt[6]{a^2})^5-(\sqrt[6]{a^2})^4\sqrt[6]{b^3}+(\sqrt[6]{a^2})^3(\sqrt[6]{b^3})^2-(\sqrt[6]{a^2})^2(\sqrt[6]{b^3})^3+(\sqrt[6]{a^2})(\sqrt[6]{b^3})^4-(\sqrt[6]{b^3})^5\}}$$

$$=\frac{(\sqrt[3]{a})^5-(\sqrt[3]{a})^4\sqrt{b}+ab-(\sqrt[3]{a})^2(\sqrt{b})^3+(\sqrt[3]{a})b^2-(\sqrt{b})^5}{a^2-b^3}.$$

$$\frac{1}{\sqrt[3]{a}-\sqrt[3]{b}+\sqrt[3]{c}}=\frac{(\sqrt[3]{a})^2+(\sqrt[3]{b})^2+(\sqrt[3]{c})^2+(\sqrt[3]{b})\sqrt[3]{c}-(\sqrt[3]{c})\sqrt[3]{a}+(\sqrt[3]{a})\sqrt[3]{b}}{a-b+c+3\sqrt[3]{abc}}$$

and this may be dealt with as $\frac{a}{b+c\sqrt[3]{d}}$.

PELICOTETICS

PART LAST

ALGEBRA

CHAPTER IV

THE PASSAGE FROM ARITHMETIC TO ALGEBRA

323. Of the fundamental laws of operational equivalence on which all Arithmetic hinges those into which additions and subtractions alone enter are

$$\begin{cases} a+b=b+a & \text{(i)} \\ a+(b+c)=a+b+c & \text{(ii)} \end{cases}$$

$$\begin{cases} a-b=a+c-(b+c) & \text{(iij)} \\ a-b-c=a-c-b & \text{(iv)} \\ a-(b+c)=a-c-b & \text{(v)} \end{cases}$$

$$\begin{cases} a+b-c=a-c+b & \text{(vj)} \\ a+(b-c)=a+b-c & \text{(vij)} \\ a-(c-b)=a+b-c & \text{(viij)} \end{cases}$$

where a b c are understood to be such numerical quantities as by the definitions of + – give the statements meaning. If however a b c be any numerical quantities while the first two of the statements always have meaning the rest sometimes have and sometimes have not. Because a not< b just when $a+c$ not< $b+c$ and a not< $b+c$ just when as shown in art. 179 after a not< c $a-c$ not< b and therefore further and ∵ $b+c=c+b$ a not< c $a-c$ not< b just when a not$<b$ $a-b$ not$<c$ in each of the statements (iij) (iv) (v) the expressions between which the mark = lies have at any one time either both meaning or both no meaning. But because when c not$>a+b$ it may happen that $c>a$ that $c>b$ that $c<b$ in each of the statements (vj) (vij) (viij) one of the two equated expressions may have meaning while the other has none.

When one only of the members of any of the equivalences (vj) (vij) (viij) is meaningless let trial be made to give it a meaning on the principle that the equivalence is to hold and that any meanings to be given to + − whenever these marks are as yet meaningless are to grow out naturally from the meanings which they already have. Let two adjoining portions OA AB of an endless straight line represent severally two magnitudes which in reference to a common unit magnitude are expressed numerically by a b so that the portion OB represents a magnitude expressed numerically by $a+b$. And taking (vj) first let a portion OC cut off close to O from OB represent a magnitude which in reference to the unit magnitude is expressed numerically by c and which is not greater than the magnitude represented by OB so that CB represents a magnitude expressed numerically by $a+b-c$. If OC not$>OA$ CA represents a magnitude expressed numerically by $a-c$ and therefore CB made up of CA AB a magnitude expressed numerically by $a-c+b$. Hence $a+b-c=a-c+b$. But if $OC>OA$ $a-c$ and therefore also $a-c+b$ is wholly unmeaning. Yet now as AC represents a magnitude expressed numerically by $c-a$ and AB is made up of AC CB the hint is easy that by taking $a-c$ when $a<c$ to mean the expression of a something which lessens by so much as is expressed by $c-a$ any magnitude put to it of the same kind as, and not less than, the magnitude expressed numerically by $c-a$ and $a-c+b$ the numerical expression of the magnitude got by thus putting to it a magnitude expressed numerically by b the magnitude represented by CB is still expressed numerically by $a-c+b$ and therefore $a+b-c=a-c+b$.

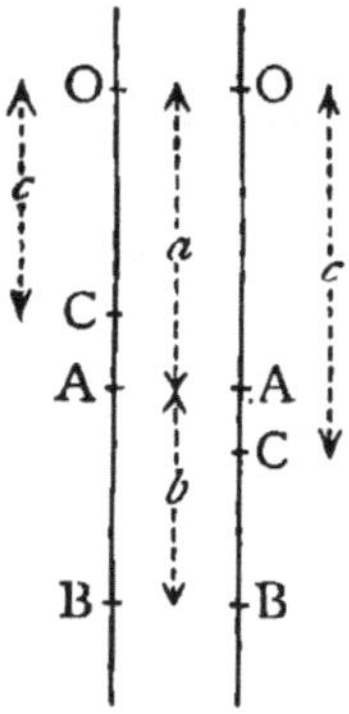

Again taking (vij) (viij) let a magnitude expressed numerically by c and not greater than the magnitude represented by OB be represented by a portion of straight line $C'B$ cut off from OB close to the end B so that OC' represents a magnitude expressed nu-

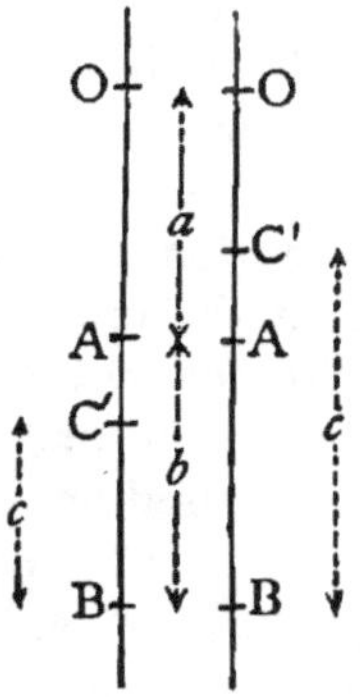

merically by $a+b-c$. Then if $C'B$ not$>$ AB AC' represents a magnitude expressed numerically by $b-c$ and therefore OC' a magnitude expressed numerically by $a+(b-c)$. Hence $a+b-c=a+(b-c)$. But if $C'B>AB$ $b-c$ and therefore too $a+(b-c)$ is utterly meaningless. Yet when $c>b$ by making $b-c$ mean the expression of a something which lessens by so much as is expressed numerically by $c-b$ any magnitude to which it is put of the same kind as, and not less than, a magnitude expressed numerically by $c-b$ and $a+(b-c)$ the numerical expression of the magnitude got by so putting it to a magnitude expressed numerically by a the magnitude represented by OC' is then also expressed numerically by $a+(b-c)$ and hence $a+b-c=a+(b-c)$. Once more if CB not$<$ AB $C'A$ represents a magnitude expressed numerically by $c-b$ and therefore and because OA is made up of OC' CA as parts OC' a magnitude expressed numerically by $a-(c-b)$. Hence $a+b-c=a-(c-b)$. But if $C'B<AB$ $c-b$ is without meaning and therefore so is $a-(c-b)$. Yet since then AC' represents a magnitude expressed numerically by $b-c$ and OC' is made up of the parts OA AC' by taking $c-b$ when $c<b$ to mean the expression of a something which greatens by so much as is expressed by $b-c$ any magnitude from which it is taken away of the same kind as the magnitude expressed numerically by $b-c$ and $a-(c-b)$ the numerical expression of the magnitude got by thus taking it away from a magnitude expressed numerically by a the magnitude represented by OC' is expressed by $a-(c-b)$ and therefore $a+b-c=a-(c-b)$.

Thus then in each of the operational equivalences (vj) (vij) (viij) whenever one of the equated members ceases to have meaning and the other does not according to the arithmetical definitions of the symbols used meaning may be given to the unmeaning member so as still to keep up the equivalence by first taking $u-(v+u)$ to be the expression of a something which on the one hand decreases by so much as is expressed numerically by v any magnitude of the same kind as, and not less than, the magnitude expressed by v either which is put to it or to which it is put and on the other hand increases by the same amount any magnitude of the said kind from which it is taken away and secondly making the marks $+$ and $-$ if not otherwise defined to stand for "changed by the putting thereto of" and "changed by the taking away therefrom of" severally.

The something here symbolized by $u-(v+u)$ through which a balance of meaning is brought about between the two members of each of the three equivalences dealt with is defined only by what happens when it undergoes certain operations that lead to strictly arithmetical results and are equivalent to strictly arithmetical operations. Whether or not therefore it have any existence apart from the operations thus undergone it may at least be used as an arithmetical tool. To get for it a symbol free from the nonessential element u make c a in the equivalence (v) and by the principle above used let

$$a-(b+a)=a-a-b, \text{ that is } 0-b \text{ or } -b.$$

Whence $-b$ which hitherto has no meaning is henceforth made to symbolize the same as $a-(b+a)$. For the sake of contrast make b 0 in the operational equivalence (i) and

$$a+0=0+a \quad \text{or} \quad a=+a.$$

Def. That which $-v$ symbolizes if v be a numerical quantity is called a MINUS QUANTITY and a numerical quantity when contrasted therewith is called a PLUS QUANTITY. Also $+v$ is read "PLUS v" $-v$ is read "MINUS v" and the numerical quantity v is called the ABSOLUTE VALUE both of $+v$ and of $-v$.

324. The thing symbolized by $-v$ has often a distinct existence. For example if a pounds be gained by one set of transactions and b pounds be lost by another set $a-b$ pounds are gained or $b-a$ pounds are lost by the two sets according as a is not less or not greater than b. But $a-b$ pounds may be said to be gained in both cases if a gain of $-c$ pounds be understood to mean a loss of c pounds and $b-a$ pounds may be said to be lost in both cases if a loss of $-c$ pounds be understood to mean a gain of c pounds.

Again a man now a years old was b years ago $a-b$ years old if a be not less than b. But if a be less than b the man may still be said to have been b years ago $a-b$ years old if $-c$ years old be taken to mean c years before birth. It may likewise be always said that b years ago was $b-a$ years before the man's birth if $-c$ years before birth be taken to mean c years after.

As a third example let O P Q be three parallel planes of which O Q are on the same side of P and a yards b yards severally distant therefrom. If $a>b$ Q is $a-b$ yards distant from O on the same side as P but if $a<b$ Q is $b-a$ yards distant from O on the opposite side. Yet Q never fails to be at once $a-b$ yards from O

on the P side and $b-a$ yards from O on the opposite side if only $-c$ yards from O on either side mean c yards from O on the other.

And generally whenever the difference between two magnitudes has a certain quality or a clean contrary one according as the one magnitude or the other is the greater the sign + written before the numerical quantity which expresses the difference may serve to mark either quality and then the sign – so written serves to mark the other. There is clearly nothing in any former use made of the signs + – to hinder them from standing as sheer prefixes for these relations of contrariety that magnitudes of the same kind may bear to one another.

325. Let it now be sought to generalize to the utmost the whole symbolic language and all the operations laws theorems and methods of Arithmetic according to

THE SYMBOLIZATION EXTENSION PRINCIPLE

That while always abiding unswervingly by whatever meanings are up to any time given to symbols all further meanings to be given are to spring up as straight as may be from, and to be bound by as close ties as may be to, those meanings and to this end that all operations of which the symbols are alike are to be named alike and are as much as may be to fulfil laws of operational equivalence of which the symbolic statements are alike.

ALGEBRA is the science built in this way on Arithmetic.

326. From what is done in art. 323 there comes by the symbolization extension principle the

Def. The (algebraic) SUBTRACTION FROM A NUMERICAL QUANTITY OF A GREATER NUMERICAL QUANTITY is the subtracting arithmetically from the latter the former and writing the sign – before the remainder.

327. As soon as a minus quantity is held to exist for its own sake and to be no longer only a handmaid of Arithmetic each of the equivalences (iij) (iv) (v) (viij) makes a wider assertion than has yet been proved. As to (iij) when $a<b \because b-a=b+c-(a+c)$ it is still true that

$$a-b=a+c-(b+c).$$

Next as to (v) when a not$<c$ but $<b+c \because a=a-c+c$

$$a-(b+c)=a-c+c-(b+c) \text{ and } \therefore =a-c-b$$

by what has just been proved. Hence as to (iv) when a not< either b or c but $a-b<c$ and $a-c<b$

$$a-b-c=a-(c+b)=a-(b+c)=a-c-b.$$

Lastly as to (viij) when $c>a+b$ let $c=d+b$ then

$$a-(c-b)=a-(d+b-b)=a-\{d+(b-b)\}=a-d$$

and $a+b-c=a+b-(d+b)=a-d$ by the wider reach now proved for the law of relativity. Therefore still

$$a-(c-b)=a+b-c.$$

328. At this stage all the eight equivalences (i)—(viij) are shown true throughout their new stretch of meaning but the balance of meaning between the two members of (vj) and of (vij) is again upset the two members of (iv) and of (v) are for the first time unbalanced in meaning and both members of (iv) are without meaning when $a<b$ and $<c$. Beginning then with (v),—for (iv) has nothing in itself whereon to hang any fresh meaning but will as in art. 67 follow from (v) through (i)—, if $a<c$ let $c=x+a$ and according to the principle of art. 325 let the unmeaning member be given a meaning such that

$$a-\{b+(x+a)\} \text{ and } \therefore\ a-(b+x+a)=a-(x+a)-b,$$
$$\text{that is } -(b+x)=-x-b.$$

Whence arises the

Def. The (algebraic) SUBTRACTION FROM A MINUS QUANTITY OF A PLUS QUANTITY is the adding arithmetically to the latter's absolute value the former's and writing the sign – before the sum.

The way in which this subtraction is drawn from the subtraction of art. 326 by making $a-c-b$ equivalent operationally to $a-(b+c)$ as well when $a<c$ as when a not< c shows that in any matter where what under certain circumstances is expressed by a plus quantity x is under other circumstances expressed by a minus quantity $-x$ if in the first case the expression $x-b$ when $x<b$ be put to any use to the like use may the expression $-x-b$ be put in the other case. For example a ship that sails first a miles northward and next c miles southward is $a-c$ miles north from where it started and if it then sail b miles southward it is $a-c-b$ or $a-(b+c)$ miles north of its starting place whether $a=x+c$ and $\therefore\ a-c-b$ $=x-b=x+c-(b+c)$ or $c=x+a$ and $\therefore\ a-c-b=-x-b=a-\{b+(x+a)\}$.

329. Since then $a-(b+c)=a-c-b$ whatever numerical quantities $a\ b\ c$ are it follows just as in art. 65 that whatever numerical quantities the letters stand for

$$\begin{aligned} x-(a+b+c+\cdots+f+g+h) &= x-h-(a+b+c+\cdots+f+g) \\ &= x-h-g-(a+b+c+\cdots+f) \\ &= \text{- - - - -} \\ &= x-h-g-f-\cdots-c-b-a \end{aligned}$$

and in particular when x is o

$$-(a+b+c+\cdots+f+g+h) = -h-g-f-\cdots-c-b-a.$$

Hence too

$$\begin{aligned} x-y-(a+b+c+\cdots+f+g+h) &= x-(a+b+c+\cdots+f+g+h+y) \\ &= x-y-h-g-f-\cdots-c-b-a \end{aligned}$$

of which a particular case is

$$-y-(a+b+c+\cdots+f+g+h) = -y-h-g-f-\cdots-c-b-a.$$

So that the law still holds good of the distribution of subtraction over additions of plus quantities.

The law of the commutation of subtractions of plus quantities also holds good. For taking $a\ b\ c...f\ g\ h$ in any other order $f\ a\ g...b\ h\ c$

$$\begin{aligned} x-a-b-c-\cdots-f-g-h &= x-(h+g+f+\cdots+c+b+a) \\ &= x-(c+h+b+\cdots+g+a+f) \\ &= x-f-a-g-\cdots-b-h-c \end{aligned}$$

and particularly when x is o $-a-b-c-\cdots-f-g-h$ $=-f-a-g-\cdots-b-h-c$. The equivalence (iv) is clearly a particular case of this general law.

330. The equivalence (vj) ceases to say aught when $c>a+b$ by the second member being unmeaning. Let $c=x+(a+b)$ and if a meaning is then to be given to $a-c+b$ so as to fulfil the equivalence

$$a+b-\{x+(a+b)\} = a-\{x+(a+b)\}+b = a-(x+a+b)+b = a-(x+b+a)+b,$$
$$\text{that is}\quad -x=-(x+b)+b.$$

Hence and from one of the elements in the definition of a minus quantity in art. 323 there comes (art. 325) the

Def. The (algebraic) ADDITION TO A MINUS QUANTITY OF A PLUS QUANTITY is the taking the arithmetical difference of their absolute values and prefixing thereto the algebraic sign of the one which has the greater absolute value.

This operation by reason of the way in which it has been got may be made use of just wherever the expression $a+b-c$ is made use of. Thus a gain of a pounds a gain of b pounds and a loss of c pounds give on the whole a gain of $a+b-c$ pounds and because

$a-c$ pounds is the gain from the a gain and the c loss there is on the whole a gain of $a-c+b$ pounds whether c not$>a$ $c>a$ but not$>a+b$ or $c>a+b$.

331. Because $a+b-c=a-c+b$ whatever numerical quantities a b c are

$$x+a+b+c+\cdots+f+g+h-p=x+a+b+c+\cdots+f+g-p+h$$
$$=x+a+b+c+\cdots+f-p+g+h$$
$$= - \quad - \quad - \quad - \quad -$$
$$=x-p+a+b+c+\cdots+f+g+h.$$

$$\therefore\ x+a+b+c+\cdots+g+h-p-q-r-\cdots-v-w$$
$$\because \text{(art. 329) it } =x+a+b+c+\cdots+g+h-(w+v+\cdots r+q+p)$$
$$=x-(w+v+\cdots+r+q+p)+a+b+c+\cdots+g+h$$
$$=x-p-q-r-\cdots-v-w+a+b+c+\cdots+g+h.$$

Under this comes the particular case of x being 0

$$a+b+c+\cdots+g+h-p-q-r-\cdots-v-w$$
$$=-p-q-r-\cdots-v-w+a+b+c+\cdots+g+h.$$

Hence $-b_1-b_2+a_1+a_2+a_3-b_3-b_4-b_5+a_4+a_5-b_6+a_6-\cdots\cdots$ taken for instance of a result of any successive additions and subtractions of plus quantities

$$=a_1+a_2+a_3-b_1-b_2-b_3-b_4-b_5+a_4+a_5-b_6+a_6-\cdots$$

hence in the same way

$$=a_1+a_2+a_3+a_4+a_5-b_1-b_2-b_3-b_4-b_5-b_6+a_6-\cdots$$
$$= - \quad - \quad - \quad - \quad - \quad - \quad - \quad - \quad -$$
$$=a_1+a_2+a_3+\cdots-b_1-b_2-b_3-\cdots\cdot$$

In like manner taking the a additions and the b subtractions in any other order of succession

$$a_3-b_5-b_2-b_4+a_6+a_1+a_5-b_6-b_3+a_4+a_2-b_1+\cdots$$
$$=a_3+a_6+a_1+a_5+a_4+a_2+\cdots-b_5-b_2-b_4-b_6-b_3-b_1-\cdots$$
$$=a_1+a_2+a_3+\cdots-b_1-b_2-b_3-\cdots$$
$$\text{and } \therefore =-b_1-b_2+a_1+a_2+a_3-b_3-b_4-b_5+a_4+a_5-b_6+a_6-\cdots$$

So that the law of the commutation of additions and subtractions of plus quantities is always true.

Moreover $x-p+a+b+c+\cdots+g+h=x+a+b+c+\cdots+g+h-p$

$$=x+(a+b+c+\cdots+g+h)-p$$
$$=x-p+(a+b+c+\cdots+g+h)$$

which when x is 0 gives in particular $-p+a+b+c+\cdots+g+h$

$=-p+(a+b+c+\cdots+g+h)$. The law then of the distribution of addition over additions of plus quantities is true for addition to a minus quantity as well as for addition to a plus quantity.

332. The first member of the equivalence (vij) has no meaning when $c>a+b$ and therefore the equivalence (vij) then makes no assertion. Let $c=x+(a+b)$ and by the symbolization extension principle a meaning is to be given to $a+(b-c)$ so that

$$a+b-\{x+(a+b)\}=a+[b-\{x+(a+b)\}]=a+\{b-(x+a+b)\},$$

that is so that $-x=a+\{-(x+a)\}$.

Whence and from one of the elements in the definition of a minus quantity in art. 323 there arises the

Def. The (algebraic) ADDITION TO A PLUS QUANTITY OF A MINUS QUANTITY is the taking the arithmetical difference of their absolute values and prefixing to it the algebraic sign of the one which has the greater absolute value.

From the way in which this operation arises it follows that wherever use can be made of the expression $a+b-c$ just there may the expression $a+(b-c)$ be made use of. Thus if $O\ X\ Y\ Z$ be four parallel straight lines such that X is a feet east of O Y b feet east of X and Z c feet west of Y then Z is $a+b-c$ feet east of O and because Z is $b-c$ feet east of X Z is also $a+(b-c)$ feet east of O whatever numerical quantities $a\ b\ c$ are.

333. Since $a+(b-c)=a+b-c$ whatever numerical quantities a b c are

$$\begin{aligned}x+(a-p-q-r-\cdots-v-w)&=x+\{a-(w+v+\cdots+r+q+p)\}\\&=x+a-(w+v+\cdots+r+q+p)\\&=x+a-p-q-r-\cdots-v-w.\end{aligned}$$

Hence $x+(-b_1-b_2+a_1+a_2+a_3-b_3-b_4-b_5+a_4+a_5-\cdots)$, taking the expression here added to x as instance of a result of additions and subtractions of plus quantities,

$$\begin{aligned}&=x+(a_1+a_2+a_3+\cdots-b_1-b_2-b_3-\cdots)\\&=x+(a_1+a_2+a_3+\cdots)-b_1-b_2-b_3-\cdots\\&=x+a_1+a_2+a_3+\cdots-b_1-b_2-b_3-\cdots\\&=x-b_1-b_2+a_1+a_2+a_3-b_3-b_4-b_5+a_4+a_5-\cdots\end{aligned}$$

And so addition in all the width of meaning now given to it still fulfils the law of distribution over additions and subtractions of plus quantities.

334. The element in the definition of a minus quantity got from equivalence (viij) gives according to art. 325 the

Def. The (algebraic) SUBTRACTION FROM A PLUS QUANTITY OF A MINUS QUANTITY is the adding arithmetically to the first the other's absolute value.

To take again the example in art. 332 of four parallel straight lines O X Y Z of which X is a feet east of O Y b feet east of X and Z c feet west of Y Z is $c-b$ feet west of X and therefore $a-(c-b)$ feet east of O even though $c<b$.

335. Because equivalence (viij) holds for all numerical quantities

$$\begin{aligned} x-(a-p-q-r-\cdots-v-w) &= x-\{a-(w+v+\cdots+r+q+p)\} \\ &= x+(w+v+\cdots+r+q+p)-a \\ &= x+w+v+\cdots+r+q+p-a. \end{aligned}$$

Hence $x-(-b_1-b_2+a_1+a_2+a_3-b_3-b_4-b_5+a_4+\cdots-b_\beta)$, taking the expression here subtracted from x to exemplify any result of additions and subtractions of plus quantities,

$$\begin{aligned} &= x-(a_1+a_2+a_3+\cdots-b_1-b_2-b_3-\cdots-b_\beta) \\ &= x+b_\beta+\cdots+b_3+b_2+b_1-(a_1+a_2+a_3+\cdots) \\ &= x+b_\beta+\cdots+b_3+b_2+b_1-\cdots-a_3-a_2-a_1 \\ &= x+b_\beta+\cdots-a_4+b_5+b_4+b_3-a_3-a_2-a_1+b_2+b_1. \end{aligned}$$

So that subtraction throughout the whole breadth of meaning which it now has fulfils the law of distribution over additions and subtractions of plus quantities.

336. The operation of adding to a minus quantity a plus quantity by reason of the laws now proved holds out a ready means of doing what amounts to distinguishing between desubtraction and sursubtraction as in arts. 60, 104, 145. Since on the one hand $k+b-b = k+(b-b)$ or k while on the other $-b+(b+k)=-b+b+k$ or k $a-b$ may be taken to symbolize the same as $a \llcorner b$ and $-b+a$ the same as $a \urcorner b$. Then if x be any numerical quantity operated on and a any numerical quantity operating there comes the system of inversions

Result of Operation	Result of the Inverse Operation
$x+a$	$x+a-a=x+(a-a)=x$
$x-a$	$x-a+a=x$
$a+x$	$-a+(a+x)=-a+a+x=x$
$-a+x$	$a+(-a+x)=a-a+x=x$
$a-x$	$-(a-x)+a=x-a+a=x$
$-x+a$	$a-(-x+a)=a-a+x=x$

Whence calling addition to a minus quantity of a plus quantity *minus addition of* or *minus addition to* according as the latter quantity or the absolute value of the former quantity is deemed the numerical quantity operated on each of the operations below is the inverse of the operation paired with it

Addition To and *Subtraction From*
Addition Of and *Minus Addition Of*
Subtraction Of and *Minus Addition To*

If in an endless unclosed line $X'OX$ two points A B be severally a yards b yards distant from O and both on the same side as X the distance from B to A in the direction toward X is expressed either numerically or algebraically by $-b+a$ in reference to a yard as unit while a point in $X'OX$ distant b yards from A on the same side as X' is $a-b$ yards distant from O in the direction OX. Likewise a man a years old was b years old $-b+a$ years ago and b years ago was $a-b$ years old.

337. Although through the meanings given in turn to $a-b$ $-a-b$ $-a+b$ $a+(-b)$ $a-(-b)$ whatever numerical quantities a b are there is now no unmeaning expression in any of the equivalences (i)—(viij) whatever numerical quantities a b c are and all these equivalences are now thoroughly fulfilled yet there are arithmetical consequences of these equivalences which with the widened meaning of the symbols both say more than has been proved and have in them unmeaning expressions. The statement that

$$a-b+(c-d)=a-b+c-d$$

is arithmetically, that is if a not$<b$ and c not$<d$, just the law (vij). If a not$<b$ and $c<d$ this statement is still (vij) but taken in the wider sense now (art. 332) understood. If $a<b$ and c not$<d$ there is still (art. 330) something said but a thing not yet proved. What is then said may however be thus proved:—

$$a-b+(c-d) \text{ (art. 330) } = a+(c-d)-b = a+c-d-b \text{ and } \therefore \text{ (art. 331) } = a-b+c-d.$$

Lastly if $a<b$ and $c<d$ $a-b+(c-d)$ has no meaning and therefore the statement says nothing. But then if $b=x+a$ and $d=y+c$ a meaning is by the symbolization extension principle to be given so that

$$\begin{aligned} a-(x+a)+\{c-(y+c)\} &= a-(x+a)+c-(y+c) \\ &= -x+c-c-y \text{ (art. 329)} \\ &= c-c-x-y \text{ (art. 331)}, \end{aligned}$$

that is $\quad -x+(-y)=-x-y.$

And hence arises the

Def. The (algebraic) ADDITION TO A MINUS QUANTITY OF A MINUS QUANTITY is the subtracting algebraically from the former the latter's absolute value.

This operation may be used for getting $a-b+(c-d)$ just wherever and only so far as $a-b+c-d$ when $a<b$ and $c<d$ expresses anything. Thus gains of $a-b$ pounds and $c-d$ pounds amount together to a gain of $a-b+(c-d)$ pounds whatever numerical quantities $a\ b\ c\ d$ may be. Also a rise of $a-b$ feet followed by a rise of $c-d$ feet gives always a rise from first to last of $a-b+(c-d)$ feet.

338. Because now $a-b+(c-d)=a-b+c-d$ whatever numerical quantities $a\ b\ c\ d$ are

$$x-y+(a-p-q-r-\cdots-v-w)=x-y+\{a-(w+v+\cdots+r+q+p)\}$$
$$=x-y+a-(w+v+\cdots+r+q+p)$$
$$=x-y+a-p-q-r-\cdots-v-w$$

of which a particular case is

$$-y+(a-p-q-r-\cdots-v-w)=-y+a-p-q-r-\cdots-v-w.$$

Hence $x-y+(-b_1-b_2+a_1+a_2+a_3-b_3-b_4-b_5+a_4+a_5-\cdots)$, taking the expression here added to $x-y$ for example of a result of any additions and subtractions of plus quantities,

$$=x-y+(a_1+a_2+a_3+\cdots-b_1-b_2-b_3-\cdots)$$
$$=x-y+(a_1+a_2+a_3+\cdots)-b_1-b_2-b_3-\cdots$$
$$=x-y+a_1+a_2+a_3+\cdots-b_1-b_2-b_3-\cdots$$
$$=x-y-b_1-b_2+a_1+a_2+a_3-b_3-b_4-b_5+a_4+a_5-\cdots.$$

In particular $-y+(-b_1-b_2+a_1+a_2+a_3-b_3-b_4-b_5+a_4+\cdots)$

$$=-y-b_1-b_2+a_1+a_2+a_3-b_3-b_4-b_5+a_4+\cdots.$$

Whence and from art. 333 the law of the distribution of addition over additions and subtractions of plus quantities holds universally.

339. The symbolic statement

$$a-b-(d-c)=a-b+c-d$$

a not$<b$ is the same as (viij) either in the arithmetical sense if d not$<c$ or in the further sense now (arts. 323, 334) understood if $d<c$. But if $a<b$ the statement either says what has not been proved if d not$<c$ or says nothing whatever if $d<c$. If $a<b$ and d not$<c$

$$a-b-(d-c)=a-(d-c+b)\ \text{(art. 328) and } \therefore \text{ (art. 335) } =a-b+c-d.$$

If $a<b$ and $d<c$ $a-b-(d-c)$ is an unmeaning expression but is

to be given meaning by the symbolization extension principle so that if $b=x+a$ and $c=y+d$

$$a-(x+a)-\{d-(y+d)\}=a-(x+a)+(y+d)-d=-x+y+d-d=d-d-x+y,$$

$$\text{that is} \quad -x-(-y)=-x+y.$$

Whence arises the

Def. The (algebraic) SUBTRACTION FROM A MINUS QUANTITY OF A MINUS QUANTITY is the adding algebraically to the former quantity the latter quantity's absolute value.

This operation may be put to use for getting $a-b-(d-c)$ just wherever the expression $a-b+c-d$ when $a<b$ and $d<c$ can be made use of. Thus a gain of $a-b$ pounds and a loss of $d-c$ pounds whatever numerical quantities $a\ b\ d\ c$ are amount to a gain of $a-b-(d-c)$ pounds. Also a rise of $a-b$ feet followed by a fall of $d-c$ feet gives on the whole always a rise of $a-b-(d-c)$ feet.

340. Because now $a-b-(d-c)=a-b+c-d$ whatever numerical quantities $a\ b\ d\ c$ are

$$\begin{aligned} x-y-(a-p-q-r-\cdots-v-w) &= x-y-\{a-(w+v+\cdots+r+q+p)\} \\ &= x-y+(w+v+\cdots+r+q+p)-a \\ &= x-y+w+v+\cdots+r+q+p-a \end{aligned}$$

and in particular

$$-y-(a-p-q-r-\cdots-v-w)=-y+w+v+\cdots+r+q+p-a.$$

Hence $x-y-(-b_1-b_2+a_1+a_2+a_3-b_3-b_4-b_5+a_4+\cdots-b_\beta)$, taking the expression here subtracted from $x-y$ for instance of a result of any additions and subtractions with plus quantities,

$$\begin{aligned} &= x-y-(a_1+a_2+a_3+\cdots-b_1-b_2-b_3-\cdots-b_\beta) \\ &= x-y+b_\beta+\cdots+b_3+b_2+b_1-(a_1+a_2+a_3+\cdots) \\ &= x-y+b_\beta+\cdots+b_3+b_2+b_1-\cdots-a_3-a_2-a_1 \\ &= x-y+b_\beta+\cdots-a_4+b_5+b_4+b_3-a_3-a_2-a_1+b_2+b_1 \end{aligned}$$

of which a particular case is

$$-y-(-b_1-b_2+a_1+a_2+\cdots-b_\beta)=-y+b_\beta+\cdots-a_2-a_1+b_2+b_1.$$

Whence and from art. 335 the universality is shown of the law of the distribution of subtraction over additions and subtractions of plus quantities.

341. The extensions of meaning thus far bestowed on the marks $+\,-$ sweep away all meaninglessness from any symbolic expression wherein $+\,-$ are the only symbols of operation and besides make every symbolic statement of what is conditionally a law of arithmetical additions and subtractions a statement of what is unconditionally a law of like named algebraic operations.

Every one of the eight results got by adding to or subtracting from a plus or minus quantity $\pm u$ a plus or minus quantity $\pm v$ is equivalent to one or other of the four simplest results $u+v$ $-u+v$ $u-v$ $-u-v$ as follows:—

$+u+(+v)=u+v$	$+u-(+v)=u-v$
$-u+(+v)=-u+v$	$-u-(+v)=-u-v$
$+u+(-v)=u-v$	$+u-(-v)=u+v$
$-u+(-v)=-u-v$	$-u-(-v)=-u+v$

Of these the particular cases when u is 0 are

$+(+v)=+v$	$-(+v)=-v$
$+(-v)=-v$	$-(-v)=+v.$

342. The laws of operational equivalence that rule in additions and subtractions of plus quantities rule in additions and subtractions of plus and minus quantities generally. If the upper or the lower of the signs $+\ -$ prefixed to any letter be used throughout

$$\begin{aligned}\pm a-(\pm b) &= \pm a+(\pm c\mp c)\mp b\\ &= \pm a\pm c\mp c\mp b\\ &= \pm a\pm c-(\pm b\pm c)\\ &= \pm a+(\pm c)-\{\pm b+(\pm c)\}.\end{aligned}$$

And so the law of relativity in subtraction is always true. The law answering to this in the operation having its result symbolized generally by $-(\pm b)+(\pm a)$ is

$$\begin{aligned}-(\pm b)+(\pm a) &= -(\pm b)-(\pm c)+(\pm c)+(\pm a)\\ &= \mp b\mp c\pm c\pm a\\ &= -(\pm c\pm b)+(\pm c\pm a)\\ &= -\{\pm c+(\pm b)\}+\{\pm c+(\pm a)\}.\end{aligned}$$

For proof of the commutation laws taking dotted reaches to mark any additions and subtractions of plus and minus quantities or the operationally equivalent additions and subtractions of plus quantities

$$\begin{aligned}\cdots\cdot\pm a\cdots\pm(-g)\cdots\pm(-p)\cdots\pm(-v)\cdots\cdot &=\cdots\cdot\pm a\cdots\cdot\mp g\cdots\cdot\mp p\cdots\mp v\cdots\cdot\\ &=\cdots\cdot\mp p\cdots\cdot\mp v\cdots\cdot\pm a\cdots\cdot\mp g\cdots\cdot\\ &=\cdots\pm(-p)\cdots\pm(-v)\cdots\pm a\cdots\pm(-g)\cdots\cdot\end{aligned}$$

So that in any additions and subtractions of plus and minus quantities an addition or a subtraction of a plus quantity may change place with an addition or a subtraction of a minus quantity and an addition or a subtraction of a minus quantity may change place with an addition or a subtraction of a minus quantity without changing the result. But any arrangement of things may be turned into any other arrangement sought of the things by making the first things that are not alike in the two arrangements change places in the first arrangement then doing the same in the new arrangement got with the first things that are not alike in it and in the sought arrangement and so on. Hence the law of the commutation of additions of subtractions and of additions and subtractions holds universally for plus and minus quantities.

Since the result of any additions and subtractions of plus and minus quantities is either a plus or a minus quantity the laws of distribution may be proved as follows:—

$$\begin{aligned} \pm a+\{\pm b\pm(-v)\cdots\} &= \pm a+(\pm b\mp v\cdots) \\ &= \pm a\pm b\mp v\cdots \\ &= \pm a\pm b\pm(-v)\cdots\cdots \\ \pm a-\{\pm b\pm(-v)\cdots\} &= \pm a-(\pm b\mp v\cdots) \\ &= \pm a\cdots \pm v\mp b \\ &= \pm a\cdots \mp(-v)\mp b. \end{aligned}$$

Special cases of these general laws are

$$+\{+(-v)\}=+(-v) \qquad -\{+(-v)\}=-(-v)$$
$$+\{-(-v)\}=-(-v) \qquad -\{-(-v)\}=+(-v).$$

343. Any plus quantity may by adding to it a plus quantity or by subtracting from it a minus quantity be changed into any greater plus quantity and inversely any plus quantity may by subtracting from it a plus quantity or by adding to it a minus quantity be changed into any less plus quantity. On this ground then after the manner of, and quite of a piece with all that has been done through, the symbolization extension principle that one of any two different plus or minus quantities may henceforth be called the ALGEBRAICALLY GREATER into which the other can be changed by adding to it a plus quantity or by subtracting from it a minus quantity and which therefore inversely by subtracting from it a plus quantity or by adding to it a minus quantity can be changed into the other. Hence any plus quantity is algebraically greater than any minus quantity and of any two minus quantities that have not the same absolute value the algebraically greater is the one with the less absolute value. Thus the terms of the doubly endless series

$$- \; - \; - \; -4 \quad -3 \quad -2 \quad -1 \quad 0 \quad 1 \quad 2 \quad 3 \quad 4 \; - \; - \; -$$

are in ascending order of algebraic greatness. When one of two plus or minus quantities is algebraically greater than the other this other must of course be called ALGEBRAICALLY LESS than the first. Also two minus quantities can be called EQUAL to one another only when they have the same absolute value.

The common final result got by algebraically adding in succession all but one of any plus or minus quantities in whatever order taken to that one and the successive results got is called the ALGEBRAIC SUM of the quantities. The result symbolized by $\pm a-(\pm b)$ is called the ALGEBRAIC REMAINDER of the subtraction from $\pm a$ of $\pm b$ and the result symbolized by $-(\pm b)+(\pm a)$ the ALGEBRAIC DIFFERENCE by which $\pm a$ differs from $\pm b$ or the ALGEBRAIC EXCESS of $\pm a$ over $\pm b$.

From the axiomatic principles of all magnitudinal relationship in arts. 6—16 it follows as in art. 143 that plus or minus quantities are equal precisely when the algebraic sums are equal got by severally adding either to them or them to any equal plus or minus quantities and also precisely when the algebraic remainders are equal got by severally subtracting either from them or them from any equal plus or minus quantities. As to inequalities too it follows likewise that the greater algebraically the plus or minus quantity algebraically added to, and the less algebraically the plus or minus quantity algebraically subtracted from, a plus or minus quantity the more is this quantity algebraically increased.

344. The fundamental laws of operational equivalence in Arithmetic that have to do with multiplications either alone or along with additions and subtractions only are

$$ab = ba \quad - \; - \; - \; - \quad \text{(ix)}$$

$$abc = (ab)c \quad - \; - \; - \quad \text{(x)}$$

$$\left.\begin{aligned}(a+b)c &= ac+bc\\ c(a+b) &= ca+cb\end{aligned}\right\} \quad - \; - \quad \text{(xj)}$$

$$\left.\begin{aligned}(a-b)c &= ac-bc\\ c(a-b) &= ca-cb\end{aligned}\right\} \quad - \; - \quad \text{(xij)}$$

Of these the last two marked (xij) are the only ones that ever cease to have arithmetical meaning and they only when $a<b$. But when $a<b$ the members $ac-bc$ $ca-cb$ of (xij) have been already given a meaning and so by the symbolization extension principle furnish a handle for giving meaning to the other members $(a-b)c$ $c(a-b)$. Beginning then with the equivalence $(a-b)c = ac-bc$ and making b $x+a$

$$\{a-(x+a)\}c = ac-(x+a)c = ac-(xc+ac)$$

$$\text{or} \quad (-x)c = -xc.$$

Whence springs the

Def. The (algebraic) MULTIPLICATION BY A MINUS QUANTITY OF A PLUS QUANTITY is the multiplying arithmetically by the former's absolute value the latter and writing the sign − before the product.

The use of this operation is to make $(a-b)c$ of the same breadth of meaning as $ac-bc$. Thus a man walking eastward at the rate of c miles an hour who passed a certain place a hours ago was b hours ago both $(a-b)c$ miles and $ac-bc$ miles east of the place whether a not$<b$ or $a<b$.

345. From the definition of $(-x)c$ and because thence $(a-b)c = ac-bc$ as well when $a<b$ as when a not$<b$ there follow in fulfilment as much as may be of multiplicational laws the same as (ix) (x) (xj) (xij)

$$(1)\quad (-f)bc = -fbc = -(fb)c = (-fb)c = \{(-f)b\}c.$$

$$(2)\quad (-u+v)w = (v-u)w = vw-uw = -uw+vw.$$

$$(3)\left\{\begin{aligned} &(-f+b)c = -fc+bc = (-f)c+bc.\\ &\{a+(-g)\}c = (a-g)c = ac-gc = ac+(-gc) = ac+(-g)c.\\ &\{-f+(-g)\}c = (-f-g)c = \{-(g+f)\}c = -(g+f)c = -(gc+fc)\\ &\quad = -fc-gc = -fc+(-gc) = (-f)c+(-g)c.\end{aligned}\right.$$

$$(4)\left\{\begin{aligned} &(-h)(a+b) = -h(a+b) = -(ha+hb) = -hb-ha = -ha-hb\\ &\quad = -ha+(-hb) = (-h)a+(-h)b.\end{aligned}\right.$$

$$(5)\quad (-u-v)w = -uw-vw \text{ as shown in (3) above.}$$

$$(6)\left\{\begin{aligned} &(-f-b)c = -fc-bc = (-f)c-bc.\\ &\{a-(-g)\}c = (a+g)c = ac+gc = ac-(-gc) = ac-(-g)c.\\ &\{-f-(-g)\}c = (-f+g)c = -fc+gc = -fc-(-gc) = (-f)c-(-g)c.\end{aligned}\right.$$

$$(7)\left\{\begin{aligned} &\text{If } a \text{ not}< b\ \ (-h)(a-b) = -h(a-b) = -(ha-hb) = hb-ha\\ &\quad = -ha+hb = -ha-(-hb) = (-h)a-(-h)b.\end{aligned}\right.$$

346. Again the equivalence $c(a-b) = ca-cb$ when b is $x+a$ becomes

$$c\{a-(x+a)\} = ca-c(x+a) = ca-(cx+ca)$$
$$\text{or } c(-x) = -cx.$$

Whence springs the

Def. The (algebraic) MULTIPLICATION BY A PLUS QUANTITY OF A MINUS QUANTITY is the multiplying arithmetically by the former the latter's absolute value and writing the sign − before the product.

The use of this operation is to give $c(a-b)$ the same breadth of meaning as $ca-cb$. Thus a ship sailing eastward at the rate of a

knots in a current running westward at the rate of b knots is $c(a-b)$ nautical miles and also $ca-cb$ east of where it was c hours ago as well if $a<b$ as if a not$<b$. The knots of a ship's logline are so adjusted that the number of them run out in half a minute is the number of nautical miles an hour at which the ship is going through the water.

347. From the definition of $c(-x)$ and because thence $c(a-b)$ $=ca-cb$ even though $a<b$

$$(1)\ (-f)b=-fb=-bf=b(-f).$$

$$(2)\ \begin{cases} a(-g)c=a(-gc)=-agc=-(ag)c=(-ag)c=\{a(-g)\}c. \\ ab(-h)=a(-bh)=-abh=-(ab)h=(ab)(-h). \end{cases}$$

And either in the same way as in art. 345 or from what is there shown by (1)

$$(3)\ (a+b)(-h)=a(-h)+b(-h). \qquad (4)\ w(-u+v)=-wu+wv.$$

$$(5)\ \begin{cases} c(-f+b)=c(-f)+cb. \quad c\{a+(-g)\}=ca+c(-g). \\ c\{-f+(-g)\}=c(-f)+c(-g). \end{cases}$$

(6) If a not$<b$ $(a-b)(-h)=a(-h)-b(-h)$. (7) $w(-u-v)=-wu-wv$.

$$(8)\ \begin{cases} c(-f-b)=c(-f)-cb. \quad c\{a-(-g)\}=ca-c(-g). \\ c\{-f-(-g)\}=c(-f)-c(-g). \end{cases}$$

348. The first of the two equivalences (xij) is arithmetically the same as

$$(a-b)(c-d)=a(c-d)-b(c-d).$$

It is by making the first of (xij) still the same as this when c not$<d$ and $a<b$ that the definition arises of multiplying by a minus quantity a plus quantity. When $c<d$ and a not$<b$ the equivalence is the law (6) proved in art. 347. When $c<d$ and $a<b$ $(a-b)(c-d)$ is an unmeaning expression and therefore the equivalence says nothing at all, but by the symbolization extension principle a meaning is then to be given to $(a-b)(c-d)$ so that if $b=x+a$ and $d=y+c$

$$\begin{aligned}\{a-(x+a)\}\{c-(y+c)\} &= a\{c-(y+c)\}-(x+a)\{c-(y+c)\}=a(-y)-(x+a)(-y) \\ &= -ay-\{-(x+a)y\}=-ay+(xy+ay)=-ay+(ay+xy) \\ &= -ay+ay+xy \\ &\text{or } (-x)(-y)=+xy.\end{aligned}$$

And hence there arises the

Def. The (algebraic) MULTIPLICATION BY A MINUS QUANTITY OF A MINUS QUANTITY is the arithmetically multiplying by the former's absolute value the latter's and writing the sign + before the product.

This operation gives the same reach of meaning and use to the expressions $(a-b)(c-d)$ $a(c-d)-b(c-d)$. For example a ship which sailing eastward at the rate of c knots in a current setting westward at the rate of d knots passed a certain place a hours ago was b hours ago east of the place $(a-b)(c-d)$ or $a(c-d)-b(c-d)$ nautical miles whatever numerical quantities a b c d are.

The statement

$$(a-b)(c-d)=(a-b)c-(a-b)d$$

if a not$<b$ is the second of the equivalences (xij) taken either arithmetically if c not$<d$ or as giving the definition (art. 346) of the multiplication by a plus quantity of a minus quantity if $c<d$ and if $a<b$ is either the law (7) proved in art. 345 if c not$<d$ or what may be proved thus if $c<d$:—If $b=x+a$ and $d=y+c$

$$(a-b)c-(a-b)d=(-x)c-(-x)(y+c)=-xc-\{-x(y+c)\}=-xc+(xy+xc)$$
$$=-xc+(xc+xy)=-xc+xc+xy=(-x)(-y)=(a-b)(c-d).$$

349. From the extensions of meaning now given to operations of multiplication and because through those extensions the operational equivalences

$$a(c-d)-b(c-d)=(a-b)(c-d)=(a-b)c-(a-b)d$$

hold whatever numerical quantities a b c d are

$$(1)\quad (-f)(-g)=fg=gf=(-g)(-f).$$

$$(2)\quad\left\{\begin{array}{l} a(-g)(-h)=agh=(ag)h=(-ag)(-h)=\{a(-g)\}(-h).\\ (-f)b(-h)=(-f)(-bh)=fbh=(fb)h=(-fb)(-h)=\{(-f)b\}(-h).\\ (-f)(-g)c=(-f)(-gc)=fgc=(fg)c=\{(-f)(-g)\}c.\\ (-f)(-g)(-h)=(-f)gh=-fgh=-(fg)h=(fg)(-h)\\ \quad=\{(-f)(-g)\}(-h).\end{array}\right.$$

$$(3)\quad (-u+v)(-w)=(v-u)(-w)=v(-w)-u(-w)=-u(-w)+v(-w).$$

$$(4)\quad\left\{\begin{array}{l} (-f+b)(-h)=-f(-h)+b(-h)=-(-fh)+b(-h)=fh+b(-h)\\ \quad=(-f)(-h)+b(-h).\\ \{a+(-g)\}(-h)=(a-g)(-h)=a(-h)-g(-h)=a(-h)-(-gh)\\ \quad=a(-h)+gh=a(-h)+(-g)(-h).\\ \{-f+(-g)\}(-h)=(-f-g)(-h)=\{-(g+f)\}(-h)=(g+f)h\\ \quad=gh+fh=fh+gh=(-f)(-h)+(-g)(-h).\end{array}\right.$$

(5) Either as in (3) or thence by (1) above and (1) of art. 347

$$(-w)(-u+v)=-(-w)u+(-w)v.$$

(6) Either as in (4) or therefrom by (1) and (1) art. 347

$$(-h)(-f+b) = (-h)(-f)+(-h)b. \quad (-h)\{a+(-g)\} = (-h)a+(-h)(-g).$$
$$(-h)\{-f+(-g)\} = (-h)(-f)+(-h)(-g).$$

$$(7) \begin{cases} (-u-v)(-w) = \{-(v+u)\}(-w) = (v+u)w = (u+v)w \\ = -(-uw)-(-vw) = -u(-w)-v(-w). \end{cases}$$

$$(8) \begin{cases} (-f-b)(-h) = -f(-h)-b(-h) = -(-fh)-b(-h) \\ = fh-b(-h) = (-f)(-h)-b(-h). \\ \{a-(-g)\}(-h) = (a+g)(-h) = a(-h)+g(-h) = a(-h)+(-gh) \\ = a(-h)-gh = a(-h)-(-g)(-h). \\ \{-f-(-g)\}(-h) = (-f+g)(-h) = (-f)(-h)+g(-h) \\ = (-f)(-h)+(-gh) = (-f)(-h)-gh = (-f)(-h)-(-g)(-h). \end{cases}$$

(9) Either as in (7) or thence by (1) and (1) art. 347

$$(-w)(-u-v) = -(-w)u-(-w)v.$$

(10) Either in the same way as in (8) or from (8) by (1) and (1) art. 347

$$(-h)(-f-b) = (-h)(-f)-(-h)b. \quad (-h)\{a-(-g)\} = (-h)a-(-h)(-g).$$
$$(-h)\{-f-(-g)\} = (-h)(-f)-(-h)(-g).$$

350. With the widened meaning of multiplication the laws of operational equivalence (ix) (x) (xj) (xij) are (arts. 345, 347, 349) true when *a b c* are any plus or minus quantities as well as when *a b c* are any numerical quantities. The general laws therefore of which these laws are the simplest special cases must follow from these for plus or minus quantities in the same way as in arts. 39, 41, 52, 54, 75, 77, if only it be true here as in multiplications with whole numbers there that the products are equal made by multiplying by equal plus or minus quantities equal plus or minus quantities.

Now two plus or minus quantities are equal precisely when the products are equal made by multiplying either them by or by them any equal plus or minus quantities that are not 0. For in multiplying any two equal plus or minus quantities either into or by other plus or minus quantities it is precisely when these others have the same absolute value that the products have the same absolute value and precisely when these others are either both plus quantities or both minus quantities that the products are either both plus or both minus.

Hence the laws of commutation colligation and distribution are fulfilled by all multiplications with plus or minus quantities.

351. In every plus or minus quantity there are two elements quite independent of one another the QUANTITATIVE element marked by the absolute value and the DESIGNATIVE element marked by the algebraic sign + or –. And the operation of multiplication has two parts the quantitative part and the designative part which have to do severally with the quantitative and with the designative element of the multiplicand or of the multiplier according as the one or the other is held to be the object operated on. These two parts of algebraic multiplication may be done in either of the two orders possible as shown in the following special cases of the laws (2) (5) of art. 345 (4) (7) of art. 347 (3) (5) (7) (9) of art. 349.

$$(+u)(+v) = \begin{Bmatrix} +u(+v) \\ +(+u)v \end{Bmatrix} = ++uv = +uv.$$

$$(-u)(+v) = \begin{Bmatrix} -u(+v) = -+uv \\ +(-u)v = +-uv \end{Bmatrix} = -uv.$$

$$(+u)(-v) = \begin{Bmatrix} +u(-v) = +-uv \\ -(+u)v = -+uv \end{Bmatrix} = -uv.$$

$$(-u)(-v) = \begin{Bmatrix} -u(-v) \\ -(-u)v \end{Bmatrix} = --uv = +uv.$$

Here what was before written $+(-c)$ to mark it the special case of $\pm a+(\pm b-c)$ when $a=0$ and $b=0$ is written $+-c$ as the bracket is now needless and the same is done with $+(+c)$ $-(+c)$ $-(-c)$.

If a b be made to stand for any plus or minus quantities it holds universally that

$$(+a)(+b) = +ab \quad (-a)(+b) = -ab \quad (+a)(-b) = -ab \quad (-a)(-b) = +ab.$$

For by the laws of distribution of multiplication of and by of addition and of subtraction if a b c d stand for any plus or minus quantities

$$(c+a)(d\pm b) = c(d\pm b)+a(d\pm b) = cd\pm cb+(ad\pm ab) = cd\pm cb+ad\pm ab$$

$$(c-a)(d\pm b) = c(d\pm b)-a(d\pm b) = cd\pm cb-(ad\pm ab) = cd\pm cb\mp ab-ad$$

even when c is 0 and d is 0.

The product of a multiplication expresses algebraically in reference to the plus unit magnitude what the multiplier expresses algebraically in reference to that as plus unit magnitude which the multiplicand expresses algebraically in reference to the plus unit magnitude. Also what the product expresses has to what the multiplicand expresses the same relation in respect of both quantity and designation as what the multiplier expresses in reference to any plus unit magnitude has to that plus unit magnitude. Thus if what is expressed algebraically by *a* in one scale of magnitude be expressed algebraically by +1 in another what is expressed algebraically by *b* in this other scale is expressed algebraically by *ba* in the first.

352. The rest of the fundamental arithmetical laws of operational equivalence deal with divisions and are

$$\frac{a}{b}=\frac{ac}{bc} \quad - - - - - \quad \text{(xiij)}$$

$$\frac{\left(\frac{a}{b}\right)}{c}=\frac{\left(\frac{a}{c}\right)}{b} \quad - - - - - \quad \text{(xiv)}$$

$$\frac{\left(\frac{a}{b}\right)}{c}=\frac{a}{cb} \quad - - - - - \quad \text{(xv)}$$

$$\frac{ab}{c}=a\frac{b}{c} \quad - - - - - \quad \text{(xvj)}$$

$$\frac{ab}{c}=\frac{a}{c}b \quad - - - - - \quad \text{(xvij)}$$

$$\frac{ab}{c}=\frac{a}{\left(\frac{c}{b}\right)} \quad - - - - - \quad \text{(xviij)}$$

$$\frac{a+b}{c}=\frac{a}{c}+\frac{b}{c} \quad - - - - - \quad \text{(xix)}$$

$$\frac{a-b}{c}=\frac{a}{c}-\frac{b}{c} \quad - - - - - \quad \text{(xx)}$$

These yield of themselves no ground whereon to rear by the symbolization extension principle a meaning for $\frac{a}{b}$ when either or each of the letters *a b* stands for a minus quantity. But since (art. 350) all plus or minus quantities are equal which if multiplied severally into a certain plus or minus quantity else than 0 give for product a certain plus or minus quantity else than 0 a meaning is now bestowed on $\frac{a}{b}$ thoroughly at one (arts. 114, 299) with all former meanings and indeed led straight up to therefrom by the

Def. The symbol $\frac{a}{b}$ stands for the plus or minus quantity which if multiplied into the plus or minus quantity b gives a product equal to the plus or minus quantity a. And the plus or minus quantity symbolized by $\frac{a}{b}$ is called the QUOTIENT OF THE DIVISION OF a BY b.

Hence the operation of Division has now meaning and use of the same width as multiplication and if u v be any numerical quantities

$$\frac{+u}{+v}=+\frac{u}{v} \qquad \frac{-u}{+v}=-\frac{u}{v} \qquad \frac{+u}{-v}=-\frac{u}{v} \qquad \frac{-u}{-v}=+\frac{u}{v}.$$

For $\frac{u}{v}$ is the numerical quantity which multiplied arithmetically into v gives a product equal to u and it is only a plus quantity which multiplied algebraically into a plus quantity produces a plus quantity or into a minus quantity a minus and only a minus quantity which multiplied into a plus quantity produces a minus quantity or into a minus quantity a plus. More generally if a b be any plus or minus quantities

$$\frac{+a}{+b}=+\frac{a}{b} \qquad \frac{-a}{+b}=-\frac{a}{b} \qquad \frac{+a}{-b}=-\frac{a}{b} \qquad \frac{-a}{-b}=+\frac{a}{b}.$$

For $\frac{a}{b}b=a$ by the definition of $\frac{a}{b}$ and (art. 351) $\left(\pm\frac{a}{b}\right)(+b)=\pm\frac{a}{b}b$ $\left(\pm\frac{a}{b}\right)(-b)=\mp\frac{a}{b}b$.

In a division the dividend and the divisor refer to the same plus unit magnitude and that which the dividend expresses algebraically is expressed algebraically by the quotient in reference to that as plus unit magnitude which the divisor expresses algebraically. Also what the quotient expresses in reference to any plus unit magnitude has to that plus unit magnitude the same relation in respect both of quantity and of designation as what the dividend expresses has to what the divisor expresses each in reference to a common plus unit magnitude. Thus if what is expressed algebraically by a in one scale of magnitude be expressed algebraically by $+1$ in another what is expressed algebraically by b in the first scale is expressed algebraically by $\frac{b}{a}$ in the other.

353. From the definition of algebraic division (art. 352) and because (art. 350) two plus or minus quantities are equal precisely

when the products are equal made by multiplying either them severally by or by them severally equal plus or minus quantities that are not o it follows that two plus or minus quantities are equal precisely when the quotients are equal got from dividing either them severally by or by them severally any equal plus or minus quantities that are not o. For if a a' be any plus or minus quantities and b b' any plus or minus quantities but o that are equal to one another $a=a'$ precisely when $\frac{a}{b}b=\frac{a'}{b'}b'$ and therefore precisely when $\frac{a}{b}=\frac{a'}{b'}$ also $a=a'$ precisely when $ba'=b'a$, that is $\left(\frac{b}{a}a\right)a'$ $=\left(\frac{b'}{a'}a'\right)a$, therefore precisely when $\frac{b}{a}aa'=\frac{b'}{a'}a'a$ and therefore, $\because$ $aa'=a'a$, precisely when $\frac{b}{a}=\frac{b'}{a'}$.

Hence if throughout arts. 116, 118, 120, 122, 124, 127, 130, the letters be taken to stand for any plus or minus quantities instead of for any numbers the laws there proved for numbers become proved for plus or minus quantities generally and hence all the laws into which divisions enter whether of relativity of commutation of colligation or of distribution hold for all plus or minus quantities.

354. The Laws of Operational Equivalence that rule all Additions Subtractions Multiplications and Divisions are now shown to be the same for plus or minus quantities as for numerical quantities. It is besides shown that two plus or minus quantities are equal precisely when equal sums are got in adding either to them or them to any equal plus or minus quantities precisely when equal remainders are got in subtracting either from them or them from any equal plus or minus quantities precisely when equal products are made in multiplying either them by or by them any equal plus or minus quantities but o and precisely when equal quotients arise in dividing either them by or by them any equal plus or minus quantities but o. Now it is just these very laws and tests of equality that throughout their arithmetical stretch lead to all results of Arithmetic beyond those got immediately from the several operations. Hence all general processes methods and theorems dealing with numerical quantities are only special cases of like processes methods and theorems dealing universally with plus or minus quantities that have severally the very same proofs. Indeed the broad algebraic proof is often shorter than the narrow arithmetical inasmuch as the sundry cases which may befall in the latter are embraced at one sweep in the former.

355. The fundamental power laws or laws of indices are

$$a^n a^m = a^{m+n} \qquad\qquad (a^n)^m = a^{mn}$$

	If m not< n	If m not> n
$\left(\frac{1}{a}\right)^n a^m =$	a^{m-n}	$\left(\frac{1}{a}\right)^{-m+n}$
$\frac{1}{a^n} a^m =$	a^{m-n}	$\frac{1}{a^{n-m}}$
$\left.\begin{matrix}\frac{a^m}{a^n} \\ a^m\frac{1}{a^n}\end{matrix}\right\} =$	a^{-n+m}	$\frac{1}{a^{-m+n}}$
$a^m\left(\frac{1}{a}\right)^n =$	a^{-n+m}	$\left(\frac{1}{a}\right)^{n-m}$

$$a^n b^n = (ab)^n \qquad\qquad \frac{a^n}{b^n} = \left(\frac{a}{b}\right)^n.$$

In these m n are whole numbers as hitherto but since (art. 43) the meaning of the word *power* must always answer to the meaning of the word *multiplication* a b are now understood to be any plus or minus quantities.

Since if m be any whole number and n any whole number but 1 $\frac{1}{n}$ and $\frac{m}{n}$ are by definition such that $n\frac{1}{n} = 1$ and $m\frac{1}{n} = \frac{m}{n}$ the unmeaning symbols $a^{\frac{1}{n}}$ $a^{\frac{m}{n}}$ are by the symbolization extension principle (art. 325) to be given a meaning such that after the second of the five fundamental index laws

$$(a^{\frac{1}{n}})^n = a^{n\frac{1}{n}} = a \qquad (a^{\frac{1}{n}})^m = a^{m\frac{1}{n}} = a^{\frac{m}{n}}.$$

Had then multiplications to do only with numerical quantities $a^{\frac{1}{n}}$ would symbolize the same thing as $\sqrt[n]{a}$ or the nth root of a. But minus quantities enter into multiplications as well as plus quantities. A minus quantity $-v$ has the same second power v^2 as the plus quantity $+v$ of the same absolute value. Besides other things than even plus or minus quantities may hereafter be found to undergo or result from operations called in some extended sense of the term *multiplications*. Whence the following definitions have to be made.

Def. Any object or any result of an arithmetical or algebraic operation is called an ALGEBRAIC QUANTITY.

Def. If a be any algebraic quantity and m n any two whole numbers any algebraic quantity of which the nth power is equal to a is symbolized by $a^{\frac{1}{n}}$ and called indifferently AN nTH ROOT OF a and the $\frac{1}{n}$TH POWER OF a and the mth power of what $a^{\frac{1}{n}}$ symbolizes is symbolized by $a^{\frac{m}{n}}$ and called the $\frac{m}{n}$TH POWER OF a.

If m be a multiple rn of n $a^{\frac{rn}{n}} = (a^{\frac{1}{n}})^{rn}$ by definition and $\therefore$ $= \{(a^{\frac{1}{n}})^n\}^r$ if $a^{\frac{1}{n}}$ be a plus or minus quantity or whatever else fulfils the second index law and then therefore by definition $= a^r$.

356. If the result of a certain operation performed on anything a be symbolized by fa then the result of the same operation performed on fa is symbolized by ffa the result of the same operation performed on ffa is symbolized by $fffa$ and so on. If further the symbol f have no meaning by itself apart from a symbol of something operated on, or in other words be a symbol of pure operation, the result ffa may be symbolized by f^2a the result $fffa$ by f^3a and generally the final result of n successive operations f performed severally on a and the successive results got may be handily symbolized by f^na. The operation thus compounded of the n simple f operations and symbolized by f^n may be called the nTH POWER of the operation symbolized by f. Then f^nf^ma means the very same as $f^{m+n}a$ and $f^t \ldots f^pf^nf^ma$ the very same as $f^{m+n+p+\cdots+t}a$ for there is here nothing answering to the law of the colligation of multiplications. Hence too $(f^n)^ma$ means the very same as $f^{mn}a$.

Thus if a be a plus or minus quantity since $+$ and $-$ are symbols purely operational the result symbolized by $++\cdots+++a$ with $+$ written n times may be symbolized by $(+)^na$ and the result symbolized by $--\cdots---a$ with $-$ written n times may be symbolized by $(-)^na$. Hence $(+)^na = +a$ $(-)^na = +a$ or $-a$ according as n is even or odd

$$(+)^{t'}(-)^t \ldots (+)^{n'}(-)^n(+)^{m'}(-)^ma = (-)^t \ldots (-)^n(-)^ma = (-)^{m+n+\cdots+t}a.$$

If m n be whole numbers else than 1 $f^{\frac{1}{n}}a$ and $f^{\frac{m}{n}}a$ may be given meanings such that

$$(f^{\frac{1}{n}})^na = f^{n\frac{1}{n}}a = fa \qquad (f^{\frac{1}{n}})^ma = f^{m\frac{1}{n}}a = f^{\frac{m}{n}}a.$$

And so $f^{\frac{1}{n}}$ symbolizes any operation whose nth power yields the same result as the operation f and $f^{\frac{m}{n}}$ the mth power of the operation symbolized by $f^{\frac{1}{n}}$.

357. If a b be numerical quantities m n r s whole numbers of which neither n nor s is 1 and heed be taken only of arithmetical roots

$$a^{\frac{m}{n}} = (\sqrt[n]{a})^m = \{(\sqrt[t]{\sqrt[n]{a}})^t\}^m \text{ (if } t \text{ be any whole number but 1)}$$

$$= (\sqrt[nt]{a})^{mt} = a^{\frac{mt}{nt}}$$

$a^{\frac{r}{s}} a^{\frac{m}{n}} = (\sqrt[ss']{a})^{rs'} (\sqrt[nn']{a})^{mn'}$ (s' n' being any whole numbers and if so taken that $ss' = nn'$)

$$= (\sqrt[nn']{a})^{mn'+rs'} = a^{\frac{m}{n}+\frac{r}{s}}$$

$(a^{\frac{r}{s}})^{\frac{m}{n}} = \{\sqrt[n]{(\sqrt[s]{a})^r}\}^m = \{\sqrt[nn']{(\sqrt[sr']{a})^{rr'}}\}^{mn'}$ (n' r' being any whole numbers and if taken so that $nn' = rr'$)

$$= (\sqrt[sr']{a})^{mn'} = a^{\frac{mn'}{sr'}} = a^{\frac{m}{n}\frac{r}{s}}$$

$$\left(\frac{1}{a}\right)^{\frac{r}{s}} a^{\frac{m}{n}} \left\{ \begin{array}{l} \left(\text{if } \frac{m}{n} \text{ not} < \frac{r}{s}\right) = \left(\frac{1}{a}\right)^{\frac{r}{s}} a^{\frac{m}{n}-\frac{r}{s}+\frac{r}{s}} = \left(\frac{1}{a}\right)^{\frac{r}{s}} a^{\frac{r}{s}} a^{\frac{m}{n}-\frac{r}{s}} = \left\{\left(\frac{1}{a}\right)^{\frac{r}{s}} a^{\frac{r}{s}}\right\} a^{\frac{m}{n}-\frac{r}{s}} \\ \qquad = \left\{\left(\sqrt[s]{\frac{1}{a}}\right)\sqrt[s]{a}\right\}^r a^{\frac{m}{n}-\frac{r}{s}} = \left(\sqrt[s]{\frac{1}{a}a}\right)^r a^{\frac{m}{n}-\frac{r}{s}} = a^{\frac{m}{n}-\frac{r}{s}} \\ \left(\text{if } \frac{m}{n} \text{ not} > \frac{r}{s}\right) = \left(\frac{1}{a}\right)^{\frac{m}{n}+\left(-\frac{m}{n}+\frac{r}{s}\right)} a^{\frac{m}{n}} = \left\{\left(\frac{1}{a}\right)^{-\frac{m}{n}+\frac{r}{s}} \left(\frac{1}{a}\right)^{\frac{m}{n}}\right\} a^{\frac{m}{n}} \\ \qquad = \left(\frac{1}{a}\right)^{-\frac{m}{n}+\frac{r}{s}} \left(\frac{1}{a}\right)^{\frac{m}{n}} a^{\frac{m}{n}} = \left(\frac{1}{a}\right)^{-\frac{m}{n}+\frac{r}{s}} \end{array} \right.$$

$$\frac{1}{a^{\frac{r}{s}}} a^{\frac{m}{n}} \left\{ \begin{array}{l} \left(\text{if } \frac{m}{n} \text{ not} < \frac{r}{s}\right) = \frac{1}{a^{\frac{r}{s}}} a^{\frac{r}{s}} a^{\frac{m}{n}-\frac{r}{s}} = \left(\frac{1}{a^{\frac{r}{s}}} a^{\frac{r}{s}}\right) a^{\frac{m}{n}-\frac{r}{s}} = a^{\frac{m}{n}-\frac{r}{s}} \\ \left(\text{if } \frac{m}{n} \text{ not} > \frac{r}{s}\right) = \frac{1}{a^{\frac{m}{n}} a^{\frac{r}{s}-\frac{m}{n}}} a^{\frac{m}{n}} = \left(\frac{1}{a^{\frac{r}{s}-\frac{m}{n}}} \frac{1}{a^{\frac{m}{n}}}\right) a^{\frac{m}{n}} = \frac{1}{a^{\frac{r}{s}-\frac{m}{n}}} \frac{1}{a^{\frac{m}{n}}} a^{\frac{m}{n}} = \frac{1}{a^{\frac{r}{s}-\frac{m}{n}}} \end{array} \right.$$

$$\frac{a^{\frac{m}{n}}}{a^{\frac{r}{s}}} = \left\{ \begin{array}{l} \left(\text{if } \frac{m}{n} \text{ not} < \frac{r}{s}\right) \frac{a^{-\frac{r}{s}+\frac{m}{n}} a^{\frac{r}{s}}}{a^{\frac{r}{s}}} = a^{-\frac{r}{s}+\frac{m}{n}} = a^{-\frac{r}{s}+\frac{m}{n}} a^{\frac{r}{s}} \frac{1}{a^{\frac{r}{s}}} \\ \qquad = \left(a^{-\frac{r}{s}+\frac{m}{n}} a^{\frac{r}{s}}\right) \frac{1}{a^{\frac{r}{s}}} \\ \left(\text{if } \frac{m}{n} \text{ not} > \frac{r}{s}\right) \frac{a^{\frac{m}{n}}}{a^{-\frac{m}{n}+\frac{r}{s}} a^{\frac{m}{n}}} = \frac{1}{a^{-\frac{m}{n}+\frac{r}{s}}} = \left(a^{\frac{m}{n}} \frac{1}{a^{\frac{m}{n}}}\right) \frac{1}{a^{-\frac{m}{n}+\frac{r}{s}}} \\ \qquad = a^{\frac{m}{n}} \frac{1}{a^{\frac{m}{n}}} \frac{1}{a^{-\frac{m}{n}+\frac{r}{s}}} = a^{\frac{m}{n}} \frac{1}{a^{-\frac{m}{n}+\frac{r}{s}} a^{\frac{m}{n}}} \end{array} \right\} = a^{\frac{m}{n}} \frac{1}{a^{\frac{r}{s}}}$$

$$a^{\frac{m}{n}}\left(\frac{1}{a}\right)^{\frac{r}{s}}\left\{\begin{aligned}&\left(\text{if } \frac{m}{n} \text{ not} < \frac{r}{s}\right) = \left(a^{-\frac{r}{s}+\frac{m}{n}}a^{\frac{r}{s}}\right)\left(\frac{1}{a}\right)^{\frac{r}{s}} = a^{-\frac{r}{s}+\frac{m}{n}}(\sqrt[s]{a})^r\left(\sqrt[s]{\frac{1}{a}}\right)^r\\&\qquad = a^{-\frac{r}{s}+\frac{m}{n}}\left(\sqrt[s]{a\frac{1}{a}}\right)^r = a^{-\frac{r}{s}+\frac{m}{n}}\\&\left(\text{if } \frac{m}{n} \text{ not} > \frac{r}{s}\right) = a^{\frac{m}{n}}\left(\frac{1}{a}\right)^{\frac{m}{n}}\left(\frac{1}{a}\right)^{\frac{r}{s}-\frac{m}{n}} = \left\{(\sqrt[n]{a})^m\left(\sqrt[n]{\frac{1}{a}}\right)^m\right\}\left(\frac{1}{a}\right)^{\frac{r}{s}-\frac{m}{n}}\\&\qquad = \left(\sqrt[n]{a\frac{1}{a}}\right)^m\left(\frac{1}{a}\right)^{\frac{r}{s}-\frac{m}{n}} = \left(\frac{1}{a}\right)^{\frac{r}{s}-\frac{m}{n}}\end{aligned}\right.$$

$$a^{\frac{m}{n}}b^{\frac{m}{n}} = (\sqrt[n]{a})^m(\sqrt[n]{b})^m = \{(\sqrt[n]{a})\sqrt[n]{b}\}^m = (\sqrt[n]{ab})^m = (ab)^{\frac{m}{n}}$$

$$\frac{a^{\frac{m}{n}}}{b^{\frac{m}{n}}} = \frac{(\sqrt[n]{a})^m}{(\sqrt[n]{b})^m} = \left(\frac{\sqrt[n]{a}}{\sqrt[n]{b}}\right)^m = \left(\sqrt[n]{\frac{a}{b}}\right)^m = \left(\frac{a}{b}\right)^{\frac{m}{n}}.$$

So then the fundamental and therefore all the other index laws hold for the arithmetical values of any fractional indexed powers of numerical quantities.

358. An arithmetical meaning can now be given to the symbol a^x when a is any numerical quantity and x is any incommensurable numerical quantity. For n being any whole number there is a whole number m such that

$$x > \frac{m}{n} \text{ but } < \frac{m+1}{n}$$

and leaving unheeded all but arithmetical roots

$$\text{if } a > 1 \quad -a^{\frac{m}{n}}+a^{\frac{m+1}{n}} = -a^{\frac{m}{n}}+a^{\frac{1}{n}}a^{\frac{m}{n}} = (-1+a^{\frac{1}{n}})a^{\frac{m}{n}}$$

$$\text{if } a < 1 \quad -a^{\frac{m+1}{n}}+a^{\frac{m}{n}} = -a^{\frac{m+1}{n}}+\left(\frac{1}{a}\right)^{\frac{1}{n}}a^{\frac{m+1}{n}} = \left\{-1+\left(\frac{1}{a}\right)^{\frac{1}{n}}\right\}a^{\frac{m+1}{n}}.$$

But (arts. 173, 300) however small a given numerical quantity κ may be n may be taken so great that

$$(1+\kappa)^n > \left\{\begin{matrix} a \text{ if } a > 1 \\ \frac{1}{a} \text{ if } a < 1\end{matrix}\right. \text{ and then } \left.\begin{matrix} -1+a^{\frac{1}{n}} \text{ if } a > 1 \\ -1+\left(\frac{1}{a}\right)^{\frac{1}{n}} \text{ if } a < 1\end{matrix}\right\} < \kappa.$$

Besides if of any magnitude expressed numerically by x in reference to a unit magnitude $\frac{\mu}{\nu}$ $\frac{\mu+\theta}{\nu}$ be any pair of intercepting fractions relative to that unit magnitude (art. 241) n may be taken so great that each of the fractions $\frac{m}{n}$ $\frac{m+1}{n}$ is greater than $\frac{\mu}{\nu}$ and less than $\frac{\mu+\theta}{\nu}$ and then

$$\text{if } a > 1 \quad a^{\frac{m}{n}} < a^{\frac{\mu+\theta}{\nu}} \text{ and if } a < 1 \quad a^{\frac{m+1}{n}} < a^{\frac{\mu}{\nu}}.$$

Hence $a^{\frac{m}{n}}$ may be made to differ from $a^{\frac{m+1}{n}}$ by less than any given numerical quantity λ if n be taken first so great that both $\frac{m}{n}$ and $\frac{m+1}{n}$ is greater than $\frac{\mu}{\nu}$ and less than $\frac{\mu+\theta}{\nu}$ and further so great that $-1+a^{\frac{1}{n}} < \frac{\lambda}{a^{\frac{\mu+\theta}{\nu}}}$ or $-1+\left(\frac{1}{a}\right)^{\frac{1}{n}} < \frac{\lambda}{a^{\frac{\mu}{\nu}}}$ according as a is greater or less than 1. What therefore $a^{\frac{m}{n}}$ $a^{\frac{m+1}{n}}$ become nearer to than by any given numerical quantity when n is taken great enough is what a^x must be taken to mean if no heed be given to any other roots than arithmetical ones.

Since the arithmetical value of $1^{\frac{m}{n}}$ and of $1^{\frac{m+1}{n}}$ is 1 whatever whole number n may be the arithmetical value of 1^x is 1. As to arithmetical values too while $a^{\frac{1}{n}}$ is ever greater or less than 1 according as a is greater or less than 1 yet (arts. 173, 300) if κ be any given numerical quantity less than 1 n may be taken so great that if $a > 1$ $(1+\kappa)^n > a$ and if $a < 1$ $(1-\kappa)^n < a$ and therefore that $a^{\frac{1}{n}}$ in either case differs from 1 by less than any given numerical quantity however small so that here again a^0 taken arithmetically must stand for 1.

The arithmetical values of powers with incommensurable indices follow the same laws as the arithmetical values of the powers with commensurable indices that approximate to them and through which alone they are known.

359. If when f is a purely operational symbol and x is an incommensurable numerical quantity greater than $\frac{m}{n}$ and less than $\frac{m+1}{n}$ it can be shown that corresponding values of $f^{\frac{m}{n}}a$ $f^{\frac{m+1}{n}}a$ come endlessly near to one another as n becomes endlessly great the common value thus endlessly neared must be held to be the corresponding value of what $f^x a$ symbolizes.

360. The symbol a^0 has already been often used as a symbol for 1 when a is a numerical quantity and it is by taking a^0 to symbolize 1 when a is a plus or minus quantity that the third of the five fundamental index laws as written in art. 355 is made true when

$m = n$. The same meaning is got for a^0 by the symbolization extension principle from the first index law of art. 355 since to make this law true even when m is 0

$$a^n a^0 = a^{0+n} = a^n$$

and +1 is the only plus or minus quantity which multiplied by any plus or minus quantity a^n gives this same plus or minus quantity a^n for product.

The unmeaning symbols a^{-1} and $a^{-\alpha}$ are now as far as may be to be likewise given meanings when a is any plus or minus quantity and α is any numerical quantity so that according to the first and second fundamental index laws

$$aa^{-1} = a^{-1+1} = a^0 = 1 \qquad (a^{-1})^\alpha = a^{\alpha(-1)} = a^{-\alpha}.$$

And thus arises the

Def. If a be any plus or minus quantity a^0 stands for +1 a^{-1} for $\frac{1}{a}$ and $a^{-\alpha}$ if α be a commensurable numerical quantity for $\left(\frac{1}{a}\right)^\alpha$. If a be any numerical quantity and α any incommensurable numerical quantity the arithmetical meaning of $a^{-\alpha}$ is the arithmetical meaning of $\left(\frac{1}{a}\right)^\alpha$. What $a^{-\alpha}$ symbolizes is called the $-\alpha$TH POWER OF a.

Hence if t be *ten* the numbers denoted decimally by - - - 0·001 0·01 0·1 1 10 100 1000 - - - are now severally symbolized by - - - t^{-3} t^{-2} t^{-1} $t^{\pm 0}$ t t^2 t^3 - - -

361. If f stand for a pure operation and a for anything operated on f^0 may be given a meaning such that

$$f^n f^0 a = f^{0+n} a = f^n a.$$

So that $f^0 a$ symbolizes the same as a and f^0 no performance of the operation f. Then f^{-1} and $f^{-\alpha}$ when α is any numerical quantity may be given meanings such that

$$ff^{-1}a = f^{-1+1}a = f^0 a = a \qquad (f^{-1})^\alpha a = f^{\alpha(-1)}a = f^{-\alpha}a.$$

So that f^{-1} symbolizes any operation which is such that if the result of its performance on anything a were operated on by f the result would always be a and $f^{-\alpha}$ symbolizes the same as $(f^{-1})^\alpha$.

Thus to say that x is anything such that $fx = a$ is just to say that $x = f^{-1}a$. But any operation which if performed on the result of a certain operation always gives back as result the object first operated on is called an INVERSE operation and hence f^{-1} symbolizes any inverse of the operation symbolized by f.

362. If when $\alpha\ \beta$ are whole numbers $a\ b$ be plus or minus quantities but if when $\alpha\ \beta$ are other numerical quantities than whole numbers $a\ b$ be numerical quantities and the arithmetical values of powers of numerical quantities be alone heeded

$$a^{-\beta}a^{\alpha} = \left(\frac{1}{a}\right)^{\beta} a^{\alpha} = \left\{\begin{matrix} (\text{if } \alpha \text{ not} < \beta)\ a^{\alpha-\beta} \\ (\text{if } \alpha \text{ not} > \beta)\ \left(\frac{1}{a}\right)^{-\alpha+\beta} = a^{-(-\alpha+\beta)} = a^{-\beta+\alpha} \end{matrix}\right\} = a^{\alpha+(-\beta)}$$

$$a^{\beta}a^{-\alpha} = a^{\beta}\left(\frac{1}{a}\right)^{\alpha} \left\{\begin{matrix} (\text{if } \alpha \text{ not} < \beta) = \left(\frac{1}{a}\right)^{\alpha-\beta} = a^{-(\alpha-\beta)} = a^{\beta-\alpha} \\ (\text{and at once if } \alpha \text{ not} > \beta) \end{matrix}\right\} = a^{-\alpha+\beta}$$

$$a^{-\beta}a^{-\alpha} = \left(\frac{1}{a}\right)^{\beta}\left(\frac{1}{a}\right)^{\alpha} = \left(\frac{1}{a}\right)^{\alpha+\beta} = a^{-(\alpha+\beta)} = a^{-\beta-\alpha} = a^{-\alpha-\beta} = a^{-\alpha+(-\beta)}$$

$$(a^{-\beta})^{\alpha} = \left\{\left(\frac{1}{a}\right)^{\beta}\right\}^{\alpha} = \left(\frac{1}{a}\right)^{\alpha\beta} = a^{-\alpha\beta} = a^{\alpha(-\beta)}$$

$$(a^{\beta})^{-\alpha} = \left(\frac{1}{a^{\beta}}\right)^{\alpha} = \left\{\left(\frac{1}{a}\right)^{\beta}\right\}^{\alpha} = a^{-\alpha\beta} = a^{(-\alpha)\beta}$$

$$(a^{-\beta})^{-\alpha} = \left\{\frac{1}{\left(\frac{1}{a}\right)^{\beta}}\right\}^{\alpha} = \left[\left\{\frac{1}{\left(\frac{1}{a}\right)}\right\}^{\beta}\right]^{\alpha} = (a^{\beta})^{\alpha} = a^{\alpha\beta} = a^{(-\alpha)(-\beta)}$$

$$\left(\frac{1}{a}\right)^{-\beta} a^{\alpha} = a^{\beta}a^{\alpha} = a^{\alpha+\beta} \left\{\begin{matrix} = a^{\alpha-(-\beta)} \text{ and also} \\ = \left(\frac{1}{a}\right)^{-(\alpha+\beta)} = \left(\frac{1}{a}\right)^{-\beta-\alpha} = \left(\frac{1}{a}\right)^{-\alpha-\beta} = \left(\frac{1}{a}\right)^{-\alpha+(-\beta)} \end{matrix}\right.$$

$$\left(\frac{1}{a}\right)^{\beta} a^{-\alpha} = a^{-\beta}a^{-\alpha} = a^{-\alpha-\beta} \text{ and also } = \left(\frac{1}{a}\right)^{\beta}\left(\frac{1}{a}\right)^{\alpha} = \left(\frac{1}{a}\right)^{\alpha+\beta} = \left(\frac{1}{a}\right)^{-(-\alpha)+\beta}$$

$$\left(\frac{1}{a}\right)^{-\beta} a^{-\alpha} = a^{\beta}a^{-\alpha} = a^{-\alpha-(-\beta)} \text{ and also} = \left(\frac{1}{a}\right)^{-\beta}\left(\frac{1}{a}\right)^{\alpha} = \left(\frac{1}{a}\right)^{-(-\alpha)+(-\beta)}$$

$$\frac{1}{a^{-\beta}}a^{\alpha} = \frac{1}{\left(\frac{1}{a}\right)^{\beta}}a^{\alpha} = a^{\beta}a^{\alpha} = a^{\alpha-(-\beta)} \text{ and also} = \frac{1}{a^{-\beta}}\,\frac{1}{(a^{\alpha})^{-1}} = \frac{1}{a^{-\alpha}a^{-\beta}} = \frac{1}{a^{-\beta-\alpha}}$$

$$\frac{1}{a^{\beta}}a^{-\alpha} = (a^{\beta})^{-1}a^{-\alpha} = a^{-\alpha-\beta} \text{ and also} = \frac{1}{a^{\beta}}\,\frac{1}{(a^{-\alpha})^{-1}} = \frac{1}{a^{\alpha}a^{\beta}} = \frac{1}{a^{\beta-(-\alpha)}}$$

$$\frac{1}{a^{-\beta}}a^{-\alpha} = (a^{-\beta})^{-1}a^{-\alpha} = a^{-\alpha-(-\beta)} \text{ and also} = \frac{1}{a^{\alpha}a^{-\beta}} = \frac{1}{a^{-\beta-(-\alpha)}}$$

$$\frac{a^\alpha}{a^{-\beta}} = \frac{a^\alpha(a^{-\beta})^{-1}}{a^{-\beta}(a^{-\beta})^{-1}} = a^\alpha(a^{-\beta})^{-1} = a^{-(-\beta)+\alpha} \text{ and also } = \frac{a^\alpha(a^\alpha)^{-1}}{a^{-\beta}(a^\alpha)^{-1}} = \frac{1}{a^{-\alpha+(-\beta)}}$$

$$\frac{a^{-\alpha}}{a^{\beta}} = \frac{a^{-\alpha}(a^{\beta})^{-1}}{a^{\beta}(a^{\beta})^{-1}} = a^{-\alpha}(a^{\beta})^{-1} = a^{-\beta+(-\alpha)} \text{ and also } = \frac{a^{-\alpha}(a^{-\alpha})^{-1}}{a^{\beta}(a^{-\alpha})^{-1}} = \frac{1}{a^{-(-\alpha)+\beta}}$$

$$\frac{a^{-\alpha}}{a^{-\beta}} = \frac{a^{-\alpha}(a^{-\beta})^{-1}}{a^{-\beta}(a^{-\beta})^{-1}} = a^{-\alpha}(a^{-\beta})^{-1} = a^{-(-\beta)+(-\alpha)} \text{ and also } = \frac{a^{-\alpha}(a^{-\alpha})^{-1}}{a^{-\beta}(a^{-\alpha})^{-1}} = \frac{1}{a^{-(-\alpha)+(-\beta)}}$$

$$a^\alpha\left(\frac{1}{a}\right)^{-\beta} = a^\alpha a^\beta = a^{-(-\beta)+\alpha} \text{ and also } = \left(\frac{1}{a}\right)^{-\alpha}\left(\frac{1}{a}\right)^{-\beta} = \left(\frac{1}{a}\right)^{-\beta-\alpha}$$

$$a^{-\alpha}\left(\frac{1}{a}\right)^{\beta} = a^{-\alpha}a^{-\beta} = a^{-\beta+(-\alpha)} \text{ and also } = \left(\frac{1}{a}\right)^{\alpha}\left(\frac{1}{a}\right)^{\beta} = \left(\frac{1}{a}\right)^{\beta-(-\alpha)}$$

$$a^{-\alpha}\left(\frac{1}{a}\right)^{-\beta} = a^{-\alpha}a^{\beta} = a^{-(-\beta)+(-\alpha)} \text{ and also } = \left(\frac{1}{a}\right)^{\alpha}\left(\frac{1}{a}\right)^{-\beta} = \left(\frac{1}{a}\right)^{-\beta-(-\alpha)}$$

$$a^{-\alpha}b^{-\alpha} = \left(\frac{1}{a}\right)^{\alpha}\left(\frac{1}{b}\right)^{\alpha} = \left(\frac{1}{a}\,\frac{1}{b}\right)^{\alpha} = \left(\frac{1}{ba}\right)^{\alpha} = (ba)^{-\alpha} = (ab)^{-\alpha}$$

$$\frac{a^{-\alpha}}{b^{-\alpha}} = \frac{\left(\frac{1}{a}\right)^{\alpha}}{\left(\frac{1}{b}\right)^{\alpha}} = \left\{\frac{\left(\frac{1}{a}\right)}{\left(\frac{1}{b}\right)}\right\}^{\alpha} = \left(\frac{b}{a}\right)^{\alpha} = \left(\frac{a}{b}\right)^{-\alpha}.$$

The fundamental index laws then and therefore too all the index laws hold for those powers of plus or minus quantities whereof the indices are plus or minus whole numbers and for the arithmetical values of those powers of numerical quantities whereof the indices are plus or minus quantities.

Moreover if a α β be the same as above

$$a^{\alpha-\beta} = \left(\frac{1}{a}\right)^{-(\alpha-\beta)} = \left(\frac{1}{a}\right)^{\beta-\alpha} = \left(\frac{1}{a}\right)^{-\alpha+\beta} \text{ and also } = \frac{1}{a^{\beta-\alpha}}$$

$$a^{-\beta+\alpha} = \left(\frac{1}{a}\right)^{-(-\beta+\alpha)} = \left(\frac{1}{a}\right)^{-\alpha+\beta} = \left(\frac{1}{a}\right)^{\beta-\alpha} \text{ and also } = \frac{1}{a^{-\alpha+\beta}}.$$

363. If a be any numerical quantity but 0 and 1 and n be any whole number but 0 magnitudes $A_{\frac{1}{n}}$ $A_{\frac{2}{n}}$ $A_{\frac{3}{n}}$ ... that are severally expressed in reference to a unit magnitude A_0 by the arithmetical values of $a^{\frac{1}{n}}$ $a^{\frac{2}{n}}$ $a^{\frac{3}{n}}$ ... (arts. 358, 173, 300) either if $a > 1$ go on getting ever greater and greater and at length become greater than any given magnitude however great of the same kind as A_0 or if $a < 1$ go on getting ever less and less and at length become less than any given magnitude however small of the same kind as A_0.

Moreover if $\frac{m}{n}$ $\frac{m+1}{n}$ be any pair of simple fractions intercepting one magnitude relative to another (art. 358) n may be taken so great that the magnitudes $A_{\frac{m}{n}}$ $A_{\frac{m+1}{n}}$ severally expressed in reference to A_0 as unit by the arithmetical values of $a^{\frac{m}{n}}$ $a^{\frac{m+1}{n}}$ differ from one another by less than any given magnitude of the same kind as A_0. Hence there is always some numerical quantity α commensurable or incommensurable which makes the arithmetical value of a^{α} express in reference to A_0 as unit either any magnitude A_{α} greater than A_0 if $a>1$ or any magnitude A_{α} less than A_0 if $a<1$. In like manner there is always some commensurable or incommensurable numerical quantity α which makes the arithmetical value of $a^{-\alpha}$ express in reference to A_0 as unit either any magnitude $A_{-\alpha}$ less than A_0 if $a>1$ or any magnitude $A_{-\alpha}$ greater than A_0 if $a<1$. And so those powers of any numerical quantity else than 0 or 1 whose indices are plus or minus quantities have arithmetical values that cover the whole field of numerical quantity.

If one ratio be the multiplicate compound ratio of the nth degree of another this other is in turn called the SUBMULTIPLICATE COMPOUND RATIO OF THE nTH DEGREE of the first. The ratio represented by the arithmetical value of $a^{\frac{m}{n}}$ then is the multiplicate compound ratio of the mth degree of the submultiplicate compound ratio of the nth degree of the ratio represented numerically by a and is called shortly the $m$plicate compound ratio of the sub$n$plicate compound ratio of the same or still more shortly the $\frac{m}{n}$plicate compound ratio. By a stretch of this way of naming ratios the ratios represented by the arithmetical values of a^{α} and $a^{-\alpha}$ are severally called the QUANTUPLICATE COMPOUND RATIO OF THE αTH DEGREE or shortly the αplicate compound ratio and the QUANTUPLICATE COMPOUND INVERSE RATIO OF THE αTH DEGREE or shortly the $-\alpha$plicate compound ratio of the ratio represented numerically by a and generally the ratio represented by the arithmetical value of $a^{\pm\alpha}$ is called the COMPLICATE COMPOUND RATIO OF THE $\pm\alpha$TH DEGREE or shortly the $\pm\alpha$plicate compound ratio of the ratio represented numerically by a. With like stretch of meaning the indices of plus or minus indexed powers of a are called the LOGARITHMS,—that is λόγων ἀριθμοὶ *numbers of ratios* or *ratio numbers*—, of the arithmetical values of those several powers in the system whose BASE is a. Whence the

Def. The LOGARITHM of a numerical quantity to any given numerical quantity but 0 and 1 as BASE is the plus or minus index of that power of the latter numerical quantity whose arithmetical value is equal to the former. And the logarithm to the base a of b is symbolized by $\log_a b$.

So far then as a power's arithmetical value only is heeded $\log_a b$ is just that which makes $a^{\log_a b}$ equal to b.

364. If a be any unchanging numerical quantity but 0 and 1 and x be a changing plus or minus quantity made to pass in unbroken ascending order of algebraic greatness over the whole range of minus and plus quantity from algebraically less than any given minus quantity however algebraically small through 0 up to greater than any given plus quantity however great the arithmetical value of a^x is at the same time made to pass in unbroken order over the whole range of numerical quantity either if $a>1$ from 0 through 1 up to greater than any given numerical quantity however great or if $a<1$ from greater than any given numerical quantity however great through 1 down to 0. Hence as to arithmetical values of such powers of numerical quantities as have plus or minus indices

$a^x = a^y$ precisely when $x = y$

$a^x = a^0$ or 1 . . . $x = 0$ and $\therefore \log_a 1$ is 0

$a^x = a$ . . . $x = 1$ and $\therefore \log_a a$ is 1.

Def. What is greater than any given numerical quantity however great is called an ENDLESSLY GREAT NUMERICAL QUANTITY or INFINITY and is symbolized by ∞.

Hence if $a > 1$ $a^{\infty} = \infty$ $a^{-\infty} = 0$ and $\therefore \log_a \infty$ is ∞ $\log_a 0$ is $-\infty$

if $a < 1$ $a^{\infty} = 0$ $a^{-\infty} = \infty$ and $\therefore \log_a 0$ is ∞ $\log_a \infty$ is $-\infty$.

365. The following are the laws of operational equivalence that have to do with logarithms. In the proofs the arithmetical values of powers are alone heeded.

If p be any numerical quantity and $a\ b\ c \ldots f\ g$ be any other numerical quantities than 0 or 1 $p = a^{\log_a p}$ and likewise $= b^{\log_b p}$ $= (a^{\log_a b})^{\log_b p} = a^{(\log_b p)\log_a b}$. Therefore (art. 364)

$$(1)\quad \log_a p = (\log_b p) \log_a b.$$

Hence $\log_a p = (\log_g p) \log_a g = (\log_g p)(\log_f g) \log_a f = \cdots$

$$= (\log_g p)(\log_f g) \ldots (\log_c d)(\log_b c) \log_a b.$$

If x be any plus or minus quantity

$p^x = a^{\log_a p^x}$ and likewise $= (a^{\log_a p})^x = a^{x \log_a p}$. Therefore

$$(2)\quad \log_a p^x = x \log_a p.$$

If $p\ q\ r \ldots u\ v$ be any numerical quantities

$qp = a^{\log_a qp}$ and likewise $= a^{\log_a q}\, a^{\log_a p} = a^{\log_a p + \log_a q}$. Therefore

$$(3)\quad \log_a qp = \log_a p + \log_a q.$$

Hence

$\log_a vu \ldots rqp = \log_a ut \ldots rqp + \log_a v = \log_a t \ldots rqp + \log_a u + \log_a v = \cdots$

$$= \log_a p + \log_a q + \log_a r + \cdots + \log_a u + \log_a v.$$

Lastly $\dfrac{p}{q} = a^{\log_a \frac{p}{q}}$ and likewise $= \dfrac{a^{\log_a p}}{a^{\log_a q}} = a^{-\log_a q + \log_a p}$. Therefore

$$(4)\quad \log_a \frac{p}{q} = -\log_a q + \log_a p.$$

366. COR. The first of the four fundamental logarithmic laws of art. 365 becomes when p is a

$$1 = (\log_b a) \log_a b \quad \text{or} \quad \log_a b = (\log_b a)^{-1}.$$

367. Leaving unheeded all but the arithmetical values of powers if any numerical quantity a greater than 1 be taken the base of a system of logarithms and n be any whole number but 0 the terms of the series

$$\text{-}\quad \text{-}\quad \text{-}\quad a^{-\frac{3}{n}}\quad a^{-\frac{2}{n}}\quad a^{-\frac{1}{n}}\quad a^{\pm\frac{0}{n}}\quad a^{\frac{1}{n}}\quad a^{\frac{2}{n}}\quad a^{\frac{3}{n}}\quad \text{-}\quad \text{-}\quad \text{-}$$

backward and forward from $a^{\pm\frac{0}{n}}$ or 1 to any length may be found to any sought degree of nearness and by taking n great enough two consecutive terms anywhere taken may be made to differ from one another by less than any given numerical quantity but 0. Moreover if h be a numerical quantity less than 1 any index lying between the two consecutive indices either $\dfrac{m}{n}\ \dfrac{m+1}{n}$ or $-\dfrac{m}{n}\ \dfrac{-m+1}{n}$ either $= \dfrac{m}{n} + h\dfrac{1}{n}$ or $= -\dfrac{m}{n} + h\dfrac{1}{n}$ and

$$a^{\frac{m}{n}+h\frac{1}{n}} = a^{\frac{m}{n}} + \{-1 + (a^{\frac{1}{n}})^h\} a^{\frac{m}{n}} \qquad a^{-\frac{m}{n}+h\frac{1}{n}} = a^{-\frac{m}{n}} + \{-1 + (a^{\frac{1}{n}})^h\} a^{-\frac{m}{n}}.$$

Now if c be any numerical quantity and $\kappa\ \nu$ be any whole numbers each greater than 1

$$-1 + c^{\frac{\kappa}{\nu}} = (1 + c^{\frac{1}{\nu}} + c^{\frac{2}{\nu}} + \cdots + c^{\frac{\kappa-1}{\nu}})(-1 + c^{\frac{1}{\nu}})$$

$$-1 + c = (1 + c^{\frac{1}{\nu}} + c^{\frac{2}{\nu}} + \cdots + c^{\frac{\nu-1}{\nu}})(-1 + c^{\frac{1}{\nu}})$$

$$\therefore\ -1+c^{\frac{\kappa}{\nu}} = \frac{1+c^{\frac{1}{\nu}}+c^{\frac{2}{\nu}}+\dots+c^{\frac{\kappa-1}{\nu}}}{1+c^{\frac{1}{\nu}}+c^{\frac{2}{\nu}}+\dots+c^{\frac{\nu-1}{\nu}}}(-1+c).$$

If $c>1$ each of the numerical quantities $c^{\frac{1}{\nu}}\ c^{\frac{2}{\nu}}\dots c^{\frac{\kappa-1}{\nu}}>1$ and $<c^{\frac{\kappa}{\nu}}$ and

$$\therefore\ -1+c^{\frac{\kappa}{\nu}}>\frac{\kappa}{\nu c}(-1+c) \text{ and } <\frac{\kappa c^{\frac{\kappa}{\nu}}}{\nu}(-1+c).$$

Hence too and $\because\ \frac{\kappa}{\nu c}(-1+c)=\frac{\kappa}{c\nu}(-1+c)=\left(\frac{\kappa}{\nu}\frac{1}{c}\right)(-1+c)=\frac{\kappa}{\nu}\frac{1}{c}(-1+c)$

$=\frac{\kappa}{\nu}(-c^{-1}+1)$ and $\frac{\kappa c^{\frac{\kappa}{\nu}}}{\nu}(-1+c)=\left(\frac{\kappa}{\nu}c^{\frac{\kappa}{\nu}}\right)(-1+c)=\frac{\kappa}{\nu}c^{\frac{\kappa}{\nu}}(-1+c)$

$$-1+c^{\frac{\kappa}{\nu}}>\frac{\kappa}{\nu}(-c^{-1}+1) \text{ and } <\frac{\kappa}{\nu}c^{\frac{\kappa}{\nu}}(-1+c).$$

Wherefore and because c^h when h is incommensurable is only known as the common value to which $c^{\frac{\kappa}{\nu}}\ c^{\frac{\kappa+1}{\nu}}$ become endlessly near when ν becomes endlessly great and κ is always taken so that $h>\frac{\kappa}{\nu}$ and $<\frac{\kappa+1}{\nu}$

$$-1+(a^{\frac{1}{n}})^h>h\{-(a^{\frac{1}{n}})^{-1}+1\} \text{ and } <h(a^{\frac{1}{n}})^h(-1+a^{\frac{1}{n}})$$

$$\therefore >h(-1+a^{\frac{1}{n}})-ha^{-\frac{1}{n}}(-1+a^{\frac{1}{n}})^2 \text{ and } <h(-1+a^{\frac{1}{n}})+h(-1+a^{h\frac{1}{n}})(-1+a^{\frac{1}{n}})$$

and much more therefore $\because\ a^{-\frac{1}{n}}<1$ and $a^{h\frac{1}{n}}<a^{\frac{1}{n}}$

$$-1+(a^{\frac{1}{n}})^h>h(-1+a^{\frac{1}{n}})-h(-1+a^{\frac{1}{n}})^2 \text{ and } <h(-1+a^{\frac{1}{n}})+h(-1+a^{\frac{1}{n}})^2.$$

But as shown in art. 358 n may be taken so great that $-1+a^{\frac{1}{n}}$ is less than any given numerical quantity but 0 and much more $(-1+a^{\frac{1}{n}})^2$. If then n be great

$$-1+(a^{\frac{1}{n}})^h=h(-1+a^{\frac{1}{n}}) \text{ nearly}$$

and hence the logarithm $\frac{m}{n}+h\frac{1}{n}$ of a numerical quantity k which $>a^{\frac{m}{n}}$ but $<a^{\frac{m+1}{n}}$ may be found very nearly by making h such that

$$k=a^{\frac{m}{n}}+\{h(-1+a^{\frac{1}{n}})\}a^{\frac{m}{n}}=a^{\frac{m}{n}}+h(-1+a^{\frac{1}{n}})a^{\frac{m}{n}}, \text{ that is } h=\frac{-a^{\frac{m}{n}}+k}{(-1+a^{\frac{1}{n}})a^{\frac{m}{n}}}.$$

Likewise the logarithm of a numerical quantity k greater than $a^{-\frac{m}{n}}$ but less than $a^{\frac{-m+1}{n}}$

$$=-\frac{m}{n}+\frac{-a^{-\frac{m}{n}}+k}{(-1+a^{\frac{1}{n}})a^{-\frac{m}{n}}}\frac{1}{n} \text{ nearly.}$$

In the same way if λ λ' be the logarithms of the less and the greater severally of two numerical quantities a^{λ} $a^{\lambda'}$

$$-1+a^{h(-\lambda+\lambda')}$$

$$>h(-1+a^{-\lambda+\lambda'})-h(-1+a^{-\lambda+\lambda'})^2 \text{ and } <h(-1+a^{-\lambda+\lambda'})+h(-1+a^{-\lambda+\lambda'})^2$$

and hence if $a^{-\lambda+\lambda'}$ be very nearly equal to 1 the intermediate numerical quantity $a^{\lambda+h(-\lambda+\lambda')}$ having the intermediate logarithm $\lambda+h(-\lambda+\lambda')$

$$=a^{\lambda}+\{-1+a^{h(-\lambda+\lambda')}\}a^{\lambda}=a^{\lambda}+h(-1+a^{-\lambda+\lambda'})a^{\lambda} \text{ very nearly}$$
$$\text{and this } =a^{\lambda}+h(-a^{\lambda}+a^{\lambda'}).$$

368. The expression $h(-1+a^{-\lambda+\lambda'})a^{\lambda}$ approximately equal (art. 367) to the small change $-a^{\lambda}+a^{\lambda+h(-\lambda+\lambda')}$ of a numerical quantity a^{λ} arising from a small change $h(-\lambda+\lambda')$ of the numerical quantity's logarithm λ would be made very simple if the base a of the system of logarithms could be so chosen as to have the small change $-1+a^{-\lambda+\lambda'}$ of 1 arising from a small change $-\lambda+\lambda'$ of 1's logarithm 0 equal to $-\lambda+\lambda'$ the small change of logarithm. But if z be a plus or minus quantity of absolute value less than 1 and a power's arithmetical value be alone heeded to make a such that

$$-1+a^{z}=z \quad \text{or} \quad a=(1+z)^{\frac{1}{z}}$$

would be to make a change with z. All that can be done then toward the end sought is to make that numerical quantity the base of a system of logarithms to which $(1+z)^{\frac{1}{z}}$ becomes endlessly near as z becomes endlessly near to 0 or to which in other words if e be a plus or minus quantity of absolute value greater than 1 $\left(1+\frac{1}{e}\right)^{e}$ becomes endlessly near as e's absolute value becomes endlessly great.

Let ν be any whole number greater than 1 and κ any whole number but 0 then putting E E' for $1+\frac{1}{e}$ when e is $\frac{\kappa}{\nu}$ $\frac{\kappa+1}{\nu}$ severally

$$-E^{\frac{\kappa}{\nu}}+E'^{\frac{\kappa+1}{\nu}}=(-1+E^{\frac{1}{\nu}})E^{\frac{\kappa}{\nu}}-(-E'^{\frac{\kappa+1}{\nu}}+E^{\frac{\kappa+1}{\nu}})$$

$$=\frac{1}{1+E^{\frac{1}{\nu}}+E^{\frac{2}{\nu}}+\cdots+E^{\frac{\nu-1}{\nu}}}(-1+E)E^{\frac{\kappa}{\nu}}-\frac{E'^{\frac{\kappa}{\nu}}+E'^{\frac{\kappa-1}{\nu}}E^{\frac{1}{\nu}}+E'^{\frac{\kappa-2}{\nu}}E^{\frac{2}{\nu}}+\cdots+E^{\frac{\kappa}{\nu}}}{E'^{\frac{\nu-1}{\nu}}+E'^{\frac{\nu-2}{\nu}}E^{\frac{1}{\nu}}+E'^{\frac{\nu-3}{\nu}}E^{\frac{2}{\nu}}+\cdots+E^{\frac{\nu-1}{\nu}}}(-E'+E)$$

$$=\frac{\nu}{(\kappa+1)\kappa}\left\{\frac{(\kappa+1)E^{\frac{\kappa}{\nu}}}{1+E^{\frac{1}{\nu}}+\cdots+E^{\frac{\nu-1}{\nu}}}-\frac{E'^{\frac{\kappa}{\nu}}+E'^{\frac{\kappa-1}{\nu}}E^{\frac{1}{\nu}}+\cdots+E^{\frac{\kappa}{\nu}}}{E'^{\frac{\nu-1}{\nu}}+E'^{\frac{\nu-2}{\nu}}E^{\frac{1}{\nu}}+\cdots+E^{\frac{\nu-1}{\nu}}}\right\}.$$

But $E^{\frac{\kappa}{\nu}}>E'^{\frac{\kappa-i}{\nu}}E^{\frac{i}{\nu}}$ if i be any whole number less than κ and $E'^{\frac{i}{\nu}}>1$ if i be any whole number greater than 0

$$\therefore\ (\kappa+1)E^{\frac{\kappa}{\nu}}>E'^{\frac{\kappa}{\nu}}+E'^{\frac{\kappa-1}{\nu}}E^{\frac{1}{\nu}}+\cdots+E^{\frac{\kappa}{\nu}}$$

$$1+E^{\frac{1}{\nu}}+\cdots+E^{\frac{\nu-1}{\nu}}<E'^{\frac{\nu-1}{\nu}}+E'^{\frac{\nu-2}{\nu}}E^{\frac{1}{\nu}}+\cdots+E^{\frac{\nu-1}{\nu}}$$

$$\text{and }\therefore\ \frac{(\kappa+1)E^{\frac{\kappa}{\nu}}}{1+E^{\frac{1}{\nu}}+\cdots+E^{\frac{\nu-1}{\nu}}}>\frac{E'^{\frac{\kappa}{\nu}}+E'^{\frac{\kappa-1}{\nu}}E^{\frac{1}{\nu}}+\cdots+E^{\frac{\kappa}{\nu}}}{E'^{\frac{\nu-1}{\nu}}+E'^{\frac{\nu-2}{\nu}}E^{\frac{1}{\nu}}+\cdots+E^{\frac{\nu-1}{\nu}}}.$$

Therefore $E'^{\frac{\kappa+1}{\nu}}>E^{\frac{\kappa}{\nu}}$. And $\left(1+\frac{1}{e}\right)^{e}$ if e be an incommensurable numerical quantity is only known as greater than $E^{\frac{\kappa}{\nu}}$ and less than $E'^{\frac{\kappa+1}{\nu}}$ when $e>\frac{\kappa}{\nu}$ and $<\frac{\kappa+1}{\nu}$ however great ν is. Hence if e be a numerical quantity $\left(1+\frac{1}{e}\right)^{e}$ gets ever greater and greater as e becomes greater and greater.

Again if c be a numerical quantity greater than 1 and ν a whole number greater than 1

$$-1+c^{\frac{1}{\nu}}=\frac{1}{1+c^{\frac{1}{\nu}}+c^{\frac{2}{\nu}}+\cdots+c^{\frac{\nu-1}{\nu}}}(-1+c)\ \text{ and }\therefore>\frac{1}{\nu c}(-1+c)\ \text{ or }\ \frac{1}{\nu}\left(-\frac{1}{c}+1\right)$$

$$\therefore\ c>\left\{1+\frac{1}{\nu}\left(-\frac{1}{c}+1\right)\right\}^{\nu}.$$

Making then $c\ \frac{n+1}{n}$ where n is any whole number but 0

$$\left\{1+\frac{1}{\nu}\left(-\frac{n}{n+1}+1\right)\right\}^{\nu}<\frac{n+1}{n}\ \text{ and }\therefore\ \left\{1+\frac{1}{(n+1)\nu}\right\}^{(n+1)\nu}<\left(1+\frac{1}{n}\right)^{n+1}.$$

And as $\left(1+\frac{1}{5}\right)^{5+1}=2{\cdot}985984$ it follows that however great ν is $\left(1+\frac{1}{6\nu}\right)^{6\nu}<2{\cdot}985984$. Hence $\left(1+\frac{1}{e}\right)^{e}$ if e be a numerical quantity although it always increases with e yet never becomes greater than some fixed numerical quantity less than 3. This fixed numerical quantity of course $>\left(1+\frac{1}{1}\right)^1$ or 2.

Since then when e is made to increase endlessly through the whole range of numerical quantity $\left(1+\frac{1}{e}\right)^e$ constantly increases yet so as never to be greater than a fixed numerical quantity greater than 2 and less than 3, however great a given numerical quantity e may be a whole number n has only to be taken greater than e for making $\left(1+\frac{1}{n}\right)^n$ greater than $\left(1+\frac{1}{e}\right)^e$ and therefore the numerical quantity which $\left(1+\frac{1}{e}\right)^e$ when e is a numerical quantity endlessly nears as e endlessly increases is that which $\left(1+\frac{1}{n}\right)^n$ when n is a whole number endlessly nears as n endlessly increases. Now

$$\left(1+\frac{1}{n}\right)^n = 1+\frac{n}{\lfloor 1}\frac{1}{n}+\frac{(n-1)n}{\lfloor 2}\left(\frac{1}{n}\right)^2+\frac{(n-2)(n-1)n}{\lfloor 3}\left(\frac{1}{n}\right)^3+\cdots$$

$$\cdots+\frac{\{n-(n-1)\}\ldots(n-2)(n-1)n}{\lfloor n}\left(\frac{1}{n}\right)^n \text{ and } \therefore$$

$$= 1+\frac{1}{\lfloor 1}+\frac{1-\frac{1}{n}}{\lfloor 2}+\frac{\left(1-\frac{2}{n}\right)\left(1-\frac{1}{n}\right)}{\lfloor 3}+\cdots+\frac{\left(1-\frac{i-2}{n}\right)\cdots\left(1-\frac{2}{n}\right)\left(1-\frac{1}{n}\right)}{\lfloor i-1}$$

$$+\left\{\frac{\left(1-\frac{i-1}{n}\right)\cdots\left(1-\frac{2}{n}\right)\left(1-\frac{1}{n}\right)}{\lfloor i}+\frac{\left(1-\frac{i}{n}\right)\cdots\left(1-\frac{2}{n}\right)\left(1-\frac{1}{n}\right)}{\lfloor i+1}+\cdots+\frac{\left(1-\frac{n-1}{n}\right)\cdots\left(1-\frac{2}{n}\right)\left(1-\frac{1}{n}\right)}{\lfloor n}\right\}$$

i being any whole number greater than 0 and less than n. When n becomes endlessly great and i is taken a fixed whole number each of the numbers $\frac{1}{n}\ \frac{2}{n}\cdots\frac{i}{n}$ becomes endlessly near to 0 therefore each of the numbers $1-\frac{1}{n}\ \ 1-\frac{2}{n}\cdots 1-\frac{i}{n}$ becomes endlessly near to 1 therefore as in art. 295 the product $\left(1-\frac{i}{n}\right)\cdots\left(1-\frac{2}{n}\right)\left(1-\frac{1}{n}\right)$ becomes endlessly near to 1^i which is 1 therefore as in art. 297 the quotient

$\dfrac{\left(1-\frac{i}{n}\right)\cdots\left(1-\frac{2}{n}\right)\left(1-\frac{1}{n}\right)}{\lfloor i+1}$ becomes endlessly near to $\dfrac{1}{\lfloor i+1}$ and therefore as in art. 291 the sum $1+\dfrac{1}{\lfloor 1}+\dfrac{1-\frac{1}{n}}{\lfloor 2}+\cdots+\dfrac{\left(1-\frac{i-2}{n}\right)\cdots\left(1-\frac{2}{n}\right)\left(1-\frac{1}{n}\right)}{\lfloor i-1}$ becomes endlessly near to $1+\dfrac{1}{\lfloor 1}+\dfrac{1}{\lfloor 2}+\cdots+\dfrac{1}{\lfloor i-1}$. Further however great n may be

$$\frac{\left(1-\frac{i-1}{n}\right)\cdots\left(1-\frac{2}{n}\right)\left(1-\frac{1}{n}\right)}{\lfloor i}+\cdots+\frac{\left(1-\frac{n-1}{n}\right)\cdots\left(1-\frac{2}{n}\right)\left(1-\frac{1}{n}\right)}{\lfloor n}$$

$$<\frac{1}{\lfloor i}+\frac{1}{\lfloor i+1}+\cdots+\frac{1}{\lfloor n} \text{ or this's operational equivalent}$$

$$\left\{1+\frac{1}{i+1}+\frac{1}{(i+1)(i+2)}+\cdots+\frac{1}{(i+1)\ldots(n-1)n}\right\}\frac{1}{\lfloor i}.$$

And $$1+\frac{1}{i+1}+\frac{1}{(i+1)(i+2)}+\cdots+\frac{1}{(i+1)\ldots(n-1)n}$$

$$<1+\frac{1}{i+1}+\frac{1}{(i+1)^2}+\cdots+\frac{1}{(i+1)^{-i+n}} \quad \text{which}$$

$$=1+\frac{1}{i+1}+\left(\frac{1}{i+1}\right)^2+\cdots+\left(\frac{1}{i+1}\right)^{-i+n}=\frac{1-\left(\frac{1}{i+1}\right)^{-i+n+1}}{1-\frac{1}{i+1}}$$

$$\text{and } \therefore <\frac{1}{1-\frac{1}{i+1}} \text{ or } 1+\frac{1}{i}.$$

$$\therefore \frac{\left(1-\frac{i-1}{n}\right)\cdots\left(1-\frac{2}{n}\right)\left(1-\frac{1}{n}\right)}{\lfloor i}+\cdots+\frac{\left(1-\frac{n-1}{n}\right)\cdots\left(1-\frac{2}{n}\right)\left(1-\frac{1}{n}\right)}{\lfloor n}$$

$$<\left(1+\frac{1}{i}\right)\frac{1}{\lfloor i} \text{ or } \frac{1}{\lfloor i}+\frac{1}{i}\frac{1}{\lfloor i}.$$

Hence when n becomes endlessly great $\left(1+\frac{1}{n}\right)^n$ becomes endlessly near to a numerical quantity greater than $1+\dfrac{1}{\lfloor 1}+\dfrac{1}{\lfloor 2}+\cdots+\dfrac{1}{\lfloor i-1}$ by less than $\frac{1}{\underline{|i}}+\frac{1}{i}\frac{1}{\underline{|i}}$. In like manner by taking $i+1$ instead of i the numerical quantity endlessly neared by $\left(1+\frac{1}{n}\right)^n$ as n becomes endlessly great is greater than $1+\frac{1}{\underline{|1}}+\frac{1}{\underline{|2}}+\cdots+\frac{1}{\underline{|i}}$ by less than $\frac{1}{\underline{|i+1}}+\frac{1}{i+1}\frac{1}{\underline{|i+1}}$. Therefore as n becomes endlessly great $\left(1+\frac{1}{n}\right)^n$ endlessly nears what is greater than $1+\frac{1}{\underline{|1}}+\frac{1}{\underline{|2}}+\cdots+\frac{1}{\underline{|i}}$ by less than $\frac{1}{i}\frac{1}{\underline{|i}}$.

But in the same way if $i\ i'$ be any other whole numbers than 0

$$1+\frac{1}{\underline{|1}}+\frac{1}{\underline{|2}}+\cdots+\frac{1}{\underline{|i}}+\cdots+\frac{1}{\underline{|i+i'}}$$

$$=1+\frac{1}{\underline{|1}}+\cdots+\frac{1}{\underline{|i-1}}+\left\{1+\frac{1}{i+1}+\cdots+\frac{1}{(i+1)\ldots(i+i'-1)(i+i')}\right\}\frac{1}{\underline{|i}}$$

$$\text{and } \therefore <1+\frac{1}{\underline{|1}}+\frac{1}{\underline{|2}}+\cdots+\frac{1}{\underline{|i}}+\frac{1}{i}\frac{1}{\underline{|i}}$$

however great i' may be. And i may be taken so great as to make $\frac{1}{i}\frac{1}{\underline{|i}}$ less than any given numerical quantity but 0 however small. There is therefore a fixed numerical quantity greater than $1+\frac{1}{\underline{|1}}$ or 2 and less than $1+\frac{1}{\underline{|1}}+\frac{1}{1}\frac{1}{\underline{|1}}$ or 3 which by taking i' ever greater and greater $1+\frac{1}{\underline{|1}}+\frac{1}{\underline{|2}}+\cdots+\frac{1}{\underline{|i+i'}}$ increases ever more and more nearly up to and from which by taking i' great enough it may be made to differ by less than any given numerical quantity but 0. Therefore that fixed numerical quantity is what

$$1+\frac{1}{\underline{|1}}+\frac{1}{\underline{|2}}+\frac{1}{\underline{|3}}+\cdots$$

symbolizes and can only be known as what

$$>1+\frac{1}{\underline{|1}}+\frac{1}{\underline{|2}}+\cdots+\frac{1}{\underline{|i}} \quad \text{and} <1+\frac{1}{\underline{|1}}+\frac{1}{\underline{|2}}+\cdots+\frac{1}{\underline{|i}}+\frac{1}{i}\frac{1}{\underline{|i}}$$

however great a whole number i may be taken.

Hence $\left(1+\frac{1}{e}\right)^e$ when e becomes an endlessly great numerical quantity becomes endlessly near to $1+\frac{1}{\underline{|1}}+\frac{1}{\underline{|2}}+\cdots$.

Still further since if e be a minus quantity such that $-e>1$

$$\left(1+\frac{1}{e}\right)^e=\left(1-\frac{1}{-e}\right)^{--e}=\left(\frac{1}{1-\frac{1}{-e}}\right)^{-e}=\left(\frac{-e-1+1}{-e-1}\right)^{-e}$$

$$=\left(1+\frac{1}{-e-1}\right)\left(1+\frac{1}{-e-1}\right)^{-e-1}$$

and $1+\frac{1}{-e-1}$ becomes endlessly near to 1 when $-e$ becomes an endlessly great numerical quantity $\left(1+\frac{1}{e}\right)^e$ when e becomes endlessly near to $-\infty$ becomes endlessly near to what $\left(1+\frac{1}{-e-1}\right)^{-e-1}$ becomes endlessly near to when $-e-1$ becomes an endlessly great numerical quantity, that is to $1+\frac{1}{\underline{|1}}+\frac{1}{\underline{|2}}+\cdots$.

Def. The numerical quantity $1+\frac{1}{\underline{|1}}+\frac{1}{\underline{|2}}+\cdots$ is symbolized by the Greek letter ϵ and the system of logarithms having the base ϵ is called NAPIER'S or NAPIERIAN from JOHN NAPIER of MERCHISTON the inventor of Logarithms having first chosen that system.

As to the arithmetical values of powers then ϵ is what $\left(1+\frac{1}{e}\right)^e$ endlessly nears as e endlessly nears either $+\infty$ or $-\infty$ and what $(1+z)^{\frac{1}{z}}$ endlessly nears as z endlessly nears, whether by decrease down to or by algebraic increase up to, 0. Hence if a be any numerical quantity but 1

$$\because\ \frac{1}{z}(-1+a^z)=\left\{(\log_\epsilon a)\frac{1}{\log_\epsilon a}\right\}\frac{1}{z}(-1+a^z)=(\log_\epsilon a)\left(\frac{1}{\log_\epsilon a}\frac{1}{z}\right)(-1+a^z)$$

$$=(\log_\epsilon a)\frac{1}{z\log_\epsilon a}\frac{1}{\left(\frac{1}{-1+a^z}\right)}=(\log_\epsilon a)\frac{1}{\frac{1}{-1+a^z}\log_\epsilon a^z}$$

$$=(\log_\epsilon a)\frac{1}{\log_\epsilon\{1+(-1+a^z)\}^{\frac{1}{-1+a^z}}}$$

and $-1+a^z$ endlessly nears 0 as z does $\frac{1}{z}(-1+a^z)$ endlessly nears $\log_e a$ as z endlessly nears 0. And hence if z be very nearly 0 $\frac{1}{z}(-1+a^z)$ very nearly $= \log_e a$ or $-1+a^z = z \log_e a$ very nearly. So that ϵ is precisely such a numerical quantity as makes $\frac{1}{z}(-1+\epsilon^z)$ become endlessly near to 1 as z endlessly nears 0 and $-1+\epsilon^z$ be very nearly equal to z when z is very nearly equal to 0.

The numerical quantity ϵ is incommensurable. For if ϵ were a simple fraction $\frac{m}{n}$ where n is a whole number greater than 1 it would follow that

$$\frac{m}{n} = 1+\frac{1}{\underline{|1}}+\frac{1}{\underline{|2}}+\cdots+\frac{1}{\underline{|n}}+\left(\frac{1}{\underline{|n+1}}+\frac{1}{\underline{|n+2}}+\cdots\right),$$

$$\frac{m}{n}\underline{|n} = \left\{1+\frac{1}{\underline{|1}}+\cdots+\frac{1}{\underline{|n}}+\left(\frac{1}{\underline{|n+1}}+\cdots\right)\right\}\underline{|n}$$

$$= \underline{|n}+\frac{1}{\underline{|1}}\underline{|n}+\cdots+\frac{1}{\underline{|n}}\underline{|n}+\left\{\frac{1}{n+1}\,\frac{1}{\underline{|n}}+\frac{1}{(n+1)(n+2)}\,\frac{1}{\underline{|n}}+\cdots\right\}\underline{|n},$$

$$-\left\{\underline{|n}+\left(\frac{1}{\underline{|1}}\underline{|1}\right).2.3\ldots(n-1)n+\left(\frac{1}{\underline{|2}}\underline{|2}\right).3.4\ldots(n-1)n+\cdots+\frac{1}{\underline{|n}}\underline{|n}\right\}+\left(\frac{m}{n}n\right)\underline{|n-1}$$

$$= \frac{1}{n+1}\,\frac{1}{\underline{|n}}\underline{|n}+\frac{1}{(n+1)(n+2)}\,\frac{1}{\underline{|n}}\underline{|n}+\cdots,$$

$$-\{\underline{|n}+2.3\ldots(n-1)n+3.4\ldots(n-1)n+\cdots+n+1\}+m\underline{|n-1}$$

$$= \frac{1}{n+1}+\frac{1}{(n+1)(n+2)}+\cdots.$$

But $-\{\underline{|n}+2.3\ldots(n-1)n+\cdots+1\}+m\underline{|n-1}$ is a whole number which because equal to else than 0 is not 0 and

$$\frac{1}{n+1}+\frac{1}{(n+1)(n+2)}+\cdots < \frac{1}{n+1}+\frac{1}{(n+1)^2}+\cdots \text{ which}$$

$$= \frac{1}{n+1}\left\{1+\frac{1}{n+1}+\left(\frac{1}{n+1}\right)^2+\cdots\right\} = \frac{1}{n+1}\,\frac{1}{1-\frac{1}{n+1}} = \frac{1}{n}.$$

And therefore a whole number not 0 would be equal to a numerical quantity greater than 0 and less than 1.

A decimally denoted number approximating to ϵ may be found as below to any sought degree of nearness.

		−		+	
$1 =$	1				
$\frac{1}{\lfloor 1} =$	1				
$\frac{1}{\lfloor 2} =$	0·5				
$\frac{1}{\lfloor 3} = \frac{1}{3} \times \frac{1}{\lfloor 2}$ and nearly =	166666 67	4	3		
$\frac{1}{\lfloor 4} = \frac{1}{4} \times \frac{1}{\lfloor 3}$ - - -	41666 67	4	3		
$\frac{1}{\lfloor 5} = \frac{1}{5} \times \frac{1}{\lfloor 4}$ - - -	8333 33			3	4
$\frac{1}{\lfloor 6} = \frac{1}{6} \times \frac{1}{\lfloor 5}$ - - -	1388 89	2	1		
$\frac{1}{\lfloor 7} = \frac{1}{7} \times \frac{1}{\lfloor 6}$ - - -	198 41			2	3
$\frac{1}{\lfloor 8} = \frac{1}{8} \times \frac{1}{\lfloor 7}$ - - -	24 80			1	2
$\frac{1}{\lfloor 9} = \frac{1}{9} \times \frac{1}{\lfloor 8}$ - - -	2 76	5	4		
$\frac{1}{\lfloor 10} = \frac{1}{10} \times \frac{1}{\lfloor 9}$ - - -	28	5	4		
$\frac{1}{\lfloor 11} = \frac{1}{11} \times \frac{1}{\lfloor 10}$ - - -	3	5	4		
		25	19	6	9
		6	9		
	2·718281 84	19	10		

The two double columns here to the right are for setting down in one of them against any such row of digits as denotes only approximately a number the two digits which written singly in the digit place next following the row's last digit denote severally those numbers that severally subtracted from, or severally added to, the number denoted by the row as may happen give the nearest numbers with one more digit in their denoting rows which are less and greater severally than the number approximated to. As for instance $\frac{1}{\lfloor 3} > 0{\cdot}166666\,67 - 0{\cdot}000000\,004$ and

$< 0{\cdot}166666\,67 - 0{\cdot}000000\,003$ $\frac{1}{\lfloor 5} > 0{\cdot}008333\,33 + 0{\cdot}000000\,003$ and

$< 0{\cdot}008333\,33 + 0{\cdot}000000\,004$. It thus comes out that

$$1+\frac{1}{\lfloor 1}+\frac{1}{\lfloor 2}+\cdots+\frac{1}{\lfloor 11} > 2{\cdot}71828184-0{\cdot}000000019 \text{ or } 2{\cdot}718281821$$
$$\text{and} < 2{\cdot}71828184-0{\cdot}00000001 \text{ or } 2{\cdot}71828183.$$

Hence and $\because \frac{1}{11}\times\frac{1}{\lfloor 11} > 0{\cdot}000000002$ and $< 0{\cdot}000000003$

$\epsilon > 2{\cdot}718281821$ and $< 2{\cdot}718281833$ and therefore is $2{\cdot}7182818$ right as far as the 7th decimal place.

369. If 10 the base of the system of notation be made the base of a system of logarithms

$$\log_{10}k(10)^x = \log_{10}(10)^x + \log_{10}k = x + \log_{10}k$$

k being any numerical quantity and x any plus or minus quantity. In particular then if k be taken any decimally denoted number greater than 1 and less than 10 so therefore that $\log_{10}k > 0$ and < 1 and x be taken any plus or minus whole number so therefore that by the principle of digit knitting $k(10)^x$ is just equal to any number denoted decimally by the same row of digits as k in all but the decimal dot's place the logarithms of numbers denoted decimally by digit rows the same in all but the decimal dot's places are all equal to the sums got by adding to plus or minus whole numbers the same numerical quantity less than 1. Thus if $\log_{10}7{\cdot}386425$ be $0{\cdot}8684343$ right to the 7th decimal place to the same degree of nearness $\log_{10}73864{\cdot}25$ is $4+0{\cdot}8684343$ $\log_{10}0{\cdot}00007386425$ is $-5+0{\cdot}8684343$ and the like.

Def. When a logarithm is viewed as the sum got by adding to a plus or minus whole number a numerical quantity less than 1 the plus or minus whole number is called the CHARACTERISTIC and the numerical quantity less than 1 the MANTISSA of the logarithm.

Of all numbers then denoted by the same row of digits heedless of where the decimal dot is the logarithms have a common mantissa. This along with the characteristic of a decimally denoted number's logarithm being at once known from the number of places that the leftmost significant digit of the number's denoting row is to left or right of the unit's digit marks out that system of logarithms whose base is 10 as the fittest for common use in calculating with decimally denoted numbers.

Def. The system having the base 10 and taken characteristic and mantissa wise is called the COMMON system of logarithms.

By an extension of the common decimal notation the final sum got by adding severally numbers that can be severally denoted by digits in order to the right of a decimal dot to a minus decimally denoted whole number and the successive sums got is denoted by writing in a row what denotes the whole number with the sign $-$ written over close before the decimal dot and the digits in their places after. Hence by the law of the distribution of addition to a minus quantity over additions of plus quantities (art. 331) a logarithm with a minus characteristic is operationally equivalent to what is denoted by the characteristic's decimally denoted absolute value with the sign $-$ over it written close before a decimal dot and that part of the decimally denoted mantissa which is to the right of the decimal dot written in the same row close after. Thus the logarithm above $-5+0\cdot8684343=\bar{5}\cdot8684343$ in the same kind of way as $4+0\cdot8684343=4\cdot8684343$.

In a table of logarithms of the common system the mantissæ right up to a certain decimal place but with the unit's digit 0 and the decimal dot left out are given of the logarithms of all numbers denoted by dotless digit rows of a certain number of digits each and the logarithm of a number $\nu+h(-\nu+\nu')$ between one tabular number ν and the next greater one ν' is to be found as nearly as may be from what is shown in art. 367 that

$$\log_{10}\{\nu+h(-\nu+\nu')\}=\log_{10}\nu+h(-\log_{10}\nu+\log_{10}\nu') \text{ nearly.}$$

Since this approximate operational equivalence holds the more nearly the nearer $h(-\nu+\nu')$ is to 0 h can always be found to at least the degree of nearness to which if ν'' be the next greater number in the table to ν' the table shows that

$$-\log_{10}\nu+\log_{10}\nu''=2(-\log_{10}\nu+\log_{10}\nu')$$

or what is the same $-\log_{10}\nu+\log_{10}\nu'=-\log_{10}\nu'+\log_{10}\nu''$.

An intermediate number $\nu+h(-\nu+\nu')$ whose logarithm is $\log_{10}\{\nu+h(-\nu+\nu')\}$ is in like manner to be found as nearly as may be from what is the same as before to wit that

$$h=\frac{-\log_{10}\nu+\log_{10}\{(\nu+h(-\nu+\nu')\}}{-\log_{10}\nu+\log_{10}\nu'} \text{ nearly.}$$

The results of operations whereinto logarithms with minus characteristics enter may always be got by the common notational processes if only the minus quantities that those logarithms are be denoted decimally. Also if a minus logarithm $-\alpha$ result whose absolute value α is less than a whole number n but greater than the next less whole number $n-1$

$$-\alpha=-n+n-\alpha=-n+(n-\alpha)$$

and so the mantissa $n-z$ may be found and compared with the mantissæ set down in a table.

In dealing with logarithms having minus characteristics special notational processes may often however be handily used resting on the laws of operational equivalence that have to do with plus and minus quantities much as the common notational processes rest on the laws of operational equivalence that have to do with numbers. For instance

$$\bar{2}\cdot 47+3\cdot 28+\bar{5}\cdot 51=-2+3-5+(0\cdot 47+0\cdot 28+0\cdot 51)=-2+3-5+1+0\cdot 26=\bar{3}\cdot 26$$

$$3\cdot 59-\bar{4}\cdot 67=3\cdot 59-0\cdot 67+4=3+4+0\cdot 59-0\cdot 67=6\cdot 92$$

$$\bar{5}\cdot 27-2\cdot 54=-5-2-1+(1+0\cdot 27-0\cdot 54)=\bar{8}\cdot 73$$

$$7\times\bar{2}\cdot 72=-7\times 2+7\times 0\cdot 72=-14+5\cdot 04=\bar{9}\cdot 04$$

$$\frac{1}{7}\times\bar{9}\cdot 04=\frac{1}{7}\{-7\times 2+(14+\bar{9}\cdot 04)\}=-\left(\frac{1}{7}\times 7\right)\times 2+\frac{1}{7}\times 5\cdot 04=\bar{2}\cdot 72.$$

370. The following are some of the uses to which logarithms and the arithmetical values of fractional incommensurable and minus indexed powers of numerical quantities may be put.

A sum of p pounds laid out at compound interest at the rate of r pounds a pound a year (art. 153) becomes in a whole number n of years $p(1+r)^n$ pounds. Here each year's interest is due at the year's end and is then joined to the principal sum after which the whole bears interest for the next year in the same way at the same rate. As to what happens between the beginning and the end of any year nothing whatever is settled so that the interest due after the beginning may be none at all until quite the end if only it then suddenly start into what the rate of r pounds a pound gives. Since however there is no reason either why money should bear interest at one time rather than another or with the same rate of interest why money should bear more interest throughout one length of time than another equal length the only right way is for money always to bear interest for interest as soon as borne to begin to bear interest and for all equal sums of money at the same interest rate to bear equal amounts of interest in all equal portions of time. If then when r pounds a pound a year is the interest rate the year be cut into any whole number ν of equal parts and the interest on 1 pound in $\frac{1}{\nu}$ year be q pounds $(1+q)^\nu=1+r$, or $1+q=(1+r)^{\frac{1}{\nu}}$ paying heed to arithmetical roots only, and in any whole number κ of $\frac{1}{\nu}$ths of a year p pounds becomes

$$p(1+q)^{\kappa} \text{ or } p(1+r)^{\frac{\kappa}{\nu}} \text{ pounds.}$$

Hence (art. 358) if α be any numerical quantity commensurable or incommensurable p pounds becomes in α years $p(1+r)^{\alpha}$ pounds. Still further

$$\because \{p(1+r)^{-\alpha}\}(1+r)^{\alpha}=p(1+r)^{-\alpha}(1+r)^{\alpha}=p(1+r)^{\alpha-\alpha}=p$$

$p(1+r)^{-\alpha}$ pounds is the sum of money which becomes in α years p pounds or in other words the present worth of p pounds due α years hence. On the whole therefore if x be a plus or minus quantity a sum of money which laid out at interest at the yearly rate of r pounds a pound is now p pounds is x years hence changed into $p(1+r)^{x}$ pounds.

When p r x are given either exactly or to a certain degree of nearness as decimally denoted numbers $p(1+r)^{x}$ may be found by help of a table of common logarithms as a decimally denoted number to the degree of nearness that the particular table used and the degree of nearness of the numbers allow from the logarithmic operational equivalence

$$\log_{10} p(1+r)^{x} = x\log_{10}(1+r)+\log_{10} p$$

$$\text{or} \quad p(1+r)^{x}=\log_{10}^{-1}\{x\log_{10}(1+r)+\log_{10} p\}$$

and α being x's absolute value $\alpha\log_{10}(1+r)$ may be found in like manner from the equivalence

$$\alpha\log_{10}(1+r)=\log_{10}^{-1}\{\log_{10}^{2}(1+r)+\log_{10}\alpha\}.$$

If the rate of interest be such that the interest in each $\frac{1}{\nu}$th of a year is at the rate of ρ pounds a pound a year simple interest 1 pound becomes in 1 year (art. 153)

$$\left(1+\frac{1}{\nu}\rho\right)^{\nu} \text{ and } \therefore =\left\{\left(1+\frac{1}{\nu}\rho\right)^{\nu}\right\}^{\rho\frac{1}{\rho}}=\left[\left\{\left(1+\frac{1}{\nu}\rho\right)^{\nu}\right\}^{\frac{1}{\rho}}\right]^{\rho}$$

$$=\left\{\left(1+\frac{1}{\nu}\rho\right)^{\frac{1}{\rho}\nu}\right\}^{\rho}=\left\{\left(1+\frac{1}{\nu}\rho\right)^{\frac{1}{\frac{1}{\nu}\rho}}\right\}^{\rho}$$

which (art. 368) when ν is taken endlessly great and therefore $\frac{1}{\nu}\rho$ an endlessly small numerical quantity approaches endlessly near to ϵ^{ρ}. Wherefore if the rate of r pounds a pound a year at which money bears interest be such as would yield simple interest at the rate of ρ pounds a pound a year

$$1+r=\epsilon^{\rho} \quad \text{or} \quad r=-1+\epsilon^{\rho} \quad \rho=\log_{\epsilon}(1+r).$$

Since $-1+\left(1+\frac{1}{\nu}\rho\right)^{\nu} > \nu\frac{1}{\nu}\rho$, which $=\left(\nu\frac{1}{\nu}\right)\rho$ or ρ, however great a whole number ν is $r>\rho$.

The rate of interest in pounds per pound per annum at which in x years p pounds is changed into p' pounds

$$=-1+\log_{10}^{-1}\frac{1}{x}(\log_{10}p'-\log_{10}p)$$

$$=\begin{cases}-1+\log_{10}^{-1}\{\log_{10}(\log_{10}p'-\log_{10}p)-\log_{10}x\} & \text{if } x \text{ be plus}\\ -1+\log_{10}^{-1}\{\log_{10}(\log_{10}p-\log_{10}p')-\log_{10}(-x)\} & \text{if } x \text{ be minus.}\end{cases}$$

The algebraic expression in reference to a coming year as $+1$ of the time in which p pounds is turned into p' pounds by money bearing interest at the rate of r pounds a pound a year

$$=\frac{\log_{10}p'-\log_{10}p}{\log_{10}(1+r)}$$

$$=\begin{cases}\log_{10}^{-1}\{-\log_{10}^{2}(1+r)+\log_{10}(\log_{10}p'-\log_{10}p)\} & \text{if } p' \text{ not} < p\\ -\log_{10}^{-1}\{-\log_{10}^{2}(1+r)+\log_{10}(\log_{10}p-\log_{10}p')\} & \text{if } p' \text{ not} > p.\end{cases}$$

If in x years the population of a place change from h heads to h' the mean rate of increase in heads a head a year,— after the same manner as money put out to interest changes in all but that the mean rate of increase per head of population may be a minus quantity not algebraically less than -1 as well as a plus quantity—,

$$=-1+\log_{10}^{-1}\frac{1}{x}\log_{10}\frac{1}{h}h' = -1+\log_{10}^{-1}\frac{1}{x}(\log_{10}h'-\log_{10}h).$$

A magnitude that has to a magnitude expressed numerically by a in reference to a unit magnitude the same ratio as one magnitude has to another expressed severally by numerical quantities b c in reference to a common unit magnitude is in reference to the first unit magnitude expressed numerically by $\frac{b}{c}a$ which

$$=\log_{10}^{-1}\{\log_{10}a+(-\log_{10}c+\log_{10}b)\}=\log_{10}^{-1}(\log_{10}a-\log_{10}c+\log_{10}b).$$

If in c grains of a certain compound there be s grains of a certain substance in 1 grain of the compound there are $\frac{1}{c}s$ grains of the substance and therefore in c' grains $c'\frac{1}{c}s$ or its equivalent

$$\log_{10}^{-1}(\log_{10}s-\log_{10}c+\log_{10}c').$$

A railway train that passes over m miles in h hours is going at the mean rate of $h^{-1}m$ miles an hour and this

$$=\log_{10}^{-1}(\log_{10}m+\log_{10}h^{-1})=\log_{10}^{-1}\{\log_{10}m+(-1)\log_{10}h\}$$
$$=\log_{10}^{-1}(\log_{10}m-\log_{10}h).$$

371. If a be any plus or minus quantity and n any whole number but 0 $(+)^n a=+a$ so that (art. 356) $+a$ is a value of $(+)^{\frac{1}{n}}a$ and the operation symbolized by + is as may happen either the operation, or one of the operations of which any one is, symbolized by $(+)^{\frac{1}{n}}$. Likewise $(-)^n a=+a$ if n be even and $-a$ if n be odd so that the operation − if n be even is one of the operations of which any one is symbolized by $(+)^{\frac{1}{n}}$ and if n be odd is either the operation, or one of the operations of which any one is, symbolized by $(-)^{\frac{1}{n}}$. An operation affecting like these only the designation of a plus or minus quantity must now be handled which if symbolized by j is such that $j^2a=-a$ and of which therefore $(-)^{\frac{1}{2}}$ is the symbol. This new operation then is just any operation which performed on a gives such a result that the same operation if performed on it would give as result $-a$. Let $x'Ox$ be a straight line stretching away endlessly from a point O on each side and let α be any numerical quantity but 0. From the straight line Ox ended at O but endless toward x cut off close to O a portion OA which in reference to some ended straight line taken as unit is expressed numerically by α and from Ox' ended at O but endless toward x' cut off close to O an equal portion OA' which therefore is also expressed numerically in reference to the unit straight line by α. Through the point O draw in any plane passing through $x'Ox$ a straight line $y'Oy$ stretching away endlessly on each side of O toward y' y cutting at right angles $x'Ox$ and cut off close to O from Oy OB and from Oy' OB' each equal to OA or OA' and therefore each expressed numerically in reference to the unit line by α. If the algebraic expression of the distance from O in the direction Ox of A be $+\alpha$ the algebraic expression of the distance from O in the direction Ox of A' (art. 324) is $-\alpha$. Now OA viewed as a straight line lying in $x'Ox$ may be shifted into where OB viewed as lying in $y'Oy$ is by turning OA through a right angle in the plane of the straight lines $x'x$ $y'y$ the way round marked by the order $xyx'y'xy$.... and if OA shifted into where OB is were further turned through a right angle in the same plane the same way round OA would be shifted into where OA' viewed as lying in xOx' is. The

operation then of turning through a right angle in the plane of $x'x$ $y'y$ the $xyx'y'x\ldots$ way round is an operation which performed on the distance from O to A along Ox gives as result the distance from O to B along Oy and performed on the distance from O to B along Oy gives as result the distance from O to A' along Ox'. Likewise the same operation performed on the distance from O to A' along Ox' gives the distance from O to B' along Oy' as result and performed on this result gives as new result the distance from O to A along Ox. Hence $(-)^{\frac{1}{2}}$ may be taken to symbolize this operation but since everything has reference to the initial or primary direction Ox just as the distance from O to A' along Ox' is called the distance from O in the direction Ox of A' expressed algebraically by $-\alpha$ so must the distance from O of B along Oy be called the distance from O in the direction Ox of B expressed algebraically by $(-)^{\frac{1}{2}}\alpha$ and the distance from O of B' along Oy' the distance from O in the direction Ox of B' expressed algebraically by $(-)^{\frac{1}{2}}(-\alpha)$. A straight line may be shifted from any of the positions OA OB OA' OB' OA OB - - - into the next following by turning it in the plane of $x'x$ $y'y$ the $xyx'y'x\ldots$ way round through $1+n\times 4$ times a right angle if n be any whole number as well as through a right angle and a turning to that amount may be taken as the operation symbolized by $(-)^{\frac{1}{2}}$ instead of the other. After the same manner too since a straight line may be shifted from the position OA to the position OB' from OB' to OA' from OA' to OB and from OB to OA by a turn through thrice a right angle in the plane of $x'x$ $y'y$ the $xyx'y'x\ldots$ way round the shifting thiswise may be taken to be what $(-)^{\frac{1}{2}}$ symbolizes but then it is the distance from O in the direction Ox of B' that is expressed algebraically by $(-)^{\frac{1}{2}}\alpha$ and from O in the direction Ox of B by $(-)^{\frac{1}{2}}(-\alpha)$. This last may be brought about as well by making the angle turned through be $3+n\times 4$ times a right angle. Moreover since Oy may be on either side of $x'Ox$ and Oy' on the other all that is said of B B' and the way round in the order $xyx'y'x\ldots$ may equally be said severally of B' B and the way round in the order $xy'x'yx\ldots$.

On the whole therefore it is precisely a distance from O expressed numerically by α along a straight line bisecting an angle bounded by Ox and Ox' that in the plane of that angle may be expressed algebraically either by $(-)^{\frac{1}{2}}\alpha$ or by $(-)^{\frac{1}{2}}(-\alpha)$ in reference to a distance from O expressed numerically by 1 along Ox as $+1$ and because when any straight line drawn from O bisects an angle beginning at Ox and ending at Ox' the straight line drawn in the same straight line but in the contrary direction bisects the equal angle in the same plane beginning at Ox' going the same way round and ending at Ox if a distance from O along the former straight line be expressed by $(-)^{\frac{1}{2}}a$ an equal distance along the latter is expressed by $(-)^{\frac{1}{2}}(-a)$. Hence and because every angle at O in the plane of $x'x$ $y'y$ bounded by Ox Ox' is bisected by and only by either Oy or Oy' it is precisely either of the distances from O of B along Oy and from O of B' along Oy' that in the plane of $x'x$ $y'y$ may be expressed algebraically by $(-)^{\frac{1}{2}}a$ and then the other is expressed algebraically by $(-)^{\frac{1}{2}}(-a)$.

Of magnitudes generally those only can be dealt with in the foregoing way that have about them something answering to, and therefore representable by, directions not only contrary but at right angles to one another of straight lines in a plane.

372. If in an endless straight line $x'Ox$ the algebraic expression of the distance from a point O in the direction Ox of a point A be a and the algebraic expression of the distance from A in the direction Ax of a point B be b the algebraic expression of the distance from O in the direction Ox of B (arts. 292, 330, 332, 337) is $a+b$. On this ground is raised (art. 325) the

Def. If a b be any plus or minus quantities and j stand for $(-)^{\frac{1}{2}}$, in a plane through an endless straight line $x'Ox$ where the algebraic expression of the distance from a point O in the direction Ox of a point A is a and the algebraic expression of the distance from A in the direction Ax of a point B is jb the symbol $a+jb$ stands for the algebraic expression of the distance from O in the direction Ox of B.

And as a further generalization in the same way the following definitions are made.

Def. Two endless straight lines $x'Ox$ $x'Px$ are said to be CONDIRECTIONATE which are either in the same straight line or parallel to one another and have the portions endlessly away toward x x on one side of any straight line that cuts them and therefore the portions endlessly away toward x' x' on the other side.

Def. If a b a' b' a'' b''.... be any plus or minus quantities and j stand for $(-)^{\frac{1}{2}}$, in a plane through an endless straight line $x'Ox$ where $a+jb$ expresses algebraically the distance from O in the direction Ox of a point P and an endless straight line $y'Oy$ cuts at right angles $x'Ox$ so that Oy or Oy' is on the same side of $x'Ox$ as P according as b is a plus quantity or a minus if through P an endless straight line $x'Px$ be drawn condirectionate with $x'Ox$ and an endless straight line $y'Py$ condirectionate with $y'Oy$ and P' be the point whose distance from P in the direction Px is expressed algebraically by $a'+jb'$ on the understanding that P' is on the same side of $x'Px$ as Py or Py' according as b' is a plus quantity or a minus then $a+jb+(a'+jb')$ symbolizes the algebraic expression of the distance from O in the direction Ox of P'. Also if through P' an endless straight line $x'P'x$ be drawn condirectionate with $x'Ox$ or $x'Px$ and an endless straight line $y'P'y$ condirectionate with $y'Oy$ or $y'Py$ and P'' be the point whose distance from P' in the direction $P'x$ is expressed algebraically by $a''+jb''$ on the understanding that P'' is on the same side of $x'P'x$ as $P'y$ or $P'y'$ according as b'' is a plus quantity or a minus then $a+jb+(a'+jb')+(a''+jb'')$ symbolizes the algebraic expression of the distance from O in the direction Ox of P''. And so on.

The straight line $x'Ox$ because it cuts the straight line $y'Oy$ cuts the parallel, or the same, straight line $y'Py$. Let then N be the point where $x'Ox$ cuts $y'Py$ and let N' be the point where likewise $x'Ox$ cuts $y'P'y$. Because the distance from O in the direction Ox of N is expressed algebraically by a and the distance from N in the direction Nx of N' by a' the distance from O in the direction Ox of N' is expressed algebraically by $a+a'$. Also because the distance from $x'Ox$ in the direction Oy of $x'Px$ is expressed algebraically by b and the distance from $x'Px$ in the direction Py of $x'P'x$ by b' the distance from $x'Ox$ in the direction $N'y$ of $x'P'x$ is expressed algebraically by $b+b'$. Therefore the distance from O in the direction Ox of P' is expressed algebraically by $a+a'+j(b+b')$ and therefore this expression and $a+jb+(a'+jb')$ symbolize the same thing. In like manner $a+a'+a''+j(b+b'+b'')$ symbolizes the same as $a+jb+(a'+jb')+(a''+jb'')$. And so on.

373. With regard to any direction chosen while a plus or minus quantity can only express a distance in either that or the contrary direction $a+(-)^{\frac{1}{2}}b$ if a b be plus or minus quantities can express a distance in any direction whatever in a certain plane and hence the

Def. What $a+(-)^{\frac{1}{2}}b$ symbolizes when a is any plus or minus quantity and b any plus or minus quantity but o is called a DITENSIVE QUANTITY and by way of contrast a plus or minus quantity is then called PROTENSIVE. Also a is called the PROTENSIVE ELEMENT in $a+(-)^{\frac{1}{2}}b$ and the other element $(-)^{\frac{1}{2}}b$ without which $a+(-)^{\frac{1}{2}}b$ would not be ditensive at all is called the PURELY DITENSIVE ELEMENT.

374. *Def.* Two straight distances OP QR are said to be CONDIRECTIONATE when if endlessly produced beyond P R severally toward x x and beyond O Q severally toward x' x' the endless straight lines $x'OPx$ $x'QRx$ are condirectionate.

Def. Ditensive quantities are said to be EQUAL which in reference to condirectionate and equal unit distances and with $(-)^{\frac{1}{2}}$ meaning the same operation express condirectionate and equal distances.

375. PROP. *Ditensive quantities are equal precisely when their protensive elements are equal and their purely ditensive elements equal.*

For a b a' b' being protensive quantities and j one meaning of $(-)^{\frac{1}{2}}$ if OP OP' be the directed distances starting at a common point O in an endless straight line $x'Ox$ that are severally expressed by $a+jb$ $a'+jb'$ in reference to a common unit distance in the direction Ox (art. 374) $a+jb=a'+jb'$ precisely when P P' are the same point. But if N N' be the points, whose several distances from O in the direction Ox are expressed algebraically by a a', where endless condirectionate straight lines $y'Py$ $y'P'y$ through P P' severally cut at right angles $x'Ox$ P is P' precisely when N is N' and NP is $N'P'$ which again happen precisely when $a=a'$ and $b=b'$. Therefore $a+jb=a'+jb'$ precisely when $a=a'$ and $b=b'$.

376. PROP. *Quantities protensive or ditensive are equal precisely when the sums are equal got by severally adding either to them or them to any quantities protensive or ditensive that are equal to one another.*

Making j stand for one of the operations (art. 371) indifferently symbolized by $(-)^{\frac{1}{2}}$ and the other letters for protensive quantities let $a+jb$ $a'+jb'$ be any two quantities and $c+jd$ $c'+jd'$ any two quantities that are equal to one another and therefore (art. 375) such that $c=c'$ and $d=d'$. Then $a+jb=a'+jb'$ precisely when $a=a'$ and $b=b'$ whereof (arts. 143, 300, 343) the former happens precisely when either $a+c=a'+c'$ or $c+a=c'+a'$ and the latter precisely when either $b+d=b'+d'$ or $d+b=d'+b'$. Wherefore $a+jb=a'+jb'$ precisely when either $a+c+j(b+d)=a'+c'+j(b'+d')$, that is (art. 372) $a+jb+(c+jd)=a'+jb'+(c'+jd')$, or $c+a+j(d+b)=c'+a'+j(d'+b')$, that is $c+jd+(a+jb)=c'+jd'+(a'+jb')$.

377. PROP. *The laws of operational equivalence in additions that hold for protensive quantities hold for quantities protensive or ditensive.*

If a b a' b' a'' b'' be any protensive quantities and j symbolize any one meaning of $(-)^{\frac{1}{2}}$, by the definitions of art. 372 and the laws for protensive quantities of the commutation of additions and the distribution of addition over additions,

$$a+jb+(a'+jb')=a+a'+j(b+b')=a'+a+j(b'+b)=a'+jb'+(a+jb).$$

$$a+jb+(a'+jb')+(a''+jb'')=a+a'+a''+j(b+b'+b'')$$
$$=a+(a'+a'')+j\{b+(b'+b'')=a+jb+\{a'+a''+j(b'+b'')\}$$
$$=a+jb+\{a'+jb'+(a''+jb'')\}.$$

The same may be otherwise shown. For if OP PQ QR be directed distances in a plane severally expressed by u v w any three quantities protensive or ditensive OQ is the directed distance expressed by $u+v$ PR is the directed distance expressed by $v+w$ therefore OR is the directed distance expressed at once by $u+v+w$ and by $u+(v+w)$ and therefore $u+v+w=u+(v+w)$. Again if a directed distance OL be taken condirectionate with and equal to PQ the directed distance LQ is condirectionate with and equal to OP OL is expressed by v and LQ by u therefore OQ is expressed not only by $u+v$ but also by $v+u$ and therefore $u+v=v+u$.

From the fundamental laws of additions thus shown true for all quantities whether protensive or ditensive the general laws follow by the proposition of art. 376 in the same way as for whole numbers in arts. 33, 34.

Of these general laws special cases are

$$a+jb=jb+a$$
$$a+jb+(a'+jb')=a+jb+a'+jb'=a+a'+jb+jb'$$
$$=a+a'+(jb+jb')=a+a'+j(b+b').$$

378. The remainder $a-b$ got by subtracting from any protensive quantity a any protensive quantity b is precisely that protensive

quantity which by adding to it b gives a sum equal to a. On this by the symbolization extension principle is grounded the

Def. If a b a' b' be any protensive quantities and j be any one and the same meaning of $(-)^{\frac{1}{2}}$ the symbol $a+jb-(a'+jb')$ stands for the protensive or ditensive quantity which by adding to it $a'+jb'$ gives a sum equal to $a+jb$.

If then x be the protensive and jy the purely ditensive element of the protensive or ditensive quantity that $a+jb-(a'+jb')$ symbolizes $x+jy+(a'+jb')$, that is (art. 372) $x+a'+j(y+b')$, $=a+jb$. Therefore (art. 375) $x+a'=a$ and $y+b'=b$ therefore $x=a-a'$ and $y=b-b'$ and therefore

$$a+jb-(a'+jb')=a-a'+j(b-b').$$

Also if OQ PQ be directed distances severally expressed by $a+jb$ $a'+jb'$ the directed distance OP is expressed by $a+jb-(a'+jb')$ or $a-a'+j(b-b')$. In particular when O is Q,— which happens precisely when $a+jb=0$ or $a=0$ $b=0$—, the directed distance QP is expressed by $-(a'+jb')$, that is $-a'+j(-b')$. And when further a' is 0 $-jb'=j(-b')$.

Hence it is with a ditensive quantity v as with a protensive that if v or $+v$ be the algebraic expression of a distance in one direction $-v$ is the algebraic expression of an equal distance in the contrary direction.

Hence also and because if ja symbolize the result of performing on any protensive quantity a any one operation symbolized by $(-)^{\frac{1}{2}}$ all the results of performing severally on a all the operations symbolized by $(-)^{\frac{1}{2}}$ are (art. 371) precisely ja and $j(-a)$ $(-)^{\frac{1}{2}}a$ is either ja or $-ja$. Since the symbols j and $-$ are here purely operational the operation whose result on a is $-ja$ is, after the manner of art. 356, all that can be understood by the operation $-j$ so that $-ja$ and $(-j)a$ are only two symbols for one thing.

If OQ OL be directed distances severally expressed by $a+jb$ $a'+jb'$ the directed distance LO is expressed by $-(a'+jb')$ and therefore (art. 372) the directed distance LQ is expressed by $-(a'+jb')+(a+jb)$.

379. PROP. *Quantities protensive or ditensive are equal precisely when the remainders are equal got by severally subtracting either from them or them from any quantities protensive or ditensive that are equal to one another.*

If j be any one $(-)^{\frac{1}{2}}$ operation and a b c d a' b' c' d' be protensive quantities such that $c+jd=c'+jd'$, or (art. 375) that $c=c'$ and

$d=d'$, $a+jb=a'+jb'$ precisely when $a=a'$ and $b=b'$ of which the former happens precisely when either $a-c=a'-c'$ or $c-a=c'-a'$ and the latter precisely when either $b-d=b'-d'$ or $d-b=d'-b'$. Therefore $a+jb=a'+jb'$ precisely when either $a-c+j(b-d)$ $=a'-c'+j(b'-d')$ or $c-a+j(d-b)=c'-a'+j(d'-b')$, that is (art. 378) precisely when either $a+jb-(c+jd)=a'+jb'-(c'+jd')$ or $c+jd-(a+jb)$ $=c'+jd'-(a'+jb')$.

380. PROP. *The laws of subtractional equivalence that hold for protensive quantities hold for protensive or ditensive quantities.*

Let u v w be any three quantities protensive or ditensive. If OQ PQ QR be directed distances in one plane severally expressed by u v w the directed distance OR is expressed by $u+w$ the directed distance PR by $v+w$ and the directed distance OP by both $u-v$ and $u+w-(v+w)$.

$$\therefore\ u-v=u+w-(v+w).$$

If OQ OL MO be three directed distances in a plane severally expressed by u v w the directed distance MQ is expressed by $w+u$ the directed distance ML by $w+v$ and the directed distance LQ by both $-v+u$ and $-(w+v)+(w+u)$.

$$\therefore\ -v+u=-(w+v)+(w+u).$$

If OQ PQ RP be directed distances in a plane severally expressed by u v w the directed distance OP is expressed by $u-v$ the directed distance RQ by $w+v$ and the directed distance OR by both $u-v-w$ and $u-(w+v)$.

$$\therefore\ u-v-w=u-(w+v).$$

$$\text{And } \therefore\ u-v-w=u-(w+v)=u-(v+w)=u-w-v.$$

If OQ PQ PR be directed distances in a plane expressed by u v w severally and RK be a directed distance taken condirectionate with and equal to PQ then QK is condirectionate with and equal to PR the directed distance OP is expressed by $u-v$ the directed distance OK by $u+w$ and the directed distance OR by both $u-v+w$ and $u+w-v$.

$$\therefore\ u-v+w=u+w-v.$$

Lastly if OP PQ RQ be directed distances in a plane expressed severally by u v w the directed distance OQ is expressed by $u+v$ the directed distance PR by $v-w$ the directed distance RP by

$w-v$ and the directed distance OR at once by $u+v-w$ by $u+(v-w)$ and by $u-(w-v)$.

$$\therefore\ u+(v-w)=u+v-w=u-(w-v).$$

These laws may too be proved as follows where j is any one and the same operation of which the second power is $-$ and the other letters are protensive quantities.

$$a+jb-(c+jd)=a-c+j(b-d)=a+e-(c+e)+j\{b+f-(d+f)\}$$
$$=a+e+j(b+f)-\{c+e+j(d+f)\}=a+jb+(c+jf)-\{c+jd+(e+jf)\}.$$
$$-(c+jd)+(a+jb)=-c+a+j(-d+b)=-(e+c)+(e+a)+j\{-(f+d)+(f+b)\}$$
$$=-\{e+c+j(f+d)\}+\{e+a+j(f+b)\}=-\{c+jf+(c+jd)\}+\{e+jf+(a+jb)\}.$$
$$a+jb-\{c+jd+(e+jf)\}=a+jb-\{c+e+j(d+f)\}=a-(c+e)+j\{b-(d+f)\}$$
$$=a-e-c+j(b-f-d)=a-e+j(b-f)-(c+jd)=a+jb-(e+jf)-(c+jd).$$
$$a+jb-(c+jd)-(e+jf)=a-c-e+j(b-d-f)=a-e-c+j(b-f-d)$$
$$=a+jb-(e+jf)-(c+jd).$$
$$a+jb+(c+jd)-(e+jf)=a+c-e+j(b+d-f)=a-e+c+j(b-f+d)$$
$$=a+jb-(e+jf)+(c+jd).$$
$$\left.\begin{array}{l}a+jb+\{c+jd-(e+jf)\}=a+jb+\{c-e+j(d-f)\}=a+(c-e)+j\{b+(d-f)\}\\ a+jb-\{e+jf-(c+jd)\}=a+jb-\{e-c+j(f-d)\}=a-(e-c)+j\{b-(f-d)\}\end{array}\right\}$$
$$=a+c-e+j(b+d-f)=a+c+j(b+d)-(e+jf)=a+jb+(c+jd)-(e+jf).$$

From these the simplest cases of those laws of operational equivalence that have to do with subtractions either alone or only along with additions the other cases follow by art. 379's proposition in the same way as in arts. 65, 67, 69, 71, 73.

Under these general laws are embraced as particulars

$$a-jb=a+(-jb)=a+j(-b)=-jb+a$$
$$a+jb-(c+jd)=a+jb-jd-c=a-c+jb-jd=a-c+(jb-jd)=a-c+j(b-d).$$

381. By reason of the definitions now given of Addition and Subtraction and of the laws of operational equivalence now shown to be fulfilled by these operations the system of Inverses given in art. 336 for numerical quantities holds true for all quantities protensive or ditensive.

382. In the other operations yet to follow with ditensive quantities the straight line as a bare magnitude and that straight line's direction which as bound up together in a directed distance any

ditensive quantity serves algebraically to express have to be specially dealt with. Inasmuch as these answer to what the quantitative and designative elements of a protensive quantity severally express the symbolization extension principle leads to calling the straight line's numerical expression the ABSOLUTE VALUE or QUANTITATIVE ELEMENT of the ditensive quantity and whatever expression can be got for the straight line's direction the DESIGNATIVE or DIRECTIONAL ELEMENT. For finding the latter element the matter of the next some arts. is needed but the former element is at once given by the proposition—

Third proportionates to any straight line and the several sides about the right angle of a right angled triangle are together equal to a third proportionate to that straight line and the side over against the right angle.

Let ABC be a triangle right angled at C V any straight line and X Y Z three straight lines such that

$$X : BC = BC : V \qquad Y : CA = CA : V$$
$$Z : AB = AB : V.$$

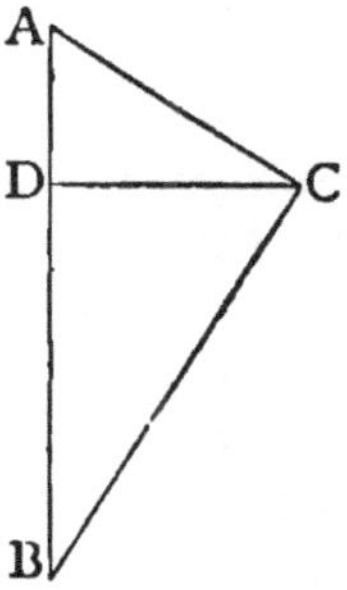

From C draw CD perpendicular to AB and therefore (EUCLID Bk. vi Pr. 8) cutting ABC into the two triangular parts ACD BCD similar to the whole triangle ABC and to one another. Since then from the similar triangles CBA DBC

$$CB : BA = DB : BC$$

X BC AB are three straight lines and DB BC V other three which taken two and two in cross order have severally the same ratios. Therefore (art. 269)

$$X : AB = DB : V.$$

In like manner $\qquad Y : AB = AD : V.$

$\therefore$ (art. 277) $(X, Y) : AB = (AD, DB) : V$, that is $= AB : V$.

$\therefore$ (art. 254) $(X, Y) : AB = Z : AB$ and $\therefore$ (art. 261) $(X, Y) = Z$.

If now BC CA AB be severally expressed numerically in reference to V as unit by a b c (art. 295) X Y Z are severally expressed numerically in reference to V as unit by a^2 b^2 c^2 and therefore (art. 291) (X, Y) is expressed numerically in reference to V as unit by a^2+b^2.

$$\therefore\ a^2+b^2 = c^2.$$

Hence and because a^2 b^2 a^2+b^2 are all plus quantities whether either a or b be a plus quantity or a minus the absolute value of any ditensive quantity $a+(-)^{\frac{1}{2}}b$ is $\sqrt{(a^2+b^2)}$.

383. *Def.* An ANGLE is the amount of opening out in a plane between two straight lines that have a common end.

Since the amount of opening out may be any whatever the bounding straight lines of an angle may be in the same straight line and portions of an angle may overlap one another.

Def. A PERIGON is the angle without any overlapping bounded by two straight lines lying in the same straight line upon the same side of their common end.

A straight line being everywise alike upon all sides everywhere throughout is in any plane through it anglewise alike upon both sides at any point in it and hence half a perigon or a HEMIPERIGON is the unoverlapping angle bounded by two straight lines lying in the same straight line upon opposite sides of their common end. A right angle is both one-half of a hemiperigon or a HEMISEMIPERIGON and one-fourth of a perigon.

Angles like every kind of magnitude are estimated and expressed numerically in reference to some one angle chosen for unit by the numerical quantities representing their several ratios to that unit angle. And like any kind of magnitude having certain counterwardnesses of run reach or stretch angular distances in the same plane at the same point from straight lines ending at that point are expressed algebraically by plus quantities or by minus according as they go one way round the point or the contrary way.

If ω be the algebraic expression of an angle in reference to a perigon as unit the algebraic expression the same way round of any angle beginning at the same first bounding line and ending at the same last is $\omega+i$ or $i+\omega$ where i is some plus or minus whole number or 0.

384. Let a straight line stretching away endlessly from a point O turn round in a plane from a certain initial position Ox into any other position Or through an angle expressed algebraically by ω. Produce xO endlessly to x' and rO endlessly to r'. From the point O in the endless straight line $x'Ox$ draw on that side of $x'Ox$ upon which are the portions close to Ox of such angles as have plus algebraic expressions an endless straight line Oy at right angles to $x'Ox$. And produce yO endlessly to y'. Because two straight lines which meet if endlessly produced either lie wholly together or meet in only one point and there cut one another the endless straight line $y'Oy$ meets in O only and there cuts $x'Ox$ and therefore Oy Oy' are on opposite sides of $x'Ox$. For the same reason $r'Or$ either lies wholly in $x'Ox$ or meets in O only and has Or Or' on opposite

sides of $x'Ox$. Likewise $r'Or$ either lies wholly in $y'Oy$ or meets in O only and has Or Or' on opposite sides of $y'Oy$. In $r'Or$ take points P P' anywhere but at O and through P P' draw severally endless straight lines $y'Py$ $y'P'y$ condirectionate with $y'Oy$. Let N N' be the points where $x'Ox$ because it cuts $y'Oy$ cuts $y'Py$ $y'P'y$ severally. Let r r' be the algebraic expressions of the distances from O in the direction Or of P P' severally q q' the algebraic expressions of the distances from O in the direction Ox of N N' severally and p p' the algebraic expressions of the several distances from N in the direction Ny of P and from N' in the direction $N'y$ of P' all in reference to a common unit straight line. If $r'Or$ be in $x'Ox$ N is P and N' is P' therefore either $q=r$ and $q'=r'$ or $q=-r$ and $q'=-r'$ and further $p=0$ and $p'=0$ wherefore $\frac{q}{r}=\frac{q'}{r'}$ each being either $+1$ or -1 and $\frac{p}{r}=\frac{p'}{r'}$ each being 0. If $r'Or$ be in $y'Oy$ N N' are both O therefore $q=0$ $q'=0$ and either $p=r$ $p'=r'$ or $p=-r$ $p'=-r'$ wherefore $\frac{q}{r}=\frac{q'}{r'}$ each being 0 and $\frac{p}{r}=\frac{p'}{r'}$ each being either $+1$ or -1. If $r'Or$ be neither in $x'Ox$ nor in $y'Oy$ N O P and N' O P' are each three points not in one straight line therefore NOP $N'OP'$ are triangles of these the angles NOP $N'OP'$ are equal being either the same angle or vertically opposite angles and the right angles ONP $ON'P'$ are equal therefore the triangles NOP $N'OP'$ are similar therefore $NO:OP=N'O:OP'$ and $NP:PO=N'P':P'O$ moreover according as P P' are on the same or contrary sides of O are N N' on the same or contrary sides of O and P P' on the same or contrary sides of $x'Ox$ and therefore according as r r' have like or contrary signs have q q' like or contrary signs and p p' like or contrary signs on the whole therefore $\frac{q}{r}=\frac{q'}{r'}$ and $\frac{p}{r}=\frac{p'}{r'}$. Hence each of the quotients $\frac{q}{r}$ $\frac{p}{r}$ abides ever the same wherever else in $r'Or$ than at O the point P may be taken.

Def. The quotients $\frac{q}{r}$ $\frac{p}{r}$ are called severally the COSINE and the SINE of the angle algebraically expressed by ω and are severally written shortly $\cos\omega$ $\sin\omega$.

Since (art. 382) $q^2+p^2=r^2$ $1=\frac{q^2+p^2}{r^2}=\frac{q^2}{r^2}+\frac{p^2}{r^2}=\left(\frac{q}{r}\right)^2+\left(\frac{p}{r}\right)^2$, that is

$$(\cos\omega)^2+(\sin\omega)^2=1.$$

385. Let $x'Ox$ $y'Oy$ be two doubly endless straight lines cutting

one another at right angles and taking a point A in Ox anywhere but at O let a straight line equal to OA turn round in the plane of $x'x\ y'y$ about one end O through the right angle xOy from the position OA along Ox to a position OD along Oy and passing through first a position OB between OA and OD and then a position OC between OB and OD so that the right angle AOD is made up of the angles $AOB\ BOD$ as parts the angle BOD of the angles $BOC\ COD$ as parts and therefore

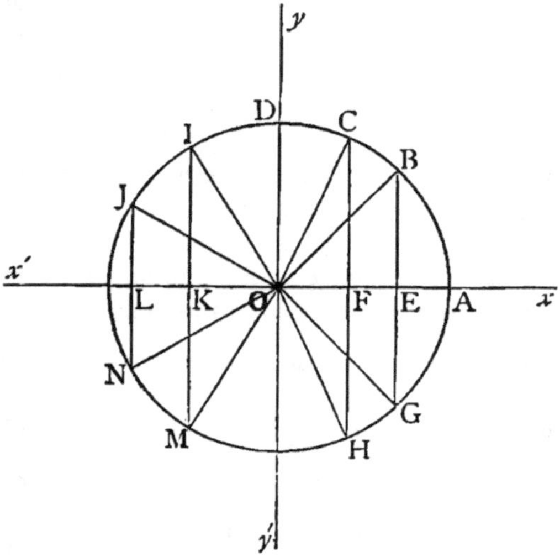

AOD of $AOB\ BOC\ COD$ as parts. From O as center at the distance OA describe a circle which since $OA\ OB\ OC\ OD$ are all equal passes through the points $B\ C\ D$. From $B\ C$ of which neither is in $x'Ox$ for neither is at O the only point common to either OB or OC and $x'Ox$ draw perpendiculars $BE\ CF$ to $x'Ox$. Because the angle xOB is part of and therefore less than a right angle the point E is on the side of O toward x therefore OEB is a triangle and this triangle being right angled at E has the side OB over against the right angle greater than either of the sides $OE\ EB$ bounding the right angle. In the same way F is on the x side of $O\ OFC$ is a triangle and because this triangle is right angled at $F\ OC$ is greater than either OF or FC. The points $E\ F$ being at less distances $OE\ OF$ from O the center of the circle BCD than $B\ C$ severally and therefore than any points in the circumference are within the circle therefore $BE\ CF$ if produced cut the circumference of BCD on the opposite sides of $E\ F$ to $B\ C$ severally. Produce BE endlessly to G and CF endlessly to H and let $G\ H$ be the points where they so cut the circumference. Since it is only in $E\ F$ that $x'Ox$ meets the straight lines $BG\ CH$ severally neither G nor H is in $x'Ox$. Join $OG\ OH$ and hence and because neither E nor F is at $O\ OEG\ OFH$ are triangles. Because too O is in $x'Ox$ but not at either E or F the only points common to $x'Ox$ and $BG\ CH$ severally $OBG\ OCH$ are triangles. Moreover *EB EG* being upon opposite sides of *E* are on opposite sides of *x'Ox* that cuts *BG* in *E* therefore *OB OG* on the same sides severally of *x'Ox* as *EB EG* are on opposite sides of *x'Ox* therefore the angles *EOB EOG* are upon opposite sides of *x'Ox* and therefore the angles *EOB EOG* make up as parts the whole angle *BOG*. Likewise the angles *FOC FOH* are upon opposite sides of *x'Ox* and make up as parts the whole angle *COH*. Now the straight line *x'Ox* passing through the center *O* of the circle *ADG* because it cuts at right angles the straight line *BG* in the circle not passing through the center bisects *BG* and *OB OG* are equal being straight lines drawn from the center of a circle to the circumference therefore in the triangles *EOB EOG* the two sides *EO OB* are equal severally to the two sides *EO OG* and the base *EB* is equal to the base *EG* therefore the angle *EOB* is equal to the angle *EOG* and therefore the straight line *Ox* bisects the angle *BOG*. In like manner *Ox* bisects the straight line *CH* and the angle *COH*. Again because the angle *COF* (art. 8) is greater than its own part the angle *BOE* the angle *COH* the former's double (art. 89) is greater than the angle *BOG* the latter's double hence the triangles *OCH OBG* have the two sides *CO OH* equal severally to the two sides *BO OG* but the angle *COH* greater than the angle *BOG* therefore the base *CH* is greater than the base *BG* and therefore (art. 91) *CH*'s half *CF* is greater than *BG*'s half *BE*. Since then of the straight lines *CH BG* in the circle *ACB CH* is greater than *BG* *CH*'s distance *OF* from the center is less than *OE BG*'s and hence and because *OF OE* have a common end *O* and lie in the same straight line *Ox* upon the same side of *O OF* is a part of *OE*.

If the turning straight line equal to *OA* instead of stopping at the position *OD* go on through the right angle *yOx'* to a position along *Ox'* passing meanwhile through a position *OI* between *Oy* and *Ox'* and afterward through a position *OJ* between *OI* and *Ox'* so that not only is *yOx'* made up of the part angles *yOI IOx'* the angle *IOx'* of the part angles *IOJ JOx'* and therefore *yOx'* of the parts *yOI IOJ JOx'* but taking in the right angled turn through *xOy* the angle *AOJ* is made up of the part angles *AOI IOJ* and the hemiperigon bounded by *Ox Ox'* upon the same side of *x'Ox* as *Oy*,—which call shortly *Y*—, of the part angles *AOI IOx'* and therefore too of the parts *AOI IOJ JOx'* then as before *I J* are in the circumference of *ADG* but neither of them in *x'Ox* perpendiculars *IK JL* to *x'Ox* have feet *K L* on the *x'* side of *O* within *ADG* and if produced cut *ADG*'s circumference in some points *M N* on the other side of $x'Ox$ than I J are on joining OM ON OKI OKM are equal and similar triangles upon opposite sides of $x'Ox$ so are OLJ OLN IK MK JL NL are each less than OA IK is greater than JL and OK is a part of OL.

In like manner if the turning line pass on over the right angle $x'Oy'$ through a position ON between Ox' and Oy' and then a position OM between ON and Oy' so that the angle bounded by Ox ON made up of Y and the angle $x'ON$ is part of the angle bounded by Ox OM made up of Y and the angle $x'OM$ and this angle made up of Y and $x'OM$ part of the angle bounded by Ox Oy' made up of Y and $x'Oy'$ or of the three right angles xOy yOx' $x'Oy'$ everything said as to I J M N holds as to M N I J. And also taking OH OG for successive positions passed through by the turning line in making the fourth right angled turn from Oy' to Ox everything said as to B C G H holds as to G H B C. If after coming to Ox the turning line goes on further the angles turned through from the first position OA begin to overlap and then the same set of positions are passed through as before in the same order and so on to any amount of turning of the line number of overlaps of the angle turned through and runs round the set of positions.

If a straight line equal to OA turned round in the same plane about one end O but the contrary way to wit from Ox to Oy' from Oy' to Ox' and so on the same positions are passed through but in the reverse order.

Hence if ω be the algebraic expression in reference to a perigon as unit of the angular distance from Ox the $xyx'y'x$...way round of a straight line OP equal to OA expressed numerically by c in reference to a unit line in reference to which also as unit a expresses algebraically the distance from O in the direction Ox of the point Q where an endless straight line $y'Py$ drawn through P condirectionate to $y'Oy$ cuts $x'Ox$ and b expresses algebraically the distance from Q in the direction Qy of P as ω increases constantly from 0 through $\frac{1}{4}$ to $\frac{1}{2}$ from $\frac{1}{2}$ through $\frac{3}{4}$ to 1 from 1 through $\frac{5}{4}$ to $\frac{3}{2}$ and so on a correspondingly decreases constantly from c through 0 to $-c$ increases constantly from $-c$ through 0 to $+c$ decreases constantly from $+c$ through 0 to $-c$ and so on and therefore $\cos\omega$ which $=\frac{a}{c}$ correspondingly decreases constantly from 1 through 0 to -1 increases constantly from -1 through 0 to $+1$ decreases constantly from +1 through 0 to −1 and so on. The same changes also happen correspondingly as ω decreases constantly from 0 through $-\frac{1}{4}$ to $-\frac{1}{2}$ from $-\frac{1}{2}$ through $-\frac{3}{4}$ to −1 from −1 through $-\frac{5}{4}$ to $-\frac{3}{2}$ and so on. Likewise as ω either constantly increases from $-\frac{1}{4}$ through 0 to $+\frac{1}{4}$ from $+\frac{1}{4}$ through $\frac{1}{2}$ to $\frac{3}{4}$ from $\frac{3}{4}$ through 1 to $\frac{5}{4}$ and so on or constantly decreases from $-\frac{1}{4}$ through $-\frac{1}{2}$ to $-\frac{3}{4}$ from $-\frac{3}{4}$ through −1 to $-\frac{5}{4}$ from $-\frac{5}{4}$ through $-\frac{3}{2}$ to $-\frac{7}{4}$ and so on b correspondingly increases constantly from $-c$ through 0 up to $+c$ decreases constantly from $+c$ through 0 down to $-c$ increases constantly from $-c$ through 0 up to $+c$ and so on and therefore $\sin\omega$ which $=\frac{b}{c}$ correspondingly increases constantly from −1 through 0 up to +1 decreases constantly from +1 through 0 down to −1 increases constantly from −1 through 0 up to +1 and so on.

386. It is precisely all the algebraically expressed angles beginning at Ox and ending at either OB or OG (art. 385) that have cosines equal to the cosine of any one of these and if ω be the algebraic expression of any one ending at either of the bounding lines OB OG $-\omega$ is the algebraic expression of the equal one ending at the other. The same may likewise be said of all the algebraically expressed angles beginning at Ox and ending at either Oy or Oy' and of all beginning at Ox and ending at either OI or OM. Therefore all the algebraically expressed angles that have cosines equal to $\cos\omega$ are precisely all that can be expressed by either $\omega+i$ or $-\omega+i$ if i be a plus or minus whole number or 0. Since each of the straight lines passing through E O K at right angles to $x'Ox$ cuts the circumference of the circle ADG at points equidistant from and on opposite sides of $x'Ox$

$$\sin(-\omega+i)=-\sin(\omega+i')$$

if i' be any plus or minus whole number or 0.

387. If the angles IOK BOE of art. 385 be equal the triangles OKI OEB have the angles at O K equal severally to the angles at O E and the sides OI OB over against equal angles equal therefore the other sides are equal severally to the other sides to wit KI to EB and OK to OE. And the angles IOK BOE are equal

precisely when the angles xOI $x'OB$ are equal for each two are the remainders got by taking from Y the other two. If then ω be the algebraic expression in reference to a perigon as unit of any angle beginning at Ox and ending at either of the lines OB OI $\frac{1}{2}-\omega$ is the algebraic expression of the corresponding angle beginning at Ox and ending at the other, that is of the angle whose ending bounder is at an angular distance from Ox' the opposite way round expressed by ω. In like manner if the angles MOK GOE be equal KM is equal to EG and OK to OE the angles MOK GOE are equal precisely when the angles xOM $x'OG$ are equal and if ω express algebraically the angular distance from Ox the xyx'... way round of either of the straight lines OG OM and therefore also the angular distance from Ox' the $x'yx$... way round of the other $\frac{1}{2}-\omega$ expresses algebraically the angular distance from Ox the xyx'... way round of this other. Hence and from what would likewise happen if each of the ending bounding lines were along Oy or Oy' or if one of them were along Ox and the other along Ox' all the algebraically expressed angles that have sines equal to $\sin\omega$ are precisely all that can be expressed by either $i'+\omega$ or $i'+\left(\frac{1}{2}-\omega\right)$ and therefore by $\frac{i}{2}+(-)^i\omega$ if i' i be plus or minus whole numbers or 0. Since BG IM cut $x'Ox$ on opposite sides of O

$$\cos\left\{\frac{i}{2}+(-)^i\omega\right\}=(-)^i\cos\omega.$$

388. Since (art. 386) $\cos\omega=\cos\omega'$ precisely when $\omega'=i'\pm\omega$ and (art. 387) $\sin\omega=\sin\omega'$ precisely when $\omega'=\frac{i}{2}+(-)^i\omega$ it follows that $\cos\omega=\cos\omega'$ and $\sin\omega=\sin\omega'$ precisely when $\omega'=i'\pm\omega=\frac{i}{2}+(-)^i\omega$. But $\omega'=i'\pm\omega=\frac{i}{2}+(-)^i\omega$ precisely when if i be a plus or minus even number $i''\times 2$ either $i'=i''$ and $\omega'=i'+\omega$ or $\omega=\frac{1}{2}(-i''+i')$ and $\omega'=i''+\omega$ and if i be a plus or minus odd number $i''\times 2+1$ either $i'=\frac{i''\times 2+1}{2}$, that is the plus or minus even number $i'\times 2$ is equal to the plus or minus odd number $i''\times 2+1$ which can never be, or $\omega=\frac{1}{2}\left(-i'+\frac{i''\times 2+1}{2}\right)$ and $\omega'=i'+\omega$. Hence (art. 383) $\cos\omega=\cos\omega'$

and $\sin\omega = \sin\omega'$ precisely when the straight lines are condirectionate whose angular distances from the initial line Ox are expressed by $\omega\ \omega'$.

389. If as in art. 385 a straight line OP expressed numerically in reference to a unit line by c be at an angular distance from Ox the xyx'... way round expressed algebraically in reference to a perigon as unit by ω and through P an endless straight line $y'Py$ condirectionate with $y'Oy$ cut $x'Ox$ at a point Q whose distance from O in the direction Ox is expressed algebraically in reference to the unit line by a and an endless straight line $x'Px$ condirectionate with $x'Ox$ cut $y'Oy$ at a point R whose distance from O in the direction Oy is expressed algebraically in reference to the same unit line by b then OP is at an angular distance from Oy the yxy'... way round expressed algebraically by $\frac{1}{4}-\omega$ the distance from R in the direction Rx of P is expressed algebraically by a and the distance from Q in the direction Qy of P is expressed algebraically by b. Hence

$$\cos\left(\frac{1}{4}-\omega\right)=\frac{b}{c}=\sin\omega \quad\text{and}\quad \sin\left(\frac{1}{4}-\omega\right)=\frac{a}{c}=\cos\omega.$$

390. From the theorems,—or Laws of Operational Equivalence as they may be called since cos. sin. are symbols of operation—, proved in arts. 386, 387, 389, these others follow:—

$$\cos\left(\omega+\frac{1}{4}\right)=\sin\left\{\frac{1}{4}-\left(\omega+\frac{1}{4}\right)\right\}=\sin(-\omega)=-\sin\omega.$$

$$\sin\left(\omega+\frac{1}{4}\right)=\cos\left\{\frac{1}{4}-\left(\omega+\frac{1}{4}\right)\right\}=\cos(-\omega)=\cos\omega.$$

$$\cos\left(\omega+\frac{1}{2}\right)=-\cos\left\{\frac{1}{2}-\left(\omega+\frac{1}{2}\right)\right\}=-\cos\omega.$$

$$\sin\left(\omega+\frac{1}{2}\right)=\sin\left\{\frac{1}{2}-\left(\omega+\frac{1}{2}\right)\right\}=-\sin\omega.$$

$$\cos\left(\omega+\frac{3}{4}\right)=-\cos\left\{\frac{1}{2}-\left(\omega+\frac{3}{4}\right)\right\}=-\cos\left(-\frac{1}{4}-\omega\right)=-\cos\left(\omega+\frac{1}{4}\right)=\sin\omega.$$

$$\sin\left(\omega+\frac{3}{4}\right)=-\sin\left\{1-\left(\omega+\frac{3}{4}\right)\right\}=-\sin\left(\frac{1}{4}-\omega\right)=-\cos\omega.$$

$$\cos\left(\omega-\frac{1}{4}\right)=\cos\left\{-\left(\omega-\frac{1}{4}\right)\right\}=\cos\left(\frac{1}{4}-\omega\right)=\sin\omega.$$

$$\sin\left(\omega-\frac{1}{4}\right)=-\sin\left\{-\left(\omega-\frac{1}{4}\right)\right\}=-\cos\omega.$$

$$\cos\left(\omega-\frac{1}{2}\right)=\cos\left(\frac{1}{2}-\omega\right)=-\cos\omega.$$

$$\sin\left(\omega-\frac{1}{2}\right)=-\sin\left(\frac{1}{2}-\omega\right)=-\sin\omega.$$

$$\cos\left(\omega-\frac{3}{4}\right)=\cos\left(\frac{3}{4}-\omega\right)=\cos\left(-\omega+\frac{3}{4}\right)=\sin(-\omega)=-\sin\omega.$$

$$\sin\left(\omega-\frac{3}{4}\right)=-\sin\left(-\omega+\frac{3}{4}\right)=\cos\omega.$$

And the cosine or sine of any angle expressed by a plus or minus quantity is equal to plus or minus the cosine or sine of some angle expressed by a plus quantity not greater than $\frac{1}{4}$ in reference to a perigon as unit. For if α be the numerical expression of any angle in reference to a perigon as unit $\cos(-\alpha)=\cos\alpha$ and $\sin(-\alpha)=-\sin\alpha$. Now α either is less than 1 is a whole number i and then $\cos i=\cos 0=1$ $\sin i=\sin 0=0$ or is equal to $i+\beta$ β being less than 1. Again β either is less than $\frac{1}{2}$ is $\frac{1}{2}$ and then $\cos\frac{1}{2}=-\cos 0$ $\sin\frac{1}{2}=\sin 0$ or is equal to $\frac{1}{2}+\gamma$ γ being less than $\frac{1}{2}$. And lastly γ either is less than $\frac{1}{4}$ is $\frac{1}{4}$ or $=\frac{1}{4}+\delta$ δ being less than $\frac{1}{4}$. Then first $\cos(i+\beta)=\cos\beta$ $\sin(i+\beta)=\sin\beta$ secondly $\cos\left(\frac{1}{2}+\gamma\right)=-\cos\gamma$ $\sin\left(\frac{1}{2}+\gamma\right)=-\sin\gamma$ and thirdly $\cos\left(\frac{1}{4}+\delta\right)=-\sin\delta$ $\sin\left(\frac{1}{4}+\delta\right)=\cos\delta$.

Def. The COMPLEMENT of an angle is that angle of which the algebraic expression if added to the algebraic expression of the angle gives for sum the numerical expression of a right angle. And the SUPPLEMENT of an angle is that angle of which the algebraic expression if added to the algebraic expression of the angle gives for sum the numerical expression of a hemiperigon.

If ω x be the algebraic expressions severally of an angle and its complement in reference to a perigon as unit $\omega+x=\frac{1}{4}$ and $\therefore$ $x=-\omega+\frac{1}{4}$. Likewise $-\omega+\frac{1}{2}$ is the algebraic expression of the supplement of the angle expressed algebraically by ω. And

$$\cos\left(-\omega+\frac{1}{4}\right)=\sin\omega \qquad \sin\left(-\omega+\frac{1}{4}\right)=\cos\omega$$

$$\cos\left(-\omega+\frac{1}{2}\right)=-\cos\omega \qquad \sin\left(-\omega+\frac{1}{2}\right)=\sin\omega.$$

391. In a plane let there be sundry endless straight lines $A'OA$ passing through a point O and having anywhere in it a point P_1, $B'P_1B$ passing through P_1 and having anywhere in it a point P_2, $C'P_2C$ passing through P_2 and having anywhere in it a point P_3, and so on. Through O draw in the plane any two endless straight lines $x'Ox$ $y'Oy$ cutting one another at right angles and through P_1 P_2 ... severally draw endless straight lines $x'P_1x$ $x'P_2x$ ... condirectionate with $x'Ox$ and $y'P_1y$ $y'P_2y$ ... condirectionate with $y'Oy$. Let N_1 N_2 ... be the several points where $x'Ox$ because it cuts at right angles $y'Oy$ cuts at right angles $y'P_1y$ $y'P_2y$ ... and M_1 M_2 ... the several points where $y'Oy$ because it cuts at right angles $x'Ox$ cuts at right angles $x'P_1x$ $x'P_2x$ ... Then if

θ_1 express algebraically the angular distance from Ox the xyx'... way round of OA

θ_2 — — — — — — — — — — — P_1x — — — — P_1B

θ_3 — — — — — — — — — — — P_2x — — — — P_2C

— — — — —

and in reference to a common unit line

r_1 express algebraically the distance from O in the direction OA of P_1

r_2 — — — — — P_1 — — P_1B - P_2

r_3 — — — — — P_2 — — P_2C - P_3

— — — — —

the several distances from O in the direction Ox of N_1 N_2 N_3 ... or the several perpendicular distances from $y'Oy$ toward that side of $y'Oy$ on which is Ox of P_1 P_2 P_3 ... are expressed algebraically by

$$(\cos\theta_1)r_1 \quad (\cos\theta_1)r_1+(\cos\theta_2)r_2 \quad (\cos\theta_1)r_1+(\cos\theta_2)r_2+(\cos\theta_3)r_3 \; - - -$$

and the several distances from O in the direction Oy of M_1 M_2 M_3 ... or the several perpendicular distances from $x'Ox$ toward that side of $x'Ox$ on which is Oy of P_1 P_2 P_3 ... by

$$(\sin\theta_1)r_1 \quad (\sin\theta_1)r_1+(\sin\theta_2)r_2 \quad (\sin\theta_1)r_1+(\sin\theta_2)r_2+(\sin\theta_3)r_3 \; - - -.$$

This which springs at once out of the definition (art. 384) of the cosine and sine of an algebraically expressed angle is called the PRINCIPLE OF ALGEBRAIC PROJECTION from the foot of the perpendicular to a straight line from a point being called the point's

PROJECTION on the straight line and the portion of an endless straight line intercepted between perpendiculars to it from the ends of an ended straight line the PROJECTION on the endless straight line of the ended.

392. Let a straight line stretching away endlessly from a point O turn round about O in a plane from an initial position Ox into any position OA through an angle expressed algebraically by ω and then let it turn round in the same plane from the position OA into any position OB through an angle expressed algebraically by ψ so that on the whole it has turned round from the first position Ox to the last position OB through an angle expressed algebraically by $\omega+\psi$ and so therefore that ω ψ are of the same or contrary algebraic signs according as the turnings are the same or contrary ways round. Produce endlessly xO toward x' AO toward A' and BO toward B'. In the endless straight line $B'OB$ take a point P anywhere but at O and through P draw an endless straight line $C'PC$ cutting the endless straight line $A'OA$ perpendicularly at Q and having the portion QC on that side of $A'OA$ by turning toward which from OA such angles as have plus algebraic expressions begin. Through Q draw an endless straight line $x'Qx$ condirectionate with the endless straight line $x'Ox$. Since a straight line stretching away endlessly from Q if it were to turn round about Q first from the position Qx into the position QA through precisely the intermediate positions that are condirectionate with the intermediate positions passed through in the turning from Ox to OA and then from the position QA into the position QC through the right angle AQC would on the whole turn round from Qx to QC through an angle expressed algebraically by $\omega+\rho$ expressing numerically a right angle by ρ this $\omega+\rho$ expresses algebraically the angular distance from Qx of QC the way round that plus expressed angles are turned through. And if r express algebraically the distance from O in the direction OB of P $(\cos\psi)r$ expresses algebraically the distance from O in the direction OA of Q and $(\sin\psi)r$ the distance from Q in the direction QC of P. Hence by the principle of algebraic projection (art. 391) the distance from O in the direction Ox of the point where $x'Ox$ is cut perpendicularly by a straight line drawn through P is expressed algebraically by

$$(\cos\omega)(\cos\psi)r+\{\cos(\omega+\rho)\}(\sin\psi)r$$

$$\text{which } =\{(\cos\omega)\cos\psi-(\sin\omega)\sin\psi\}r$$

and the perpendicular distance from $x'Ox$ of P toward that side of

$x'Ox$ upon which plus expressed angles at O bounded by Ox begin is expressed algebraically by

$$(\sin \omega)(\cos \psi)r + \{\sin (\omega+\rho)\}(\sin \psi)r$$

$$\text{which} = \{(\sin \omega) \cos \psi + (\cos \omega) \sin \psi\} r.$$

$$\therefore \cos (\omega+\psi) = (\cos \omega) \cos \psi - (\sin \omega) \sin \psi$$

$$\sin (\omega+\psi) = (\sin \omega) \cos \psi + (\cos \omega) \sin \psi.$$

Whence and because $\omega-\psi = \omega+(-\psi)$

$$\cos (\omega-\psi) = (\cos \omega) \cos \psi + (\sin \omega) \sin \psi$$

$$\sin (\omega-\psi) = (\sin \omega) \cos \psi - (\cos \omega) \sin \psi.$$

393. After the manner of all multiplications hitherto the symbolization extension principle gives the

Def. If j be any one meaning of $(-)^{\frac{1}{2}}$ and a_1 b_1 a_2 b_2 be any protensive quantities $(a_2+jb_2)(a_1+jb_1)$ symbolizes the algebraic expression in reference to the unit distance drawn in the primary direction of the directed distance expressed algebraically by a_2+jb_2 in reference to that directed distance as primarily directed unit distance which is expressed algebraically by a_1+jb_1 in reference to the primarily directed unit distance.

Let Ox be the primary direction OV the unit distance lying along Ox and OP the directed distance expressed algebraically by a_1+jb_1 in reference to the primarily directed unit distance OV. Produce xO endlessly to x' and draw in the plane of reference an endless straight line $y'Oy$ cutting at right angles $x'Ox$ with Oy on that side of $x'Ox$ on which P is when b_1 is a plus quantity else than 0. Produce OP endlessly to r and PO endlessly to r' and draw in the plane of reference an endless straight line $s'Os$ cutting at right angles $r'OPr$ with Os on that side of $r'OPr$ which makes the ..$s'rsr'$.. way round O the same as the ..$y'xyx'$.. way. Let OQ be the directed distance which in reference to OP as a unit distance lying along Or as a primary direction is expressed algebraically by a_2+jb_2 and through Q draw an endless straight line $s'Qs$ condirectionate with $s'Os$ and therefore cutting perpendicularly $r'Or$ in some point N. Because j symbolizes the same operation in a_2+jb_2 as in a_1+jb_1 Q is on the side of N toward s or toward s' according as b_2 is a plus or a minus quantity. If then θ_1 be the algebraic expression in reference to a perigon as unit of the angular distance from Ox the xyx'... way round of Or and c_1 be at once the absolute value of a_1+jb_1 the numerical expression of OP in reference to OV as unit and the algebraic expression in reference to OV as unit of the distance from O in the direction Or of P $a_1=(\cos\theta_1)c_1$ $b_1=(\sin\theta_1)c_1$ $c_1=\sqrt{(a_1^2+b_1^2)}$ the distance from O in the direction Or of N being expressed algebraically by a_2 in reference to OP as unit is expressed algebraically by a_2c_1 in reference to OV as unit and the distance from N in the direction Ns of Q being expressed algebraically by b_2 in reference to OP as unit is expressed algebraically by b_2c_1 in reference to OV as unit. Hence by the principle of algebraic projection (art. 391) the distance from O in the direction Ox of the point where a straight line through Q cuts perpendicularly $x'Ox$ is expressed algebraically in reference to OV as unit by

$$(\cos\theta_1)a_2c_1+\left\{\cos\left(\theta_1+\frac{1}{4}\right)\right\}b_2c_1 \text{ which } =(\cos\theta_1)a_2c_1-(\sin\theta_1)b_2c_1$$

and the perpendicular distance from $x'Ox$ of Q toward that side of $x'Ox$ on which Oy is is expressed algebraically in reference to OV as unit by

$$(\sin\theta_1)a_2c_1+\left\{\sin\left(\theta_1+\frac{1}{4}\right)\right\}b_2c_1 \text{ which } =(\sin\theta_1)a_2c_1+(\cos\theta_1)b_2c_1.$$

$$\therefore\ (a_2+jb_2)(a_1+jb_1)=(\cos\theta_1)a_2c_1-(\sin\theta_1)b_2c_1+j\{(\sin\theta_1)a_2c_1+(\cos\theta_1)b_2c_1\}$$

and $\therefore$ too $=a_2(\cos\theta_1)c_1-b_2(\sin\theta_1)c_1+j\{a_2(\sin\theta_1)c_1+b_2(\cos\theta_1)c_1\}$, that is $a_2a_1-b_2b_1+j(a_2b_1+b_2a_1)$.

If θ_2 be the algebraic expression in reference to a perigon as unit of the angular distance from Or the rsr'... way round of OQ and c_2 be the numerical expression of OQ in reference to OP as unit $a_2=(\cos\theta_2)c_2$ $b_2=(\sin\theta_2)c_2$ $c_2=\sqrt{(a_2^2+b_2^2)}$ $\therefore$ $(\cos\theta_1)a_2c_1$ $=(\cos\theta_1)\{(\cos\theta_2)c_2\}c_1=(\cos\theta_1)(\cos\theta_2)c_2c_1$ and the like for the other like products $\therefore$ $(\cos\theta_1)a_2c_1-(\sin\theta_1)b_2c_1+j\{(\sin\theta_1)a_2c_1+(\cos\theta_1)b_2c_1\}$

$$=(\cos\theta_1)(\cos\theta_2)c_2c_1-(\sin\theta_1)(\sin\theta_2)c_2c_1$$
$$+j\{(\sin\theta_1)(\cos\theta_2)c_2c_1+(\cos\theta_1)(\sin\theta_2)c_2c_1\}$$
$$=\{(\cos\theta_1)\cos\theta_2\}c_2c_1-\{(\sin\theta_1)\sin\theta_2\}c_2c_1$$
$$+j[\{(\sin\theta_1)\cos\theta_2\}c_2c_1+\{(\cos\theta_1)\sin\theta_2\}c_2c_1]$$
$$=\{(\cos\theta_1)\cos\theta_2-(\sin\theta_1)\sin\theta_2\}c_2c_1+j\{(\sin\theta_1)\cos\theta_2+(\cos\theta_1)\sin\theta_2\}c_2c_1$$

$$\text{and } \therefore\ (a_2+jb_2)(a_1+jb_1)=\{\cos(\theta_1+\theta_2)\}c_2c_1+j\{\sin(\theta_1+\theta_2)\}c_2c_1.$$

Special cases of these are

$$(a_2+jb_2)a_1=a_2a_1+jb_2a_1 \qquad (jb_2)a_1=jb_2a_1$$
$$(\cos\theta_1+j\sin\theta_1)c_1=(\cos\theta_1)c_1+j(\sin\theta_1)c_1$$
$$(a_2+jb_2)jb_1=-b_2b_1+ja_2b_1=ja_2b_1-b_2b_1 \qquad a_2jb_1=ja_2b_1$$
$$(jb_2)jb_1=-b_2b_1=j^2b_2b_1=jb_2\,ib_1$$

$$a_2(a_1+jb_1) = a_2a_1+ja_2b_1 \qquad (jb_2)(a_1+jb_1) = -b_2b_1+jb_2a_1$$

$$(a_2+jb_2)(a_1-jb_1) = (a_2+jb_2)\{a_1+j(-b_1)\} = a_2a_1-b_2(-b_1)+j\{a_2(-b_1)+b_2a_1\}$$

$$= a_2a_1+b_2b_1+j(-a_2b_1+b_2a_1) = a_2a_1+b_2b_1+j(b_2a_1-a_2b_1)$$

$$= a_2a_1+b_2b_1-j(a_2b_1-b_2a_1)$$

$$(a_2-jb_2)(a_1+jb_1) = a_2a_1+b_2b_1+j(a_2b_1-b_2a_1)$$

$$(a_2-jb_2)(a_1-jb_1) = a_2a_1-b_2b_1-j(a_2b_1+b_2a_1).$$

If the operation j be taken to operate on any ditensive quantity in the same way as on either a protensive quantity or a ditensive quantity with no protensive element it changes the quantity into another expressing a directed distance equal to the directed distance before expressed but turned through a right angle the way round that plus expressed angular distances stretch. Then since

$$(j1)\{(\cos\theta_1)c_1+j(\sin\theta_1)c_1\} = -(\sin\theta_1)c_1+j(\cos\theta_1)c_1$$

$$= \left\{\cos\left(\theta_1+\frac{1}{4}\right)\right\}c_1+j\left\{\sin\left(\theta_1+\frac{1}{4}\right)\right\}c_1$$

$$j\{(\cos\theta_1)c_1+j(\sin\theta_1)c_1\} = (j1)\{(\cos\theta_1)c_1+j(\sin\theta_1)c_1\}$$

$$\text{or } j(a_1+jb_1) = (j1)(a_1+jb_1).$$

The ditensive quantity $\cos\theta_1+j\sin\theta_1$ expresses algebraically a unit distance so directed as to be at an angular distance from Ox the $xyx'\ldots$ way round expressed algebraically by θ_1 and therefore serves to mark out the direction of the directed distance expressed algebraically by $(\cos\theta_1)c_1+j(\sin\theta_1)c_1$ or $(\cos\theta_1+j\sin\theta_1)c_1$. The expression $\cos\theta_1+j\sin\theta_1$ or $\dfrac{a_1}{\sqrt{(a_1^2+b_1^2)}}+j\dfrac{b_1}{\sqrt{(a_1^2+b_1^2)}}$ is then (art. 382) the directional or designative element or factor of the ditensive quantity $(\cos\theta_1+j\sin\theta_1)c_1$ or a_1+jb_1.

Def. Θ_θ is used as a short symbol for $\cos\theta+j\sin\theta$.

Since $\Theta_{\frac{1}{4}}=j1$ and $(j1)(a_1+jb_1)=j(a_1+jb_1)$ multiplying by $\Theta_{\frac{1}{4}}$ does the same as operating on with j and therefore $\Theta_{\frac{1}{4}}$ may be used instead of j. Likewise $\Theta_{\frac{1}{2}}=-1$ $\Theta_0=+1$ and $(-1)(a_1+jb_1)=-(a_1+jb_1)$ $(+1)(a_1+jb_1)=+(a_1+jb_1)$ so that $\Theta_{\frac{1}{2}}$ may be used instead of $-$ and Θ_0 instead of $+$.

From the definition of Θ_θ and what is above shown $\Theta_{\theta_1}c_1$ stands indifferently for $(\cos\theta_1+j\sin\theta_1)c_1$ $(\cos\theta_1)c_1+j(\sin\theta_1)c_1$ a_1+jb_1. Hence

$$(\Theta_{\theta_2}c_2)\Theta_{\theta_1}c_1 = \Theta_{\theta_1+\theta_2}c_2c_1$$

and special cases of this being $c_2\Theta_{\theta_1}c_1 = \Theta_{\theta_1}c_2c_1$ $\Theta_{\theta_2}\Theta_{\theta_1}c_1 = \Theta_{\theta_1\theta_2}c_1$ $\Theta_{\theta_2}\Theta_{\theta_1} = \Theta_{\theta_1+\theta_2}$ it follows that

$$(\Theta_{\theta_2}c_2)\Theta_{\theta_1}c_1 = \Theta_{\theta_2}c_2\Theta_{\theta_1}c_1 = \Theta_{\theta_2}\Theta_{\theta_1}c_2c_1 = (\Theta_{\theta_2}\Theta_{\theta_1})c_2c_1 = \Theta_{\theta_1+\theta_2}c_2c_1.$$

The directed distance expressed algebraically by the product of a multiplication has then to the directed distance expressed algebraically by the multiplicand each in reference to a common primarily directed unit distance the same relation in regard to quantity and designation as the directed distance expressed algebraically by the multiplier in reference to any directed distance as primarily directed unit has to this directed distance and if a_1+jb_1 express algebraically in one plan of directed distances what $+1$ expresses algebraically in another a_2+jb_2 expresses algebraically in this other plan what $(a_2+jb_2)(a_1+jb_1)$ expresses algebraically in the first.

394. PROP. *Quantities protensive or ditensive are equal precisely when the products are equal got from severally multiplying either them by or by them any quantities protensive or ditensive that are equal to one another but none of them* 0.

For in multiplying severally quantities protensive or ditensive either by or into others that are equal to one another and of which none is 0 it is precisely when the quantitative factors are equal that the products have equal quantitative factors and precisely when the directional factors are equal that the products have equal directional factors.

395. PROP. *The laws of operational equivalence that have to do with multiplications are the same for quantities in general as for protensive quantities.*

Let $\Theta_\omega c$ $\Theta_\psi d$ $\Theta_\chi e$ be any three protensive or ditensive quantities c d e being numerical quantities

$$(\Theta_\psi d)\Theta_\omega c = \Theta_{\omega+\psi} dc = \Theta_{\psi+\omega} cd = (\Theta_\omega c)\Theta_\psi d.$$

$$(\Theta_\chi e)(\Theta_\psi d)\Theta_\omega c = (\Theta_\chi e)\Theta_{\omega+\psi} dc = \Theta_{\omega+\psi+\chi} edc = \Theta_{\omega+(\psi+\chi)}(cd)c$$
$$= (\Theta_{\psi+\chi} ed)\Theta_\omega c = \{(\Theta_\chi e)\Theta_\psi d\}\Theta_\omega c.$$

If $\Theta_\omega c+\Theta_\psi d = \Theta_\theta g$ where

$$g = \surd[\{(\cos\omega)c+(\cos\psi)d\}^2+\{(\sin\omega)c+(\sin\psi)d\}^2]$$
$$= \surd[c^2+d^2+2\{\cos(-\omega+\psi)\}cd]$$
$$(\Theta_\omega c+\Theta_\psi d)\Theta_\chi e = (\Theta_\theta g)\Theta_\chi e = \Theta_{\chi+\theta} ge$$

$$= \left[\begin{array}{l}(\cos\chi)\dfrac{(\cos\omega)c+(\cos\psi)d}{g}-(\sin\chi)\dfrac{(\sin\omega)c+(\sin\psi)d}{g} \\ \quad +j\left\{(\sin\chi)\dfrac{(\cos\omega)c+(\cos\psi)d}{g}+(\cos\chi)\dfrac{(\sin\omega)c+(\sin\psi)d}{g}\right\}\end{array}\right] gc$$

$$= \Theta_{\chi+\omega} ce+\Theta_{\chi+\psi} de = (\Theta_\omega c)\Theta_\chi e+(\Theta_\psi d)\Theta_\chi c.$$

Wherefore if the unit angle be a perigon

$$(\Theta_\omega c-\Theta_\psi d)\Theta_\chi e=(\Theta_\omega c+\Theta_{\psi+\frac{1}{2}}d)\Theta_\chi e=(\Theta_\omega c)\Theta_\chi e+\Theta_{\chi+(\psi+\frac{1}{2})}de$$

$$=(\Theta_\omega c)\Theta_\chi e+\Theta_{\chi+\psi+\frac{1}{2}}de=(\Theta_\omega c)\Theta_\chi e-\Theta_{\chi+\psi}de=(\Theta_\omega c)\Theta_\chi e-(\Theta_\psi d)\Theta_\chi e.$$

Either in the same way as the last two or from them by the commutation law first of all proved

$$(\Theta_\chi e)(\Theta_\omega c+\Theta_\psi d)=(\Theta_\chi e)\Theta_\omega c+(\Theta_\chi e)\Theta_\psi d$$

$$(\Theta_\chi e)(\Theta_\omega c-\Theta_\psi d)=(\Theta_\chi e)\Theta_\omega c-(\Theta_\chi e)\Theta_\psi d.$$

From the fundamental laws of multiplications thus proved the general laws follow by help of art. 394's proposition in the same way as (arts. 108, 350) for other kinds of multiplication.

396. As with other divisions so now is there made by the symbolization extension principle the

Def. If when a b $a_{,}$ $b_{,}$ are any protensive quantities but $a_{,}$ $b_{,}$ not both 0 and j is any one meaning of $(-)^{\frac{1}{2}}$ $a+jb$ $a_{,}+jb_{,}$ refer to the same primarily directed unit distance the symbol $\dfrac{a+jb}{a_{,}+jb_{,}}$ stands for the algebraic expression in reference to the directed distance as primarily directed unit distance that $a_{,}+jb_{,}$ algebraically expresses of the directed distance that $a+jb$ algebraically expresses.

As with all former quotients then $\dfrac{a+jb}{a_{,}+jb_{,}}$ is precisely the quantity protensive or ditensive which is such that

$$\frac{a+jb}{a_{,}+jb_{,}}(a_{,}+jb_{,})=a+jb.$$

Hence if $\Theta_\omega c=a+jb$ and $\Theta_{\omega_{,}}c_{,}=a_{,}+jb_{,}$ where c $c_{,}$ are numerical quantities and ω $\omega_{,}$ refer to a common unit angle

$$\frac{a+jb}{a_{,}+jb_{,}}=\frac{\Theta_\omega c}{\Theta_{\omega_{,}}c_{,}}=\Theta_{-\omega_{,}+\omega}\frac{c}{c_{,}}=\frac{\{(\cos\omega_{,})c_{,}\}(\cos\omega)c+\{(\sin\omega_{,})c_{,}\}(\sin\omega)c}{c_{,}^{2}}$$

$$+j\frac{-\{(\sin\omega_{,})c_{,}\}(\cos\omega)c+\{(\cos\omega_{,})c_{,}\}(\sin\omega)c}{c_{,}^{2}}$$

$$=\frac{a_{,}a+b_{,}b}{a_{,}^{2}+b_{,}^{2}}+j\frac{-b_{,}a+a_{,}b}{a_{,}^{2}+b_{,}^{2}}.$$

This may be otherwise got. For the common primarily directed unit distance to which $a+jb$ $a_{,}+jb_{,}$ refer is expressed by $\Theta_{-\omega_{,}}c_{,}$ or $a_{,}-jb_{,}$ in reference to some distance in the $\omega_{,}$ expressed direction as unit and the directed distances severally expressed by $a+jb$ $a_{,}+jb_{,}$ in reference to the one unit are severally expressed in reference to the other by $(a+jb)(a_{,}-jb_{,})$ or $aa_{,}+bb_{,}+j(-ab_{,}+ba_{,})$ and $(a_{,}+jb_{,})(a_{,}-jb_{,})$ or $a_{,}^2+b_{,}^2$.

$$\therefore \frac{a+jb}{a_{,}+jb_{,}} = \frac{aa_{,}+bb_{,}+j(-ab_{,}+ba_{,})}{a_{,}^2+b_{,}^2} = \frac{aa_{,}+bb_{,}}{a_{,}^2+b_{,}^2}+j\frac{-ab_{,}+ba_{,}}{a_{,}^2+b_{,}^2}.$$

In a division the directed distance expressed by the dividend has to the directed distance expressed by the divisor each in reference to a common primarily directed unit distance the same relation in regard to quantity and designation as the directed distance expressed by the quotient in reference to any primarily directed unit distance has to this primarily directed unit distance and if the directed distance expressed by $a_{,}+jb_{,}$ in one plan of directed distances be expressed by $+1$ in another $\frac{a+jb}{a_{,}+jb_{,}}$ expresses in this other plan the directed distance that $a+jb$ expresses in the first.

397. In the same way as in arts. 353, 143, but with the wider meanings in which words and symbols are now understood is shown the

PROP. *Quantities protensive or ditensive are equal precisely when the quotients are equal got from dividing either them by or by them any other quantities but* 0 *protensive or ditensive that are equal to one another.*

In the same way likewise as in art. 353 or arts. 116, 118, 120, 122, 124, 127, 130, follows the

PROP. *The laws of operational equivalence that have to do with divisions are the same for quantities generally as for protensive quantities.*

398. In all Additions Subtractions Multiplications and Divisions the tests of equality (arts. 376, 379, 394, 397) and the laws of operational equivalence (arts. 377, 380, 395, 397) are the same for protensive or ditensive quantities as for protensive quantities. All the general theorems methods and processes having to do with these operations therefore that are made out for protensives are made out in the very same way for protensives or ditensives.

399. If $\omega_1\ \omega_2 \ldots \omega_n$ be algebraic expressions of angles in reference to a common unit angle and $c_1\ c_2 \ldots c_n$ be numerical quantities (art. 393)

$$(\Theta_{\omega_n}c_n)\ldots(\Theta_{\omega_3}c_3)(\Theta_{\omega_2}c_2)\Theta_{\omega_1}c_1 = (\Theta_{\omega_n}c_n)\ldots(\Theta_{\omega_3}c_3)\Theta_{\omega_1+\omega_2}c_2c_1$$

$$= (\Theta_{\omega_n}c_n)\ldots(\Theta_{\omega_4}c_4)\Theta_{\omega_1+\omega_2+\omega_3}c_3c_2c_1 = \text{- - -} = \Theta_{\omega_1+\omega_2+\omega_3+\ldots+\omega_n}c_n\ldots c_3c_2c_1.$$

If in particular $\omega_1=\omega_2=---=\omega_n$ and each $=\omega$ and $c_1=c_2=---=c_n$ and each $=c$

$$(\Theta_\omega c)^n=\Theta_{n\omega}c^n.$$

400. If ω θ be algebraic expressions of angles in reference to a perigon as unit n a whole number else than 0 and c k numerical quantities (art. 374) $\Theta_\omega c=(\Theta_\theta k)^n=\Theta_{n\theta}k^n$ precisely when $\Theta_\omega=\Theta_{n\theta}$ and $c=k^n$ and therefore (arts. 302, 375, 388) precisely when $k=\sqrt[n]{c}$ and $n\theta=\omega+i$ i being some plus or minus whole number or 0. Hence and because $n\theta=\omega+i$ precisely when $\theta=\frac{1}{n}(\omega+i)=\frac{1}{n}\omega+\frac{1}{n}i=\frac{1}{n}\omega+\frac{i}{n}$

$$(\Theta_\omega c)^{\frac{1}{n}}=\Theta_{\frac{1}{n}\omega+\frac{i}{n}}\sqrt[n]{c}.$$

Now the values $\Theta_{\frac{1}{n}\omega+\frac{i'}{n}}$ $\Theta_{\frac{1}{n}+\omega\frac{i_,}{n}}$ of $\Theta_{\frac{1}{n}\omega+\frac{i}{n}}$ can be equal only when $\frac{1}{n}\omega+\frac{i'}{n}=\frac{1}{n}\omega+\frac{i_,}{n}+i''$ i'' being some plus or minus whole number or 0 therefore only when $-\left(\frac{1}{n}\omega+\frac{i_,}{n}\right)+\left(\frac{1}{n}\omega+\frac{i'}{n}\right)=-\left(\frac{1}{n}\omega+\frac{i_,}{n}\right)+\left(\frac{1}{n}\omega+\frac{i_,}{n}+i''\right)$ therefore again only when $\frac{-i_,+i'}{n}=i''$, that is (art. 194) only when $i_,$ i' either are equal or differ by some multiple of n. Wherefore if i be made in turn 0 1 2 3 ... $n-1$ the n values got of $\Theta_{\frac{1}{n}\omega+\frac{i}{n}}$ all differ but if i be made in turn n $n+1$ $n+2$ ... the same n different values are got over and over again in the same order and if i be made in turn -1 -2 -3 ... those same values are still got over and over again but in backward order. Hence any algebraic quantity $\Theta_\omega c$ has precisely n different nth roots and these are

$$\Theta_{\frac{1}{n}\omega}\sqrt[n]{c}\quad \Theta_{\frac{1}{n}\omega+\frac{1}{n}}\sqrt[n]{c}\quad \Theta_{\frac{1}{n}\omega+\frac{2}{n}}\sqrt[n]{c}\quad - - - \quad \Theta_{\frac{1}{n}\omega+\frac{n-1}{n}}\sqrt[n]{c}.$$

Since $\Theta_{\frac{1}{n}\omega+\frac{i}{n}}k=\Theta_{\frac{i}{n}}\Theta_{\frac{1}{n}\omega}k$ and $\Theta_{\frac{i}{n}}=\Theta_{i\frac{1}{n}}=\Theta^i_{\frac{1}{n}}$ the n different values of $(\Theta_\omega c)^{\frac{1}{n}}$ are both $\Theta_{\frac{1}{n}\omega}\sqrt[n]{c}$ $\Theta_{\frac{1}{n}}\Theta_{\frac{1}{n}\omega}\sqrt[n]{c}$ $\Theta_{\frac{2}{n}}\Theta_{\frac{1}{n}\omega}\sqrt[n]{c}$ - - - $\Theta_{\frac{n-1}{n}}\Theta_{\frac{1}{n}\omega}\sqrt[n]{c}$ and $\Theta_{\frac{1}{n}\omega}\sqrt[n]{c}$ $\Theta_{\frac{1}{n}}\Theta_{\frac{1}{n}\omega}\sqrt[n]{c}$ $\Theta^2_{\frac{1}{n}}\Theta_{\frac{1}{n}\omega}\sqrt[n]{c}$ - - - $\Theta^{n-1}_{\frac{1}{n}}\Theta_{\frac{1}{n}\omega}\sqrt[n]{c}$.

In particular if c be 1 and ω 0 the n nth roots of $+1$ or if (art. 393) viewed operationally of $+$ are Θ_0 $\Theta_{\frac{1}{n}}$ $\Theta_{\frac{2}{n}}$ $\Theta_{\frac{3}{n}}$ - - - $\Theta_{\frac{n-1}{n}}$ of which the first $\Theta_0=\Theta_{\frac{n}{n}}$. And by taking i any consecutive n terms of the series ... −3 −2 −1 0 1 2 3 ... the same round of values of $\Theta_{\frac{1}{n}\omega+\frac{i}{n}}$ is got though not always beginning with the same value. Therefore all the n values of $(\Theta_\omega c)^{\frac{1}{n}}$ are the products made by multiplying any one of them by the several values of $(+1)^{\frac{1}{n}}$ or $(+)^{\frac{1}{n}}1$. Further if ν be any whole number less than and prime to n since the value $\Theta_{\frac{i'\nu}{n}}$ of $(+)^{\frac{1}{n}}1$ is equal to the value $\Theta_{\frac{i\nu}{n}}$ precisely when $\frac{(-i+i')\nu}{n}$ and therefore $\frac{\nu(-i+i')}{n}$ is 0 or a plus or minus whole number and this happens precisely when $-i+i'$ is either 0 or (art. 210) a plus or minus multiple of n all the n values of $(+)^{\frac{1}{n}}1$ are $\Theta_{\frac{\nu}{n}}$ $\Theta_{\frac{2\nu}{n}}$ $\Theta_{\frac{3\nu}{n}}$ - - - $\Theta_{\frac{n\nu}{n}}$.

Thus the 2 second roots of $+c$ or $\Theta_0 c$ are

$$\Theta_0\sqrt{c} \quad \Theta_{\frac{1}{2}}\sqrt{c} \quad \text{or} \quad +\sqrt{c} \quad -\sqrt{c}$$

of $-c$ or $\Theta_{\frac{1}{2}}c$ $\quad \Theta_{\frac{1}{4}}\sqrt{c}$ $\Theta_{\frac{3}{4}}\sqrt{c}$ or $(-)^{\frac{1}{2}}\sqrt{c}$ $-(-)^{\frac{1}{2}}\sqrt{c}$

of $(-)^{\frac{1}{2}}c$ or $\Theta_{\frac{1}{4}}c$ $\quad \Theta_{\frac{1}{8}}\sqrt{c}$ $\Theta_{\frac{5}{8}}\sqrt{c}$

$$\text{or} \quad \left\{\sqrt{\frac{1}{2}}+(-)^{\frac{1}{2}}\sqrt{\frac{1}{2}}\right\}\sqrt{c} \quad \left\{-\sqrt{\frac{1}{2}}-(-)^{\frac{1}{2}}\sqrt{\frac{1}{2}}\right\}\sqrt{c}$$

and of $-(-)^{\frac{1}{2}}c$ or $\Theta_{\frac{3}{4}}c$ $\quad \Theta_{\frac{3}{8}}\sqrt{c}$ $\Theta_{\frac{7}{8}}\sqrt{c}$

$$\text{or} \quad \left\{-\sqrt{\frac{1}{2}}+(-)^{\frac{1}{2}}\sqrt{\frac{1}{2}}\right\}\sqrt{c} \quad \left\{\sqrt{\frac{1}{2}}-(-)^{\frac{1}{2}}\sqrt{\frac{1}{2}}\right\}\sqrt{c}.$$

The 3 3rd roots of $+c$ are $\Theta_0\sqrt[3]{c}$ $\Theta_{\frac{1}{3}}\sqrt[3]{c}$ $\Theta_{\frac{2}{3}}\sqrt[3]{c}$ or

$$\sqrt[3]{c} \quad \left\{-\frac{1}{2}+(-)^{\frac{1}{2}}\frac{\sqrt{3}}{2}\right\}\sqrt[3]{c} \quad \left\{-\frac{1}{2}-(-)^{\frac{1}{2}}\frac{\sqrt{3}}{2}\right\}\sqrt[3]{c}$$

of $-c$ $\quad \Theta_{\frac{1}{6}}\sqrt[3]{c}$ $\Theta_{\frac{1}{2}}\sqrt[3]{c}$ $\Theta_{\frac{5}{6}}\sqrt[3]{c}$ or

$$\left\{\frac{1}{2}+(-)^{\frac{1}{2}}\frac{\sqrt{3}}{2}\right\}\sqrt[3]{c} \quad -\sqrt[3]{c} \quad \left\{\frac{1}{2}-(-)^{\frac{1}{2}}\frac{\sqrt{3}}{2}\right\}\sqrt[3]{c}$$

of $(-)^{\frac{1}{2}}c$ $\quad \Theta_{\frac{1}{12}}\sqrt[3]{c}$ $\Theta_{\frac{5}{12}}\sqrt[3]{c}$ $\Theta_{\frac{3}{4}}\sqrt[3]{c}$ or

$$\left\{\frac{\sqrt{3}}{2}+(-)^{\frac{1}{2}}\frac{1}{2}\right\}\sqrt[3]{c} \quad \left\{-\frac{\sqrt{3}}{2}+(-)^{\frac{1}{2}}\frac{1}{2}\right\}\sqrt[3]{c} \quad -(-)^{\frac{1}{2}}\sqrt[3]{c}$$

and of $-(-)^{\frac{1}{2}}c$ $\quad \Theta_{\frac{1}{4}}\sqrt[3]{c}$ $\Theta_{\frac{7}{12}}\sqrt[3]{c}$ $\Theta_{\frac{11}{12}}\sqrt[3]{c}$ or

$$(-)^{\frac{1}{2}}\sqrt[3]{c} \quad \left\{-\frac{\sqrt{3}}{2}-(-)^{\frac{1}{2}}\frac{1}{2}\right\}\sqrt[3]{c} \quad \left\{\frac{\sqrt{3}}{2}-(-)^{\frac{1}{2}}\frac{1}{2}\right\}\sqrt[3]{c}.$$

Because generally

$$\cos\omega = \left(\cos\frac{1}{2}\omega\right)^2 - \left(\sin\frac{1}{2}\omega\right)^2 = 2\left(\cos\frac{1}{2}\omega\right)^2 - 1 = 1 - 2\left(\sin\frac{1}{2}\omega\right)^2$$

and specially $\cos\frac{1}{8} = \sin\frac{1}{8} = \sqrt{\frac{1}{2}}$ and $\cos\frac{1}{16}$ $\sin\frac{1}{16}$ are both plus $\cos\frac{1}{16} = \sqrt{\frac{1}{2}\left(\sqrt{\frac{1}{2}}+1\right)}$ $\sin\frac{1}{16} = \sqrt{\frac{1}{2}\left(-\sqrt{\frac{1}{2}}+1\right)}$. Putting then f g for these severally—

The four values of $(+c)^{\frac{1}{4}}$ are $+\sqrt[4]{c}$ $(-)^{\frac{1}{2}}\sqrt[4]{c}$ $-\sqrt[4]{c}$ $-(-)^{\frac{1}{2}}\sqrt[4]{c}$ of $(-c)^{\frac{1}{4}}$

$$\left\{\sqrt{\frac{1}{2}}+(-)^{\frac{1}{2}}\sqrt{\frac{1}{2}}\right\}\sqrt[4]{c} \qquad \left\{-\sqrt{\frac{1}{2}}+(-)^{\frac{1}{2}}\sqrt{\frac{1}{2}}\right\}\sqrt[4]{c}$$

$$\left\{-\sqrt{\frac{1}{2}}-(-)^{\frac{1}{2}}\sqrt{\frac{1}{2}}\right\}\sqrt[4]{c} \qquad \left\{\sqrt{\frac{1}{2}}-(-)^{\frac{1}{2}}\sqrt{\frac{1}{2}}\right\}\sqrt[4]{c}$$

of $\{(-)^{\frac{1}{2}}c\}^{\frac{1}{4}}$ $\{f+(-)^{\frac{1}{2}}g\}\sqrt[4]{c}$ $\{-g+(-)^{\frac{1}{2}}f\}\sqrt[4]{c}$ $\{-f-(-)^{\frac{1}{2}}g\}\sqrt[4]{c}$ $\{g-(-)^{\frac{1}{2}}f\}\sqrt[4]{c}$

of $\{-(-)^{\frac{1}{2}}c\}^{\frac{1}{4}}$ $\{g+(-)^{\frac{1}{2}}f\}\sqrt[4]{c}$ $\{-f+(-)^{\frac{1}{2}}g\}\sqrt[4]{c}$ $\{-g-(-)^{\frac{1}{2}}f\}\sqrt[4]{c}$ $\{f-(-)^{\frac{1}{2}}g\}\sqrt[4]{c}$.

401. If ω be the algebraic expression of an angle in reference to a perigon as unit m n any other whole numbers than 0 and c a numerical quantity

$$(\Theta_\omega c)^{\frac{m}{n}} = \{(\Theta_\omega c)^{\frac{1}{n}}\}^m = \{\Theta_{\frac{1}{n}(\omega+i)}\sqrt[n]{c}\}^m = \Theta_{\frac{m}{n}(\omega+i)}(\sqrt[n]{c})^m = \Theta_{\frac{m}{n}\omega+\frac{m}{n}i}(\sqrt[n]{c})^m$$

i being a plus or minus whole number.

If μ ν express the multiples that m n are severally of their greatest common measure $\frac{m}{n}i = \frac{\mu}{\nu}i = \frac{\mu i}{\nu}$. Let an unfractional division of μi by ν give a plus or minus whole quotient q and a plus or minus whole remainder r (which may be 0) of absolute value less than ν. Then $\frac{\mu i}{\nu} = \frac{r+q\nu}{\nu} = \frac{r}{\nu}+q$ and

$$\Theta_{\frac{m}{n}\omega+\frac{m}{n}i} = \Theta_{\frac{m}{n}\omega+\left(\frac{r}{\nu}+q\right)} = \Theta_{\frac{m}{n}\omega+\frac{r}{\nu}+q} = \Theta_{\frac{m}{n}\omega+\frac{r}{\nu}}.$$

But different values i, i' of i give the same value of r precisely when $\frac{\mu(-i,+i')}{\nu}$ is a plus or minus whole number and since μ is prime to ν this happens (art. 210) precisely when $-i,+i'$ is either 0 or a plus or minus multiple of ν. Hence the values 0 1 2 3 4 ... $\nu-1$

of i all give different values of r and hence $(\Theta_\omega c)^{\frac{m}{n}}$ has precisely the ν different values

$$\Theta_{\frac{m}{n}\omega}(\sqrt[n]{c})^m \quad \Theta_{\frac{m}{n}\omega+\frac{1}{\nu}}(\sqrt[n]{c})^m \quad \Theta_{\frac{m}{n}\omega+\frac{2}{\nu}}(\sqrt[n]{c})^m \; - \; - \; - \; \Theta_{\frac{m}{n}\omega+\frac{\nu-1}{\nu}}(\sqrt[n]{c})^m.$$

As in art. 400 these ν different values of $(\Theta_\omega c)^{\frac{m}{n}}$ are the products of multiplications of any one of them by each of the ν νth roots of $+1$.

402. *Def.* The symbol $\overline{\underline{c^d}}$ if c be a numerical and d a protensive quantity stands for the arithmetical value of c^d.

Thus $\overline{\underline{c^{\frac{m}{n}}}}$ symbolizes the same as $(\sqrt[n]{c})^m$.

403. If d be an incommensurable numerical quantity and so known only by this that when n any whole number but 0 is chosen at pleasure there is always a whole number m such that

$$d > \frac{m}{n} \text{ and } < \frac{m+1}{n}$$

the symbol $(\Theta_\omega c)^d$ wherein c is a numerical quantity and ω is the algebraic expression of an angle in reference to a perigon as unit can only get meaning through corresponding values $\Theta_{\frac{m}{n}(\omega+i)}(\sqrt[n]{c})^m$ $\Theta_{\frac{m+1}{n}(\omega+i)}(\sqrt[n]{c})^{m+1}$ of $(\Theta_\omega c)^{\frac{m}{n}}$ $(\Theta_\omega c)^{\frac{m+1}{n}}$ severally. But those corresponding values express algebraically directed distances which when n is taken endlessly great have numerical expressions endlessly near $\overline{\underline{c^d}}$ and make with the primary direction angles having algebraic expressions endlessly near $d(\omega+i)$. Wherefore $(\Theta_\omega c)^d$ must be taken to mean

$$\Theta_{d(\omega+i)}\overline{\underline{c^d}}$$

i being taken any plus or minus whole number.

Since two values $\Theta_{d(\omega+i_,)}$ $\Theta_{d(\omega+i')}$ of $\Theta_{d(\omega+i)}$ are equal precisely when $d(\omega+i') = d(\omega+i_,)+i''$ where i'' is a plus or minus whole number and this is precisely when $d(-i_,+i') = i''$ or the incommensurable d is equal to the plus or minus commensurable $\dfrac{i''}{-i_,+i'}$ it follows that the endless values of $\Theta_{d(\omega+i)}\overline{\underline{c^d}}$ got by making i in turn $\ldots -3 \; -2 \; -1 \; 0 \; 1 \; 2 \; 3 \ldots$ are all different.

Since as n becomes endlessly great $\sqrt[n]{c}$ becomes endlessly near 1 and $\frac{1}{n}(\omega+i)$ whatever unchanging value be given to i endlessly near 0 $(\Theta_\omega c)^0$ must be taken to symbolize Θ_0 or $+1$.

404. If c e be numerical quantities and ω be the algebraic expression of an angle in reference to a perigon as unit $(\Theta_\omega c)^{-1}$ is just as in art. 360 to be taken to symbolize $\frac{1}{\Theta_\omega c}$ and $(\Theta_\omega c)^{-e}$ $\{(\Theta_\omega c)^{-1}\}^e$ or $\left(\frac{1}{\Theta_\omega c}\right)^e$.

Since then (art. 396) $\frac{1}{\Theta_\omega c} = \Theta_{-\omega}\frac{1}{c} = \Theta_{-\omega}c^{-1}$

$$(\Theta_\omega c)^{-e} = (\Theta_{-\omega}c^{-1})^e = \Theta_{e(-\omega+i)}\underline{\overline{(c^{-1})^e}} = \Theta_{-e\omega+ei}\underline{\overline{c^{-e}}}$$
$$= \Theta_{(-e)\omega+(-e)(-i)}\underline{\overline{c^{-e}}} = \Theta_{(-e)(\omega+-i)}\underline{\overline{c^{-e}}}$$

where i or $-i$ is any plus or minus whole number.

Hence if e be a whole number n $(\Theta_\omega c)^{-n}$ has the single value $\Theta_{-n\omega}c^{-n}$ if e be a simple fraction $\frac{m}{n}$ with the terms m n prime to one another $(\Theta_\omega c)^{-\frac{m}{n}}$ has the n different values of $\Theta_{-\frac{m}{n}\omega+\frac{i}{n}}\underline{\overline{c^{-\frac{m}{n}}}}$ got by making i in turn 0 1 2 3...$n-1$ and if e be an incommensurable numerical quantity d $(\Theta_\omega c)^{-d}$ has the endless different values of $\Theta_{-d\omega+di}\underline{\overline{c^{-d}}}$ got by making i in turn ...-3 -2 -1 0 1 2 3....

It here comes out that all the different values of $(\Theta_\omega c)^a$ if a be any protensive quantity are precisely all the different values which can be got of $\Theta_{a(\omega+i)}\underline{\overline{c^a}}$ or $\Theta_{a\omega+ai}\underline{\overline{c^a}}$ by making i in turn all plus or minus whole numbers.

405. After the meaning and the values come the laws of protensive indexed powers of quantities protensive or ditensive. If ω ψ be algebraic expressions of angles in reference to a perigon as unit c d numerical quantities a a protensive quantity and i h g plus or minus whole numbers taken at will

$$(\Theta_\psi d)^a(\Theta_\omega c)^a = \{\Theta_{a(\psi+i)}\underline{\overline{d^a}}\}\Theta_{a(\omega+h)}\underline{\overline{c^a}} = \Theta_{a(\omega+h)+a(\psi+i)}\underline{\overline{d^a}}\,\underline{\overline{c^a}} = \Theta_{a\{\omega+\psi+(h+i)\}}\underline{\overline{(dc)^a}}$$
$$\{(\Theta_\psi d)\Theta_\omega c\}^a = (\Theta_{\omega+\psi}dc)^a = \Theta_{a(\omega+\psi+g)}\underline{\overline{(dc)^a}}$$

and by making g h i all plus or minus whole numbers $h+i$ is made all plus or minus whole numbers so that $\Theta_{a\{\omega+\psi+(h+i)\}}$ has the very same set of values as $\Theta_{a(\omega+\psi+g)}$.

$$\therefore\ (\Theta_\psi d)^a(\Theta_\omega c)^a = \{(\Theta_\psi d)\Theta_\omega c\}^a.$$

Also $$\frac{(\Theta_\psi d)^a}{(\Theta_\omega c)^a} = \frac{\Theta_{a(\psi+i)}\underline{\overline{d^a}}}{\Theta_{a(\omega+h)}\underline{\overline{c^a}}} = \Theta_{-a(\omega+h)+a(\psi+i)}\frac{\underline{\overline{d^a}}}{\underline{\overline{c^a}}} = \Theta_{a\{-\omega+\psi+(-h+i)\}}\underline{\overline{\left(\frac{d}{c}\right)^a}}$$

$$\text{and} \quad \left(\frac{\Theta_\psi d}{\Theta_\omega c}\right)^a = \left(\Theta_{-\omega+\psi}\frac{d}{c}\right)^a = \Theta_{a(-\omega+\psi+g)}\overline{\underline{\left(\frac{d}{c}\right)^a}}$$

$$\therefore \quad \frac{(\Theta_\psi d)^a}{(\Theta_\omega c)^a} = \left(\frac{\Theta_\psi d}{\Theta_\omega c}\right)^a.$$

406. If b be a protensive quantity and all else as before

$$(\Theta_\omega c)^b(\Theta_\omega c)^a = (\Theta_{b\omega+bi}\overline{\underline{c^b}})\Theta_{a\omega+ah}\overline{\underline{c^a}} = \Theta_{a\omega+ah+(b\omega+bi)}\overline{\underline{c^b}}\,\overline{\underline{c^a}}$$

$$= \Theta_{(a+b)\omega+ah+bi}\overline{\underline{c^{a+b}}} = \Theta_{(a+b)\omega+(ah+bi)}\overline{\underline{c^{a+b}}}$$

$$\text{and} \quad (\Theta_\omega c)^{a+b} = \Theta_{(a+b)\omega+(a+b)g}\overline{\underline{c^{a+b}}}.$$

Hence and because $ah+bi=(a+b)g$ whenever $i=h=g$ the different values of $(\Theta_\omega c)^{a+b}$ are all among the different values of $(\Theta_\omega c)^b(\Theta_\omega c)^a$.

If either a or b be a plus or minus whole number any value of $(a+b)\omega+ah+bi$ differs by a plus or minus whole number from the value of $(a+b)\omega+(a+b)g$ got by giving g the same value as either i in the former case or h in the latter and therefore all the different values of $(\Theta_\omega c)^b(\Theta_\omega c)^a$ are then precisely all the different values of $(\Theta_\omega c)^{a+b}$.

If a b be simple fractions $\frac{m}{n}$ $\frac{r}{s}$ each expressed in the least terms (art. 401) $(\Theta_\omega c)^{\frac{m}{n}} = \Theta_{\frac{m}{n}\omega+\frac{h}{n}}\overline{\underline{c^{\frac{m}{n}}}}$ $(\Theta_\omega c)^{\frac{r}{s}} = \Theta_{\frac{r}{s}\omega+\frac{i}{s}}\overline{\underline{c^{\frac{r}{s}}}}$ and therefore

$$(\Theta_\omega c)^{\frac{r}{s}}(\Theta_\omega c)^{\frac{m}{n}} = \Theta_{(\frac{m}{n}+\frac{r}{s})\omega+(\frac{h}{n}+\frac{i}{s})}\overline{\underline{c^{\frac{m}{n}+\frac{r}{s}}}}.$$

Let $n=\nu\kappa$ and $s=\sigma\kappa$ where κ is the greatest common measure of n and s then $\frac{h}{n}+\frac{i}{s}=\frac{h\sigma+i\nu}{\nu\sigma\kappa}$ and it has now to be found whether or no h i can be taken such plus or minus whole numbers as to make $h\sigma+i\nu$ any given plus or minus whole number f. Since any multiple $l\sigma$ of σ is a multiple of ν precisely when the equal σl is and since σ is prime to ν this (art. 210) is precisely when l is a multiple of ν unfractional divisions of σ 2σ $3\sigma\ldots(\nu-1)\sigma$ severally by ν are all inexact. And since the divisions of the less $k\sigma$ and of the greater $k'\sigma$ of any two of these can give the same remainder only when $-k\sigma+k'\sigma$ or the equal $(-k+k')\sigma$ is a multiple of ν which as before is precisely when $-k+k'$ is a multiple of ν the remainders of the $\nu-1$ unfractional divisions are all different. Those remainders then are just all the whole numbers but 0 less than ν and among them therefore is 1. In the division that gives the remainder 1 let $h'\sigma$ be the dividend and i' the quotient so that $1 = h'\sigma - i'\nu$ and therefore $f = f(h'\sigma - i'\nu) = (fh')\sigma - (fi')\nu$. Hence $h\sigma + i\nu = f$ precisely when $h\sigma + i\nu = (fh')\sigma - (fi')\nu$ therefore precisely when $(i+fi')\nu = (-h+fh')\sigma$ or

$$i+fi' = \frac{(-h-fh')\sigma}{\nu} = \frac{\sigma(-h+fh')}{\nu}$$

therefore again as before precisely when $-h+fh'$ is a multiple $t\nu$ of ν and at the same time $i+fi' = \sigma t = t\sigma$ or is the same multiple of σ and therefore at last precisely when

$$h = fh' - t\nu \quad \text{and} \quad i = t\sigma - fi'$$

t being any plus or minus whole number taken at pleasure. Any plus or minus whole number f then being taken at pleasure

$$(\Theta_\omega c)^{\frac{r}{s}}(\Theta_\omega c)^{\frac{m}{n}} = \Theta_{(\frac{m}{n}+\frac{r}{s})\omega + \frac{f}{\nu\sigma\kappa}} \overline{\underline{c^{\frac{m}{n}+\frac{r}{s}}}}$$

and hence has precisely the $\nu\sigma\kappa$ different values got by making f in turn 0 1 2...$\nu\sigma\kappa-1$. On the other hand $\frac{m}{n}+\frac{r}{s} = \frac{m\sigma+r\nu}{\nu\sigma\kappa}$. And $m\sigma+r\nu$ is prime to ν since any common measure of them measures (art. 197) $r\nu$ therefore (art. 200) $m\sigma$ or σm therefore (210) m which being prime to $\nu\kappa$ or $\kappa\nu$ is (art. 198) prime to ν. Also $m\sigma+r\nu$ is prime to σ. Hence (art. 207) $m\sigma+r\nu$ is prime to $\nu\sigma$ and therefore (art. 210) all the common measures of $m\sigma+r\nu$ and $(\nu\sigma)\kappa$ or $\nu\sigma\kappa$ are precisely all the common measures of $m\sigma+r\nu$ and κ. Let then λ be the greatest common measure of $m\sigma+r\nu$ and κ and let $\kappa = \kappa'\lambda$. Because then λ is the greatest common measure of $m\sigma+r\nu$ and $\nu\sigma\kappa$ and $\nu\sigma\kappa = \nu\sigma\kappa'\lambda = (\nu\sigma\kappa')\lambda$

$$(\Theta_\omega c)^{\frac{m}{n}+\frac{r}{s}} = \Theta_{(\frac{m}{n}+\frac{r}{s})\omega + \frac{g}{\nu\sigma\kappa'}} \overline{\underline{c^{\frac{m}{n}+\frac{r}{s}}}}$$

and has therefore precisely the $\nu\sigma\kappa'$ different values got by making g in turn 0 1 2...$\nu\sigma\kappa'-1$. Thus it is only those values of $(\Theta_\omega c)^{\frac{r}{s}}(\Theta_\omega c)^{\frac{m}{n}}$ got by making f a plus or minus multiple of λ that are values of $(\Theta_\omega c)^{\frac{m}{n}+\frac{r}{s}}$ and therefore it is only when λ is 1 that the several values are the same of $(\Theta_\omega c)^{\frac{r}{s}}(\Theta_\omega c)^{\frac{m}{n}}$ and of $(\Theta_\omega c)^{\frac{m}{n}+\frac{r}{s}}$.

If a be $\frac{m}{n}$ and b $-\frac{r}{s}$ then as above (art. 404)

$$(\Theta_\omega c)^{-\frac{r}{s}}(\Theta_\omega c)^{\frac{m}{n}} = \Theta_{(\frac{m}{n}-\frac{r}{s})\omega + \frac{f}{\nu\sigma\kappa}} \overline{\underline{c^{\frac{m}{n}-\frac{r}{s}}}}$$

and $m\sigma - r\nu$ is prime to $\nu\sigma$ so that if λ be now the greatest common measure of $m\sigma - r\nu$ and κ and $\kappa = \kappa'\lambda$

$$(\Theta_\omega c)^{\frac{m}{n}-\frac{r}{s}} = \Theta_{\left(\frac{m}{n}-\frac{r}{s}\right)\omega+\frac{g}{\nu\sigma\kappa'}} \overline{\underline{c^{\frac{m}{n}-\frac{r}{s}}}}.$$

The $\nu\sigma\kappa'$ different values of $(\Theta_\omega c)^{\frac{m}{n}-\frac{r}{s}}$ are therefore $\frac{\nu\sigma\kappa}{\lambda}$ of the $\nu\sigma\kappa$ different values of $(\Theta_\omega c)^{-\frac{r}{s}}(\Theta_\omega c)^{\frac{m}{n}}$. And the cases when a is $-\frac{m}{n}$ and b $\frac{r}{s}$ and when a is $-\frac{m}{n}$ and b $-\frac{r}{s}$ may be dealt with in a like way.

If a be plus or minus an incommensurable numerical quantity and b $\pm\frac{r}{s}$ $\left(a\pm\frac{r}{s}\right)\omega+ah+\frac{i}{s}$ differs from $\left(a\pm\frac{r}{s}\right)\omega+\left(a\pm\frac{r}{s}\right)g$ by a plus or minus whole number t just when $a(-g+h) = \pm\frac{r}{s}g+t-\frac{i}{s}$ and inasmuch as the incommensurable a cannot be equal to the commensurable $\dfrac{\pm\frac{r}{s}g+t-\frac{i}{s}}{-g+h}$ this is just when $-g+h=0$ and $\pm\frac{r}{s}g+t-\frac{i}{s}=0$. Hence in each of the endless sets of s each that make up all the endless different values of $(\Theta_\omega c)^{\pm\frac{r}{s}}(\Theta\ c)^a$ there is just one that is a value of $(\Theta_\omega c)^{a\pm\frac{r}{s}}$

Lastly if a b be plus or minus quantities both incommensurable in absolute value $(a+b)\omega+ah+bi$ differs by a plus or minus whole number t from $(a+b)\omega+(a+b)g$ precisely when $a(-g+h)+b(-g+i)=t$. Unless a b be such that $b=a\frac{u}{p}+\frac{v}{q}$ p q being whole numbers each else than 0 u a plus or minus whole number else than 0 prime to p and v a plus or minus whole number prime to q this is precisely when $-g+i=0$ $-g+h=0$ and $t=0$ and then the only values common to $(\Theta_\omega c)^b(\Theta_\omega c)^a$ and $(\Theta_\omega c)^{a+b}$ are those wherein $i=h=g$. But if $b=a\frac{u}{p}+\frac{v}{q}$ $a(-g+h)+b(-g+i)=t$ otherwise precisely when $b=-a\frac{-g+h}{-g+i}+\frac{t}{-g+i}$ therefore precisely when $a\left(\frac{-g+h}{-g+i}+\frac{u}{p}\right)=\frac{t}{-g+i}-\frac{v}{q}$ therefore a being incommensurable precisely when $\frac{-g+h}{-g+i}+\frac{u}{p}=0$ and $\frac{t}{-g+i}-\frac{v}{q}=0$ therefore again (art. 211) precisely when x y being any plus or minus whole numbers but 0 $-g+h=-ux$ $-g+i=px=qy$

and $t=vy$ and at length therefore if $p=p_{,}\rho$ $q=q_{,}\rho$ ρ being the greatest common measure of p and q and z be any plus or minus whole number but 0,—since $px=qy$ just when $\frac{x}{y}=\frac{1}{p}q=\frac{q}{p}=\frac{q_{,}}{p_{,}}$ and therefore just when $x=q_{,}z$ and $y=p_{,}z$—, precisely when $h=g-(uq_{,})z$ $i=g+(pq_{,})z$ or $g+(qp_{,})z$ and $t=(vp_{,})z$. The first two of these last happen precisely when $g=h+(uq_{,})z$ and $i=h+\{(u+p)q_{,}\}z$ and hence if $b=a\frac{u}{p}+\frac{v}{q}$ of the endless different values of $(\Theta_\omega c)^b(\Theta_\omega c)^a$ that have any and the same value of h not only is that one in which $i=h$ a value of $(\Theta_\omega c)^{a+b}$ but also every one in which i has one or other of the endless different values $h+(u+p)q_{,}$ $h+\{(u+p)q_{,}\}\times 2$ $h+\{(u+p)q_{,}\}\times 3$ - - - $h-(u+p)q_{,}$ $h-\{(u+p)q_{,}\}\times 2$ $h-\{(u+p)q_{,}\}\times 3$ - - -. Moreover since it is precisely when $(u+p)q_{,}=1$ that $q_{,}=1$ and $u+p=1$ and precisely when $(u+p)q_{,}=-1$ that $q_{,}=1$ and $u+p=-1$ when too in either case $q=\rho$ $p=p_{,}q=qp_{,}$ and therefore $\frac{v}{q}=\frac{vp_{,}}{p}$ where $vp_{,}$ then is any plus or minus whole number w it is precisely when either $b=a\frac{1-p}{p}+\frac{w}{p}$ without p being 1 or $b=a\frac{-1-p}{p}+\frac{w}{p}$ that all the different values of $(\Theta_\omega c)^b(\Theta_\omega c)^a$ are values of $(\Theta_\omega c)^{a+b}$.

Since in all cases $(a+b)\omega+ah+bi=(a+b)\omega+(a+b)h+b(-h+i)$

$$(\Theta_\omega c)^b(\Theta_\omega c)^a=\Theta_0^b(\Theta_\omega c)^{a+b}=(+)^b(\Theta_\omega c)^{a+b}.$$

407. If the letters stand for the same as in arts. 405, 406,

$$\left(\frac{1}{\Theta_\omega c}\right)^b(\Theta_\omega c)^a=(\Theta_\omega c)^{-b}(\Theta_\omega c)^a=(+)^{-b}(\Theta_\omega c)^{a+-b}=(+)^b(\Theta_\omega c)^{a-b}$$

$$\text{and }=\left(\frac{1}{\Theta_\omega c}\right)^b\left(\frac{1}{\Theta_\omega c}\right)^{-a}=(+)^b\left(\frac{1}{\Theta_\omega c}\right)^{-a+b}$$

$$\frac{1}{(\Theta_\omega c)^b}(\Theta_\omega c)^a=\left(\frac{1}{\Theta_\omega c}\right)^b(\Theta_\omega c)^a=(+)^b(\Theta_\omega c)^{a-b}$$

$$\therefore\text{ too}=(+)^b\left(\frac{1}{\Theta_\omega c}\right)^{-(a-b)}=(+)^b\frac{1}{(\Theta_\omega c)^{b-a}}$$

$$\frac{(\Theta_\omega c)^a}{(\Theta_\omega c)^b}=(\Theta_\omega c)^a\frac{1}{(\Theta_\omega c)^b}=(+)^a(\Theta_\omega c)^{-b+a}\text{ and }=(+)^a\frac{1}{(\Theta_\omega c)^{-a+b}}$$

$$(\Theta_\omega c)^a\left(\frac{1}{\Theta_\omega c}\right)^b=(+)^a(\Theta_\omega c)^{-b+a}\text{ and }=(+)^a\left(\frac{1}{\Theta_\omega c}\right)^{b-a}.$$

408. If a b be protensive quantities c a numerical quantity ω the algebraic expression of an angle in reference to a perigon as unit and g h i any whole numbers chosen at will

$$\{(\Theta_\omega c)^b\}^a = \{\Theta_{b(\omega+h)}\overline{c^b}\}^a = \Theta_{a\{b(\omega+h)+i\}}\overline{(\overline{c^b})^a} = \Theta_{(ab)\omega+(ab)h+ai}\overline{c^{ab}}$$

$$\text{and} \quad (\Theta_\omega c)^{ab} = \Theta_{(ab)\omega+(ab)g}\overline{c^{ab}}.$$

Any value of $(\Theta_\omega c)^{ab}$ is therefore always a value of $\{(\Theta_\omega c)^b\}^a$.

If n s be whole numbers each else than 0 and m r plus or minus whole numbers prime severally to n s

$$\left(\Theta_\omega^{\frac{r}{s}}\right)^{\frac{m}{n}} = \Theta^{\frac{m}{n}}_{\frac{r}{s}\omega+\frac{h}{s}} = \Theta_{\left(\frac{m}{n}\frac{r}{s}\right)\omega+\frac{m}{n}\frac{h}{s}+\frac{i}{n}} = \Theta_{\left(\frac{m}{n}\frac{r}{s}\right)\omega+\left(\frac{mh}{sn}+\frac{i}{n}\right)}$$

and if κ being the greatest common measure of m and s $m=m'\kappa$ and $s=s'\kappa$ $\frac{mh}{sn}=\frac{hm'}{ns'}$ $\frac{mh}{sn}+\frac{i}{n}=\frac{hm'+is'}{ns'}$ and as in art. 406 h i may be taken such that $hm'+is'$ is any plus or minus whole number f whatever so that

$$\left(\Theta_\omega^{\frac{r}{s}}\right)^{\frac{m}{n}} = \Theta_{\left(\frac{m}{n}\frac{r}{s}\right)\omega+\frac{f}{ns'}}.$$

Again if λ being the greatest common measure of n and r $n=n_,\lambda$ and $r=r_,\lambda$ $\frac{m}{n}\frac{r}{s}=\frac{mr_,}{sn_,}=\frac{m'r_,}{s'n_,}$ and because each of the plus or minus whole numbers m' $r_,$ is prime to each of the whole numbers s' $n_,$ $m'r_,$ is (art. 208) prime to $s'n_,$ so that

$$\Theta_\omega^{\frac{m}{n}\frac{r}{s}} = \Theta_{\left(\frac{m}{n}\frac{r}{s}\right)\omega+\frac{g}{s'n_,}}.$$

Therefore the $s'n_,$ different values of $(\Theta_\omega c)^{\frac{m}{n}\frac{r}{s}}$ are $\frac{s'n}{\lambda}$ of the $s'n$ or ns' different values of $\{(\Theta_\omega c)^{\frac{r}{s}}\}^{\frac{m}{n}}$.

If a be a plus or minus incommensurable numerical quantity

$$\left(\Theta_\omega^{\frac{r}{s}}\right)^a = \Theta_{\left(a\frac{r}{s}\right)\omega+a\frac{h}{s}+ai} \qquad \Theta_\omega^{a\frac{r}{s}} = \Theta_{\left(a\frac{r}{s}\right)\omega+\left(a\frac{r}{s}\right)g}$$

and $\left(a\frac{r}{s}\right)\omega+a\frac{h}{s}+ai$ differs by a plus or minus whole number t from $\left(a\frac{r}{s}\right)\omega+\left(a\frac{r}{s}\right)g$ precisely when $a\left(\frac{-rg+h}{s}+i\right)=t$ therefore a being incommensurable precisely when $\frac{-rg+h}{s}+i=0$ and $t=0$ and therefore precisely when $i=\frac{-h+rg}{s}$. Let $h=\eta+h's$ η being a whole number less than s and h' a plus or minus whole number then since the unfractional divisions by s of 0 r $r\times 2$ $r\times 3$ ... $r(s-1)$ severally all yield different remainders the division of some one $r\gamma$ of these yields the remainder η. Let $r\gamma=\eta+ks$ k being a plus or minus whole number and it is precisely when $g=\gamma+g's$ g' being a plus or minus whole number that $i=\dfrac{-h+rg}{s}=-h'+k+rg'$ for it is then only that $-h+rg$ is a plus or minus multiple of s. Hence every value of $\{(\Theta_\omega c)^{\frac{r}{s}}\}^a$ is a value of $(\Theta_\omega c)^{a\frac{r}{s}}$ just when r is ± 1.

If b be a plus or minus incommensurable numerical quantity

$$(\Theta_\omega^b)^{\frac{m}{n}}=\Theta_{\left(\frac{m}{n}b\right)\omega+\left(\frac{m}{n}b\right)h+\frac{i}{n}} \qquad \Theta_\omega^{\frac{m}{n}b}=\Theta_{\left(\frac{m}{n}b\right)\omega+\left(\frac{m}{n}b\right)g}$$

and $\left(\frac{m}{n}b\right)\omega+\left(\frac{m}{n}b\right)h+\frac{i}{n}$ differs by a plus or minus whole number t from $\left(\frac{m}{n}b\right)\omega+\left(\frac{m}{n}b\right)g$ precisely when $\frac{m}{n}b(-g+h)=t-\frac{i}{n}$. Hence there are no other values common to $\{(\Theta_\omega c)^b\}^{\frac{m}{n}}$ and $(\Theta_\omega c)^{\frac{m}{n}b}$ than those in which $h=g$ and i is a plus or minus multiple of n.

Lastly if a b be plus or minus incommensurable numerical quantities $(ab)\omega+(ab)h+ai$ differs by a plus or minus whole number t from $(ab)\omega+(ab)g$ precisely when $a\{b(-g+h)+i\}=t$. Unless a b be such that $a^{-1}=b\frac{u}{p}+\frac{v}{q}$ where p q are whole numbers other than 0 u a plus or minus whole number not 0 prime to p and v a plus or minus whole number prime to q this is precisely when $h=g$ $i=0$ and $t=0$. But if $a^{-1}=b\frac{u}{p}+\frac{v}{q}$ $a\{b(-g+h)+i\}=t$ further precisely when $a^{-1}=b\frac{-g+h}{t}+\frac{i}{t}$ therefore precisely when $b\left(-\frac{-g+h}{t}+\frac{u}{p}\right)=\frac{i}{t}-\frac{v}{q}$ therefore again b being incommensurable precisely when $-\frac{-g+h}{t}+\frac{u}{p}=0$ and $\frac{i}{t}-\frac{v}{q}=0$ and therefore precisely when x y being plus or minus whole numbers other than 0 $-g+h=ux$ $t=px=qy$ and $i=vy$. Since $px=qy$ precisely when $\frac{x}{y}=\frac{1}{p}q=\frac{q}{p}=\frac{q_{\prime}}{p_{\prime}}$ $p_{\prime}$ $q_{\prime}$ being the whole numbers other than 0 that severally express what multiples p q are of ρ their greatest common measure $x=q_{\prime}z$ and $y=p_{\prime}z$ z being a plus or minus whole number else than 0 and therefore each of the above relations and sets of relations holds precisely when $g=h-(uq_{\prime})z$ $i=(vp_{\prime})z$ and $t=(p_{\prime}q_{\prime}\rho)z$. Hence it is precisely any such value of $\{(\Theta_\omega c)^b\}^a$ as has h any value and i either 0 or $(vp_,)z$ that is a value of $(\Theta_\omega c)^{ab}$ the value to wit which has g either h or $h-(uq_,)z$. Moreover since $vp_,$ is 1 or -1 precisely when $p_,$ is 1 and v either 1 or -1 and then $p=\rho \quad q=q_,p=pq_,$ and $\frac{u}{p}=\frac{uq_,}{pq_,}=\frac{w}{q}$ where w is any plus or minus whole number but 0 all the values of $\{(\Theta_\omega c)^b\}^a$ and of $(\Theta_\omega c)^{ab}$ are the same precisely when $\frac{1}{a}=b\frac{w}{q}\pm\frac{1}{q}$.

Since in every case $\Theta_{(ab)\omega+(ab)h+ai}=\Theta_0^a\,\Theta_{(ab)\omega+(ab)h}$

$$\{(\Theta_\omega c)^b\}^a=\Theta_0^a\,(\Theta_\omega c)^{ab}=(+)^a(\Theta_\omega c)^{ab}.$$

CHAPTER V

EQUATIONS
OF THE FIRST AND SECOND DEGREES

409. *Def.* AN EQUATION is a statement that two quantities are equal. And these quantities are called the equation's MEMBERS TERMS or SIDES.

There are as many kinds of equations as meanings of the word *equal.* Thus an equation may be one of identity or sameness as $2a+b=2a+b$ of simple result stating as $4+5-6=3$ of definition of a symbol as $3x=x+x+x$ or of operational equivalence as $3x+2x=5x$. But the following is what chiefly bears the name *equation.*

Def. An EQUATION OF SPECIFIC RELATION is one which having one or each member either a symbolized quantity or an expression with one or more symbolized quantities in it marks out these quantities to be not any whatever but certain only.

Def. When of the quantities that enter into an equation or set of equations of specific relation one or more are taken as given or known and the rest as sought or unknown it is called SOLVING the equation or set of equations to find precisely what operations performed on the given quantities give the sought.

410. If the letters stand for any numbers whole or fractional any numerical quantities commensurable or incommensurable any protensive quantities plus or minus or any algebraic quantities protensive or ditensive

$$a+b+c+\cdots+g=p+q+\cdots+v$$

precisely when any one of the following equations holds

$$a+b+c+\cdots+g-b=p+q+\cdots+v-b$$
$$a+b-b+c+\cdots+g=p-b+q+\cdots+v$$
$$a+(b-b)+c+\cdots+g=p-b+q+\cdots+v$$
$$a+c+\cdots+g=p-b+q+\cdots+v.$$

Since $p+q=p-(-q)$ $p+q+\cdots+v$ may be taken as the result of any series of successive additions and subtractions and likewise $a+b+c+\cdots+g$. Hence an equation expressing the equality of results of additions and subtractions holds precisely when another equation holds got therefrom by changing either an addition of a quantity anywhere on one side into a subtraction of that quantity anywhere on the other or a subtraction of a quantity anywhere on one side into an addition of that quantity anywhere on the other. Each of these changings is called TRANSPOSITION. Of course in either series of successive additions and subtractions each or any of the quantities operated with may be o.

411. If $a'\ b'\ c' \ldots g'$ be any quantities and $a\ b\ c \ldots g\ p$ any quanties but o the equation

$$\frac{a'}{a}+\frac{b'}{b}+\frac{c'}{c}+\cdots+\frac{g'}{g}=0$$

holds precisely when severally

$$\left(\frac{a'}{a}+\frac{b'}{b}+\frac{c'}{c}+\cdots+\frac{g'}{g}\right)p=0$$

$$\frac{a'}{a}p+\frac{b'}{b}p+\frac{c'}{c}p+\cdots+\frac{g'}{g}p=0.$$

Further if $a'=\alpha'a_{,}$ $a=\alpha a_{,}$ $b'=\beta'b_{,}$ $b=\beta b_{,}$ $c'=\gamma'c_{,}$ $c=\gamma c_{,}$ - - - $g'=\theta'g_{,}$ $g=\theta g_{,}$ then $\frac{a'}{a}=\frac{\alpha'}{\alpha}$ $\frac{b'}{b}=\frac{\beta'}{\beta}$ $\frac{c'}{c}=\frac{\gamma'}{\gamma}$ - - - $\frac{g'}{g}=\frac{\theta'}{\theta}$ and if $\alpha_{,}$ $\beta_{,}$ $\gamma_{,}\ldots$ $\theta_{,}$ being quantities such that $\alpha\alpha_{,}=\beta\beta_{,}=\gamma\gamma_{,}=---=\theta\theta_{,}$ p be taken equal to each of these products $\frac{\alpha'}{\alpha}\alpha\alpha_{,}=\left(\frac{\alpha'}{\alpha}\alpha\right)\alpha_{,}=\alpha'\alpha_{,}$ and so for the rest and the equation then holds precisely when

$$\alpha'\alpha_{,}+\beta'\beta_{,}+\gamma'\gamma_{,}+\cdots+\theta'\theta_{,}=0.$$

This process is called CLEARING AN EQUATION OF FRACTIONS. When $\alpha'\ \alpha\ \beta'\ \beta \ldots \theta'\ \theta$ are all whole p is most simply taken equal to the simplest common multiple of $\alpha\ \beta \ldots \theta$. Were it not for the condition that p is not o the equation $\alpha'\alpha_{,}+\beta'\beta_{,}+\cdots+\theta'\theta_{,}=0$ might be fulfilled when the equation $\frac{a'}{a}+\frac{b'}{b}+\cdots+\frac{g'}{g}=0$ is not. The better to meet that condition fractions may be cleared off one by one instead of all at once. Thus for getting rid of $\frac{b'}{b}$ or the equivalent $\frac{\beta'}{\beta}$ p may be taken β and then $\frac{a'}{a}+\frac{b'}{b}+\frac{c'}{c}+\cdots+\frac{g'}{g}=0$ precisely when

$$\frac{\alpha'}{\alpha}\beta+\frac{\beta'}{\beta}\beta+\cdots+\frac{\theta'}{\theta}\beta=0 \quad\text{or}\quad \frac{\alpha'\beta}{\alpha}+\beta'+\frac{\gamma'\beta}{\gamma}+\cdots+\frac{\theta'\beta}{\theta}=0.$$

If an equation $\frac{1}{\alpha}\alpha'+\frac{1}{\beta}\beta'+\cdots+\frac{1}{\theta}\theta'=0$ be given the more immediate way to clear from fractions is that this holds precisely when severally

$$p\left(\frac{1}{\alpha}\alpha'+\frac{1}{\beta}\beta'+\cdots+\frac{1}{\theta}\theta'\right)=0 \qquad p\frac{1}{\alpha}\alpha'+p\frac{1}{\beta}\beta'+\cdots+p\frac{1}{\theta}\theta'=0$$

and if $p=\alpha_{,}\alpha=\beta_{,}\beta=---=\theta_{,}\theta$ $\because$ $(\alpha_{,}\alpha)\frac{1}{\alpha}\alpha'=\alpha_{,}\alpha\frac{1}{\alpha}\alpha'=\alpha_{,}\left(\alpha\frac{1}{\alpha}\right)\alpha'$ and so on precisely when

$$\alpha_{,}\alpha'+\beta_{,}\beta'+\cdots+\theta_{,}\theta'=0.$$

412. In art. 411 none of the quantities $a_{,}$ α $b_{,}$ β .... $g_{,}$ θ is 0 because it is there understood that none of the quantities a b...g is 0. But it may happen that some values of the quantity or quantities held as unknown in the equation $\frac{a'}{a}+\frac{b'}{b}+\cdots+\frac{g'}{g}=0$ make some of the quantities a b ... g to be 0.

If $a'=\alpha'a_{,}$ and $a=\alpha a_{,}$ always and some values of the unknowns give $a_{,}$ the value 0 then for these values $\frac{\alpha' a_{,}}{\alpha a_{,}}$ takes the utterly unmeaning shape $\frac{0}{0}$ and therefore cannot be said to be equal to $\frac{\alpha'}{\alpha}$. As in all other uses of the quotient symbol so now by the symbolization extension principle $\frac{u}{v}$ is still to be held such that $\frac{u}{v}v=u$ and therefore in particular when u is 0 and v 0 such that

$$\frac{0}{0}\times 0=0.$$

But it is precisely any quantity whatever that multiplied into 0 gives the product 0 and hence $\frac{0}{0}$ must be taken to symbolize any quantity be it what it may. So too $\frac{u}{0}$ when u is not 0 must be held such that

$$\frac{u}{0}\times 0=u.$$

But there is no quantity whatever that multiplied into o can give a product anything else than o. Yet since however great an absolute value any given quantity k may have v may (arts. 172, 300) be taken of so small an absolute value that the absolute value of $\frac{1}{v}$ is greater than the absolute value of $\frac{1}{u}k$ and therefore (arts. 91, 89, 300) that the absolute value of $u\frac{1}{v}$ or of the equivalent $\frac{u}{v}$ is greater than the absolute value of $u\frac{1}{u}k$ or of the equivalent k v may be taken so near o that $\frac{u}{v}$ has an absolute value greater than any given numerical quantity however great. Hence $\frac{u}{0}$ when u is any quantity but o can only be taken to mean what has an endlessly great absolute value and what therefore $\Theta_{\theta}\times\infty$ θ being the algebraic expression of an angle otherwise symbolizes.

In like manner $\frac{1}{0}\times 0$ and $\frac{1}{0}u$ are to be held such as severally fulfil the relations

$$0\times\frac{1}{0}\times 0=0 \qquad 0\times\frac{1}{0}u=u.$$

And hence $\frac{1}{0}\times 0$ is any quantity and $\frac{1}{0}u$ u being any quantity but o what has an endlessly great absolute value.

If then in the equation $\frac{a'}{a}+\frac{b'}{b}+\frac{c'}{c}+\cdots+\frac{g'}{g}=0$ some values of some of the unknowns make $\frac{b'}{b}$ one of the quantities $\frac{a'}{a}$ $\frac{b'}{b}$ $\frac{c'}{c}\cdots\frac{g'}{g}$ become $\frac{0}{0}$ and none of the rest of the shape $\frac{u}{0}$ with u not o the equation holds just when

$$\frac{b'}{b}=-\frac{a'}{a}-\frac{g'}{g}-\cdots-\frac{c'}{c}$$

and here the first member being simply anything which multiplied into o gives o for product may be taken as equal to the other member whether this other be of the shape $\frac{0}{0}$ or some fixed quantity. Hence those values fulfil the equation.

If again some values of some of the unknowns make $\frac{b'}{b}\ \frac{0}{0}$ and one and only one $\frac{c'}{c}$ of the other quotients $\frac{u}{0}$ where u is not 0 the member $\frac{b'}{b}$ is still what multiplied into 0 gives a product 0 but the member $-\frac{a'}{a}-\frac{g'}{g}-\cdots-\frac{c'}{c}=-\frac{a'}{a}-\frac{g'}{g}-\cdots-\frac{e'}{e}+\left(-\cdots-\frac{c'}{c}\right)$ and therefore such that $\left(-\frac{a'}{a}-\frac{g'}{g}-\cdots-\frac{c'}{c}\right)\times 0=\left(-\frac{a'}{a}-\cdots\right)\times 0-\frac{e'}{e}\times 0+\left(-\cdots-\frac{c'}{c}\right)\times 0$ and $\therefore =-\frac{c'}{c}\times 0$ which is not 0. Hence the values of the unknowns do not in this case fulfil the equation.

If for some values of some of the unknowns $\frac{b'}{b}$ takes the shape $\frac{0}{0}$ and of the other quotients several $\frac{a'}{a}\ \frac{d'}{d}\cdots\frac{f'}{f}$ and only these take the shape of $\frac{u}{0}$ u being another quantity than 0 $-\frac{a'}{a}-\frac{g'}{g}-\cdots-\frac{c'}{c}=-\frac{a'}{a}-\frac{f'}{f}-\cdots-\frac{d'}{d}-\frac{g'}{g}-\cdots-\frac{c'}{c}$ and a single quotient $\frac{s}{t}$ must (arts. 132, 354, 398) be found equal to $-\frac{a'}{a}-\frac{f'}{f}-\cdots-\frac{d'}{d}$ for other values of the unknowns then according as $\frac{s}{t}$ takes the shape $\frac{u}{0}$ or not for the values of the unknowns is the equation not fulfilled or fulfilled by these values.

Likewise values of the unknowns that make only one quotient $\frac{u}{0}$ u being not 0 do not fulfil the equation and values that make several of the quotients of the shape $\frac{u}{0}$ fulfil the equation precisely when the algebraic sum of those several quotients for other values of the unknowns is equal to a quotient which for the values takes the shape $\frac{0}{0}$ since if the several quotients were $\frac{b'}{b}\ \frac{d'}{d}\cdots\frac{f'}{f}$ the equation holds precisely when severally

$$\frac{a'}{a}+\frac{c'}{c}+\cdots+\frac{g'}{g}+\left(\frac{b'}{b}+\frac{d'}{d}+\cdots+\frac{f'}{f}\right)=0$$

$$\frac{b'}{b}+\frac{d'}{d}+\cdots+\frac{f'}{f}=-\frac{g'}{g}-\cdots-\frac{c'}{c}-\frac{a'}{a}.$$

The cases when one or more of the quantities summed in an equation take the shape $\frac{1}{0}\times 0$ $0\times\frac{1}{0}$ $\frac{1}{0}\times u$ or $u\frac{1}{0}$ u being else than 0 are to be dealt with in a like way.

413. PROP. *A product is equal to* 0 *precisely when one or other of its factors is equal to* 0.

For if any one of the factors of a product be 0 the product is 0 and if each of the factors be else than 0 the product is else than 0.

414. Equations are CLEARED OF ROOTS as in the following instances. Let an equation free from root symbols be sought expressing the very same relation as the equation $p^{\frac{1}{2}}+q^{\frac{1}{2}}=0$. If a x be any values of $p^{\frac{1}{2}}$ $x^2=p$ and $a^2=p$ $\therefore$ $x^2=a^2$ and this is precisely when severally $x^2-a^2=0$ $(x-a)(x+a)=0$. But (art. 413) $(x-a)(x+a)=0$ precisely when either $x-a=0$ or $x+a=0$ which severally are precisely when $x=a$ $x=-a$. Wherefore if a be any value of $p^{\frac{1}{2}}$ the only other value is $-a$. This is at one with what was found in art. 400. Likewise if b be any value of $q^{\frac{1}{2}}$ the only other is $-b$. Hence the equation $p^{\frac{1}{2}}+q^{\frac{1}{2}}=0$ holds precisely when one or other of the four equations holds

$$a+b=0 \qquad -a+b=0 \qquad a-b=0 \qquad -a-b=0.$$

And since $-a+b=0$ precisely when $a-b=0$ and $-a-b=0$ precisely when $a+b=0$ $p^{\frac{1}{2}}+q^{\frac{1}{2}}=0$ precisely when either $a+b=0$ or $a-b=0$ and therefore (art. 413) precisely when severally

$$(a+b)(a-b)=0 \qquad a^2-b^2=0, \text{ that is } p-q=0.$$

The equations $p^{\frac{1}{2}}+q^{\frac{1}{2}}=0$ $p-q=0$ then express the very same relation.

Inasmuch as $p-q=(p^{\frac{1}{2}}+q^{\frac{1}{2}})(p^{\frac{1}{2}}-q^{\frac{1}{2}})=(p^{\frac{1}{2}}-q^{\frac{1}{2}})(p^{\frac{1}{2}}+q^{\frac{1}{2}})$ the equation $p-q=0$ may be held as got from the equation $p^{\frac{1}{2}}+q^{\frac{1}{2}}=0$ by means of the factor $p^{\frac{1}{2}}-q^{\frac{1}{2}}$. An expression used in this way to get from an equation with one or more root symbols another equation expressing the same relation but without those root symbols is called an UNROOTING FACTOR of the equation.

The equation $p^{\frac{1}{2}}-q^{\frac{1}{2}}=0$ may be shown in the same way to express the same relation as the equation $p-q=0$ and hence the equation $p^{\frac{1}{2}}-q^{\frac{1}{2}}=0$ expresses the same relation as the equation $p^{\frac{1}{2}}+q^{\frac{1}{2}}=0$. Equations having the same one or more root symbols are called CONJUGATE which singly express the same relation as the same other equation free from those root symbols.

Either of the equations $k+p^{\frac{1}{2}}=0$ $k-p^{\frac{1}{2}}=0$ in like manner expresses the same relation as the equation $(k+p^{\frac{1}{2}})(k-p^{\frac{1}{2}})=0$ or the equation $k^2-p=0$.

For finding the rootless equation that expresses the same relation as the equation $p^{\frac{1}{2}}+q^{\frac{1}{2}}+r^{\frac{1}{2}}=0$ let $p^{\frac{1}{2}}$'s two values be a and $-a$ $q^{\frac{1}{2}}$'s two b and $-b$ and $r^{\frac{1}{2}}$'s c and $-c$ then $p^{\frac{1}{2}}+q^{\frac{1}{2}}+r^{\frac{1}{2}}=0$ precisely when one or other of these eight expressions $=0$

$$\begin{array}{cccc} a+b+c & a+b-c & a-b+c & a-b-c \\ -a+b+c & -a+b-c & -a-b+c & -a-b-c \end{array}$$

and since the last four in order are severally 0 precisely when the first four in backward order are this happens precisely when $(a+b+c)(a+b-c)(a-b+c)(a-b-c)$ or the operational equivalent

$$(a^2)^2+(b^2)^2+(c^2)^2-2b^2c^2-2c^2a^2-2a^2b^2=0,$$

$$\text{that is}\quad p^2+q^2+r^2-2qr-2rp-2pq=0.$$

The same end is gained by clearing away the roots one by one and then it may chance that more roots than one are cleared away at one stroke. Thus $p^{\frac{1}{2}}+q^{\frac{1}{2}}+r^{\frac{1}{2}}=0$ precisely when $(p^{\frac{1}{2}}+q^{\frac{1}{2}}+r^{\frac{1}{2}})(p^{\frac{1}{2}}+q^{\frac{1}{2}}-r^{\frac{1}{2}})$ or

$$p+q+2p^{\frac{1}{2}}q^{\frac{1}{2}}-r=0$$

and therefore precisely when $(p+q-r+2p^{\frac{1}{2}}q^{\frac{1}{2}})(p+q-r-2p^{\frac{1}{2}}q^{\frac{1}{2}})$ or $p^2+q^2+r^2-2qr-2rp-2pq=0$. Since

$$(a+b-c)(a-b+c)(-a+b+c)$$
$$=a(b^2+c^2-a^2)+b(c^2+a^2-b^2)+c(a^2+b^2-c^2)-2abc$$

the equation $p^{\frac{1}{2}}(q+r-p)+q^{\frac{1}{2}}(r+p-q)+r^{\frac{1}{2}}(p+q-r)-2p^{\frac{1}{2}}q^{\frac{1}{2}}r^{\frac{1}{2}}=0$ is conjugate to, and therefore expresses the same relation as, the equation $p^{\frac{1}{2}}+q^{\frac{1}{2}}+r^{\frac{1}{2}}=0$.

Again a $-a$ being the two second roots of p and b $-b$ the two second roots of q $p^{\frac{3}{2}}$ has the two values a^3 $-a^3$ and $q^{\frac{3}{2}}$ the two b^3 $-b^3$. Hence $p^{\frac{3}{2}}+q^{\frac{3}{2}}=0$ precisely when severally $(a^3+b^3)(a^3-b^3)=0$ $(a^3)^2-(b^3)^2=0$ $(a^2)^3-(b^2)^3=0$ and at length $p^3-q^3=0$. Or since $p^{\frac{3}{2}}+q^{\frac{3}{2}}=(p^{\frac{1}{2}}+q^{\frac{1}{2}})(p-p^{\frac{1}{2}}q^{\frac{1}{2}}+q)$ and $p^{\frac{1}{2}}+q^{\frac{1}{2}}=0$ just when $p-q=0$ and $p-p^{\frac{1}{2}}q^{\frac{1}{2}}+q=0$ just when $(p+q)^2-(p^{\frac{1}{2}}q^{\frac{1}{2}})^2$ or the equivalent $p^2+pq+q^2=0$ $p^{\frac{3}{2}}+q^{\frac{3}{2}}=0$ just when $(p-q)(p^2+pq+q^2)$ or the equivalent $p^3-q^3=0$. Moreover the equations $p^{\frac{3}{2}}+q^{\frac{3}{2}}=0$ $p^{\frac{3}{2}}-q^{\frac{3}{2}}=0$ are conjugate relative to the equation $p^3-q^3=0$.

To find the equation without roots that expresses the same relation as the equation $p^{\frac{1}{3}}+q^{\frac{1}{3}}=0$ let x a be any values of $p^{\frac{1}{3}}$ so that $x^3=p$ $a^3=p$ and $\therefore$ $x^3=a^3$ or $x^3-a^3=0$. Now

$$x^3-a^3=(x-a)(x^2+xa+a^2)=(x-a)\left\{\left(x+\frac{1}{2}a\right)^2+\left(-\frac{1}{4}+1\right)a^2\right\}$$

$$=(x-a)\left[\left(x+\frac{1}{2}a\right)^2-\left\{(-)^{\frac{1}{2}}\frac{\sqrt{3}}{2}a\right\}^2\right]$$

$$=(x-a)\left[x+\left\{\frac{1}{2}+(-)^{\frac{1}{2}}\frac{\sqrt{3}}{2}\right\}a\right]\left[x+\left\{\frac{1}{2}-(-)^{\frac{1}{2}}\frac{\sqrt{3}}{2}\right\}a\right]$$

and therefore $x^3=a^3$ precisely when (art. 413) one or other of the equations is true

$$x-a=0 \qquad x+\left\{\frac{1}{2}+(-)^{\frac{1}{2}}\frac{\sqrt{3}}{2}\right\}a=0 \qquad x+\left\{\frac{1}{2}-(-)^{\frac{1}{2}}\frac{\sqrt{3}}{2}\right\}a=0.$$

Hence if a be any value of $p^{\frac{1}{3}}$ the only other values (as in art. 400) are $-\left\{\frac{1}{2}+(-)^{\frac{1}{2}}\frac{\sqrt{3}}{2}\right\}a$ $-\left\{\frac{1}{2}-(-)^{\frac{1}{2}}\frac{\sqrt{3}}{2}\right\}a$. If either of the last two be symbolized by λa the other $=\lambda^2 a$ and $\lambda^3=1$. Likewise if b be any value of $q^{\frac{1}{3}}$ all the three values are b λb $\lambda^2 b$. The equation $p^{\frac{1}{3}}+q^{\frac{1}{3}}=0$ therefore holds precisely when one or other of these equations holds to wit

$$a+b=0 \quad a+\lambda b=0 \quad a+\lambda^2 b=0 \quad \lambda a+b=0 \quad \lambda a+\lambda b=0$$
$$\lambda a+\lambda^2 b=0 \quad \lambda^2 a+b=0 \quad \lambda^2 a+\lambda b=0 \quad \lambda^2 a+\lambda^2 b=0.$$

But $\lambda a+b=\lambda a+\lambda^3 b=\lambda(a+\lambda^2 b)$ $\lambda^2 a+b=\lambda^2(a+\lambda b)$ $\lambda^2 a+\lambda b=\lambda^2(a+\lambda^2 b)$ $\lambda a+\lambda b=\lambda(a+b)$ $\lambda a+\lambda^2 b=\lambda(a+\lambda b)$ $\lambda^2 a+\lambda^2 b=\lambda^2(a+b)$ and neither λ nor λ^2 is 0 so that one or other of the nine equations holds precisely when one or other of the first three of them holds. Therefore $p^{\frac{1}{3}}+q^{\frac{1}{3}}=0$ precisely when

$$(a+b)(a+\lambda b)(a+\lambda^2 b)=0$$

and therefore $\because$ $(a+b)(a+\lambda b)(a+\lambda^2 b)=(a+b)(a^2-ab+b^2)=a^3+b^3$ precisely when $p+q=0$.

In like manner each of the equations $p^{\frac{1}{3}}+\lambda q^{\frac{1}{3}}=0$ $p^{\frac{1}{3}}+\lambda^2 q^{\frac{1}{3}}=0$ and therefore too the equation $(p^{\frac{1}{3}}+\lambda q^{\frac{1}{3}})(p^{\frac{1}{3}}+\lambda^2 q^{\frac{1}{3}})=0$ or $p^{\frac{2}{3}}-p^{\frac{1}{3}}q^{\frac{1}{3}}+q^{\frac{2}{3}}=0$ holds precisely when $p+q=0$.

In the same way $p^{\frac{2}{3}}+q^{\frac{2}{3}}=0$ precisely when one or other of the expressions a^2+b^2 $a^2+\lambda^2 b^2$ $a^2+\lambda^4 b^2$ or $a^2+\lambda b^2$ is 0 therefore precisely when $(a^2)^3+(b^2)^3$ or $(a^3)^2+(b^3)^2=0$ and therefore precisely when $p^2+q^2=0$. And $p^{\frac{4}{3}}-p^{\frac{2}{3}}q^{\frac{2}{3}}+q^{\frac{4}{3}}=0$ precisely when $p^2+q^2=0$.

Also either of the equations $p^{\frac{1}{3}}-q^{\frac{1}{3}}=0$ $p^{\frac{2}{3}}+p^{\frac{1}{3}}q^{\frac{1}{3}}+q^{\frac{2}{3}}=0$ holds just when $p-q=0$ and either of the equations $p^{\frac{2}{3}}-q^{\frac{2}{3}}=0$ $p^{\frac{4}{3}}+p^{\frac{2}{3}}q^{\frac{2}{3}}+q^{\frac{4}{3}}=0$ just when $p^2-q^2=0$. Moreover both $p^{\frac{4}{3}}+q^{\frac{4}{3}}=0$ and $p^{\frac{8}{3}}-p^{\frac{4}{3}}q^{\frac{4}{3}}+q^{\frac{8}{3}}=0$ just when $p^4+q^4=0$ and both $p^{\frac{4}{3}}-q^{\frac{4}{3}}=0$ and $p^{\frac{8}{3}}+p^{\frac{4}{3}}q^{\frac{4}{3}}+q^{\frac{8}{3}}=0$ just when $p^4-q^4=0$.

For getting the equation clear of roots that expresses the same relation as the equation $p^{\frac{1}{3}}+q^{\frac{1}{3}}+r^{\frac{1}{3}}=0$ let $p^{\frac{1}{3}}$'s three values be a λa $\lambda^2 a$ $q^{\frac{1}{3}}$'s b λb $\lambda^2 b$ and $r^{\frac{1}{3}}$'s c λc $\lambda^2 c$ then $p^{\frac{1}{3}}+q^{\frac{1}{3}}+r^{\frac{1}{3}}=0$ precisely when one or other of the nine expressions is 0

$$a+b+c \qquad a+b+\lambda c \qquad a+b+\lambda^2 c \qquad a+\lambda b+c \qquad a+\lambda b+\lambda c$$

$$a+\lambda b+\lambda^2 c \qquad a+\lambda^2 b+c \qquad a+\lambda^2 b+\lambda c \qquad a+\lambda^2 b+\lambda^2 c.$$

For any one of the nine got from these by putting λa for a and any one of the nine got by putting $\lambda^2 a$ for a is 0 just when some one of these is. As for instance $\lambda^2 a+\lambda b+c=\lambda^2 a+\lambda^4 b+\lambda^3 c=\lambda^2(a+\lambda^2 b+\lambda c)$ and therefore is 0 just when $a+\lambda^2 b+\lambda c$ is. The product of those nine expressions is operationally equivalent to the product of the products of severally the first three of them the next three and the last three and therefore putting for $a^3+b^3+c^3$ $3a^2b$ $3ab^2$ severally f g h

$$=(f+g+h)(f+\lambda g+\lambda^2 h)(f+\lambda^2 g+\lambda h)=(f+g+h)(f^2+g^2+h^2-gh-hf-fg)$$

$$=f^3+g^3+h^3-3fgh=(a^3+b^3+c^3)^3-27a^3b^3c^3.$$

Hence $p^{\frac{1}{3}}+q^{\frac{1}{3}}+r^{\frac{1}{3}}=0$ precisely when $(p+q+r)^3-27pqr=0$ or $\frac{1}{3}(p^3+q^3+r^3)+p^2(q+r)+q^2(r+p)+r^2(p+q)-7pqr=0$. Hence too the equation $p^{\frac{1}{3}}+q^{\frac{1}{3}}+r^{\frac{1}{3}}=0$ is true just when any one of the following equations is

$$(p^{\frac{1}{3}}+q^{\frac{1}{3}})^2-(p^{\frac{1}{3}}+q^{\frac{1}{3}})r^{\frac{1}{3}}+r^{\frac{2}{3}}=0 \qquad (p^{\frac{1}{3}}+q^{\frac{1}{3}}+r^{\frac{1}{3}})\{(p^{\frac{1}{3}}+q^{\frac{1}{3}})^2-(p^{\frac{1}{3}}+q^{\frac{1}{3}})r^{\frac{1}{3}}+r^{\frac{2}{3}}\}=0$$

$$(p^{\frac{1}{3}}+q^{\frac{1}{3}})^3+r=0 \qquad (p^{\frac{1}{3}}+\lambda q^{\frac{1}{3}})^3+r=0 \qquad (p^{\frac{1}{3}}+\lambda^2 q^{\frac{1}{3}})^3+r=0$$

$$\{(p^{\frac{1}{3}}+\lambda q^{\frac{1}{3}})^3+r\}\{(p^{\frac{1}{3}}+\lambda^2 q^{\frac{1}{3}})^3+r\}=0$$

$$(p+q+r)^2-3(p+q+r)(p^{\frac{2}{3}}q^{\frac{1}{3}}+p^{\frac{1}{3}}q^{\frac{2}{3}})+9(p^{\frac{4}{3}}q^{\frac{2}{3}}-pq+p^{\frac{2}{3}}q^{\frac{4}{3}})=0$$

$$\{p+q+r+3(p^{\frac{2}{3}}q^{\frac{1}{3}}+p^{\frac{1}{3}}q^{\frac{2}{3}})\}\left\{\begin{matrix}(p+q+r)^2-3(p+q+r)(p^{\frac{2}{3}}q^{\frac{1}{3}}+p^{\frac{1}{3}}q^{\frac{2}{3}}) \\ +9(p^{\frac{4}{3}}q^{\frac{2}{3}}-pq+p^{\frac{2}{3}}q^{\frac{4}{3}})\end{matrix}\right\}=0$$

$$p+q+r-3p^{\frac{1}{3}}q^{\frac{1}{3}}r^{\frac{1}{3}}=0 \qquad (p^{\frac{1}{3}}+q^{\frac{1}{3}}+r^{\frac{1}{3}})(p^{\frac{2}{3}}+q^{\frac{2}{3}}+r^{\frac{2}{3}}-q^{\frac{1}{3}}r^{\frac{1}{3}}-r^{\frac{1}{3}}p^{\frac{1}{3}}-p^{\frac{1}{3}}r^{\frac{1}{3}})=0$$

$$p^{\frac{2}{3}}+q^{\frac{2}{3}}+r^{\frac{2}{3}}-q^{\frac{1}{3}}r^{\frac{1}{3}}-r^{\frac{1}{3}}p^{\frac{1}{3}}-p^{\frac{1}{3}}q^{\frac{1}{3}}=0$$

$$(p+q+r)^2+3(p+q+r)(pqr)^{\frac{1}{3}}+9(pqr)^{\frac{2}{3}}=0.$$

The equation $p^{\frac{1}{2}}+q^{\frac{1}{3}}=0$ is fulfilled precisely when if a $-a$ be $p^{\frac{1}{2}}$'s two values and b λb $\lambda^2 b$ $q^{\frac{1}{3}}$'s three one or other is 0 of the expressions $a+b$ $a+\lambda b$ $a+\lambda^2 b$ $-a+b$ $-a+\lambda b$ $-a+\lambda^2 b$ and therefore $\because$ $(a+b)(a+\lambda b)(a+\lambda^2 b)(-a+b)(-a+\lambda b)(-a+\lambda^2 b)$
$=\{(a+b)(a+\lambda b)(a+\lambda^2 b)\}(-a+b)(-a+\lambda b)(-a+\lambda^2 b)=(a^3+b^3)(-a^3+b^3)$
$=-(a^2)^3+(b^3)^2$ and further
$=\{(a+b)(-a+b)\}\{(a+\lambda b)(-a+\lambda b)\}(a+\lambda^2 b)(-a+\lambda^2 b)$
$=(-a^2+b^2)(-a^2+\lambda^2 b^2)(-a^2+\lambda b^2)$
$=(-a^2+b^2)\{(a^2)^2+a^2b^2+b^4\}$ precisely when severally

$$-p^3+q^2=0 \qquad -p^{\frac{3}{2}}+q=0 \qquad (p^{\frac{3}{2}}+q)(-p^{\frac{3}{2}}+q)=0 \qquad p^{\frac{3}{2}}+q=0$$

$$-p+q^{\frac{2}{3}}=0 \qquad (-p+q^{\frac{2}{3}})(p^2+pq^{\frac{2}{3}}+q^{\frac{4}{3}})=0 \qquad p^2+pq^{\frac{2}{3}}+q^{\frac{4}{3}}=0.$$

If to get a rootless equation expressing the same relation as the equation $p^{\frac{1}{4}}+q^{\frac{1}{4}}=0$ x a be any values of $p^{\frac{1}{4}}$ $x^4=p$ $a^4=p$ and $\therefore$ $x^4=a^4$ which happens precisely when severally $x^4-a^4=0$ $(x^2-a^2)(x^2+a^2)=0$ $\{(x-a)(x+a)\}\{x-(-)^{\frac{1}{2}}a\}\{x+(-)^{\frac{1}{2}}a\}=0$ $(x-a)(x+a)\{x-(-)^{\frac{1}{2}}a\}\{x+(-)^{\frac{1}{2}}a\}=0$ and therefore all the values of $p^{\frac{1}{4}}$ are a $-a$ $(-)^{\frac{1}{2}}a$ $-(-)^{\frac{1}{2}}a$. If likewise b $-b$ $(-)^{\frac{1}{2}}b$ $-(-)^{\frac{1}{2}}b$ be the four values of $q^{\frac{1}{4}}$ $p^{\frac{1}{4}}+q^{\frac{1}{4}}=0$ precisely when
$(a+b)(a-b)\{a+(-)^{\frac{1}{2}}b\}\{a-(-)^{\frac{1}{2}}b\}=0$ and hence precisely when severally

$$p-q=0 \qquad (p^{\frac{1}{2}}-q^{\frac{1}{2}})(p^{\frac{1}{2}}+q^{\frac{1}{2}})=0 \qquad p^{\frac{1}{2}}+q^{\frac{1}{2}}=0 \qquad p^{\frac{1}{2}}-q^{\frac{1}{2}}=0 \qquad p^{\frac{1}{4}}-q^{\frac{1}{4}}=0$$

$$p^{\frac{1}{4}}+(-)^{\frac{1}{2}}q^{\frac{1}{4}}=0 \qquad p^{\frac{3}{4}}+p^{\frac{1}{2}}q^{\frac{1}{4}}+p^{\frac{1}{4}}q^{\frac{1}{2}}+q^{\frac{3}{4}}=0 \qquad p^{\frac{3}{4}}-p^{\frac{1}{2}}q^{\frac{1}{4}}+p^{\frac{1}{4}}q^{\frac{1}{2}}-q^{\frac{3}{4}}=0$$

and others.

As to the equation $p^{\frac{1}{5}}+q^{\frac{1}{5}}=0$ if x a be any values of $p^{\frac{1}{5}}$ $x^5=p=a^5$ and

$$x^5-a^5=(x-a)(x^4+ax^3+a^2x^2+a^3x+a^4)$$

$$=(x-a)\{(x^2+a^2)^2+(ax)(x^2+a^2)-(ax)^2\}$$

$$=(x-a)\left\{\left(x^2+a^2+\frac{1}{2}ax\right)^2-\left(1+\frac{1}{4}\right)(ax)^2\right\}$$

$$=(x-a)\left(x^2+a^2+\frac{1}{2}ax-\frac{\sqrt{5}}{2}ax\right)\left(x^2+a^2+\frac{1}{2}ax+\frac{\sqrt{5}}{2}ax\right)$$

$$=(x-a)\left[\begin{array}{l}\left(x-\frac{\sqrt{5}-1}{4}a\right)^2\\+\left\{1-\left(\frac{\sqrt{5}-1}{4}\right)^2\right\}a^2\end{array}\right]\left[\begin{array}{l}\left(x+\frac{\sqrt{5}+1}{4}a\right)^2\\+\left\{1-\left(\frac{\sqrt{5}+1}{4}\right)^2\right\}a^2\end{array}\right]$$

$$=(x-a)\{(x-ea-ga)(x-ea+ga)\}(x+fa-ha)(x+fa+ha),\text{—}$$

e f g h standing for $\frac{\sqrt{5}-1}{4}$ $\frac{\sqrt{5}+1}{4}$ $(-)^{\frac{1}{2}}\sqrt{\frac{5+\sqrt{5}}{8}}$ $(-)^{\frac{1}{2}}\sqrt{\frac{5-\sqrt{5}}{8}}$ severally—,

$$=(x-a)\{x-(e+g)a\}\{x-(e-g)a\}\{x+(f-h)a\}\{x+(f+h)a\}.$$

Hence $p^{\frac{1}{5}}$'s five values are a $\left\{\frac{\sqrt{5}-1}{4}+(-)^{\frac{1}{2}}\sqrt{\frac{5+\sqrt{5}}{8}}\right\}a$ $\left\{\frac{\sqrt{5}-1}{4}-(-)^{\frac{1}{2}}\sqrt{\frac{5+\sqrt{5}}{8}}\right\}a$ $-\left\{\frac{\sqrt{5}+1}{4}-(-)^{\frac{1}{2}}\sqrt{\frac{5-\sqrt{5}}{8}}\right\}a$ $-\left\{\frac{\sqrt{5}+1}{4}+(-)^{\frac{1}{2}}\sqrt{\frac{5-\sqrt{5}}{8}}\right\}a$ or symbolizing the 2nd of them by λa a λa $\lambda^4 a$ $\lambda^2 a$ $\lambda^3 a$. Likewise $q^{\frac{1}{5}}$'s 5 values if b be any one are b λb $\lambda^2 b$ $\lambda^3 b$ $\lambda^4 b$. Hence $p^{\frac{1}{5}}+q^{\frac{1}{5}}=0$ just when $(a+b)(a+\lambda b)(a+\lambda^2 b)(a+\lambda^3 b)(a+\lambda^4 b)=0$ and therefore just when severally

$$(p^{\frac{1}{5}}+q^{\frac{1}{5}})(p^{\frac{4}{5}}-p^{\frac{3}{5}}q^{\frac{1}{5}}+p^{\frac{2}{5}}q^{\frac{2}{5}}-p^{\frac{1}{5}}q^{\frac{3}{5}}+q^{\frac{4}{5}})=0 \qquad p+q=0$$

$$p^{\frac{4}{5}}-p^{\frac{3}{5}}q^{\frac{1}{5}}+p^{\frac{2}{5}}q^{\frac{2}{5}}-p^{\frac{1}{5}}q^{\frac{3}{5}}+q^{\frac{4}{5}}=0.$$

From the values here found of $p^{\frac{1}{5}}$ and what is found in art.400 it follows that a perigon being unit angle

$$\cos\frac{1}{5}=\frac{\sqrt{5}-1}{4}=\sin\frac{1}{20} \qquad \sin\frac{1}{5}=\sqrt{\frac{5+\sqrt{5}}{8}}=\cos\frac{1}{20}$$

$$\cos\frac{1}{10}=\frac{\sqrt{5}+1}{4}=\sin\frac{3}{20} \qquad \sin\frac{1}{10}=\sqrt{\frac{5-\sqrt{5}}{8}}=\cos\frac{3}{20}.$$

When the n different values of x that make x^n-a^n equal to 0 cannot be found in the foregoing ways use must be made of art.400 and if need be of art.416 below.

415. *Def.* If there be in an equation only one unknown quantity any such value of that unknown quantity as fulfils the equation is called a ROOT of the equation.

Def. An equation or set of equations which holds precisely when another equation or set of equations holds is said to be EQUIVALENT to that other.

Def. An equation having one member a polynome of the nth degree in one or more unknowns and the other member 0 or any equivalent equation is said to be of the nTH DEGREE in those unknowns.

416. If $x^n+p_1x^{n-1}+p_2x^{n-2}+\cdots+p_{n-1}x+p_n$ $x-a$ be polynomes in x of the nth and 1st degrees severally there are (arts. 192, 300, 354, 398) a polynome $x^{n-1}+q_1x^{n-2}+q_2x^{n-3}+\cdots+q_{n-2}x+q_{n-1}$ in x of the $(n-1)$th degree and a quantity q_n independent of x such that

$$x^n+p_1x^{n-1}+p_2x^{n-2}+\cdots+p_{n-1}x+p_n$$
$$=(x-a)(x^{n-1}+q_1x^{n-2}+q_2x^{n-3}+\cdots+q_{n-2}x+q_{n-1})+q_n.$$

If in this operational equivalence x be given the particular value a

$$a^n+p_1a^{n-1}+p_2a^{n-2}+\cdots+p_{n-1}a+p_n=q_n$$

and if further a be such that $a^n+p_1a^{n-1}+p_2a^{n-2}+\cdots+p_{n-1}a+p_n$ is 0 q_n is also 0. Hence if P_n be a polynome in x having x^n the term of highest degree and a be a root of the equation in x $P_n=0$ there is a polynome in x P_{n-1} having x^{n-1} the term of highest degree such that

$$P_n=(x-a)P_{n-1}.$$

Hence $P_n=0$ precisely when either $x-a=0$ or $P_{n-1}=0$ and hence all the roots of the equation $P_n=0$ are precisely a and all the roots of the equation $P_{n-1}=0$. In the same way if b be a root of the equation $P_{n-1}=0$ there is a polynome P_{n-2} with x^{n-2} the term of highest degree such that $P_{n-1}=(x-b)P_{n-2}$ and all the roots of the equation $P_{n-1}=0$ are precisely b and all the roots of the equation $P_{n-2}=0$. Wherefore

$$P_n=(x-a)(x-b)P_{n-2}$$

and all the roots of the equation $P_n=0$ are precisely a b and all the roots of the equation $P_{n-2}=0$. In the same way if c be a root of the equation $P_{n-2}=0$ $P_{n-2}=(x-c)P_{n-3}$ P_{n-3} being a polynome with x^{n-3} the term of highest degree

$$P_n=(x-a)(x-b)(x-c)P_{n-3}$$

and all the roots of the equation $P_n=0$ are precisely a b c and all the roots of the equation $P_{n-3}=0$. And so on until at length a polynome P_1 is come to with x the term of highest degree when if h be a root of the equation $P_1=0$ $P_1=x-h$ and h is the only root of the equation $P_1=0$ since $x-h$ is not 0 for any other value of x than h. If then in this way

$$P_n=(x-a)(x-b)\ldots(x-f)(x-g)P_1=(x-a)(x-b)\ldots(x-f)(x-g)(x-h)$$

all the roots of the equation $P_n=0$ are precisely $a\ b\ c\ldots g\ h$.

If P_n be the polynome $x^n+p_1x^{n-1}+p_2x^{n-2}+\cdots+p_{n-1}x+p_n$ this is the polynomic equivalent of $(x-a)(x-b)(x-c)\ldots(x-f)(x-g)(x-h)$ and therefore (arts. 185, 300, 354, 398)

$$p_i=(-)^i s_i$$

s_i being the sum of the products made by multiplying together every i of the n quantities $a\ b\ c\ldots f\ g\ h$.

417. An equation in x of the first degree takes one of the shapes $p_0x+p_1=0$ $xp_0+p_1=0$ p_0 being some quantity else than 0 and of these the former holds just when severally $p_0(x+p_0^{-1}p_1)=0$ $x+p_0^{-1}p_1=0$ $x=-p_0^{-1}p_1$ and the latter just when severally $(x+p_1p_0^{-1})p_0=0$ $x+p_1p_0^{-1}=0$ $x=-p_1p_0^{-1}$.

An equation in x of the second degree takes one of the shapes $p_0x^2+p_1x+p_2=0$ $x^2p_0+xp_1+p_2=0$ p_0 being else than 0 and the former of these holds just when $p_0\{x^2+(p_0^{-1}p_1)x+p_0^{-1}p_2\}=0$ or $x^2+(p_0^{-1}p_1)x+p_0^{-1}p_2=0$ and the latter just when $(x^2+xp_1p_0^{-1}+p_2p_0^{-1})p_0=0$ or $x^2+xp_1p_0^{-1}+p_2p_0^{-1}=0$. An equation of either of the shapes $x^2+px+q=0$ $x^2+xp+q=0$ holds just when severally

$$\left(x+\frac{1}{2}p\right)^2-\left(\frac{1}{2}p\right)^2+q=0 \quad \left(x+\frac{1}{2}p\right)^2-\left\{-q+\left(\frac{1}{2}p\right)^2\right\}=0$$

$$\left[x+\frac{1}{2}p-\left\{-q+\left(\frac{1}{2}p\right)^2\right\}^{\frac{1}{2}}\right]\left[x+\frac{1}{2}p+\left\{-q+\left(\frac{1}{2}p\right)^2\right\}^{\frac{1}{2}}\right]=0$$

$$\left\{x+\left[\frac{1}{2}p-\left\{-q+\left(\frac{1}{2}p\right)^2\right\}^{\frac{1}{2}}\right]\right\}\left\{x+\left[\frac{1}{2}p+\left\{-q+\left(\frac{1}{2}p\right)^2\right\}^{\frac{1}{2}}\right]\right\}=0$$

and therefore just when either $x=-\left[\frac{1}{2}p-\left\{-q+\left(\frac{1}{2}p\right)^2\right\}^{\frac{1}{2}}\right]$ or $x=-\left[\frac{1}{2}p+\left\{-q+\left(\frac{1}{2}p\right)^2\right\}^{\frac{1}{2}}\right]$ which are both included in the single expression of the two values

$$x=-\left[\frac{1}{2}p\mp\left\{-q+\left(\frac{1}{2}p\right)^2\right\}^{\frac{1}{2}}\right]=\pm\left\{-q+\left(\frac{1}{2}p\right)^2\right\}^{\frac{1}{2}}-\frac{1}{2}p=-\frac{1}{2}p\pm\left(\frac{1}{4}p^2-q\right)^{\frac{1}{2}}.$$

When q is 0 the two values of x are $\frac{1}{2}p-\frac{1}{2}p$ and $-\frac{1}{2}p-\frac{1}{2}p$ or 0 and $-p$ which come at once from the equation $x^2+px=0$ or $(x+p)x=0$. When $q=\left(\frac{1}{2}p\right)^2$ $\left\{-q+\left(\frac{1}{2}p\right)^2\right\}^{\frac{1}{2}}$ has only the single value

± 0 or 0 and the only value of x that fulfils the equation is $-\frac{1}{2}p$ which comes at once from the equation $x^2+px+\left(\frac{1}{2}p\right)^2=0$ or $\left(x+\frac{1}{2}p\right)^2=0$.

For solving an equation of any degree the one thing needed is to find polynomes of the first degree whose product is operationally equivalent to the equation's polynomic member. This however has never been done for any equation of a degree higher than the second expressed in general symbols without bringing in some condition not at all necessarily implied in the equation itself.

418. The following are examples of equation solving.

(1). $\frac{5}{3}(4x+7)+\frac{1}{2}(4-x)=20+\frac{3}{4}(5x-2)$ just when singly

$$\frac{5}{3}\times 4x+\frac{5}{3}\times 7+\left(\frac{1}{2}\times 4-\frac{1}{2}x\right)-\left(\frac{3}{4}\times 5x-\frac{3}{4}\times 2\right)-20=0$$

$$\left(\frac{5}{3}\times 4\right)x+\frac{35}{3}+2-\frac{1}{2}x+\frac{3}{2}-\left(\frac{3}{4}\times 5\right)x-20=0$$

$$\left(\frac{20}{3}-\frac{1}{2}-\frac{15}{4}\right)x-\left(20-\frac{3}{2}-2-\frac{35}{3}\right)=0$$

$$\frac{29}{12}x-\frac{29}{6}=0 \qquad \frac{29}{12}\left(x-\frac{12}{29}\times\frac{29}{6}\right)=0 \qquad x=2.$$

(2). $\frac{5x+3}{4}-\frac{4x-11}{7}=\frac{2x-9}{5}+\frac{193}{35}$ precisely when severally

$$\frac{5x}{4}+\frac{3}{4}-\left(\frac{4x}{7}-\frac{11}{7}\right)-\frac{193}{35}-\left(\frac{2x}{5}-\frac{9}{5}\right)=0$$

$$x\times\frac{5}{4}+\frac{3}{4}+\frac{11}{7}-x\times\frac{4}{7}-\frac{193}{35}+\frac{9}{5}-x\times\frac{2}{5}=0$$

$$x\left(\frac{5}{4}-\frac{4}{7}-\frac{2}{5}\right)+\left(\frac{3}{4}+\frac{11}{7}-\frac{193}{35}+\frac{9}{5}\right)=0$$

$$x\times\frac{39}{140}+-\frac{39}{28}=0 \qquad \left\{x-\frac{\left(\frac{39}{28}\right)}{\left(\frac{39}{140}\right)}\right\}\times\frac{39}{140}=0 \qquad x=5.$$

(3). $\frac{3}{2x-3}-\frac{8}{3x+2}=\frac{5}{x-1}$. Since when $2x-3=0$, that is just when $x=\frac{1}{2}\times 3=\frac{3}{2}$, both $3x+2$ and $x-1$ is not 0 the equation (art. 412) is then not fulfilled. The equation is likewise not fulfilled either when $3x+2=0$ or when $x-1=0$. Therefore the equation is fulfilled just when severally

$$\frac{3(3x+2)-8(2x-3)}{(2x-3)(3x+2)}-\frac{5}{x-1}=0$$
$$(-7x+30)(x-1)-5(2x-3)(3x+2)=0$$
$$-7x^2+37x-30+30+25x-30x^2=0$$
$$\{-37x+(37+25)\}x=0.$$

And therefore all the solutions are $x=1\frac{25}{37}$ and $x=0$.

(4). $\frac{3(2x-3)}{(2x-3)^2}-\frac{8x}{(3x+2)x}=\frac{5(x+1)(x+2)}{(x-1)(x+1)(x+2)}$ precisely when one or other holds of the equations

$$2x-3=0 \quad x=0 \quad (x+1)(x+2)=0 \quad \frac{3}{2x-3}-\frac{8}{3x+2}=\frac{5}{x-1}$$

and therefore and because the last of these is the equation (3) above the only solutions are $x=\frac{3}{2}$ $x=0$ $x=-1$ $x=-2$ and $x=1\frac{25}{37}$. The value 0 of x solves the equation twowise inasmuch as it solves both the second and the fourth of the four alternative equations.

(5). $\frac{ax}{x+b}+\frac{bx}{x+a}=a+b$ just when

$$\frac{a(x+b)-ab}{x+b}+\frac{b(x+a)-ba}{x+a}-(a+b)=0$$

and since then there is no solution when either $x+b=0$ or $x+a=0$ just when severally

$$a-\frac{ab}{x+b}+\left(b-\frac{ba}{x+a}\right)-b-a=0$$
$$(-ab)\left(\frac{1}{x+a}+\frac{1}{x+b}\right)=0$$
$$\frac{1}{x+b}+\frac{1}{x+a}=0$$
$$x+a+(x+b)=0$$
$$\cdot\,2\{x+\tfrac{1}{2}(a+b)\}=0 \qquad x=-\tfrac{1}{2}(a+b).$$

(6). $\frac{2x+5}{5x+2}+\frac{2x-5}{5x-2}=\frac{160}{221}$ just when

$$\frac{\frac{2}{5}(5x+2)+\frac{21}{5}}{5x+2}+\frac{\frac{2}{5}(5x-2)-\frac{21}{5}}{5x-2}-\frac{160}{221}=0$$

and as then no solution is got by making 0 either $5x+2$ or $5x-2$ just when singly

$$\frac{\left(\frac{21}{5}\right)}{5x+2}-\frac{\left(\frac{21}{5}\right)}{5x-2}-\left(\frac{160}{221}-\frac{2}{5}-\frac{2}{5}\right)=0$$

$$\frac{21}{5}\left(\frac{1}{5x+2}-\frac{1}{5x-2}+\frac{5}{21}\times\frac{84}{221\times5}\right)=0$$

$$\frac{5x-2-(5x+2)}{(5x+2)(5x-2)}+\frac{4}{221}=0$$

$$(-4)\left\{\frac{1}{(5x)^2-4}-\frac{1}{221}\right\}=0$$

$$221-\{(5x)^2-4\}=0 \qquad (15)^2-(5x)^2=0 \qquad (15-5x)(15+5x)=0$$

and the solutions therefore are $x=3$ and $x=-3$.

(7). $\frac{2x^2-3}{2x-3}-\frac{3x^2-4}{3x-4}=2x$ likewise precisely when singly

$$\frac{2x(x-1)+(2x-3)}{2x-3}-\frac{3x(x-1)+(3x-4)}{3x-4}-2x=0$$

$$x\left\{(x-1)\left(\frac{2}{2x-3}-\frac{3}{3x-4}\right)-2\right\}=0.$$

Therefore one solution is $x=0$ and every other is just what fulfils each of the following equations.

$$(x-1)\frac{2(3x-4)-3(2x-3)}{(2x-3)(3x-4)}-2=0$$

$$x-1-2(6x^2-17x+12)=0$$

$$-(25-35x+12x^2)=0$$

$$\left(5-\frac{7}{2}x\right)^2-\left\{-12+\left(\frac{7}{2}\right)^2\right\}x^2=0 \quad \left(5-\frac{1+7}{2}x\right)\left(5-\frac{-1+7}{2}x\right)=0$$

and so the other solutions are $x=\frac{5}{4}$ and $x=\frac{5}{3}$.

(8). $\frac{6x^2+x+3}{2x-1}+\frac{2x+3}{x+1}=3(x+2)$ just when severally

$$\frac{(3x+2)(2x-1)+5}{2x-1}+\frac{2(x+1)+1}{x+1}-(3x+6)=0$$

$$\frac{5}{2x-1}+\frac{1}{x+1}-2=0$$

$$5(x+1)+(2x-1)-2(2x-1)(x+1)=0$$

$$6+5x-4x^2=0$$

$$6+\left(\frac{5}{4}\right)^2-\left(2x-\frac{5}{4}\right)^2=0$$

and therefore just when either $x=2$ or $x=-\frac{3}{4}$.

(9). $\frac{1-x}{1+x}+\frac{2+x}{2-x}=\frac{3-x}{3+x}+\frac{4+x}{4-x}$ just when singly

$$1-\frac{2x}{1+x}+\left(1+\frac{2x}{2-x}\right)-\left(1+\frac{2x}{4-x}\right)-\left(1-\frac{2x}{3+x}\right)=0$$

$$-\frac{2x}{1+x}+\frac{2x}{3+x}+\left(\frac{2x}{2-x}-\frac{2x}{4-x}\right)=0$$

$$-\frac{4x}{3+4x+x^2}+\frac{4x}{8-6x+x^2}=0$$

$$4x\{-(8-6x+x^2)+(3+4x+x^2)\}=0 \qquad 4x\{10x-5)=0$$

and the only solutions are $x=0$ $x=\frac{1}{2}$.

(10). As to the equation $x+(x^2+a)^{\frac{1}{2}}=b$ not only has the unknown x to be found but (art.409) it is quite as much part of the solution to find what value of $(x^2+a)^{\frac{1}{2}}$ must be taken for any value found of x. To solve this equation is (art.355) just to find what $x\ y$ are so that $x+y=b$ $y^2=x^2+a$. The equation holds just when (art.410) $-b+x+(x^2+a)^{\frac{1}{2}}=0$ and therefore just when by using the unrooting factor $-b+x-(x^2+a)^{\frac{1}{2}}$ (art.414)

$$\{-b+x+(x^2+a)^{\frac{1}{2}}\}\{-b+x-(x^2+a)^{\frac{1}{2}}\}=0.$$

Whence $(-b+x)^2-(x^2+a)=0$ $b^2-a-2xb=0$ $x=\frac{1}{2}\left(b-\frac{a}{b}\right)$ and then

$$(x^2+a)^{\frac{1}{2}}=\left[\left\{\frac{1}{2}\left(b-\frac{a}{b}\right)\right\}^2+a\right]^{\frac{1}{2}} \text{ and } =-x+b=-\frac{1}{2}\left(b-\frac{a}{b}\right)+b$$

$$\text{so that } (x^2+a)^{\frac{1}{2}}=\left[\left\{\frac{1}{2}\left(\frac{a}{b}+b\right)\right\}^2\right]^{\frac{1}{2}}=\frac{1}{2}\left(\frac{a}{b}+b\right).$$

Thus

b being 1	and a 3	$x+(x^2+3)^{\frac{1}{2}}=1$	just when $x=-1$	and	$(x^2+3)^{\frac{1}{2}}=2$
- - - -1	- 3	$x+(x^2+3)^{\frac{1}{2}}=-1$	- - $x=1$	-	$(x^2+3)^{\frac{1}{2}}=-2$
- - - 1	- -3	$x+(x^2-3)^{\frac{1}{2}}=1$	- - $x=2$	-	$(x^2-3)^{\frac{1}{2}}=-1$
- - - -1	- -3	$x+(x^2-3)^{\frac{1}{2}}=-1$	- - $x=-2$	-	$(x^2-3)^{\frac{1}{2}}=1$.

The conjugate equation $-b+x-(x^2+a)^{\frac{1}{2}}=0$ or $x-(x^2+a)^{\frac{1}{2}}=b$ holds just when $x=\frac{1}{2}\left(b-\frac{a}{b}\right)$ and

$$(x^2+a)^{\frac{1}{2}}=\left[\left\{\frac{1}{2}\left(\frac{a}{b}+b\right)\right\}^2\right]^{\frac{1}{2}}=-\frac{1}{2}\left(\frac{a}{b}+b\right).$$

(11). $x+a+(x^2+2bx+c)^{\frac{1}{2}}=d$ precisely when $x+b+\{(x+b)^2+(-b^2+c)\}^2=d-a+b$ and this equation having the same shape as equation (10) but with $(x+b)$ instead of x may be solved in the same way.

The equation $x^{\frac{1}{2}}+(x+a)^{\frac{1}{2}}=b$ is the same as $x^{\frac{1}{2}}+\{(x^{\frac{1}{2}})^2+a\}^{\frac{1}{2}}=b$ and therefore like (10) gives as the solution

$$x^{\frac{1}{2}}=\frac{1}{2}\left(b-\frac{a}{b}\right) \qquad x=\left\{\frac{1}{2}\left(b-\frac{a}{b}\right)\right\}^2 \qquad (x+a)^{\frac{1}{2}}=\frac{1}{2}\left(\frac{a}{b}+b\right).$$

The equation $(x+a)^{\frac{1}{2}}+(x+b)^{\frac{1}{2}}=c$ holds just when $(x+a)^{\frac{1}{2}}+\{x+a+(-a+b)\}^{\frac{1}{2}}=c$ and therefore may be treated in the same way.

(12). $\{(ea+x)^2+c^2\}^{\frac{1}{2}}+\{(-ea+x)^2+c^2\}^{\frac{1}{2}}=2a$ precisely when severally

$$\left[\begin{matrix}-2a+\{(ea+x)^2+c^2\}^{\frac{1}{2}} \\ \qquad +\{(-ea+x)^2+c^2\}^{\frac{1}{2}}\end{matrix}\right]\left[\begin{matrix}-2a+\{(ea+x)^2+c^2\}^{\frac{1}{2}} \\ \qquad -\{(-ca+x)^2+c^2\}^{\frac{1}{2}}\end{matrix}\right]=0$$

$$[-2a+\{(ca+x)^2+c^2\}^{\frac{1}{2}}]^2-\{(-ca+x)^2+c^2\}=0$$

$$4a^2-4a\{(ea+x)^2+c^2\}^{\frac{1}{2}}+(e^2a^2+2eax+x^2+c^2)-(c^2a^2-2cax+x^2+c^2)=0$$

$$4a[a+ex-\{(ca+x)^2+c^2\}^{\frac{1}{2}}]=0$$

$$[a+ex-\{(ea+x)^2+c^2\}^{\frac{1}{2}}][a+cx+\{(ca+x)^2+c^2\}^{\frac{1}{2}}]=0$$

$$(a+ex)^2-\{(ea+x)^2+c^2\}=0 \qquad (1-e^2)a^2-c^2-(1-c^2)x^2=0$$

$$x=\left(a^2-\frac{1}{1-e^2}c^2\right)^{\frac{1}{2}}$$

and then with either value thus got of x the values to be taken of the equation's other unknowns are

$$\{(ea+x)^2+c^2\}^{\frac{1}{2}}=a+ex \qquad \{(-ea+x)^2+c^2\}^{\frac{1}{2}}=-cx+a.$$

(13). $1+x^{\frac{1}{2}}+\{(1+x)^{\frac{1}{2}}+x\}^{\frac{1}{2}}=0$ just when severally

$$(1+x^{\frac{1}{2}})^2-\{(1+x)^{\frac{1}{2}}+x\}=0 \qquad 1+2x^{\frac{1}{2}}-(1+x)^{\frac{1}{2}}=0$$

$$(1+2x^{\frac{1}{2}})^2-(1+x)=0 \qquad (4+3x^{\frac{1}{2}})x^{\frac{1}{2}}=0.$$

Hence the equation is fulfilled only when either $x^{\frac{1}{2}}=0$ and $x=0$

$(1+x)^{\frac{1}{2}} = 1$ $\{(1+x)^{\frac{1}{2}}+x\}^{\frac{1}{2}} = -1$ or $x^{\frac{1}{2}} = -\frac{4}{3}$ and $x = \frac{16}{9}$ $(1+x)^{\frac{1}{2}} = -\frac{5}{3}$ $\{(1+x)^{\frac{1}{2}}+x\}^{\frac{1}{2}} = \frac{1}{3}$.

(14). $5^{\frac{1}{2}}+[5+\{x^{\frac{1}{2}}+(x^{\frac{1}{2}}+6)^{\frac{1}{2}}\}^{\frac{1}{2}}]^{\frac{1}{2}} = 0$ precisely when singly

$$5-[5+\{x^{\frac{1}{2}}+(x^{\frac{1}{2}}+6)^{\frac{1}{2}}\}^{\frac{1}{2}}] = 0 \qquad \{x^{\frac{1}{2}}+(x^{\frac{1}{2}}+6)^{\frac{1}{2}}\}^{\frac{1}{2}} = 0$$

$$x^{\frac{1}{2}}+(x^{\frac{1}{2}}+6)^{\frac{1}{2}} = 0 \qquad x-(x^{\frac{1}{2}}+6) = 0$$

$$\left(x^{\frac{1}{2}}-\frac{1}{2}\right)^2-\left(6+\frac{1}{4}\right) = 0 \qquad \left(x^{\frac{1}{2}}-\frac{5+1}{2}\right)\left(x^{\frac{1}{2}}+\frac{5-1}{2}\right) = 0.$$

And hence the only ways of fulfilling the equation are by $x^{\frac{1}{2}} = 3$ $x = 9$ $(x^{\frac{1}{2}}+6)^{\frac{1}{2}} = -3$ and by $x^{\frac{1}{2}} = -2$ $x = 4$ $(x^{\frac{1}{2}}+6)^{\frac{1}{2}} = 2$ taking besides in each case those values of $5^{\frac{1}{2}}$ and $[5+\{x^{\frac{1}{2}}+(x^{\frac{1}{2}}+6)^{\frac{1}{2}}\}^{\frac{1}{2}}]^{\frac{1}{2}}$ that have contrary algebraic signs.

(15). $x+(2x-3)^{\frac{1}{2}} = 9$ precisely when severally

$$\{(2x-3)^{\frac{1}{2}}\}^2+2(2x-3)^{\frac{1}{2}}-15 = 0 \qquad \{(2x-3)^{\frac{1}{2}}+1\}^2-(15+1) = 0$$

$$\{(2x-3)^{\frac{1}{2}}-3\}\{(2x-3)^{\frac{1}{2}}+5\} = 0$$

and therefore precisely when either $(2x-3)^{\frac{1}{2}} = 3$ and $x = 6$ or $(2x-3)^{\frac{1}{2}} = -5$ and $x = 14$.

(16). In a like way $(7-5x+3x^2)^{\frac{1}{2}}-1-10x+6x^2 = 0$ just when singly

$$-15+(7-5x+3x^2)^{\frac{1}{2}}+2(7-5x+3x^2) = 0$$

$$-\left(\frac{1}{4}+2\times 15\right)+\left\{\frac{1}{2}+2(7-5x+3x^2)^{\frac{1}{2}}\right\}^2 = 0$$

$$(7-5x+3x^2)^{\frac{1}{2}} = \frac{1}{2}\,\frac{-1\pm 11}{2}\text{, that is either } \frac{5}{2} \text{ or } -3.$$

And $(7-5x+3x^2)^{\frac{1}{2}} = \frac{5}{2}$ only when singly $-\frac{25}{4}+7-5x+3x^2 = 0$ $3\times\frac{3}{4}-\frac{25}{4}+\left(\frac{5}{2}-3x\right)^2 = 0$ $\pm 2+\left(\frac{5}{2}-3x\right) = 0$ x is either $\frac{3}{2}$ or $\frac{1}{6}$ and $(7-5x+3x^2)^{\frac{1}{2}} = -3$ only when $x = \frac{1}{3}\,\frac{\pm 7+5}{2}$, that is either 2 or $-\frac{1}{3}$.

(17). $(x+x^{\frac{1}{2}})^{\frac{1}{2}}+(x-x^{\frac{1}{2}})^{\frac{1}{2}} = c\left(\frac{x}{x+x^{\frac{1}{2}}}\right)^{\frac{1}{2}}$. Here $x^{\frac{1}{2}}$ must be understood to have the same value throughout and although $\left(\frac{x}{v}\right)^{\frac{1}{2}}$ is (art.

405) operationally equivalent to $\frac{x^{\frac{1}{2}}}{v^{\frac{1}{2}}}$ this is only because $x^{\frac{1}{2}}$ may have any value. If then both $x^{\frac{1}{2}}$ and $(x+x^{\frac{1}{2}})^{\frac{1}{2}}$ is to have the same value as in the other parts of the equation $\left(\frac{x}{x+x^{\frac{1}{2}}}\right)^{\frac{1}{2}}$ has either the same value or values as $\frac{x^{\frac{1}{2}}}{(x+x^{\frac{1}{2}})^{\frac{1}{2}}}$ or the same as $-\frac{x^{\frac{1}{2}}}{(x+x^{\frac{1}{2}})^{\frac{1}{2}}}$. Hence taking either the upper sign throughout or the lower the equation holds just when severally

$$(x+x^{\frac{1}{2}})^{\frac{1}{2}}+(x-x^{\frac{1}{2}})^{\frac{1}{2}}\mp c\frac{x^{\frac{1}{2}}}{(x+x^{\frac{1}{2}})^{\frac{1}{2}}}=0 \qquad x+x^{\frac{1}{2}}+(x-x^{\frac{1}{2}})^{\frac{1}{2}}(x+x^{\frac{1}{2}})^{\frac{1}{2}}\mp cx^{\frac{1}{2}}=0$$

$$\{x+(1\mp c)x^{\frac{1}{2}}\}^2-\{(x-x^{\frac{1}{2}})^{\frac{1}{2}}(x+x^{\frac{1}{2}})^{\frac{1}{2}}\}^2=0 \qquad x[2(1\mp c)x^{\frac{1}{2}}+\{(1\mp c)^2+1\}]=0$$

therefore just when either $x=0$ and $x^{\frac{1}{2}}=0$ or $x^{\frac{1}{2}}=-\frac{1}{2}\left(1\mp c+\frac{1}{1\mp c}\right)$ and putting κ^2 for $1\mp c$ so that $\pm c=1-\kappa^2$ $x=\left\{\frac{1}{2}\left(\kappa^2+\frac{1}{\kappa^2}\right)\right\}^2$

$$x+\kappa^2x^{\frac{1}{2}}=\left\{-\frac{1}{2}\left(\kappa^2+\frac{1}{\kappa^2}\right)+\kappa^2\right\}\left\{-\frac{1}{2}\left(\kappa^2+\frac{1}{\kappa^2}\right)\right\}=-\left(\frac{1}{2}\right)^2\left(\kappa^2-\frac{1}{\kappa^2}\right)\left(\kappa^2+\frac{1}{\kappa^2}\right)$$

$$(x-x^{\frac{1}{2}})^{\frac{1}{2}}=\left[\left\{-\frac{1}{2}\left(\kappa^2+\frac{1}{\kappa^2}\right)-1\right\}\left\{-\frac{1}{2}\left(\kappa^2+\frac{1}{\kappa^2}\right)\right\}\right]^{\frac{1}{2}}=\frac{1}{2}\left(\kappa+\frac{1}{\kappa}\right)\left(\kappa^2+\frac{1}{\kappa^2}\right)^{\frac{1}{2}}$$

$$(x+x^{\frac{1}{2}})^{\frac{1}{2}}=\left[\left\{-\frac{1}{2}\left(\kappa^2+\frac{1}{\kappa^2}\right)+1\right\}\left\{-\frac{1}{2}\left(\kappa^2+\frac{1}{\kappa^2}\right)\right\}\right]^{\frac{1}{2}}=\frac{1}{2}\left(\kappa-\frac{1}{\kappa}\right)\left(\kappa^2+\frac{1}{\kappa^2}\right)^{\frac{1}{2}}$$

the same value of $\left(\kappa^2+\frac{1}{\kappa^2}\right)^{\frac{1}{2}}$ being taken since $x+\kappa^2x^{\frac{1}{2}}$ $=-(x-x^{\frac{1}{2}})^{\frac{1}{2}}(x+x^{\frac{1}{2}})^{\frac{1}{2}}$

$$\left(\frac{x}{x+x^{\frac{1}{2}}}\right)^{\frac{1}{2}}=\left(\frac{x^{\frac{1}{2}}}{x^{\frac{1}{2}}+1}\right)^{\frac{1}{2}}=\left\{\frac{\kappa^2+\frac{1}{\kappa^2}}{\left(\kappa-\frac{1}{\kappa}\right)^2}\right\}^{\frac{1}{2}}=\pm c^{-1}\kappa\left(\kappa^2+\frac{1}{\kappa^2}\right)^{\frac{1}{2}}.$$

If c be 1 the lower sign only can be taken and if c be -1 the upper only. The equation is fulfilled in just 5 ways unless c be either $+1$ or -1 but if c be either $+1$ or -1 in just 3.

(18). $(x+a)^{\frac{2}{3}}+b(x-a)^{\frac{2}{3}}=2c(x^2-a^2)^{\frac{1}{3}}$. Although (art. 405) $(x+a)^{\frac{1}{3}}(x-a)^{\frac{1}{3}}=\{(x+a)(x-a)\}^{\frac{1}{3}}=(x^2-a^2)^{\frac{1}{3}}$ yet this is only when each third root may have any of its three values. But when one value is taken of $(x+a)^{\frac{1}{3}}$ and one value of $(x-a)^{\frac{1}{3}}$ the resulting value of $(x+a)^{\frac{1}{3}}(x-a)^{\frac{1}{3}}$ is equal to only one of the three values of $(x^2-a^2)^{\frac{1}{3}}$ the other two values however being (art. 400 or 414) equal to the products made by multiplying that one by λ λ^2 if λ be either of the two values other than 1 of $(+)^{\frac{1}{3}}1$. Hence the right hand member of the equation is equal to $2k(x+a)^{\frac{1}{3}}(x-a)^{\frac{1}{3}}$ where k is either c $c\lambda$ or $c\lambda^2$ and hence the equation holds just when singly

$$(x+a)^{\frac{2}{3}}-2k(x+a)^{\frac{1}{3}}(x-a)^{\frac{1}{3}}+b(x-a)^{\frac{2}{3}}=0 \quad (x+a)^{\frac{1}{3}}-\{(-b+k^2)^{\frac{1}{2}}+k\}(x-a)^{\frac{1}{3}}=0$$

and therefore d standing for $(-b+k^2)^{\frac{1}{2}}+k$ just when singly

$$\{(x+a)^{\frac{1}{3}}-d(x-a)^{\frac{1}{3}}\}[(x+a)^{\frac{2}{3}}+(x+a)^{\frac{1}{3}}d(x-a)^{\frac{1}{3}}+\{d(x-a)^{\frac{1}{3}}\}^2]=0$$

$$x+a-d^3(x-a)=0 \qquad x=-\frac{1+d^3}{1-d^3}a.$$

Since k has 3 values of which each gives 2 values of d d has 3×2 values and since each of $(x+a)^{\frac{1}{3}}$'s 3 values gives with each value of d a value of $(x-a)^{\frac{1}{3}}$ there are in all $(3\times2)\times3$ or 18 ways of fulfilling the equation.

(19). $x^5+1=(x+1)^5c$ precisely when

$$(x+1)\{x^4-x^3+x^2-x+1-(x+1)^4c\}=0$$

and therefore precisely when either $x+1=0$ or severally

$$x^4+2x^2+1-(x^3+x)-x^2-(x^2+1+2x)^2c=0$$

$$(x^2+1)^2-x(x^2+1)-x^2-\{(x^2+1)^2+4x(x^2+1)+4x^2\}c=0$$

$$(x^2+1)^2(1-c)-x(x^2+1)(1+4c)-x^2(1+4c)=0.$$

If $1-c=0$ this happens precisely when $x(x^2+x+1)=0$, that is either $x=0$ or $x=\dfrac{(-3)^{\frac{1}{2}}-1}{2}$, and if $1+4c=0$ precisely when $(x^2+1)^2=0$, that is $x=(-1)^{\frac{1}{2}}$, but otherwise precisely when severally

$$\left\{(x^2+1)\frac{1-c}{1+4c}-\frac{1}{2}x\right\}^2-x^2\left(\frac{1-c}{1+4c}+\frac{1}{4}\right)=0$$

$$x^2-x\left\{\left(\frac{1-c}{1+4c}+\frac{1}{4}\right)^{\frac{1}{2}}+\frac{1}{2}\right\}\frac{1+4c}{1-c}+1=0 \qquad x=(-1+k^2)^{\frac{1}{2}}+k$$

k being $\dfrac{1}{2}\left\{\left(\dfrac{1-c}{1+4c}+\dfrac{1}{4}\right)^{\frac{1}{2}}+\dfrac{1}{2}\right\}\dfrac{1+4c}{1-c}$. The equation then is fulfilled in precisely 4 3 or 5 ways according as c is 1 $-\dfrac{1}{4}$ or neither 1 nor $-\dfrac{1}{4}$.

(20). $x^8+1=0$ just when singly

$$(x^4+1)^2-2x^4=0 \quad x^4-2^{\frac{1}{2}}x^2+1=0 \quad (x^2+1)^2-(2^{\frac{1}{2}}+2)x^2=0$$

$$x^2-(2^{\frac{1}{2}}+2)^{\frac{1}{2}}x+1=0 \quad x=\frac{1}{2}\{(2^{\frac{1}{2}}-2)^{\frac{1}{2}}+(2^{\frac{1}{2}}+2)^{\frac{1}{2}}\}.$$

419. *Def.* Two incommensurable roots of the second degree are called LIKE or UNLIKE according as either of them is or is not equal to the product made by multiplying the other by some protensive commensurable quantity.

420. PROP. *The product of two unlike incommensurable second roots of commensurable quantities cannot be equal to a commensurable quantity.*

For $x^{\frac{1}{2}}$ $y^{\frac{1}{2}}$ being incommensurable second roots of commensurable quantities x y $x^{\frac{1}{2}}y^{\frac{1}{2}}=a$ a commensurable quantity precisely when

$$x^{\frac{1}{2}}=\frac{a}{y^{\frac{1}{2}}}=\frac{\left(\frac{a}{y^{\frac{1}{2}}}\right)}{y^{\frac{1}{2}}}y^{\frac{1}{2}}=\frac{a}{y}y^{\frac{1}{2}}$$

and $\frac{a}{y}$ is commensurable.

421. PROP. *An incommensurable second root of a commensurable cannot be equal to the algebraic sum either of a commensurable and an incommensurable second root of a commensurable or of two unlike incommensurable second roots of commensurables.*

For $x^{\frac{1}{2}}=a+b^{\frac{1}{2}}$ x a b being commensurables and $x^{\frac{1}{2}}$ $b^{\frac{1}{2}}$ incommensurables just when severally $x=a^2+2ab^{\frac{1}{2}}+b$ $b^{\frac{1}{2}}=\frac{1}{2a}(-a^2+x-b)$ or just when an incommensurable is equal to a commensurable. And $x^{\frac{1}{2}}=a^{\frac{1}{2}}+b^{\frac{1}{2}}$ $a^{\frac{1}{2}}$ $b^{\frac{1}{2}}$ being unlike incommensurable second roots of commensurables just when severally $x=a+2a^{\frac{1}{2}}b^{\frac{1}{2}}+b$ $a^{\frac{1}{2}}b^{\frac{1}{2}}=\frac{1}{2}(-a+x-b)$ or the product of two unlike incommensurable second roots of commensurables is equal to a commensurable which (art. 420) is never.

422. Hence $a+b^{\frac{1}{2}}=a'+b'^{\frac{1}{2}}$ a a' b b' being commensurables and $b^{\frac{1}{2}}$ $b'^{\frac{1}{2}}$ incommensurables precisely when $a=a'$ and $b^{\frac{1}{2}}=b'^{\frac{1}{2}}$. For

otherwise $-a+a'=k$ some commensurable else than 0 and then $a'=a+k \quad b^{\frac{1}{2}}=k+b'^{\frac{1}{2}}$ which (art. 421) cannot be.

423. Prop. *To find the algebraic sum when there is any either of a commensurable and an incommensurable second root of a commensurable or of two unlike incommensurable second roots of commensurables that is equal to a second root of a given algebraic sum of a commensurable and a commensurable's incommensurable second root.*

Let s be a given commensurable and $t^{\frac{1}{2}}$ a given commensurable's incommensurable second root. Then $x=(s+t^{\frac{1}{2}})^{\frac{1}{2}}$ precisely when severally $x^2=s+t^{\frac{1}{2}}$

$$(-s+x^2)^2-t=0 \qquad s^2-t+x^4-2sx^2=0$$

$$\{-(s^2-t)^{\frac{1}{2}}+x^2\}^2-2\{s-(s^2-t)^{\frac{1}{2}}\}x^2=0$$

$$-(s^2-t)^{\frac{1}{2}}+x^2-2\left[\frac{1}{2}\{s-(s^2-t)^{\frac{1}{2}}\}\right]^{\frac{1}{2}}x=0$$

$$x=\left[\frac{1}{2}\{s+(s^2-t)^{\frac{1}{2}}\}\right]^{\frac{1}{2}}+\left[\frac{1}{2}\{s-(s^2-t)^{\frac{1}{2}}\}\right]^{\frac{1}{2}}.$$

Thus for example $(2+3^{\frac{1}{2}})^{\frac{1}{2}}=\left(\frac{3}{2}\right)^{\frac{1}{2}}+\left(\frac{1}{2}\right)^{\frac{1}{2}}$ and therefore in particular $\sqrt{(2+\sqrt{3})}=\sqrt{\frac{3}{2}}+\sqrt{\frac{1}{2}} \quad \sqrt{(2-\sqrt{3})}=\sqrt{\frac{3}{2}}-\sqrt{\frac{1}{2}}$. So $\sqrt{(7-4\sqrt{3})}=2-\sqrt{3}$ and $\sqrt{\left(\frac{5}{6}+\sqrt{\frac{2}{3}}\right)}=\sqrt{\frac{1}{2}}+\sqrt{\frac{1}{3}}$.

If instead of $(s^2-t)^{\frac{1}{2}}$ v be used

$$\{s+(s^2-v^2)^{\frac{1}{2}}\}^{\frac{1}{2}}=\left\{\frac{1}{2}(s+v)\right\}^{\frac{1}{2}}+\left\{\frac{1}{2}(s-v)\right\}^{\frac{1}{2}}.$$

424. In the same way as in art. 423 if a b be protensive quantities

$$\{a+(-)^{\frac{1}{2}}b\}^{\frac{1}{2}}=\left[\frac{1}{2}\{a+(a^2+b^2)^{\frac{1}{2}}\}\right]^{\frac{1}{2}}+\left[\frac{1}{2}\{a-(a^2+b^2)^{\frac{1}{2}}\}\right]^{\frac{1}{2}}.$$

Thus $\{1+(-)^{\frac{1}{2}}\sqrt{2}\}^{\frac{1}{2}}=(+)^{\frac{1}{2}}\left\{\sqrt{\frac{\sqrt{3}+1}{2}}+(-)^{\frac{1}{2}}\sqrt{\frac{\sqrt{3}-1}{2}}\right\}$

$$\{1+(-)^{\frac{1}{2}}\sqrt{3}\}^{\frac{1}{2}}=(+)^{\frac{1}{2}}\left\{\sqrt{\frac{3}{2}}+(-)^{\frac{1}{2}}\sqrt{\frac{1}{2}}\right\} \qquad \{(-)^{\frac{1}{2}}1\}^{\frac{1}{2}}=\{1+(-)^{\frac{1}{2}}1\}\left(\frac{1}{2}\right)^{\frac{1}{2}}$$

$$\left\{\sqrt{\frac{1}{2}}+(-)^{\frac{1}{2}}\sqrt{\frac{1}{2}}\right\}^{\frac{1}{2}}=(+)^{\frac{1}{2}}\left\{\sqrt{\frac{1}{2}\left(1+\sqrt{\frac{1}{2}}\right)}+(-)^{\frac{1}{2}}\sqrt{\frac{1}{2}\left(1-\sqrt{\frac{1}{2}}\right)}\right\}.$$

Moreover even if the s t of art. 423 be such that $s+t^{\frac{1}{2}}$ is minus still

$$(s+t^{\frac{1}{2}})^{\frac{1}{2}} = \left[\frac{1}{2}\{s+(s^2-t)^{\frac{1}{2}}\}\right]^{\frac{1}{2}} + \left[\frac{1}{2}\{s-(s^2-t)^{\frac{1}{2}}\}\right]^{\frac{1}{2}}.$$

Thus $(-7+\sqrt{13})^{\frac{1}{2}}=(-)^{\frac{1}{2}}\left(\sqrt{\frac{13}{2}}-\sqrt{\frac{1}{2}}\right)$ and $(2\sqrt{10}-7)^{\frac{1}{2}}=(-)^{\frac{1}{2}}(\sqrt{5}-\sqrt{2})$.

425. Let the polynome in x $p_0x^n+p_1x^{n-1}+p_2x^{n-2}+\cdots+p_{n-1}x+p_n$ where p_0 p_1 $p_2 \cdots p_{n-1}$ p_n are given protensive quantities be symbolized by fx. The values fa $f(a+h)$ of fx when x is given the several protensive values a $a+h$ may be made to differ from one another by a protensive quantity of less absolute value than any given numerical quantity else than 0 however small by taking h near enough to 0.

For first a product of numerical quantities some of them incommensurable is (art. 295) only known as what is endlessly neared by products of numbers taken either the same as or endlessly near to those quantities and in the same way but more simply from the needlessness of intercepting fractions a product of numbers whole or fractional is endlessly neared by products of numbers taken some of them endlessly near to and the rest if any the same as those numbers severally. Hence a product of protensive quantities is endlessly neared by products of protensive quantities taken some of them endlessly near to and therefore of the same sign as such as are not 0 of and the rest if any the same as those several protensives. In particular then the values $p_{n-i}a^i$ $p_{n-i}(a+h)^i$ of any term $p_{n-i}x^i$ of fx become endlessly near to one another as h endlessly nears 0.

Next the algebraic sum of protensive quantities is in a like way (arts. 291, 293) endlessly neared by the algebraic sum of protensive quantities taken endlessly near to those protensives severally. Hence as h endlessly nears 0 $f(a+h)$ endlessly nears fa.

426. If a be the algebraically less and b the algebraically greater of two protensive quantities that make the values fa fb of art. 425's fx one a plus quantity and the other a minus there is one protensive root at least of the equation $fx=0$ that is algebraically greater than a and algebraically less than b.

For as x is made to increase algebraically without any break from the value a up to the value b fx passes from the value fa to the value fb through values that (art. 425) everywhere lie endlessly close together in order and so too run on without any break. One value at least therefore of x is passed through which makes the value of fx pass either from being a plus quantity to being a minus or from being a minus quantity to being a plus. But in passing through any value but 0 or $\pm\infty$ fx keeps the same algebraic sign as it has and fx is $\pm\infty$ only when x is. Therefore fx changes sign only in passing through the value 0.

427. If the $p_0\ p_1 \ldots p_n$ of art. 425's fx be given plus or minus decimally denoted numbers protensive roots of the equation $fx=0$ may often be found decimally denoted to any sought degree of nearness by help of art. 190's process but with minus quantities used (art. 354) as well as plus and if need be of what is shown in art. 426. If $a_1\ a_2\ a_3 \ldots a_i$ being protensives polynomes $f_1(-a_1+x)$ $f_2(-a_2-a_1+x)$ - - - $f_i(-a_i-\cdots-a_2-a_1+x)$ be found all operationally equivalent to $fx \because -a_i-\cdots-a_1+x=-(a_1+\cdots+a_i)+x$ $a_1\ a_2 \ldots a_i$ have only to be so chosen that the last terms $f_1 0\ f_2 0$ - - - $f_i 0$ of those polynomes are all of the same algebraic sign and go on getting ever nearer and nearer to 0 for $a_1\ a_1+a_2\ a_1+a_2+a_3$ . . . to be ever closer and closer approximations to a root of the equation $fx=0$. It is handiest here as in art. 321's kindred process so to take $a_1\ a_2\ a_3 \ldots$ that their absolute values are the numbers denoted severally by the digits in order of a decimally denoted number.

By way of instance if ω be the algebraic expression of an angle in reference to a perigon as unit

$$\begin{aligned}\sin 3\omega &= \sin(2\omega+\omega) = (\sin 2\omega)\cos\omega+(\cos 2\omega)\sin\omega\\ &= 2(\sin\omega)(\cos\omega)^2+\{(\cos\omega)^2-(\sin\omega)^2\}\sin\omega\\ &= 2(\sin\omega)\{1-(\sin\omega)^2\}+\{1-2(\sin\omega)^2\}\sin\omega = 3\sin\omega-4(\sin\omega)^3\end{aligned}$$

and therefore the sines of the third parts of all algebraically expressed angles that have a given sine a are precisely such values of x as fulfil the equation

$$x^3-\frac{3}{4}x+\frac{1}{4}a=0.$$

Since all algebraically expressed angles that have sines the same as $\sin 3\omega$ are (art. 387) precisely all that are expressed by $\frac{i}{2}+(-)^i 3\omega$ i being a plus or minus whole number if $\sin 3\omega=a$ $x=\sin\left\{\frac{i}{6}+(-)^i\omega\right\}$. But i is either $i'\times 3$ $i'\times 3+1$ or $i'\times 3-1$ i' being a plus or minus whole number and

$$\sin\left\{\frac{i'\times 3}{6}+(-)^{i'\times 3}\omega\right\}=\sin\left\{\frac{i'}{2}+(-)^{i'}\omega\right\}=\sin\omega$$

$$\sin\left\{\frac{i'\times 3\pm 1}{6}+(-)^{i'\times 3\pm 1}\omega\right\}=\sin\left[\frac{i'}{2}+(-)^{i'}\left\{\pm(-)^{i'}\frac{1}{6}-\omega\right\}\right]=\sin\left\{\pm(-)^{i'}\frac{1}{6}-\omega\right\}$$

wherefore $\sin\left\{\frac{i}{6}+(-)^{i}\omega\right\}$ has no other values than $\sin\omega$ $\sin\left(\frac{1}{6}-\omega\right)$ and $\sin\left(-\frac{1}{6}-\omega\right)$ or $-\sin\left(\frac{1}{6}+\omega\right)$ got by making i in turn 0 1 and -1. Further $\sin\left(\pm\frac{1}{6}-\omega\right)=\sin\omega$ precisely when h being a plus or minus whole number $\pm\frac{1}{6}-\omega=\frac{h}{2}+(-)^{h}\omega$ which holds never if h be odd and if h be even and equal to $h'\times 2$ precisely when $\omega=-\frac{1}{2}h'\pm\frac{1}{12}$ $\sin\left(-\frac{1}{6}-\omega\right)=\sin\left(\frac{1}{6}-\omega\right)$ precisely when $-\frac{1}{6}-\omega=\frac{h}{2}+(-)^{h}\left(\frac{1}{6}-\omega\right)$ which holds never if h be even and if h be odd and equal to $h''\times 2+1$ precisely when $\omega=-\frac{1}{2}\left(h''+\frac{1}{2}\right)$ and since $-\frac{1}{2}h'\pm\frac{1}{12}$ never $=-\frac{1}{2}\left(h''+\frac{1}{2}\right)$ $\sin\omega$ $\sin\left(\frac{1}{6}-\omega\right)$ $\sin\left(-\frac{1}{6}-\omega\right)$ are never all equal. On the whole then x has precisely 2 or precisely 3 different values according as a is ± 1 or not. Since a straight line bisecting an angle of an equilateral or equiangular triangle bisects perpendicularly the side opposite $\sin\frac{1}{12}=\frac{1}{2}$ and hence making a $\frac{1}{2}$ the three roots of the equation

$$x^3-0{\cdot}75x+0{\cdot}125=0$$

are $\sin\frac{1}{36}$ $\sin\frac{5}{36}$ and $\sin-\frac{7}{36}$ or $-\sin\frac{7}{36}$. The values -1 0 $\frac{1}{2}$ 1 of x make $x^3-\frac{3}{4}x+\frac{1}{8}$ in turn $-\frac{1}{8}$ $+\frac{1}{8}$ $-\frac{1}{8}$ $+\frac{3}{8}$ and the root $\sin-\frac{7}{36}$ is to be sought between -1 and 0 the root $\sin\frac{1}{36}$ between 0 and $\frac{1}{2}$ and the root $\sin\frac{5}{36}$ between $\frac{1}{2}$ and 1.

			−0·93969262
1	0	−0·75	0·125
	−0·9	81	− 54
	−0·9	0·06	71
	− 9	1 62	
	−1·8	1·68	
	− 9		
	−2·7		
	− 3	819	− 52857
	−2·73	1·7619	18143
	− 3	828	
	−2·76	1·8447	
	− 3		
	−2·79		
	− 9	25191	− 16829019
	−2·799	1·869891	1313981
	− 9	25272	
	−2·808	1·895163	
	− 9		
	−2·817		
	− 6	1690 6	− 1138112 2
	−2·817 6	1·896853 6	175868 8
	− 6	1690 9	
	−2·818 2	1·898544 5	
	− 6		
	−2·818 8		
		25 37	− 170891 8
		1·89879 8	4977 0
		25 4	
		1·89905 2	
		6	− 3798 1
		1·8990 6	1178 9
			− 1139 4
			395
			− 380

			0·17364818
1	0	−0·75	0·125
	0·1	1	− 74
	0·1	−0·74	51
	1	2	
	0·2	−0·72	
	1		
	0·3		
	7	259	− 48587
	0·37	−0·6941	2413
	7	308	
	0·44	−0·6633	
	7		
	0·51		
	3	1539	− 1985283
	0·513	−0·661761	427717
	3	1548	
	0·516	−0·660213	
	3		
	0·519		
	6	311 8	− 395940 7
	0·519 6	−0·659901 2	31776 3
	6	312 1	
	0·520 2	−0·659589 1	
	6		
	0·520 8		
		2 08	− 26382 7
		−0·65956 8	5393 6
		2 1	
		−0·65954 7	
		4	− 5276 3
		−0·6595 4	117 3
			− 66 0
			51 3
			− 52 8

			0·76 - -
1	0	−0·75	0·125
	0·7	49	− 182
	0·7	−0·26	− 57
	7	98	
	1·4	0·72	
	7		
	2·1		
	6	1296	50976
	2·16	0·8496	− 6024
-	-	-	- -

This way of approximating to the roots of an equation is named HORNER'S METHOD after its inventor.

428. An equation of the first degree $xa+yb=c$ with two unknowns $x\ y$ does not fix the value of either unknown but only so binds together $x\ y$ that when any value is given at will to either the value of the other becomes fixed. If however another equation of the first degree $xa'+yb'=c'$ with the same unknowns holds at the same time $x\ y$ have fixed values. For since neither b nor b' is o

$$\left.\begin{matrix} xa+yb=c \\ xa'+yb'=c' \end{matrix}\right\} \text{ precisely when } y=\frac{-xa+c}{b}=\frac{-xa'+c'}{b'}$$

and $\dfrac{-xa+c}{b}=\dfrac{-xa'+c'}{b'}$ precisely when $x\left(\dfrac{a}{b}-\dfrac{a'}{b'}\right)=\dfrac{c}{b}-\dfrac{c'}{b'}$ and therefore unless $\dfrac{a}{b}=\dfrac{a'}{b'}$ precisely when $x=\dfrac{\dfrac{c}{b}-\dfrac{c'}{b'}}{\dfrac{a}{b}-\dfrac{a'}{b'}}=\dfrac{cb'-c'b}{ab'-a'b}$. If $\dfrac{a}{b}=\dfrac{a'}{b'}$

$xa+yb=c$ precisely when severally $x\dfrac{a}{b}+y=\dfrac{c}{b}$ $\ x\dfrac{a'}{b'}+y=\dfrac{c}{b}$

$xa'+yb'=\dfrac{c}{b}b'$ and hence the equations $xa+yb=c\ \ xa'+yb'=c'$ can then hold only if $\dfrac{c}{b}b'=c'$ or $\dfrac{c}{b}=\dfrac{c'}{b'}$ when indeed the second equation is no other than the first. Or since neither a nor a' is o

$$\left.\begin{matrix} xa+yb=c \\ xa'+yb'=c' \end{matrix}\right\} \text{ precisely when } x=\frac{c-yb}{a}=\frac{c'-yb'}{a'}$$

and $\dfrac{c-yb}{a}=\dfrac{c'-yb'}{a'}$ precisely when $y\left(-\dfrac{b'}{a'}+\dfrac{b}{a}\right)=-\dfrac{c'}{a'}+\dfrac{c}{a}$ and therefore,— unless $\frac{b}{a}=\frac{b'}{a'}$ when the two equations can hold only if $\frac{c}{a}=\frac{c'}{a'}$ and so only by being one and the same equation—, precisely when

$$y=\frac{-\frac{c'}{a'}+\frac{c}{a}}{-\frac{b'}{a'}+\frac{b}{a}}=\frac{-c'a+ca'}{-b'a+ba'}.$$

Of the equations $xa+yb=c$ $xa'=c'$ the latter settles x's value and then the former y's.

If neither b nor b' be 0 the equations

$$\left.\begin{matrix}ax+by=c\\ a'x+b'y=c'\end{matrix}\right\}\text{ hold precisely when } y=\frac{1}{b}(-ax+c)=\frac{1}{b'}(-a'x+c')$$

and unless $\frac{1}{b}a=\frac{1}{b'}a'$ when the equations can only hold if $\frac{1}{b}c=\frac{1}{b'}c'$ and therefore each equation is the same as the other $\frac{1}{b}(-ax+c)=\frac{1}{b'}(-a'x+c')$ precisely when $x=\frac{1}{\frac{1}{b}a-\frac{1}{b'}a'}\left(\frac{1}{b}c-\frac{1}{b'}c'\right)=\frac{1}{b'a-ba'}(b'c-bc')$.

If neither a nor a' be 0 the same equations hold precisely when $x=\frac{1}{a}(c-by)=\frac{1}{a'}(c'-b'y)$ and unless $\frac{1}{a}b=\frac{1}{a'}b'$ when the equations can only hold if $\frac{1}{a}c=\frac{1}{a'}c'$ and each equation therefore is simply a consequence of the other $\frac{1}{a}(c-by)=\frac{1}{a'}(c'-b'y)$ precisely when

$$y=\frac{1}{-\frac{1}{a'}b'+\frac{1}{a}b}\left(-\frac{1}{a'}c'+\frac{1}{a}c\right)=\frac{1}{-ab'+a'b}(-ac'+a'c).$$

429. If u_1 v_1 u_2 $v_2 \ldots u_i$ v_i be quantities such that

$$\frac{u_1}{v_1}=\frac{u_2}{v_2}=\frac{u_3}{v_3}=\cdots=\frac{u_i}{v_i}$$

then calling any one of these equals t and taking any i quantities λ_1 $\lambda_2 \ldots \lambda_i$ $u_1=tv_1$ $u_1\lambda_1=(tv_1)\lambda_1=tv_1\lambda_1$ and in like way $u_2\lambda_2=tv_2\lambda_2$ $u_3\lambda_3=tv_3\lambda_3$ - - - $u_i\lambda_i=tv_i\lambda_i$. Hence

$$u_1\lambda_1+u_2\lambda_2+\cdots+u_i\lambda_i=tv_1\lambda_1+tv_2\lambda_2+\cdots+tv_i\lambda_i=t(v_1\lambda_1+v_2\lambda_2+\cdots+v_i\lambda_i)$$

$$t=\frac{u_1\lambda_1+u_2\lambda_2+\cdots+u_i\lambda_i}{v_1\lambda_1+v_2\lambda_2+\cdots+v_i\lambda_i}.$$

If in particular the λs be such that $v_1\lambda_1+v_2\lambda_2+\cdots+v_i\lambda_i=0$ then $u_1\lambda_1+u_2\lambda_2+\cdots+u_i\lambda_i=0$.

If the λs be such that $v_1\lambda_1+v_2\lambda_2+\cdots+v_i\lambda_i$ is not 0 then since if $\frac{u_1}{v_1}\ \frac{u_2}{v_2}\cdots\frac{u_i}{v_i}$ be all equal each of them $=\frac{u_1\lambda_1+u_2\lambda_2+\cdots+u_i\lambda_i}{v_1\lambda_1+v_2\lambda_2+\cdots+v_i\lambda_i}$ and if each of them be equal to this they are all equal to one another it follows that $\frac{u_1}{v_1}=\frac{u_2}{v_2}=\frac{u_3}{v_3}=---=\frac{u_i}{v_i}$ precisely when each

$$=\frac{u_1\lambda_1+u_2\lambda_2+\cdots+u_i\lambda_i}{v_1\lambda_1+v_2\lambda_2+\cdots+v_i\lambda_i}.$$

Thus if as in art. 428 $y=\frac{-xa+c}{b}=\frac{-xa'+c'}{b'}$

$$y=\frac{(-xa+c)a'+(-xa'+c')(-a)}{ba'+b'(-a)}=\frac{ca'-c'a}{ba'-b'a}$$

and if $x=\frac{c-yb}{a}=\frac{c'-yb'}{a'}$ $\quad x=\frac{(c-yb)b'-(c'-yb')b}{ab'-a'b}=\frac{cb'-c'b}{ab'-a'b}.$

Again if $\frac{1}{v_1}u_1=\frac{1}{v_2}u_2=\frac{1}{v_3}u_3=---\frac{1}{v_i}u_i$ and any one of them be called t $\quad \lambda_1u_1=\lambda_1v_1t=(\lambda_1v_1)t \quad \lambda_2u_2=(\lambda_2v_2)t \quad ---- \quad \lambda_iu_i=(\lambda_iv_i)t$
$\lambda_1u_1+\lambda_2u_2+\cdots+\lambda_iu_i=(\lambda_1v_1+\lambda_2v_2+\cdots+\lambda_iv_i)t$ and

$$t=\frac{1}{\lambda_1v_1+\lambda_2v_2+\cdots+\lambda_iv_i}(\lambda_1u_1+\lambda_2u_2+\cdots+\lambda_iu_i)$$

with the understanding that if $\lambda_1v_1+\lambda_2v_2+\cdots+\lambda_iv_i=0$ then $\lambda_1u_1+\lambda_2u_2+\cdots+\lambda_iu_i=0$. As before too if $\lambda_1v_1+\lambda_2v_2+\cdots+\lambda_iv_i$ is not 0 then $\frac{1}{v_1}u_1=\frac{1}{v_2}u_2=\frac{1}{v_3}u_3=---=\frac{1}{v_i}u_i$ precisely when each of these products $=\frac{1}{\lambda_1v_1+\lambda_2v_2+\cdots+\lambda_iv_i}(\lambda_1u_1+\lambda_2u_2+\cdots+\lambda_iu_i)$.

Thus if as in art. 428 $y=\frac{1}{b}(-ax+c)=\frac{1}{b'}(-a'x+c')$

$$y=\frac{1}{a'b-ab'}\{a'(-ax+c)-a(-a'x+c')\}=\frac{1}{a'b-ab'}(a'c-ac')$$

and if $x=\frac{1}{a}(c-by)=\frac{1}{a'}(c'-b'y)$

$$x=\frac{1}{b'a+(-b)a'}\{b'(c-by)+(-b)(c'-b'y)\}=\frac{1}{b'a-ba'}(b'c-bc').$$

430. If $V=0$ and $V'=0$ then $\lambda V+\lambda'V'=0$ whatever quantities $\lambda\ \lambda'$ are. Again if $V=0$ and $\lambda V+\lambda'V'=0$ then $\lambda'V'=0$ and therefore if λ' be not 0 $V'=0$. So if $V'=0$ and $\lambda V+\lambda'V'=0$

then if λ be not o $V=0$. Hence if λ λ' be any quantities but o the three equations

$$V=0 \qquad V'=0 \qquad \lambda V+\lambda' V'=0$$

are such that each of them follows from the other two and therefore any two of them are equivalent to any two.

If $V=0$ and $V'=0$ then $\lambda V+\lambda' V'=0$ and $\mu V+\mu' V'=0$ λ λ' μ μ' being any quantities. Again if $\lambda V+\lambda' V'=0$ and $\mu V+\mu' V'=0$ then $\lambda(\mu V+\mu' V')+(-\mu)(\lambda V+\lambda' V')=0$ and $\mu'(\lambda V+\lambda' V')+(-\lambda')(\mu V+\mu' V')=0$ therefore $(\lambda\mu'-\mu\lambda')V'=0$ and $(\mu'\lambda-\lambda'\mu)V=0$ and therefore if $\lambda\mu'-\mu\lambda'$ or the operational equivalent $\mu'\lambda-\lambda'\mu$ be not o $V'=0$ and $V=0$. Hence if λ λ' μ μ' be any quantities such that $\lambda\mu'-\mu\lambda'$ is not o

$$\left.\begin{matrix} V=0 \\ V'=0 \end{matrix}\right\} \text{ precisely when } \left\{\begin{matrix} \lambda V+\lambda' V'=0 \\ \mu V+\mu' V'=0. \end{matrix}\right.$$

431. If then λ λ' μ μ' be any such quantities as make $\lambda\mu'-\mu\lambda'$ else than o

$$\because \lambda(ax+by-c)+\lambda'(a'x+b'y-c') = (\lambda a+\lambda' a')x+(\lambda b+\lambda' b')y-(\lambda c+\lambda' c')$$

the pair of equations $ax+by=c$ $a'x+b'y=c'$ is equivalent to the pair

$$(\lambda a+\lambda' a')x+(\lambda b+\lambda' b')y-(\lambda c+\lambda' c')=0$$
$$(\mu a+\mu' a')x+(\mu b+\mu' b')y-(\mu c+\mu' c')=0.$$

And this pair gives at once x and y if λ λ' μ μ' be so taken that

$$\lambda b+\lambda' b'=0 \qquad \lambda a+\lambda' a' = \text{not } 0 \qquad \mu a+\mu' a'=0 \qquad \mu b+\mu' b' = \text{not } 0.$$

If b be o $ax=c$ and $x=a^{-1}c$ and if b' be o $a'x=c'$. But if neither b nor b' be o $\lambda b+\lambda' b'=0$ precisely when $\lambda b'^{-1}=-\lambda' b^{-1}$ and therefore precisely when h being any quantity $\lambda=hb'$ and $\lambda'=-hb$. Also if a be o $by=c$ if a' $b'y=c'$ and if neither a be o nor a' $\mu a+\mu' a'=0$ precisely when k being any quantity $\mu=ka'$ and $\mu'=-ka$.

When λ λ' μ μ' are taken severally hb' $-hb$ ka' $-ka$

$$\lambda\mu'-\mu\lambda' = -k(\lambda a+\lambda' a') = h(\mu b+\mu' b') = hk(a'b-b'a)$$

and hence if both h and k be taken else than o the four quantities $\lambda\mu'-\mu\lambda'$ $\lambda a+\lambda' a'$ $\mu b+\mu' b'$ $a'b-b'a$ are then either all o or all not o. Moreover if neither a nor b be o $a'b-b'a=0$ precisely when d being any quantity $a'=da$ and $b'=db$ and then $a'x+b'y-c'=d(ax+by-d^{-1}c')$ so that the equations $ax+by=c$ $a'x+b'y=c'$ can only hold if $c'=dc$ when the equations are in effect one equation.

As examples of the method $x\times4-y\times5=3$ and $x\times7+y\times2=16$ precisely when the equation pairs severally hold—

$$\left.\begin{array}{r}(x\times4-y\times5-3)\times2+(x\times7+y\times2-16)\times5=0\\-(x\times4-y\times5-3)\times7+(x\times7+y\times2-16)\times4=0\end{array}\right\}$$

$$\left.\begin{array}{r}x(4\times2+7\times5)=3\times2+16\times5\\y(5\times7+2\times4)=-3\times7+16\times4\end{array}\right\}\quad\left.\begin{array}{r}x\times43=86\\y\times43=43\end{array}\right\}\quad\left.\begin{array}{r}x=2\\y=1\end{array}\right\}.$$

Any one holds precisely when any other holds of the pairs—

$$\left.\begin{array}{r}5x+12y=63\\7x+8y=53\end{array}\right\}\qquad\left.\begin{array}{r}2(5x+12y-63)-3(7x+8y-53)=0\\7(5x+12y-63)-5(7x+8y-53)=0\end{array}\right\}$$

$$\left.\begin{array}{r}(10-21)x=126-159\\(84-40)y=441-265\end{array}\right\}\qquad\left.\begin{array}{r}x=\frac{1}{11}\times33=3\\y=\frac{1}{44}\times176=4\end{array}\right].$$

The pairs following are all equivalent—

$$\left.\begin{array}{r}5x-9y=3\\3y=4\end{array}\right\}\quad\left.\begin{array}{r}5x-9y-3+3(3y-4)=0\\3\left(y-\frac{1}{3}\times4\right)=0\end{array}\right\}\quad\left.\begin{array}{r}5x-3\times5=0\\y-\frac{4}{3}=0\end{array}\right\}\quad\left.\begin{array}{r}x=3\\y=\frac{4}{3}\end{array}\right\}.$$

It is precisely when any one of the following pairs of equations holds that any other does—

$$\left.\begin{array}{r}x\times\frac{2}{3}-y\times\frac{5}{2}=\frac{16}{3}\\x\times\frac{3}{4}-y\times\frac{1}{3}=\frac{25}{24}\end{array}\right\}\qquad\left.\begin{array}{r}x\left(\frac{2}{3}\times\frac{1}{3}-\frac{3}{4}\times\frac{5}{2}\right)=\frac{16}{3}\times\frac{1}{3}-\frac{25}{24}\times\frac{5}{2}\\y\left(\frac{1}{3}\times\frac{2}{3}-\frac{5}{2}\times\frac{3}{4}\right)=-\frac{25}{24}\times\frac{2}{3}+\frac{16}{3}\times\frac{3}{4}\end{array}\right\}$$

$$\left.\begin{array}{r}x=\frac{119}{9\times16}\times\frac{9\times8}{119}=\frac{1}{2}\\y=-\frac{119}{3\times12}\times\frac{9\times8}{119}=-2\end{array}\right\}.$$

432. If $V_1=0$ $V_2=0$ $V_3=0$ then λ_1 λ_2 λ_3 μ_1 μ_2 μ_3 ν_1 ν_2 ν_3 being any quantities

$$\lambda_1V_1+\lambda_2V_2+\lambda_3V_3=0\qquad\mu_1V_1+\mu_2V_2+\mu_3V_3=0\qquad\nu_1V_1+\nu_2V_2+\nu_3V_3=0.$$

Again if these last three equations hold then

$$\mu_3(\nu_1V_1+\nu_2V_2+\nu_3V_3)-\nu_3(\mu_1V_1+\mu_2V_2+\mu_3V_3)=0$$

$$\nu_2(\mu_1V_1+\mu_2V_2+\mu_3V_3)-\mu_2(\nu_1V_1+\nu_2V_2+\nu_3V_3)=0$$

or $(\mu_3\nu_1-\mu_1\nu_3)V_1=(\mu_2\nu_3-\mu_3\nu_2)V_2\quad(\mu_1\nu_2-\mu_2\nu_1)V_1=(\mu_2\nu_3-\mu_3\nu_2)V_3$

and $\mu_2\nu_3-\mu_3\nu_2)(\lambda_1V_1+\lambda_2V_2+\lambda_3V_3)=0$

$$\text{or}\quad \lambda_1(\mu_2\nu_3-\mu_3\nu_2)\,V_1+\lambda_2(\mu_2\nu_3-\mu_3\nu_2)\,V_2+\lambda_3(\mu_2\nu_3-\mu_3\nu_2)\,V_3=0$$

$$\text{hence}\quad \lambda_1(\mu_2\nu_3-\mu_3\nu_2)\,V_1+\lambda_2(\mu_3\nu_1-\mu_1\nu_3)\,V_1+\lambda_3(\mu_1\nu_2-\mu_2\nu_1)\,V_1=0$$

$$\text{or}\quad \{\lambda_1(\mu_2\nu_3-\mu_3\nu_2)+\lambda_2(\mu_3\nu_1-\mu_1\nu_3)+\lambda_3(\mu_1\nu_2-\mu_2\nu_1)\}\,V_1=0$$

and therefore if $\lambda_1(\mu_2\nu_3-\mu_3\nu_2)+\lambda_2(\mu_3\nu_1-\mu_1\nu_3)+\lambda_3(\mu_1\nu_2-\mu_2\nu_1)$ be not 0 $V_1=0$. Doing the same with the 3rd 1st and 2nd of the same 3 equations as is here done with the 2nd 3rd and 1st and again the same with the 1st 2nd and 3rd if

$\mu_1(\nu_2\lambda_3-\nu_3\lambda_2)+\mu_2(\nu_3\lambda_1-\nu_1\lambda_3)+\mu_3(\nu_1\lambda_2-\nu_2\lambda_1)$ be not 0 $V_2=0$ and if $\nu_1(\lambda_2\mu_3-\lambda_3\mu_2)+\nu_2(\lambda_3\mu_1-\lambda_1\mu_3)+\nu_3(\lambda_1\mu_2-\lambda_2\mu_1)$ be not 0 $V_3=0$. But

$$\lambda_1(\mu_2\nu_3-\mu_3\nu_2)+\lambda_2(\mu_3\nu_1-\mu_1\nu_3)+\lambda_3(\mu_1\nu_2-\mu_2\nu_1)$$

$$=\lambda_1\mu_2\nu_3+\mu_1\nu_2\lambda_3+\nu_1\lambda_2\mu_3-\nu_1\mu_2\lambda_3-\mu_1\lambda_2\nu_3-\lambda_1\nu_2\mu_3$$

$$=\lambda_1\mu_2\nu_3-\nu_1\mu_2\lambda_3+(\mu_1\nu_2\lambda_3-\lambda_1\nu_2\mu_3)+(\nu_1\lambda_2\mu_3-\mu_1\lambda_2\nu_3)$$

$$=\mu_1(\nu_2\lambda_3-\nu_3\lambda_2)+\mu_2(\nu_3\lambda_1-\nu_1\lambda_3)+\mu_3(\nu_1\lambda_2-\nu_2\lambda_1)$$

$$=\nu_1(\lambda_2\mu_3-\lambda_3\mu_2)+\nu_2(\lambda_3\mu_1-\lambda_1\mu_3)+\nu_3(\lambda_1\mu_2-\lambda_2\mu_1).$$

Hence if $\lambda_1\ \lambda_2\ \lambda_3\ \mu_1\ \mu_2\ \mu_3\ \nu_1\ \nu_2\ \nu_3$ be any quantities such that

$$\lambda_1\mu_2\nu_3-\nu_1\mu_2\lambda_3+(\mu_1\nu_2\lambda_3-\lambda_1\nu_2\mu_3)+(\nu_1\lambda_2\mu_3-\mu_1\lambda_2\nu_3)\ \text{ is not } 0$$

$$\left.\begin{array}{l}V_1=0\\ V_2=0\\ V_3=0\end{array}\right\}\ \text{precisely when}\ \left\{\begin{array}{l}\lambda_1V_1+\lambda_2V_2+\lambda_3V_3=0\\ \mu_1V_1+\mu_2V_2+\mu_3V_3=0\\ \nu_1V_1+\nu_2V_2+\nu_3V_3=0\end{array}\right.$$

433. Hence and

$$\because\ \lambda_1(a_1x+b_1y+c_1z-d_1)+\lambda_2(a_2x+b_2y+c_2z-d_2)+\lambda_3(a_3x+b_3y+c_3z-d_3)$$

$$=(\lambda_1a_1+\lambda_2a_2+\lambda_3a_3)x+(\lambda_1b_1+\lambda_2b_2+\lambda_3b_3)y+(\lambda_1c_1+\lambda_2c_2+\lambda_3c_3)z$$
$$-(\lambda_1d_1+\lambda_2d_2+\lambda_3d_3)$$

if $\lambda_1\mu_2\nu_3-\nu_1\mu_2\lambda_3+(\mu_1\nu_2\lambda_3-\lambda_1\nu_2\mu_3)+(\nu_1\lambda_2\mu_3-\mu_1\lambda_2\nu_3)$ be not 0 the equations

$$a_1x+b_1y+c_1z=d_1\qquad a_2x+b_2y+c_2z=d_2\qquad a_3x+b_3y+c_3z=d_3$$

hold precisely when,— symbolizing $\lambda_1a_1+\lambda_2a_2+\lambda_3a_3$ for shortness by $\Sigma\lambda a$—,

$$(\Sigma\lambda a)x+(\Sigma\lambda b)y+(\Sigma\lambda c)z=\Sigma\lambda d\qquad(\Sigma\mu a)x+(\Sigma\mu b)y+(\Sigma\mu c)z=\Sigma\mu d$$
$$(\Sigma\nu a)x+(\Sigma\nu b)y+(\Sigma\nu c)z=\Sigma\nu d.$$

These last three equations severally give $x\ y\ z$ if the λs the μs and the νs be taken so that

$$\begin{array}{lll}\Sigma\lambda b=0 & \Sigma\lambda c=0 & \Sigma\lambda a=\text{not }0\\ \Sigma\mu c=0 & \Sigma\mu a=0 & \Sigma\mu b=\text{not }0\\ \Sigma\nu a=0 & \Sigma\nu b=0 & \Sigma\nu c=\text{not }0.\end{array}$$

If $b_3(-c_2)-b_2(-c_3)$ or the equivalent $b_2c_3-b_3c_2$ be not o (art. 430) $\Sigma\lambda b=0$ and $\Sigma\lambda c=0$ precisely when

$$\lambda_2(b_2c_3-c_2b_3)=\lambda_1(c_1b_3-b_1c_3) \qquad \lambda_3(c_3b_2-b_3c_2)=\lambda_1(b_1c_2-c_1b_2)$$

therefore if further neither $b_3c_1-b_1c_3$ nor $b_1c_2-b_2c_1$ be o precisely when

$$\frac{\lambda_1}{b_2c_3-b_3c_2}=\frac{\lambda_2}{b_3c_1-b_1c_3}=\frac{\lambda_3}{b_1c_2-b_2c_1}$$

and therefore precisely when t being any quantity $\lambda_1=t(b_2c_3-b_3c_2)$ $\lambda_2=t(b_3c_1-b_1c_3)$ and $\lambda_3=t(b_1c_2-b_2c_1)$. Then $\Sigma\lambda a=t\{(b_2c_3-b_3c_2)a_1+(b_3c_1-b_1c_3)a_2+(b_1c_2-b_2c_1)a_3\}$ which if t be taken else than o is o or not o according as $(b_2c_3-b_3c_2)a_1+(b_3c_1-b_1c_3)a_2+(b_1c_2-b_2c_1)a_3$ is o or not o and $\Sigma\lambda d=t\{(b_2c_3-b_3c_2)d_1+(b_3c_1-b_1c_3)d_2+(b_1c_2-b_2c_1)d_3\}$.

If $b_2c_3-b_3c_2=0$ either (1) $b_2=0$ $b_3=0$ when of the 3 given equations the 2nd and 3rd $a_2x+c_2z=d_2$ $a_3x+c_3z=d_3$ alone fix all about x z (2) $b_2=0$ $c_2=0$ when the 2nd $a_2x=d_2$ alone gives x (3) $b_2=$ not o $b_3=0$ when $c_3=0$ and the 3rd given equation $a_3x=d_3$ alone gives x or (4) $b_2=$ not o $b_3=$ not o when $\frac{1}{b_2}c_2=\frac{1}{b_3}c_3$ = some quantity e be it and the 2nd and 3rd given equations are equivalent to $a_2x+b_2(y+ez)=d_2$ and $a_3x+b_3(y+ez)=d_3$ and so alone fix all about x and $y+ez$ as two unknowns.

If $b_2c_3-b_3c_2$ $b_3c_1-b_1c_3$ be each o and b_1 b_2 b_3 each not o then $\frac{1}{b_3}c_3=\frac{1}{b_2}c_2=\frac{1}{b_1}c_1$ = some quantity f say and the given equations are severally equivalent to $a_1x+b_1(y+fz)=d_1$ $a_2x+b_2(y+fz)=d_2$ $a_3x+b_3(y+fz)=d_3$ three equations with only two unknowns x and $y+fz$.

If $b_2c_3-b_3c_2$ $b_3c_1-b_1c_3$ $b_1c_2-b_2c_1$ be each not o and $(b_2c_3-b_3c_2)a_1+(b_3c_1-b_1c_3)a_2+(b_1c_2-b_2c_1)a_3$ be o the three given equations can hold only if $(b_2c_3-b_3c_2)d_1+(b_3c_1-b_1c_3)d_2+(b_1c_2-b_2c_1)d_3$ be o then any values of the λs that make $\Sigma\lambda b$ and $\Sigma\lambda c$ both o make $\lambda_1(a_1x+b_1y+c_1z-d_1)+\lambda_2(a_2x+b_2y+c_2z-d_2)$ operationally equivalent to $-\lambda_3(a_3x+b_3y+c_3z-d_3)$ therefore each of the three equations is simply a consequence of the other two and therefore the three are in truth only two independent equations with the three unknowns x y z.

The conditions $\Sigma\mu c=0$ $\Sigma\mu a=0$ $\Sigma\mu b=$ not o and the conditions $\Sigma\nu a=0$ $\Sigma\nu b=0$ $\Sigma\nu c=$ not o are to be dealt with in the same way as the conditions $\Sigma\lambda b=0$ $\Sigma\lambda c=0$ $\Sigma\lambda a=$ not o. From all three sets of conditions it may turn out that no two of the three given equations are independent of one another.

If $\lambda_1(b_2c_3-b_3c_2)^{-1}=\lambda_2(b_3c_1-b_1c_3)^{-1}=\lambda_3(b_1c_2-b_2c_1)^{-1}=t$

$$\mu_1(c_2a_3-c_3a_2)^{-1}=\mu_2(c_3a_1-c_1a_3)^{-1}=\mu_3(c_1a_2-c_2a_1)^{-1}=u \text{ and}$$
$$\nu_1(a_2b_3-a_3b_2)^{-1}=\nu_2(a_3b_1-a_1b_3)^{-1}=\nu_3(a_1b_2-a_2b_1)^{-1}=v$$

then putting K for any one of the operational equivalents

$$(b_2c_3-b_3c_2)a_1+(b_3c_1-b_1c_3)a_2+(b_1c_2-b_2c_1)a_3$$
$$(c_2a_3-c_3a_2)b_1+(c_3a_1-c_1a_3)b_2+(c_1a_2-c_2a_1)b_3$$
$$(a_2b_3-a_3b_2)c_1+(a_3b_1-a_1b_3)c_2+(a_1b_2-a_2b_1)c_3$$
$$a_1b_2c_3-c_1b_2a_3+(b_1c_2a_3-a_1c_2b_3)+(c_1a_2b_3-b_1a_2c_3)$$
$$\mu_2\nu_3-\mu_3\nu_2=\{u(c_3a_1-c_1a_3)\}v(a_1b_2-a_2b_1)-\{u(c_1a_2-c_2a_1)\}v(a_3b_1-a_1b_3)$$
$$=uv\{(c_2a_3-c_3a_2)b_1+(c_3a_1-c_1a_3)b_2+(c_1a_2-c_2a_1)b_3\}a_1=uvKa_1$$
$$\mu_3\nu_1-\mu_1\nu_3=uvKa_2 \qquad \mu_1\nu_2-\mu_2\nu_1=uvKa_3$$
$$\lambda_1(\mu_2\nu_3-\mu_3\nu_2)+\lambda_2(\mu_3\nu_1-\mu_1\nu_3)+\lambda_3(\mu_1\nu_2-\mu_2\nu_1)$$
$$=tuvK\{(b_2c_3-b_3c_2)a_1+(b_3c_1-b_1c_3)a_2+(b_1c_2-b_2c_1)a_3\}=tuvK^2$$

and therefore if t u v be each taken else than 0

$$\lambda_1\mu_2\nu_3-\nu_1\mu_2\lambda_3+(\mu_1\nu_2\lambda_3-\lambda_1\nu_2\mu_3)+(\nu_1\lambda_2\mu_3-\mu_1\lambda_2\nu_3)$$

is or is not 0 just according as K is or is not 0.

Here follow examples. Any one is equivalent to any other of the sets of equations

$$\left.\begin{aligned}2x-3y+5z&=8\\3x-5y+8z&=12\\5x-8y+12z&=18\end{aligned}\right\}$$

$$\left.\begin{aligned}\{(-5\times12+8^2)\times2+(-8\times5+3\times12)\times3+(-3\times8+5^2)\times5\}x&=4\times8-4\times12+18\\\{(8\times5-12\times3)(-3)+(12\times2-5^2)(-5)+(5\times3-8\times2)(-8)\}y&=4\times8-12-18\\\{(-3\times8+5^2)\times5+(-5\times3+2\times8)\times8+(-2\times5+3^2)\times12\}z&=8+12-18\end{aligned}\right\}$$

$$\left.\begin{aligned}x&=2\\y&=2\\z&=2\end{aligned}\right\}.$$

As to the equations $yc+zb=a$ $za+xc=b$ $xb+ya=c$ where a b c are each not 0 since the λ factors are $-a^2t$ $(ab)t$ $(ca)t$ and so handiest when taken $-a$ b c one equation of the equivalent set is

$$(za+xc-b)b+(xb+ya-c)c-(yc+zb-a)a=0 \text{ or } x\times2bc=b^2+c^2-a^2$$

and the others may be got either in the same way as this or from this at once by symmetry of relation so that the equivalent set of equations is

$$x=\frac{b^2+c^2-a^2}{2bc} \qquad y=\frac{c^2+a^2-b^2}{2ca} \qquad z=\frac{a^2+b^2-c^2}{2ab}.$$

The equations $x-2y-3z=1$ $4y-z=11$ $y+5z=8$ are equivalent to any one of the equation sets

$$\left.\begin{array}{r}3(x-2y-3z-1)+(4y-z-11)+2(y+5z-8)=0\\ 5(4y-z-11)+(y+5z-8)=0\\ -(4y-z-11)+4(y+5z-8)=0\end{array}\right\}\quad\left.\begin{array}{r}3x-30=0\\ 21y-63=0\\ 21z-21=0\end{array}\right\}\quad\left.\begin{array}{r}x=10\\ y=3\\ z=1\end{array}\right\}.$$

The equations $3x+5y-4z=11$ $2y+7z=39$ $z=5$ hold precisely when any one holds of the sets

$$\left.\begin{array}{r}z=5\\ y=\frac{1}{2}(39-7\times5)=2\\ x=\frac{1}{3}(11+4\times5-5\times2)=7\end{array}\right\}$$

$$\left.\begin{array}{r}2(3x+5y-4z-11)-5(2y+7z-39)+43(z-5)=0\\ 2y+7z-39-7(z-5)=0\\ z=5\end{array}\right\}\quad\left.\begin{array}{r}6x-42=0\\ 2y-4=0\\ z=5\end{array}\right\}.$$

Any one holds precisely when any other does of the sets

$$\left.\begin{array}{r}x\times5+y\times3+z\times2=23\\ x\times10-y\times7+z\times4=20\\ x\times3+y\times6-z\times8=59\end{array}\right\}$$

$$\left.\begin{array}{r}x(5\times16+10\times18+3\times13)=23\times16+20\times18+59\times13\\ y(3\times2+7)=23\times2-20\\ z(2\times81-4\times21+8\times65)=23\times81-20\times21-59\times65\end{array}\right\}\quad\left.\begin{array}{r}x=\frac{1495}{299}=5\\ y=\frac{26}{13}=2\\ z=-\frac{2392}{598}=-4\end{array}\right\}.$$

In the equations $x\times8+y\times3-z\times12=18$ $x\times6+y\times4-z\times9=17$ $x\times14-y\times5-z\times21=11$ there are only two unknowns $x\times2-z\times3$ and y and $\because$ $\{(x\times2-z\times3)\times4+y\times3-18\}\times43-\{(x\times2-z\times3)\times3+y\times4-17\}\times41$ $=\{(x\times2-z\times3)\times7-y\times5-11\}\times7$ the third of the equations is only a consequence of the other two moreover these two hold just when $y=\dfrac{17\times4-18\times3}{4^2-3^2}=2$ and $x\times2-z\times3=\dfrac{18\times4-17\times3}{4^2-3^2}=3$ x and z being otherwise quite arbitrary. But the third equation had it been $x\times14-y\times5-z\times21=$ anything else than 11 would have contradicted the first two equations and therefore the three equations could then have held for no values whatever of x y z. Again of the equations $4x+5y+6z=28$ $6x-4y+5z=15$ $10x-22y+3z=-11$

$$\because\ -2(4x+5y+6z-28)+3(6x-4y+5z-15)=10x-22y+3z+11$$

each follows from the other two and hence $x\ y\ z$ are simply three quantities so linked together that any two of them have the values fixed by any two of the equations whenever a fixed value is given at random to the third. Also two equations $4x+5y+6z=28$ $6x-4y+5z=15$ and a third equation $10x-22y+3z=$ aught else than -11 are clearly incompatible.

434. Making for the greater ease expressiveness and generalizability $1_1\ 1_2\ 1_3\ldots 2_1\ \ 2_2\ldots 3_1\ldots$ symbols of quantities let $1_1 2_2 3_3 - 3_1 2_2 1_3 + (2_1 3_2 1_3 - 1_1 3_2 2_3) + (3_1 1_2 2_3 - 2_1 1_2 3_3)$ be symbolized shortly by $\lceil 1_1 2_2 3_3 \rfloor$ then (art.432) if $\lceil 1_1 2_2 3_3 \rfloor$ be not o $V_1=0$ $V_2=0$ and $V_3=0$ precisely when $1_1 V_1+1_2 V_2+1_3 V_3=0$ $2_1 V_1+2_2 V_2+2_3 V_3=0$ and $3_1 V_1+3_2 V_2+3_3 V_3=0$. Also let $1_1 2_2-2_1 1_2$ be symbolized by $\lceil 1_1 2_2 \rfloor$ then (art.430) if $1_1 2_2\rfloor$ be not o $V_1=0$ and $V_2=0$ precisely when $1_1 V_1+1_2 V_2=0$ and $2_1 V_1+2_2 V_2=0$. It is too at once clear (art.413) that $V_1=0$ precisely when $1_1 V_1=0$ if 1_1 or to use the like symbol $\lceil 1_1 \rfloor$ be not o. Now if $V_1\ V_2\ V_3\ V_4$ be each o then whatever quantities $1_1\ 1_2\ldots 4_4$ may be

$$\left.\begin{array}{r} 1_1 V_1+1_2 V_2+1_3 V_3+1_4 V_4=0 \\ 2_1 V_1+2_2 V_2+2_3 V_3+2_4 V_4=0 \\ 3_1 V_1+3_2 V_2+3_3 V_3+3_4 V_4=0 \\ 4_1 V_1+4_2 V_2+4_3 V_3+4_4 V_4=0 \end{array}\right\}.$$

And on the other hand if this last set of four equations holds then it follows from the three last of the four that

$$\lceil 3_3 4_4 \rfloor (2_1 V_1+2_2 V_2+2_3 V_3+2_4 V_4)+\lceil 4_3 2_4 \rfloor (3_1 V_1+3_2 V_2+3_3 V_3+3_4 V_4)$$
$$+\lceil 2_3 3_4 \rfloor (4_1 V_1+4_2 V_2+4_3 V_3+4_4 V_4)=0$$

or $\lceil 2_1 3_3 4_4 \rfloor V_1+\lceil 2_2 3_3 4_4 \rfloor V_2=0$ and in like ways

$$\lceil 2_1 3_4 4_2 \rfloor V_1+\lceil 2_3 3_4 4_2 \rfloor V_3=0 \qquad \lceil 2_1 3_2 4_4 \rfloor V_1+\lceil 2_4 3_2 4_3 \rfloor V_4=0.$$

But by operational equivalency $\lceil 1_1 2_2 3_3 \rfloor=\lceil 2_2 1_1 3_3 \rfloor=\lceil 1_1 3_3 2_2 \rfloor=\lceil 3_3 2_2 1_1 \rfloor$

$$=\lceil 2_1 3_2 1_3 \rfloor=\lceil 3_1 1_2 2_3 \rfloor=-\lceil 2_1 1_2 3_3 \rfloor=-\lceil 1_1 3_2 2_3 \rfloor$$
$$=\lceil 1_2 2_3 3_1 \rfloor=\lceil 1_3 2_1 3_2 \rfloor=-\lceil 1_2 2_1 3_3 \rfloor=-\lceil 1_1 2_3 3_2 \rfloor.$$

$\therefore \lceil 2_2 3_3 4_4 \rfloor V_2=-\lceil 2_3 3_4 4_1 \rfloor V_1$ $\lceil 2_2 3_3 4_4 \rfloor V_3=\lceil 2_4 3_1 4_2 \rfloor V_1$ $\lceil 2_2 3_3 4_4 \rfloor V_4=-\lceil 2_1 3_2 4_3 \rfloor V_1$.

Wherefore and because from the first of the four equations

$$1_1\lceil 2_2 3_3 4_4 \rfloor V_1+1_2\lceil 2_2 3_3 4_4 \rfloor V_2+1_3\lceil 2_2 3_3 4_4 \rfloor V_3+1_4\lceil 2_2 3_3 4_4 \rfloor V_4=0$$
$$(1_1\lceil 2_2 3_3 4_4 \rfloor-1_2\lceil 2_3 3_4 4_1 \rfloor+1_3\lceil 2_4 3_1 4_2 \rfloor-1_4\lceil 2_1 3_2 4_3 \rfloor)\,V_1=0.$$

Therefore if $1_1\lceil 2_2 3_3 4_4 \rfloor-1_2\lceil 2_3 3_4 4_1 \rfloor+1_3\lceil 2_4 3_1 4_2 \rfloor-1_4\lceil 2_1 3_2 4_3 \rfloor$ or to symbolize this as in the former cases $\lceil 1_1 2_2 3_3 4_4 \rfloor$ be not o $V_1 = 0$. In like ways if $\lceil 1_1 2_2 3_3 4_4 \rfloor$ be not o $V_2 = 0$ $V_3 = 0$ and $V_4 = 0$. On the whole then if $\lceil 1_1 2_2 3_3 4_4 \rfloor$ be not o

$$\left.\begin{array}{l} V_1 = 0 \\ V_2 = 0 \\ V_3 = 0 \\ V_4 = 0 \end{array}\right\} \text{ precisely when } \left\{\begin{array}{l} 1_1 V_1 + 1_2 V_2 + 1_3 V_3 + 1_4 V_4 = 0 \\ 2_1 V_1 + 2_2 V_2 + 2_3 V_3 + 2_4 V_4 = 0 \\ 3_1 V_1 + 3_2 V_2 + 3_3 V_3 + 3_4 V_4 = 0 \\ 4_1 V_1 + 4_2 V_2 + 4_3 V_3 + 4_4 V_4 = 0. \end{array}\right.$$

By help of this theorem may be solved a set of four equations of the first degree with four unknowns in the same way as (art. 433) a set of 3 with 3 unknowns and (art. 431) a set of 2 with 2. A like theorem may then be got for any five equations and so on.

435. If $V = PV' + Q$ then if $V = 0$ and $V' = 0$ $Q = 0$ and if $V' = 0$ and $Q = 0$ $V = 0$ so that the pair of equations $V = 0$ $V' = 0$ is equivalent to the pair $V' = 0$ $Q = 0$.

Thus $xy = 10$ and $x + y = 7$ precisely when,—
$\because xy - 10 = (x + y - 7)y - (10 - 7y + y^2)$—, $10 - 7y + y^2 = 0$ and $x + y - 7 = 0$ and therefore precisely when either $y = 2$ and $x = 5$ or $y = 5$ and $x = 2$. Again the two equations $x^2 + y^2 = 58$ $x - y = 4$,—
$\because -58 + (x^2 + y^2) = 2(x^2 - 4x - 21) - (-4 + x + y)(-4 + x - y)$—, are equivalent to the two $x^2 - 4x - 21 = 0$ $y = -4 + x$ and therefore hold just when either x is 7 and y 3 or x is -3 and y -7. Also the equations $x + y = 1$ $x^5 + y^5 = 31$

$$\because x^5 + y^5 - 31 = (x^4 - x^3y + x^2y^2 - xy^3 + y^4)(x + y - 1) \\ + \{(x^2 + y^2)^2 - (x^2 + y^2)xy - (xy)^2 - 31\}$$

and $(x^2 + y^2)^2 - (x^2 + y^2)xy - (xy)^2 - 31 = \left(x^2 + y^2 - \frac{1}{2}xy\right)^2 - \frac{5}{4}(xy)^2 - 31$

$$= \left\{(x + y + 1)(x + y - 1) + \left(1 - \frac{5}{2}xy\right)\right\}^2 - \frac{5}{4}(xy)^2 - 31$$

are equivalent to $x + y = 1$ along with

$$(x^2 + y^2)^2 - (x^2 + y^2)xy - (xy)^2 - 31 = 0 \text{ or } \left(1 - \frac{5}{2}xy\right)^2 - \frac{5}{4}(xy)^2 - 31 = 0$$

or $5(xy)^2 - 5xy - 30 = 0$ or $xy = \left\{6 + \left(\frac{1}{2}\right)^2\right\}^{\frac{1}{2}} + \frac{1}{2}$ and therefore all the solutions are (1) $x = 2$ $y = -1$ (2) $x = -1$ $y = 2$ and (3) and (4) $x = \frac{1 + (-11)^{\frac{1}{2}}}{2}$ $y = \frac{1 - (-11)^{\frac{1}{2}}}{2}$.

436. If $U=P_m\dots P_3P_2P_1$ and $V=Q_n\dots Q_3Q_2Q_1$ then (art. 413) $U=0$ and $V=0$ precisely when one or other of the factors P_1 $P_2\dots P_m$ is equal to 0 and one or other of the factors Q_1 $Q_2\dots Q_n$ is equal to 0.

Thus $x^2+y^2=34$ and $xy=15$ precisely when (art. 430)

$$x^2+y^2-34+2(xy-15)=0 \text{ and } x^2+y^2-34-2(xy-15)=0$$

$$\text{or } (x+y-8)(x+y+8)=0 \text{ and } (x-y-2)(x-y+2)=0$$

and therefore precisely when one or other holds of the equation pairs

$$\left.\begin{matrix}x+y-8=0\\x-y-2=0\end{matrix}\right\} \quad \left.\begin{matrix}x+y-8=0\\x-y+2=0\end{matrix}\right\} \quad \left.\begin{matrix}x+y+8=0\\x-y-2=0\end{matrix}\right\} \quad \left.\begin{matrix}x+y+8=0\\x-y+2=0\end{matrix}\right\},$$

that is precisely when either $x=5$ and $y=3$ $x=3$ and $y=5$ $x=-3$ and $y=-5$ or $x=-5$ and $y=-3$. In the same way generally $x^2+y^2=a$ and $xy=b$ precisely when

$$x=\frac{1}{2}\{(a+2b)^{\frac{1}{2}}+(a-2b)^{\frac{1}{2}}\} \text{ and } y=\frac{1}{2}\{(a+2b)^{\frac{1}{2}}-(a-2b)^{\frac{1}{2}}\}.$$

The equations $2x^2+3xy-4y^2=23$ $4x^2-3xy-2y^2=25$ hold (art. 430) precisely when either of them holds along with any one of these following $23(4x^2-3xy-2y^2-25)-25(2x^2+3xy-4y^2-23)=0$ $7x^2-24xy+9y^2=0$ $(x-3y)(7x-3y)=0$ and therefore,—inasmuch as $2x^2+3xy-4y^2-23=(2x+9y)(x-3y)+23(y^2-1)$ and $7^2(2x^2+3xy-4y^2-23)=(14x+27y)(7x-3y)-(5y^2+7^2)\times 23$—, precisely when either $y=1$ and $x=3$ $y=-1$ and $x=-3$, or $y=\left(-\frac{1}{5}\right)^{\frac{1}{2}}\times 7$ and $x=3\left(-\frac{1}{5}\right)^{\frac{1}{2}}$. Generally $ax^2+by^2+2cxy=f$ and $a'x^2+b'y^2+2c'xy=f'$ if neither f nor f' be 0 precisely when $ax^2+by^2+2cxy=f$ and $f'(ax^2+by^2+2cxy-f)-f(a'x^2+b'y^2+2c'xy-f')$ or the operational equivalent $(f'a-fa')x^2+(f'b-fb')y^2+2(f'c-fc')xy=0$ and $Ax^2+By^2+2Cxy=0$ if neither A nor B be 0 precisely when $\{Ax+(C+K)y\}\{By+(C+K)x\}=0$ where K is such that $K^2=C^2-AB$.

If a polynome of the second degree in x y $ay^2+(bx+c)y+(dx^2+ex+f)$ be written shortly $p_0y^2+p_1y+p_2$ and another like polynome $q_0y^2+q_1y+q_2$ then unless $p_0q_2-p_2q_0$ be 0

$$\left.\begin{matrix}p_0y^2+p_1y+p_2=0\\q_0y^2+q_1y+q_2=0\end{matrix}\right\} \text{ precisely when } \left\{\begin{matrix}(p_0q_1-q_0p_1)y+(p_0q_2-q_0p_2)=0\\\{(q_2p_0-p_2q_0)y+(q_2p_1-p_2q_1)\}y=0\end{matrix}\right.$$

or for shortness $P_1y+P_2=0$ $(P_2y+P_3)y=0$ and therefore and because P_2 being not 0 y is not 0 and $P_2(P_1y+P_2)=P_1(P_2y+P_3)+$

$(P_2^2-P_1P_3)$ precisely when $P_2y+P_3=0$ and $P_2^2-P_1P_3=0$. Of the last two equations the latter is an equation of the fourth degree for finding x and the former gives the value of y answering to any value of x.

If none of the quantities a b c be 0 the following sets of equations are all equivalent

$$\left.\begin{matrix} yz=a \\ zx=b \\ xy=c \end{matrix}\right\} \quad \left.\begin{matrix} yz-a=0 \\ (zx-b)xy+b(xy-c)=0 \\ xy-c=0 \end{matrix}\right\} \quad \left.\begin{matrix} yz-a=0 \\ x^2(yz-a)+(x^2a-bc)=0 \\ xy-c=0 \end{matrix}\right\}$$

$$\left.\begin{matrix} x=\dfrac{(abc)^{\frac{1}{2}}}{a} \\ y=\dfrac{(abc)^{\frac{1}{2}}}{b} \\ z=\dfrac{(abc)^{\frac{1}{2}}}{c} \end{matrix}\right\}.$$

The equations $x(y+z)=a$ $y(z+x)=b$ $z(x+y)=c$ hold precisely when $\frac{1}{2}\{y(z+x)+z(x+y)-x(y+z)\}$ or the equivalent $yz=\frac{1}{2}(b+c-a)$ $zx=\frac{1}{2}(c+a-b)$ $xy=\frac{1}{2}(a+b-c)$ and their solution may therefore be got in the same way. Also how $xyz=a^{-1}(y+z)=b^{-1}(z+x)=c^{-1}(x+y)$ otherwise than by having x y z severally 0 may be found in the same way since it is just then that (art. 429)
$xyz=(b+c-a)^{-1}\times 2x=(c+a-b)^{-1}\times 2y=(a+b-c)^{-1}\times 2z$ or that
$yz=2(b+c-a)^{-1}$ $zx=2(c+a-b)^{-1}$ and $xy=2(a+b-c)^{-1}$.

The equations $yz=ax$ $zx=by$ $xy=cz$ hold when $x=0$ $y=0$ $z=0$ and otherwise precisely when any one holds of the sets

$$\left.\begin{matrix} (zx-by)x+b(xy-cz)=0 \\ (yz-ax)x-(zx-by)y=0 \\ zx-by=0 \end{matrix}\right\} \quad \left.\begin{matrix} z(x^2-bc)=0 \\ by^2=ax^2 \\ zx=by \end{matrix}\right\} \quad \left.\begin{matrix} x=b^{\frac{1}{2}}c^{\frac{1}{2}} \\ y=c^{\frac{1}{2}}a^{\frac{1}{2}} \\ z=a^{\frac{1}{2}}b^{\frac{1}{2}} \end{matrix}\right\}.$$

The equations $x+y+z=a$ $yz+zx+xy=b$ $xyz=c$ hold (art. 416) just when x y z are either the roots if there be 3 of the equation in t $t^3-t^2a+tb-c=0$ two of them the roots if there be 2 of this equation and the third either of those 2 or all three the root if there be 1 only and so the ways of fulfilling the equations are 6 or 1 according as the equation in t has more than 1 root or 1. By reason of the operational equivalences

$$y+z+(z+x)+(x+y)=2(x+y+z)$$

$$(z+x)(x+y)+(x+y)(y+z)+(y+z)(z+x)=\frac{1}{2}\{3(x+y+z)^2-(x^2+y^2+z^2)\}$$

$$(y+z)(z+x)(x+y)=\frac{1}{3}\{(x+y+z)^3-(x^3+y^3+z^3)\}$$

the set of equations $x+y+z=a$ $x^2+y^2+z^2=b$ $x^3+y^3+z^3=c$ is equivalent to the set

$$\left.\begin{aligned} y+z+(z+x)+(x+y) &= 2a \\ (z+x)(x+y)+(x+y)(y+z)+(y+z)(z+x) &= \frac{1}{2}(3a^2-b) \\ (y+z)(z+x)(x+y) &= \frac{1}{3}(a^3-c) \end{aligned}\right\}$$

and therefore is fulfilled just when x y z have such values as make $y+z$ $z+x$ $x+y$ the roots or root of the equation

$$t^3-t^2\times 2a+\frac{t}{2}(3a^2-b)-\frac{1}{3}(a^3-c)=0.$$

437. Since $y^3z^3=-q^3$ and $y^3+z^3=r$ precisely when

$$y=\left\{\frac{1}{2}r+\left(\frac{1}{4}r^2+q^3\right)^{\frac{1}{2}}\right\}^{\frac{1}{3}} \text{ and } z=\left\{\frac{1}{2}r-\left(\frac{1}{4}r^2+q^3\right)^{\frac{1}{2}}\right\}^{\frac{1}{3}}$$

and further $y^3z^3+q^3=(yz+q)(y^2z^2-qyz+q^2)$ out of the 18 ways of fulfilling the equations $y^3z^3=-q^3$ $y^3+z^3=r$ if $\frac{1}{4}r^2+q^3$ be not 0 there are 6 that are the ways of fulfilling the equations $yz=-q$ $y^3+z^3=r$ and if $\frac{1}{4}r^2+q^3$ be 0 out of the 9 ways of fulfilling the former equations there are 3 that are the ways of fulfilling the latter.

Now $x^3-3(yz)x+(y^3+z^3)$

$$=\{x+(y+z)\}\left\{x-\frac{1}{2}(y+z)-\frac{(-3)^{\frac{1}{2}}}{2}(y-z)\right\}\left\{x-\frac{1}{2}(y+z)+\frac{(-3)^{\frac{1}{2}}}{2}(y-z)\right\}$$

and $x^3+3qx+r$ is operationally equivalent to $x^3-3(yz)x+(y^3+z^3)$ precisely when $yz=-q$ and $y^3+z^3=r$.

Hence if -1 τ τ^2 be the three values of $(-)^{\frac{1}{3}}1$ and g be any value of $\left\{\frac{1}{2}r+\left(\frac{1}{4}r^2+q^3\right)^{\frac{1}{2}}\right\}^{\frac{1}{3}}$ it follows that $x^3+3qx+r=0$ precisely when x has one or other of the values $-(g-g^{-1}q)$ $\tau g-\tau^2g^{-1}q$ $\tau^2g-\tau g^{-1}q$.

438. Uses may be made of equation solving such as follow.

(1). When and where are two men together that walk along a road AOB at the rates of a and b miles an hour in the direction AB if when the one is at a certain place O the other be c miles from O toward B?

At x hours after being at O the one is xa miles from O toward B and the other is then $c+xb$ miles from O toward B so that the men are together just when x is such that $xa=c+xb$ and therefore just when $x(a-b)=c$ or if $a-b$ be not 0 $x=\frac{c}{a-b}$. If $a-b$ be a plus quantity else than 0 a b being plus or minus quantities the meeting happens after the first man's being at O or before according as c is a plus or a minus quantity and since with the value of x got $xa=\frac{c}{a-b}a=c+\frac{c}{a-b}b$ the meeting happens toward B or toward A according as a c have like or unlike algebraic signs. If a b being plus or minus quantities $a-b$ be a minus quantity else than 0 the meeting happens after or before the first's O passage according as c is a minus quantity or a plus and to the B side of O or the A side according as a c have unlike or like signs. If $a-b$ be 0 the meeting is either always and everywhere to wit if c be 0 or never to wit if c be not 0. If each or any of the quantities a b c be ditensive the question has no meaning unless the road AOB be straight and then if a b c be severally $\alpha+j\alpha'$ $\beta+j\beta'$ $\gamma+j\gamma'$ where α α' β β' γ γ' are protensive quantities and j is $(-)^{\frac{1}{2}}$ since x hours is either an instant of time or some time after or before an instant (art. 375) $x\{\alpha+j\alpha'-(\beta+j\beta')\}$ or $x(\alpha-\beta)+jx(\alpha'-\beta')=\gamma+j\gamma'$ just when $x(\alpha-\beta)=\gamma$ and $x(\alpha'-\beta')=\gamma'$. Hence if $\alpha'-\beta'$ be 0 γ' must be 0 for a meeting to happen and then the men are always in a straight line either the same as or parallel to AB and therefore meet just when the straight lines drawn through them perpendicularly to AB come together as one. If $\alpha-\beta$ be 0 there can be a meeting only when γ is 0 and then the men are always in a straight line cutting perpendicularly AB and so come together just when they are at the same point in this straight line. If neither $\alpha-\beta$ nor $\alpha'-\beta'$ be 0 the men come together just if and just when the straight lines through them drawn perpendicularly to AB become one and the straight lines through them drawn in or parallel to AB become one at the very same time which is just if $\frac{\gamma}{\alpha-\beta}=\frac{\gamma'}{\alpha'-\beta'}$ and just when $x=$ either. In short the walkers can meet only by taking roads that meet and can then meet only where the roads meet.

(2). How much did a thing cost on which by selling it for a pounds there is lost just as much per cent.?

Since x pounds is expressed numerically or algebraically by $\frac{x}{100}$ in reference to 100 pounds as unit x per cent. on x pounds amounts to $\frac{x}{100}x$ pounds and therefore by giving x pounds and taking a pounds there is a loss of exactly x per cent. precisely when

$$x - \frac{x}{100}x = a.$$

This then happens just when singly

$$-a + \left\{x - \left(\frac{x}{10}\right)^2\right\} = 0 \quad -a + 5^2 - \left(-\frac{x}{10} + 5\right)^2 = 0 \quad x = \{5 + (-a+25)^{\frac{1}{2}}\} \times 10.$$

Thus if the thing were sold for 25*l.* it cost 50*l.* and this is the only case when the question has only one answer and is besides the case when the selling price is the greatest that can be for the question's conditions to be anyhow fulfilled unless indeed what is expressed by a ditensive quantity in reference to a pound as unit could be taken to mean anything. If the selling price be 24*l.* the cost price is either 60*l.* or 40*l.* If a be 0 the cost is either 100*l.* or 0. If the thing be sold for -11 pounds the meaning is that 11 pounds is paid to get rid of it and then the cost is either 110*l.* or -10*l.* of which the latter is to be understood as 10*l.* got with, instead of given for, the thing.

(3). How many books are bought for 80 shillings if by buying 4 more for the same 1 shilling a piece less would be paid?

If there be x books the price of each is $\frac{1}{x}\times 80$ shillings and therefore x is just to be such that

$$\frac{1}{x}\times 80 - 1 = \frac{1}{x+4}\times 80.$$

This holds precisely when severally

$$1 + \left(-\frac{1}{x} + \frac{1}{x+4}\right)\times 80 = 0 \quad x^2 + 4x - 4\times 80 = 0 \quad x = 2\{(80+1)^{\frac{1}{2}} - 1\}.$$

The only answer to the question then is 16. Although -20 books bought may mean 20 books sold this other root of the equation in

x gives no meaning in the rest of the question. It might very well have been that no number whatever of books met the conditions of the question. Generally all the answers to a question of this kind are precisely all such quantities that fulfil a certain equation as lie within the field of the question's view.

(4). At what prices the bushel are wheat and barley when 30 bushels of wheat and 26 of barley are together worth 10*l.* and 27 of wheat and 39 of barley together 10*l.* 19*s.*?

If a bushel of wheat be worth x shillings and a bushel of barley y shillings the question fixes x and y to be such that $30x+26y=10\times20$ and $27x+39y=10\times20+19$ and these equations hold just when severally

$$\left.\begin{array}{r}15x+13y=100\\ 9x+13y=73\end{array}\right\} \qquad \left.\begin{array}{r}15x+13y-100-(9x+13y-73=0\\ 5(9x+13y-73)-3(15x+13y-100)=0\end{array}\right\}$$

$$\left.\begin{array}{r}x=\frac{1}{6}\times27=4\frac{1}{2}\\ y=\frac{1}{2\times13}\times65=2\frac{1}{2}\end{array}\right\}.$$

(5). How many yards long is each of two pieces of cloth the one worth 17*s.* and the other 22*s.* the yard if half the first be worth 4*l.* 18*s.* more than one-third the other and the two together be worth 70*l.* 6*s.*?

Calling x yards the length of the piece at 17*s.* and y yards the length of the piece at 22*s.* x y are just such that

$$x\times17+y\times22=70\times20+6 \text{ and } -\left(\frac{1}{3}y\right)\times22+\left(\frac{1}{2}x\right)\times17=4\times20+18$$

and therefore just such that severally

$$\left.\begin{array}{r}x\times17+y\times22=1406\\ -y\times\frac{22}{3}+x\times\frac{17}{2}=98\end{array}\right\} \qquad \left.\begin{array}{r}x(17+\frac{17}{2}\times3)=1406+98\times3\\ y(22+\frac{22}{3}\times2)=1406-98\times2\end{array}\right\}$$

$$\left.\begin{array}{r}x=\dfrac{1700}{17\times\frac{5}{2}}=40\\ y=\dfrac{1210}{22\times\frac{5}{3}}=33\end{array}\right\}.$$

(6). How many dozens of wine in two several bins are there which at 56*s.* and 52*s.* the dozen would bring 73*l.* 8*s.* and at 50*s.* and 54*s.* the dozen 69*l.* 14*s.*?

If there be x dozen in the first bin and y dozen in the other x y are only known to fulfil the relations $x\times56+y\times52=73\times20+8$ $x\times50+y\times54=69\times20+14$ which hold precisely when severally

$$\left.\begin{array}{r} x\times14+y\times13-367=0 \\ (x\times14+y\times13-367)\times2-(x\times25+y\times27-697)=0 \end{array}\right\} \quad \left.\begin{array}{r} x\times14+y\times13=367 \\ x\times3-y=37 \end{array}\right\}$$

$$\left.\begin{array}{l} x=\dfrac{367+37\times13}{14+3\times13}=16 \\ y=\dfrac{367\times3-37\times14}{13\times3+14}=11 \end{array}\right\}.$$

(7). What simple fraction is it which were the numerator 1 greater would be equal to 4 and were the denominator 1 greater 3?

The numerator x and the denominator y are just whatever whole numbers fulfil the several equation pairs

$$\left.\begin{array}{l} \dfrac{x+1}{y}=4 \\ \dfrac{x}{y+1}=3 \end{array}\right\} \quad \left.\begin{array}{r} x+1=4y \\ x=3(y+1) \end{array}\right\} \quad \left.\begin{array}{r} (4-3)x=4\times3+3 \\ (4-3)y=1+3 \end{array}\right\} \quad \left.\begin{array}{r} x=15 \\ y=4 \end{array}\right\}.$$

(8). What must two lengths be so that the one may measure off into 1 more pieces than the other a length of 12 chains and that of other two lengths longer severally by 1 chain the one may measure off into 1 more pieces than the other a length of 20 chains?

Calling the several lengths x chains and y chains x y have just to fulfil the equations $\frac{12}{x}=\frac{12}{y}+1$ $\frac{20}{x+1}=\frac{20}{y+1}+1$ of which the first is equivalent to $-y^{-1}+x^{-1}=\frac{1}{12}$ and the other to

$$-(1+x)+(1+y)=\frac{1}{20}(x+1)(y+1) \quad -y^{-1}+x^{-1}=\frac{1}{20}(1+x^{-1})(1+y^{-1})$$

$$\text{or } -y^{-1}+x^{-1}-\frac{1}{12}+\frac{1}{12}=\frac{1}{20}(1+x^{-1})\left\{1+x^{-1}-\frac{1}{12}-\left(-y^{-1}+x^{-1}-\frac{1}{12}\right)\right\}.$$

These equations therefore hold just when $(1+x^{-1})\left(1+x^{-1}-\frac{1}{12}\right)=\frac{5}{3}$ and $y^{-1}=x^{-1}-\frac{1}{12}$ therefore just when $x^{-1}=-1+\left[\left\{\frac{5}{3}+\left(\frac{1}{24}\right)^{2}\right\}^{\frac{1}{2}}+\frac{1}{24}\right]$ or $-1+\frac{\pm31+1}{24}$ and $y^{-1}=-1+\frac{\pm31-1}{24}$ and therefore just when either

$x=3$ and $y=4$ or $x=-\frac{4}{9}$ and $y=-\frac{3}{7}$. Hence the lengths sought can only be 3 chains and 4.

(9). How must a straight line be cut atwo so that the whole may have to the part close to one end the same ratio as this part to the other part?

Let AB be a straight line to be so cut atwo that AB is to have to the part close to the end A the same ratio as this part to the other part. If in reference to AB as unit the part sought close to A be expressed numerically by x the other part is expressed numerically by $1-x$ and hence x has only to be such that $\frac{1}{x}=\frac{x}{1-x}$. This equation holds precisely when severally $x^2=1-x$ $x^2+x-1=0$

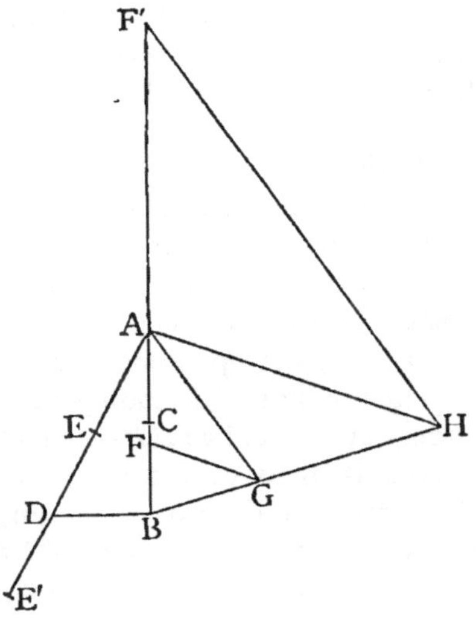

$x=\left\{1+\left(\frac{1}{2}\right)^2\right\}^{\frac{1}{2}}-\frac{1}{2}=\frac{5^{\frac{1}{2}}-1}{2}$. The only value of x then which meets the conditions of the question is $\frac{\sqrt{5}-1}{2}$ or $\sqrt{\left\{1+\left(\frac{1}{2}\right)^2\right\}}-\frac{1}{2}$ and this at once (art. 382) gives the construction:—Halve AB in C from the point B in BA draw on either side a straight line BD at right angles to BA from BD cut off close to the end B BD equal to AC or CB because D is not at B the only point common to the straight lines BA BD D is not in AB join AD then AB BD DA being three straight lines joining two and two three points A B D not in one straight line ABD is a triangle of this triangle the angle ABD being a right angle the angle BAD is less than a right angle and therefore less than the angle ABD and therefore the side AD is greater than the side BD from AD cut off close to D a part DE equal to BD because the two sides AB BD of the triangle ABD are together greater than the third side AD and BD a part of the first is equal to ED a part of the other the remainder AB is greater than the remainder AE from AB then cut off close to the end A a part

AF equal to *AE* and *AB* is cut into the parts *AF FB* so that $AB : AF = AF : FB$.

Since $-\sqrt{\left\{1+\left(\frac{1}{2}\right)^2\right\}}-\frac{1}{2}$ the other root of the equation is a minus quantity which makes equal the absolute values of $\frac{1}{x}$ $\frac{x}{1-x}$ and besides if x be the algebraic expression of the distance from *A* in the direction *AB* of a point in the endless straight line in which *AB* lies $1-x$ is the algebraic expression of the distance from *B* in the direction *BA* of the same point it follows that the minus root answers the question How must a straight line be produced beyond one end so that the line may have to the part produced the same ratio as the part produced to the whole produced line? and hence producing endlessly *AD* to *E'* from *DE'* cutting off close to *D DE'* equal to *BD* producing *BA* endlessly to *F'* and from *AF'* cutting off close to *A AF'* equal to *AE'* the straight line *AB* is produced beyond *A* to a point *F'* so that $AB : AF' = AF' : F'B$.

Both roots answer the question Where must a point be taken in an endless straight line in which an ended straight line is so that the ended straight line may have to the distance of the point from one end the same ratio as this distance to the distance of the point from the other end?

If upon, and upon one side of, *AB* a triangle *ABG* be described having the side *AG* ending at *A* equal to *AB* and the side *BG* ending at *B* equal to *AF*,— which may be done because *AB* being greater than its own part *AF AB* an equal to *AB* and *AF* are three straight lines of which any two are together greater than the third—, the angles *ABG AGB* at the base *BG* of this isosceles triangle are equal and if *FG* be joined *GBF* is a triangle because *G B F* are three points not in one straight line moreover $AB : BG = GB : BF$ and so the triangles *ABG GBF* have the angle *ABG* the same as the angle *GBF* and the sides about this common angle proportionate therefore these triangles are similar with the angle *AGB* equal to the angle *GFB* and the angle *BAG* equal to the angle *BGF*. But the angle *ABG* is equal to the angle *AGB* therefore the angle *ABG* is equal to the angle *GFB* and therefore *FG BG* the sides of the triangle *GBF* over against these equal angles are equal. Hence *AF FG GB* are all equal and the angles *ABG AGB GFB* are all equal. Further because the points *F A G* are not in one straight line *FAG* is a triangle and because *FA* is equal to FG the triangle FAG is isosceles the angles FAG FGA at the base therefore are equal and therefore are together double of the angle FAG. But because the side AF of the triangle AFG is produced to B the exterior angle GFB is equal to the offlying interior angles FAG FGA together. Therefore the angle GFB is double the angle BAG and therefore each of the angles ABG AGB GFB is double each of the angles BAG FGB. Since then ABG is an isosceles triangle having each of the angles ABG AGB at the base twice the third angle BAG all the angles of this triangle are together five times the angle BAG and hence as the angles of any triangle are altogether equal to a hemiperigon the angle BAG is $\frac{1}{5}$ a hemiperigon and therefore $\frac{1}{5}\times\frac{1}{2}$ or $\frac{1}{10}$ a perigon. Half the angle BAG is therefore $\frac{1}{2}\times\frac{1}{10}$ or $\frac{1}{20}$ perigon and since in reference to AB as unit BG is expressed by $\frac{\sqrt{5}-1}{2}$ the half of BG is in reference to AB expressed numerically by $\frac{1}{2}\times\frac{\sqrt{5}-1}{2}$ or the equal $\frac{\sqrt{5}-1}{4}$. Hence and because the straight line which bisects the vertical angle of an isosceles triangle bisects perpendicularly the base it comes out anew that $\sin\frac{1}{20}=\frac{\sqrt{5}-1}{4}$ if a perigon be unit angle.

Upon, and upon one side of, AB too a triangle ABH may be described having AH BH each equal to AF' and if $F'H$ be joined $AF'H$ $F'BH$ are triangles because the angle ABH is the angle HBF' and $AB : BH = HB : BF'$ the triangles ABH HBF' are similar $F'BH$ is an isosceles triangle each of the isosceles triangles HAB $F'BH$ has each of the base angles twice the vertical angle and each of the angles AHB $BF'H$ is $\frac{1}{10}$ perigon. If the triangles ABG ABH be upon the same side of AB the equal angles ABG ABH at the common point B having the common bounding line BA ending at B and lying in the same plane and beginning together upon the same side of BA exactly fit upon one another and therefore BG BH are in one straight line.

(10). How is a given straight line to be cut atwo for the parts to have a given mean proportionate?

If in reference to the unit straight line the straight line to be cut be expressed numerically by $2a$ the mean proportionate by b and one of the parts by $a+x$ the other part is expressed numerically by $2a-(a+x)$ and therefore by the operationally equivalent $a-x$ and $a+x$ $a-x$ have just to be such that $\frac{a+x}{b}=\frac{b}{a-x}$. This equation holds just when singly $(a+x)(a-x)=b^2$ $a^2-x^2=b^2$ $x^2=-b^2+a^2$ $x=(a^2-b^2)^{\frac{1}{2}}$. Hence the question's conditions can be met at all only when the given mean proportionate is not greater than half the given straight line to be cut the two segments of the given straight line have the greatest mean proportionate that they can have precisely when they are equal and any two segments into which a given straight line can be cut have the same mean proportionate as other two precisely when the points of section are equidistant from the middle point or what comes to the same when that one of the one two segments which is close to either end of the given straight line is equal to that one of the other two segments which is close to the other end. Hence too the construction:— Let AB be the given straight line. Bisect AB in C. If AC or CB be equal to the given mean proportionate C is the only point in which AB is cut in the way wanted. But if AC or CB be not equal to the given mean proportionate each of them is greater than it. Then from C in AB draw

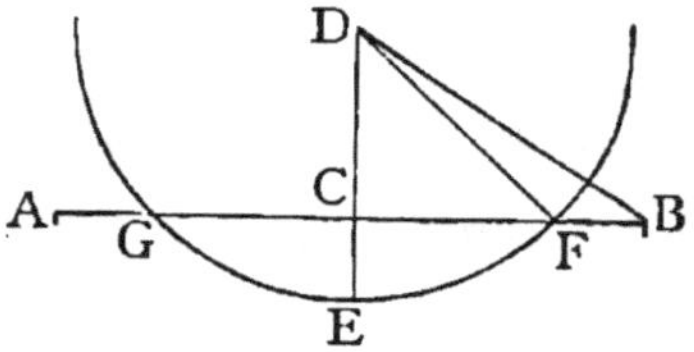

on either side CD at right angles to AB. From CD cut off close to C CD equal to the given mean proportionate. Produce DC endlessly to E and from DE cut off close to D DE equal to AC or CB. From D as center at the distance DE describe a circle FEG. Join DB. Because D is not in AB DCB is a triangle and in this triangle because the angle DCB is a right angle the angle CDB is less than a right angle and therefore less than the angle DCB therefore the side DB is greater than the side CB and therefore than DE. Since then B is at a greater distance from D the center of the circle FEG than E a point in the circumference and therefore than any point in the circumference B is without the circle. And C is within the circle being at a less distance from the center D than E or than any point in the circumference. The line CB therefore ending at the two points C B on opposite sides of the endless line the circumference of FEG is cut somewhere between C and B by the circumference. Let F be the point between C and

B where CB is so cut. Join DF then since if any of the equals AC CB DE DF be expressed numerically by a and CD by b in reference to a common unit line CF is expressed numerically in reference to the same by $\sqrt{(a^2-b^2)}$ AB is cut atwo in F so that $AF : CD = CD : FB$. If G be the point where likewise AC is cut between A and C by the circumference of FEG G cuts AB into the parts AG GB so that $AG : CD = CD : GB$. And AB can be cut nowhere else than at F and at G so that the parts may have a mean proportionate equal to CD.

If a b being still numerical quantities $b > a$ from a straight line Ox stretching away endlessly from O taken as a primary direction cut off close to O a part OA and from Ax close to A a part AQ each expressed numerically by a from the point A in Ox draw on either side a straight line AP at right angles to Ox and produce PA endlessly to P' from Ox cut off close to O a part OB expressed numerically by b and from O as center at the distance OB describe a circle. Since $a < b$ A is less distant from O than B therefore A is within the circle and the endless straight line PAP' passing through this point within the circle cuts the circumference twice once on each side of A. Let P be the point on one side of A and P' the point on the other where PAP' so cuts the circumference. Join OP PQ and OP' $P'Q$. The straight line PAP' because it meets at A and does not lie wholly together with the straight line OAx meets at A only and there cuts OAx therefore the points P P' in PAP' on opposite sides of A are on opposite sides of OAx and therefore OAP QAP are triangles upon one side of OAx and OAP' QAP' are triangles upon the other side. In the triangles OAP QAP the two sides OA AP are equal severally to the two sides QA AP and the angle OAP is equal to the angle QAP therefore the base OP is equal to the base QP. In the triangles OAP' QAP' likewise OP' is equal to QP'. Hence OP QP OP' QP' are all equal and hence the angles POQ PQO at the base of the isosceles triangle POQ are equal and also the angles $P'OQ$ $P'QO$ are equal at the base of the isosceles triangle $P'OQ$. From P draw $PB_{,}$ and from P' $P'B'$ each condirectionate and equal to OB and join BP and $QB_{,}$ BP' and QB'. Because in

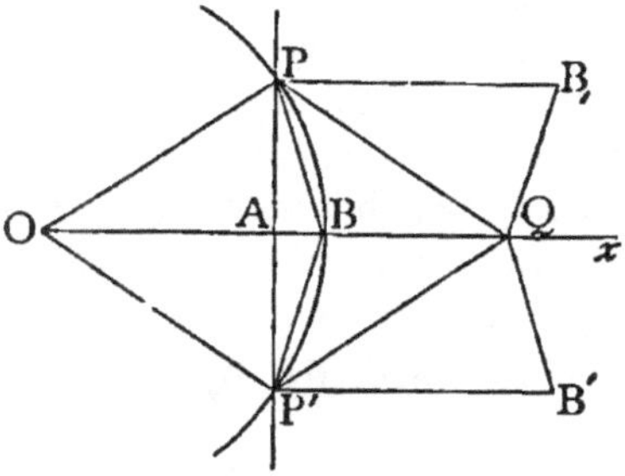

QP there is a point Q in and a point P not in OQx the straight lines QP OQx meet and do not lie wholly together therefore they meet at Q only and there cut one another and therefore the portions Qx QO of OQx upon opposite sides of Q lie on opposite sides of QP. But $PB_{,}$ being condirectionate to OB is (arts. 374, 372) on the same side of QP as Qx. Therefore $PB_{,}$ QO are on opposite sides of QP therefore the angles $QPB_{,}$ PQO that PQ makes with the parallels $PB_{,}$ Ox are alternate and therefore these angles are equal. The angles BOP $QPB_{,}$ being then each equal to the same angle PQO are equal to one another. Hence a straight line to get from the direction OB into the direction OP has to pass through the same amount of angle the same way round as to get from the direction PQ into the direction $PB_{,}$. In like manner a straight line to get from the direction OB to the direction OP' must pass through the same angular amount the same way round as to get from the direction $P'Q$ to the direction $P'B'$. The triangles OBP $PQB_{,}$ further are isosceles and equal vertical angled and therefore are similar and so likewise are the triangles OBP' $P'QB'$. On the whole then this is the way in which if in reference to the primarily directed unit distance the directed distance OQ be expressed algebraically by $2a$ the directed distances OP OB $PB_{,}$ PQ expressed severally by $a+(-)^{\frac{1}{2}}\surd(b^2-a^2)$ b b $-(-)^{\frac{1}{2}}\surd(b^2-a^2)+a$ are such that

$$a+(-)^{\frac{1}{2}}\surd(b^2-a^2)+\{-(-)^{\frac{1}{2}}\surd(b^2-a^2)+a\}=2a$$

$$\frac{a+(-)^{\frac{1}{2}}\surd(b^2-a^2)}{b}=\frac{b}{-(-)^{\frac{1}{2}}\surd(b^2-a^2)+a}$$

and the directed distances OP' OB $P'B'$ $P'Q$ expressed severally by $a-(-)^{\frac{1}{2}}\surd(b^2-a^2)$ b b $(-)^{\frac{1}{2}}\surd(b^2-a^2)+a$ such that

$$a-(-)^{\frac{1}{2}}\surd(b^2-a^2)+\{(-)^{\frac{1}{2}}\surd(b^2-a^2)+a\}=2a$$

$$\frac{a-(-)^{\frac{1}{2}}\surd(b^2-a^2)}{b}=\frac{b}{(-)^{\frac{1}{2}}\surd(b^2-a^2)+a}.$$

Another question somewhat like the above is How must a given straight line be produced so that the whole line produced and the produced part may have a given mean proportionate? Taking $2a$ b and $a+x$ severally as the numerical expressions in reference to the unit line of the given line the given mean proportionate and the whole line stretching from one end and produced beyond the other the numerical expression of the produced part is

$-2a+(a+x)$ or the equivalent $-a+x$ and therefore x has to be such that $\frac{a+x}{b}=\frac{b}{-a+x}$. This equation in x is fulfilled precisely when severally $(a+x)(-a+x)=b^2$ $-a^2+x^2=b^2$ $x=(a^2+b^2)^{\frac{1}{2}}$. The only value of x which straightway fits the case put is $\sqrt{(a^2+b^2)}$. But inasmuch as the other value $-\sqrt{(a^2+b^2)}$ makes $\frac{a+x}{b}$ $\frac{b}{-a+x}$ although both minus yet of equal absolute value $a+x$ $-a+x$ have only to be taken as the algebraic expressions of the distances from the ends in one direction of a point in the straight line in which the given straight line lies for both values to answer the question. Hence if AB be the given straight line produce AB endlessly to E beyond B and F beyond A from B in the straight line $FABE$ draw on either side BC at right angles to FE from BC cut off close to B a part BC equal to the given mean proportionate bisect AB in D join DC and from D as center at the distance DC describe a circle. The endless straight line FDE passing through D the center of the circle cuts the circumference on each side of D let E be the point where FDE cuts the circumference on the same side of D as B and F the point where FDE cuts the circumference on the same side of D as A. In the triangle DBC the side DC over against the right angle DBC is greater than the side DB over against the acute angle DCB moreover DE DF are each equal to DC being all straight lines drawn from the center to the circumference of a circle and AD is equal to DB therefore DE is greater than DB and DF than DA. Since DB DE are two unequal straight lines that have a common end D and lie in the same straight line upon the same side of D the less DB is a part of the greater DE or in other words E is in DB produced and in the same way F is in DA produced. Then E is the only point in AB produced and F the only point in BA produced such that $AE : BC = BC : BE$ and $AF : BC = BC : BF$.

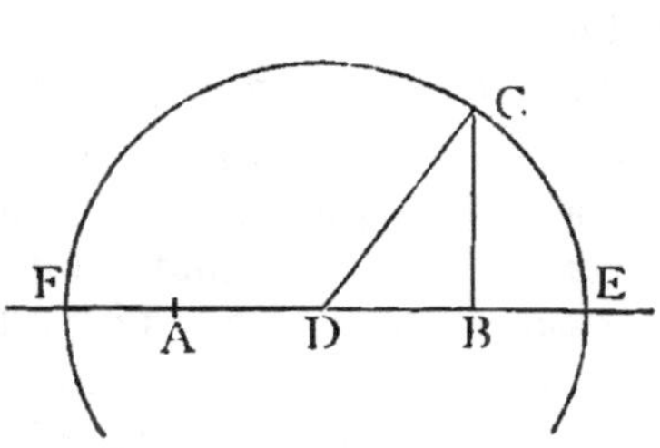

(11). What must the sides be bounding the right angle of a right angled triangle so that they may together be equal to a given straight line and the perpendicular drawn from their common end to the third side may be equal to a given straight line?

If $x\ y$ express numerically the sides bounding the right angle q the straight line to which those sides are to be together equal and p the straight line to which the perpendicular is to be equal all in reference to a common unit line $x+y=q$ and from the similarity of the parts to the whole of a right angled triangle cut atwo by the perpendicular drawn from the right angle to the side opposite $\frac{x}{\sqrt{(x^2+y^2)}}=\frac{p}{y}$. The latter of these equations is equivalent to the equation $xy=p\sqrt{(x^2+y^2)}$ and since $xy=p(x^2+y^2)^{\frac{1}{2}}$ precisely when singly $(xy)^2-p^2(x^2+y^2)=0$ $(xy)^2-p^2\{(x+y-q)(x+y+q)+q^2-2xy\}=0$ it hence follows that the values of $x\ y$ sought are precisely such of those values of $x\ y$ that fulfil the equations $x+y=q$ $(xy)^2+2p^2xy-p^2q^2=0$ as fulfil the equations $x+y=q$ $xy=p\sqrt{(x^2+y^2)}$. But $(xy)^2+2p^2xy-p^2q^2=0$ precisely when $xy=p\{(q^2+p^2)^{\frac{1}{2}}-p\}$ and therefore by the foregoing question (10) the sides bounding the right angle can only be equal severally to the two parts into which the straight line expressed by q is cut when these parts have the same mean proportionate as the straight lines severally expressed by p $\sqrt{(q^2+p^2)}-p$. There is then a triangle fulfilling the required conditions just when $p\{\sqrt{(q^2+p^2)}-p\}$ not$>\left(\frac{1}{2}q\right)^2$ or $2p^2$ not$>\left(\frac{1}{2}q\right)^2$, that is just when the straight line to which the perpendicular is to be equal is not greater than the perpendicular drawn from the vertex to the base of a right angled isosceles triangle having each of the equal sides expressed numerically by $\frac{1}{2}q$.

(12). If the sides BC CA AB of a triangle ABC be severally expressed numerically in reference to the unit line by a b c and $B_{,}BCC'$ be the endless straight line wherein BC lies what is the numerical expression of the perpendicular drawn from A to $B_{,}BCC'$ and where in $B_{,}BCC'$ is that perpendicular's foot?

Calling AF the perpendicular from A to $B_{,}BCC'$ let p express numerically AF and x algebraically the distance from B in the direction BC' of F each in reference to the unit line. Then $a-x$ expresses algebraically the distance from C in the direction $CB_{,}$ of F. Since either BFA is a right angled triangle or AF is AB and x^2 is a plus quantity whether x be a plus quantity or a minus $x^2+p^2=c^2$. Also $(a-x)^2+p^2=b^2$. Hence

$$p^2=-x^2+c^2=-(a-x)^2+b^2 \quad x=\frac{1}{2a}(c^2+a^2-b^2) \quad a-x=\frac{1}{2a}(a^2+b^2-c^2)$$

$$p=\frac{1}{2a}\surd\{-(c^2+a^2-b^2)^2+(2a)^2c^2\}=\frac{1}{2a}\surd(2b^2c^2+2c^2a^2+2a^2b^2-a^4-b^4-c^4)$$

$$=\frac{1}{2a}\surd(a+b+c)(b+c-a)(c+a-b)(a+b-c).$$

If A B C be the numerical expressions in reference to some unit angle of the angles of the triangle at the points A B C severally the numerical expression of FA is $(\sin B)c$ the algebraic expression of the distance from B toward C' of F is $(\cos B)c$ and the algebraic expression of the distance from F toward C' of C is $-(\cos B)c+a$. Hence either from the right angled triangle AFC if F be not at C or because FA is CA if F be at C

$$b^2=\{(\sin B)c\}^2+\{-(\cos B)c+a\}^2=c^2+a^2-2(\cos B)ca$$

whence the same expressions as before may be got as equals of $(\cos B)c$ or x and $(\sin B)c$ or p severally.

Since $(\sin C)b$ also expresses numerically FA $(\sin C)b=(\sin B)c$ wherefore and because any other side line of the triangle ABC may be dealt with in the same way as $B{,}BCC'$

$$\frac{1}{\sin A}a=\frac{1}{\sin B}b=\frac{1}{\sin C}c.$$

In these two equations and a third equation,—which if a perigon be the unit angle is

$$A+B+C=\frac{1}{2}-,$$

stating that the angles of a triangle are altogether equal to a hemiperigon are wrapped up all the numerical ties among a triangle's sides and angles. Each of the equals $(\sin A)^{-1}a$ $(\sin B)^{-1}b$ $(\sin C)^{-1}c$ expresses numerically the diameter of the circle circumscribed about the triangle ABC for BD being the diameter ending at B if BD be BC the angle BAC is right but if not and CD be joined the angle BCD is right and the angles BAC BDC being either in the same segment of the circle or in counter segments are either equal or supplementary.

Thus from the three fundamental equations in the numerically expressed six elements of a triangle

$$\sin A=\sin\left(-A+\frac{1}{2}\right)=\sin(B+C)=(\sin B)\cos C+(\cos B)\sin C$$

and therefore if d stand for the common value of $(\sin A)^{-1}a$ $(\sin B)^{-1}b$ $(\sin C)^{-1}c$

$(\sin A)d=(\cos B)(\sin C)d+(\cos C)(\sin B)d$ or $a=(\cos B)c+(\cos C)b$

likewise $b=(\cos C)a+(\cos A)c$ $c=(\cos A)b+(\cos B)a$ and hence

$$-a^2+b^2+c^2$$

$$=-\{(\cos B)c+(\cos C)b\}a+\{(\cos C)a+(\cos A)c\}b+\{(\cos A)b+(\cos B)a\}c$$

$$=2(\cos A)bc.$$

Again

$$\because -\cos A=\cos\left(-A+\frac{1}{2}\right)=\cos(B+C)=(\cos B)\cos C-(\sin B)\sin C$$

$$\{-\cos A+(\sin B)\sin C\}^2=(\cos B)^2(\cos C)^2=\{1-(\sin B)^2\}\{1-(\sin C)^2\}$$

$$\text{and } \therefore (\sin A)^2=(\sin B)^2+(\sin C)^2-2(\cos A)(\sin B)\sin C$$

$$\{(\sin A)d\}^2=\{(\sin B)d\}^2+\{(\sin C)d\}^2-2(\cos A)\{(\sin B)d\}(\sin C)d$$

$$a^2=b^2+c^2-2(\cos A)bc.$$

(13). If two endless straight lines cut one another how are the distances severally from each in the other's direction linked together of a point in a straight line in their plane? And when and where do two endless straight lines in a plane therefore meet?

Through a point P anywhere in the plane of two endless straight lines $x'Ox$ $y'Oy$ that cut one another draw an endless straight line $y'Py$ condirectionate with $y'Oy$ and let N be the point where $x'Ox$ because it cuts $y'Oy$ cuts $y'Py$. Let x express algebraically the distance from O in the direction Ox of N and y the distance from N in the direction Ny of P each in reference to the same unit line. Since N is the only point in $x'Ox$ whose distance from O in the direction Ox x expresses and $y'Ny$ is the only condirectionate to $y'Oy$ that can pass through N x serves to mark out $y'Ny$ among all condirectionates of $y'Oy$ and since P is the only point in $y'Ny$ whose distance from N in the direction Ny y expresses y serves to mark out P among all points in $y'Ny$. Hence $x\,y$ serve together to mark out P among all points in the plane of $x'Ox$ $y'Oy$. Further if through the same point P an endless straight line $x'Px$ be drawn condirectionate with $x'Ox$ and M be the point where $y'Oy$ because it cuts $x'Ox$ cuts $x'Px$ then OM is equal and condirectionate to NP and MP to ON and therefore y expresses algebraically the distance from O in the direction Oy of M and x the distance from M in the direction Mx of P. These protensive quantities $x\,y$ are called the COORDINATES of P relative to $x'Ox$ $y'Oy$ as AXES.

Let SL be an endless straight line in the plane of $x'Ox$ $y'Oy$. Through O draw an endless straight line $p'Op$ cutting perpendicularly SL and let Q be the point of cutting. Let p be the algebraic expression in reference to the unit line of the distance from

O in the direction Op of Q. Let ω express numerically the angle xOy and α algebraically the angular distance from Ox the $xyx'y'$ way round of Op and therefore too the angular distance from Op the contrary or $xy'x'y$ way round of Ox each in reference to a common unit angle so that $-\alpha+\omega$ expresses algebraically in reference to this unit angle the angular distance from Op the $xyx'y'$ way round of Oy and the operationally equivalent $\omega-\alpha$ the angular distance from Oy the $xy'x'y$ way round of Op. Then since Ox is angularly distant from Op the $xy'x'y$ way round by what α algebraically expresses and Ny angularly distant from Np, one of the two portions stretching away endlessly from N of an endless straight line $p'Np$ drawn through N condirectionate to $p'Op$, the same way round by what $\alpha-\omega$ algebraically expresses (art. 391) the foot of a perpendicular drawn from P to $p'Op$ if P be not in $p'Op$ or P itself if in $p'Op$ is distant from O in the direction Op by what $(\cos\alpha)x+\{\cos(\alpha-\omega)\}y$ or the operationally equivalent $(\cos\alpha)x+\{\cos(\omega-\alpha)\}y$ algebraically expresses. Now P is in SL precisely when a straight line drawn through P perpendicularly to $p'Op$ is the straight line SL and hence and symbolizing $\omega-\alpha$ by β P is in SL precisely when

$$(\cos\alpha)x+(\cos\beta)y=p.$$

Or $-p+\{(\cos\alpha)x+(\cos\beta)y\}$ expresses algebraically the distance from Q in the direction QP of the point where $p'Op$ is cut perpendicularly by a straight line drawn through P and therefore the perpendicular distance from SL toward that side on which Qp is of P and hence P is in SL precisely when this expression is equal to o. The equation $(\cos\alpha)x+(\cos\beta)y-p=0$ then inasmuch as all the pairs of protensive values of x y that fulfil it are precisely all the pairs of coordinates of points in SL serves to mark out among all points in the plane of $x'x$ $y'y$ those of them that are in the straight line SL and hence is called the EQUATION OF THE STRAIGHT LINE SL.

If SL be the same as or parallel to $x'x$ $\cos\alpha$ is o $\cos\beta$ is $\pm\sin\omega$ and the equation is $y\mp(\sin\omega)^{-1}p=0$. If SL be or be parallel to $y'y$ $\cos\alpha$ $\cos\beta$ and the equation are severally $\pm\sin\omega$ o and $x\mp(\sin\omega)^{-1}p=0$. If SL pass through O p is o and the equation $(\cos\alpha)x+(\cos\beta)y=0$. If SL meet $x'x$ at a point whose distance from O in the direction Ox is expressed algebraically by a the values a of x and o of y fulfil the equation and $\therefore$ $(\cos\alpha)a-p=0$. If SL meet $y'y$ at a distance from O in the direction Oy expressed algebraically by b y is b when x is o and $\therefore$ $(\cos\beta)b-p=0$. If SL

cut elsewhere than at O $x'x$ at a point whose coordinates are a o and $y'y$ at a point whose coordinates are o b none of the quantities a b p is o $\therefore$ $\cos\alpha=\frac{p}{a}$ $\cos\beta=\frac{p}{b}$ and the equation is equivalent to $\frac{1}{a}x+\frac{1}{b}y-1=0$ and therefore too to this's equivalent $\frac{x}{a}+\frac{y}{b}=1$.

Any equation $Ax+By+C=0$ of the first degree wherein the quantities A B C independent of x y are protensive is the equation of a straight line. For if K be any protensive quantity but o SL's equation is equivalent to $(K\cos\alpha)x+(K\cos\beta)y-Kp=0$ and this is the same as $Ax+By+C=0$ if α p K can be and be taken such that $K\cos\alpha=A$ $K\cos\beta=B-Kp=C$. Now α may (art. 385) be taken so as to have $\cos\alpha$ equal to any given protensive quantity not algebraically less than -1 and not algebraically greater than $+1$ and the only tie between $\cos\alpha$ and $\cos\beta$ is that

$$\cos\beta=\cos(\omega-\alpha)=(\cos\omega)\cos\alpha+(\sin\omega)\{-(\cos\alpha)^2+1\}^{\frac{1}{2}}$$

or
$$(\cos\alpha)^2+(\cos\beta)^2-2(\cos\omega)(\cos\alpha)\cos\beta=(\sin\omega)^2.$$

Therefore K must be taken so that $K^{-1}=\frac{\sin\omega}{\{A^2+B^2-2(\cos\omega)AB\}^{\frac{1}{2}}}$ and then $\because$ $A^2+B^2-2(\cos\omega)AB=(\sin\omega)^2A^2+\{(\cos\omega)A-B\}^2$ $=\{A-(\cos\omega)B\}^2+(\sin\omega)^2B^2$ and $\therefore$ not$<(\sin\omega)^2A^2$ and not$<(\sin\omega)^2B^2$ $\cos\alpha$ may be taken $K^{-1}A$ and $\cos\beta$ $K^{-1}B$. When then all this is done and p taken $-K^{-1}C$ the equation $Ax+By+C=0$ is the equation of SL. If $\cos\alpha$ $\cos\beta$ p marking out the direction Op and the distance from O to $p$ward of Q be the values got by taking either value of K the values got by taking the other value are $-\cos\alpha$ $-\cos\beta$ $-p$ marking out the opposite direction Op' and the distance from O to p'ward of Q. Inasmuch as $(\cos\alpha)x+(\cos\beta)y-p$ or $\left(\frac{1}{K}A\right)x+\left(\frac{1}{K}B\right)y+\frac{1}{K}C$ expresses algebraically the perpendicular distance from SL toward that side on which Qp is of any point whose coordinates are x y the equation $(\cos\alpha)x+(\cos\beta)y-p=0$ or $\left(\frac{1}{K}A\right)x+\left(\frac{1}{K}B\right)y+\frac{1}{K}C=0$ is called SL's CATHETIC equation.

Since it is precisely a straight line the same as or parallel to SL that cuts at right angles $p'Op$ two straight lines having as equations $Ax+By+C=0$ $A'x+B'y+C'=0$ are the same or parallel precisely when $\frac{A'}{A}=\frac{B'}{B}$.

If λ λ' be protensive quantities independent of x y the equation $\lambda(Ax+By+C)+\lambda'(A'x+B'y+C')=0$ or the equivalent $(\lambda A+\lambda'A')x$

$+(\lambda B+\lambda' B')y+(\lambda C+\lambda' C')=0$ is of the first degree in x y and is therefore the equation of a straight line. If further neither λ nor λ' be 0 this equation and the equations $Ax+By+C=0$ $A'x+B'y+C'=0$ are (art. 430) such that the values of x y are the same which fulfil any two of them as fulfil any other two of them and therefore unless $A'B=B'A$ the three straight lines having these as their equations are such that each passes through the point of section of the other two. By taking λ λ' (art. 431) such that $\lambda A+\lambda' A'=0$ the equation $(\lambda B+\lambda' B')y+(\lambda C+\lambda' C')=0$ is got of a straight line the same as or parallel to $x'x$ passing through the point where the straight lines having as equations $Ax+By+C=0$ $A'x+B'y+C'=0$ cut one another and therefore the y coordinate of this point and in like way is the x coordinate of the point got.

(14). What relation holds among the perpendicular distances from the endless straight lines in which a triangle's sides lie of a point in the triangle's plane?

Let ABC be a triangle whose sides BC CA AB lie severally in the endless straight lines $B_{,}BCC'$ $C_{,}CAA'$ $A_{,}ABB'$. Through any point P in the plane of ABC draw three endless straight lines $\alpha'P\alpha$ $\beta'P\beta$ $\gamma'P\gamma$ cutting perpendicularly the straight lines $B_{,}C'$ $C_{,}A'$ $A_{,}B'$ severally in the points L M N so that $L\alpha$ $M\beta$ $N\gamma$ may be on the opposite sides severally of $B_{,}C'$ $C_{,}A'$ $A_{,}B'$ to those upon which ABC is and let α β γ express algebraically the several distances from L in the direction $L\alpha$ from M in the direction $M\beta$ and from N in the direction $N\gamma$ of P. Further let λ μ ν express algebraically the several distances from B in the direction BC' of L from C in the direction CA' of M and from A in the direction AB' of N. Then A B C expressing numerically in reference to a perigon as unit the several angles CAB ABC BCA and f g h in reference to the same unit line as all the distances are referred to the several perpendiculars drawn from A to $B_{,}C'$ from B to $C_{,}A'$ and from C to $A_{,}B'$ since to pass over a $-\frac{1}{\sin C}g+\lambda$ expressed distance from C toward C' and an α expressed distance from L $L\alpha$ward is to pass to a β expressed perpendicular distance from $C_{,}A'$ $M\beta$ward

$$(\sin C)\left(-\frac{1}{\sin C}g+\lambda\right)+\left\{\sin\left(C-\frac{1}{4}\right)\right\}\alpha=\beta \text{ or } (\sin C)\lambda-(\cos C)\alpha-\beta=g$$

and since to pass over a λ expressed distance from B toward C' and an α expressed distance from L $L\alpha$ward is to pass to a $-\gamma$ expressed perpendicular distance from $A_{,}B'$ $N\gamma'$ward

$$(\sin B)\lambda + \left\{\sin\left(B+\frac{1}{4}\right)\right\}\iota = -\gamma \quad \text{or} \quad (\sin B)\lambda + (\cos B)\alpha + \gamma = 0.$$

These two equations,—

$$\because (\sin C)\cos B + (\sin B)\cos C = \sin(B+C) = \sin\left(-A+\frac{1}{2}\right) = \sin A -,$$

are equivalent to the two

$$(\sin A)\alpha + (\sin B)\beta + (\sin C)\gamma = -(\sin B)g$$
$$(\sin A)\lambda - (\cos B)(g+\beta) + (\cos C)\gamma = 0$$

of which the first shows that $(\sin A)\alpha+(\sin B)\beta+(\sin C)\gamma$ has the same value wherever P is and this value either therefore by taking P at A and at C in turn or in like manner as it has been shown the same as $-(\sin B)g$ is the same as $-(\sin A)f$ and as $-(\sin C)h$. If d be the numerical expression of the circumscribed circle's diameter $(\sin A)(\sin B)(\sin C)d = (\sin A)f = (\sin B)g = (\sin C)h$ and therefore

$$(\sin A)\alpha + (\sin B)\beta + (\sin C)\gamma + (\sin A)(\sin B)(\sin C)d = 0.$$

From the other equation $(\sin A)\lambda - (\cos B)(g+\beta) + (\cos C)\gamma = 0$ and the equations

$$(\sin B)\mu - (\cos C)(h+\gamma) + (\cos A)\alpha = 0$$
$$(\sin C)\nu - (\cos A)(f+\alpha) + (\cos B)\beta = 0$$

that must on like grounds hold it follows that

$$(\sin A)\lambda + (\sin B)\mu + (\sin C)\nu = (\cos A)f + (\cos B)g + (\cos C)h$$

and therefore and

$$\because (\cos A)(\sin B)\sin C + (\cos B)(\sin C)\sin A + (\cos C)(\sin A)\sin B$$
$$= (\cos A)(\sin B)\sin C + \{-(\cos A)^2+1\}$$
$$= (\cos A)\{(\sin B)\sin C + \cos(B+C)\} + 1 = (\cos A)(\cos B)\cos C + 1$$
$$(\sin A)\lambda + (\sin B)\mu + (\sin C)\nu = \{(\cos A)(\cos B)\cos C + 1\}d.$$

The equation in $\alpha\ \beta\ \gamma$ may be otherwise found from what is shown in question (13) above. Taking any two cutting endless straight lines $x'Ox\ y'Oy$ in the plane of the triangle ABC as coordinate axes through O draw three endless straight lines p'_1Op_1 p'_2Op_2 p'_3Op_3 cutting perpendicularly $B_{,}C'$ $C_{,}A'$ $A_{,}B'$ severally in $Q_1\ Q_2\ Q_3$ so that $Q_1p_1\ Q_2p_2\ Q_3p_3$ may be on the opposite sides severally of $B_{,}C'$ $C_{,}A'$ $A_{,}B'$ to those upon which ABC is and let $p_1\ p_2\ p_3$ express algebraically the several distances from O in the direction Op_1 of Q_1 in the direction Op_2 of Q_2 and in the direction Op_3 of Q_3. After first naming the endless portions of $y'Oy$ upon opposite sides of O Oy Oy' in such a manner that the $xyx'y'$ way round may be the same as the $A_{,}A'B_{,}B'C_{,}C'$ way round let ω express numerically the angle xOy in reference to a perigon as unit and in reference to a perigon as unit let θ_1 θ_2 θ_3 so express algebraically the angular distances from Ox the xyx' way round of Op_1 Op_2 Op_3 severally and ϕ_1 ϕ_2 ϕ_3 the angular distances from Oy the yxy' way round of the same severally that besides $\theta_1+\phi_1$ $\theta_2+\phi_2$ $\theta_3+\phi_3$ being each equal to ω $-\theta_2+\theta_3$ which $=-\left(\frac{1}{4}+\theta_2\right)+\left(\frac{1}{4}+\theta_3\right)=-\left(\theta_2+\frac{1}{4}\right)+\left(\theta_3+\frac{1}{4}\right)$ and therefore expresses algebraically the angular distance from AA' the $A'B_{,}B'$ way round of AB' may be equal to $\frac{1}{2}-A$ the numerical expression of the angle $A'AB'$ likewise $-\theta_3+\theta_1$ equal to $\frac{1}{2}-B$ and hence $-\theta_1+\theta_2$ equal to $-\frac{1}{2}-C$. Then if x y be the coordinates of P

$$-p_1+(\cos\theta_1)x+(\cos\phi_1)y=\alpha \qquad -p_2+(\cos\theta_2)x+(\cos\phi_2)y=\beta$$
$$-p_3+(\cos\theta_3)x+(\cos\phi_3)y=\gamma.$$

And $\because$
$$(\cos\theta_2)\cos\phi_3-(\cos\theta_3)\cos\phi_2=(\sin\omega)\sin(-\theta_2+\theta_3)$$
$$(\cos\theta_3)\cos\phi_1-(\cos\theta_1)\cos\phi_3=(\sin\omega)\sin(-\theta_3+\theta_1)$$
and
$$(\cos\theta_1)\cos\phi_2-(\cos\theta_2)\cos\phi_1=(\sin\omega)\sin(-\theta_1+\theta_2)$$

it follows that the equation free from x y which with any two of these equations is equivalent to all three is

$$(\sin A)(p_1+\alpha)+(\sin B)(p_2+\beta)+(\sin C)(p_3+\gamma)=0.$$

Whence $(\sin A)\alpha+(\sin B)\beta+(\sin C)\gamma$ has always the unchanging value $(\sin A)(-p_1)+(\sin B)(-p_2)+(\sin C)(-p_3)$ however P be changed. The equation itself states how the algebraic expressions $p_1+\alpha$ $p_2+\beta$ $p_3+\gamma$ are bound together of P's perpendicular distances from the several endless straight lines drawn through O condirectionate severally with $B_{,}C'$ $C_{,}A'$ $A_{,}B'$ toward the several sides whereon are Op_1 Op_2 Op_3.

439. PROP. *To explain how triangles and parallelograms are estimated numerically and algebraically.*

Let OAB be a triangle and produce AO any ended distance OU and BO any ended distance OV. Because OA OB are adjoining sides of a triangle the straight lines AOU BOV meet and are not in one straight line they therefore meet at O only and there cut one another therefore OA OU are on opposite sides of BOV and OB OV on opposite sides of AOU. Join BU and UV. Since neither of the points B V is in the same straight line with

AOU both *UOB* and *UOV* is a triangle. Let *a* be the numerical expression of *OA* in reference to *OU* as unit and *b* the numerical expression of *OB* in reference to *OV* as unit. Because the triangles *OAB OUB* have as a common height the perpendicular drawn from *B* to the endless straight line in which their bases *OA OU* lie $OAB : OUB = OA : OU$ and therefore *a* is the numerical expression of *OAB* in reference to *OUB* as unit. Again because the triangles *OUB OUV* have as a common height the perpendicular drawn from *U* to the endless straight line in which their bases *OB OV* lie $OUB : OUV = OB : OV$ and therefore *b* is the numerical expression of *OUB* in reference to *OUV* as unit. Since then *a* expresses numerically *OAB* in reference to *OUB* as unit and *b* expresses numerically *OUB* in reference to *OUV* as unit *ab* expresses numerically *OAB* in reference to *OUV* as unit. Let *OW* be a straight line equal to *OU* and lying in the same straight line and upon the same side of *O* as *OA*. Join *VW*. Because *V* is not in the same straight line with *WOU OWV* is a triangle and the triangles *OUV OWV* being upon equal bases *OU OW* and between the same parallels *UW* and a parallel to *UW* through *V* are equal. Hence *ab* is the numerical expression of *OAB* in reference to *OWV* as unit.

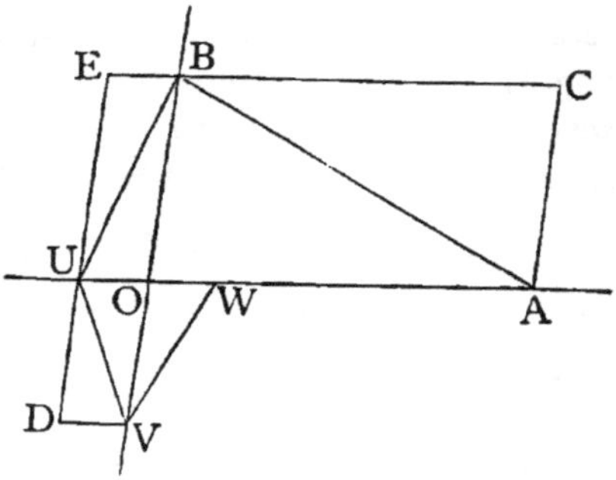

If then in particular *OV* be equal to *OU* and therefore to *OW*,—and it is always understood if not otherwise stated that the unit is the same to which the numerical expressions refer of all straight lines dealt with at the same time—, a triangle having two sides expressed numerically by *a b* is expressed numerically by *ab* in reference to an isosceles triangle as unit surface having each of the equal sides equal to the unit line and the angle bounded by them either equal or supplementary to the angle bounded by the two sides of the triangle. Moreover since the triangle *OAB* is equal to any other triangle upon a base equal to *OB* and between the endless straight line wherein *OB* lies and the parallel thereto through *A* and either of the triangles *OUV OWV* is equal to any other triangle upon a base equal to *OU* or *OW* and between the endless straight line in which *UW* lies and the parallel straight line through *V* a triangle with a side expressed numerically by *b* and a straight line expressed numerically by a drawn from the opposite angle and striking at any angles the straight line of that side is expressed numerically by ab in reference to a triangle as unit surface having a unit side and a unit straight line drawn from the angle opposite and striking at angles equal severally to those angles the straight line of the unit side. In particular if the angles be right angles at which the side lines are struck a triangle upon a b numerically expressed base and of an a numerically expressed height is numerically expressed by ab in reference to a triangle upon unit base and of unit height as unit surface.

Next let $OACB$ be a parallelogram and produce AO to any point U and BO to any point V. Because OA OB are adjoining sides of a parallelogram the straight lines AOU BOV meet and do not lie wholly together therefore AOU BOV meet at O only and there cut one another and therefore OA OU lie on opposite sides of BOV and OB OV on opposite sides of AOU. Through V a point not in AOU draw a straight line VD parallel to AOU and through U a point not in BOV draw a straight line UE parallel to BOV. Because in AOU there is a point U in and a point O not in UE,— for O is in BOV a parallel to UE—, AOU UE meet but do not lie wholly together therefore they meet at U only and there cut one another. And because UE cuts AOU one of the parallels AOU VD or AOU CB it also cuts the other VD or CB. Let then D be the point where UE cuts VD and E the point where UE cuts the straight line in which CB lies. Since OU meets OB only at O BE not at all and EU only at U the straight line OU and the line $OBEU$ made up of the three straight parts OB BE EU end at the same two points O U and only there meet one another therefore these lines shut in a portion of surface and therefore $OBEU$ is a figure. This $OBEU$ figure too being a plane figure bounded by one line wholly made up of four straight parts of which no adjoining two are in one straight line is a four sided polygon and having each opposite two of the four sides parallel is a parallelogram. In like ways is $OVDU$ a parallelogram. Let a express numerically OA in reference to OU as unit and b OB in reference to OV as unit. Because the parallelograms $OACB$ $OUEB$ are of the same height $OACB : OUEB = OA : OU$ and because the parallelograms $OUEB$ $OUDV$ are of the same height $OUEB : OUDV = OB : OV$. Hence a expresses numerically $OACB$ in reference to $OUEB$ as unit and b $OUEB$ in reference to $OUDV$ as unit. Therefore ab expresses numerically $OACB$ in reference to $OUDV$ as unit.

If the units OU OV to which a b severally refer be equal it follows in particular that a parallelogram having adjoining sides expressed numerically by a b in reference to a common unit straight line is expressed numerically by ab in reference to an equilateral parallelogram equiangular to the parallelogram and with each side equal to the unit line as unit surface. Since too $OACB$ is equal to any parallelogram upon a base equal to OB and between the parallels in which OB AC lie and $OUDV$ is equal to any parallelogram upon a base equal to OU and between the parallels in which OU VD lie a parallelogram having a side expressed numerically by b and a straight line expressed numerically by a crossing from the line of that side at any angles to the line of the side opposite is expressed numerically by ab in reference to a parallelogram as unit surface having a unit side and a unit straight line crossing from the line of the unit side at angles equal severally to those angles to the line of the side opposite. If in particular the straight lines cross at right angles a parallelogram upon a base numerically expressed by b and of a height numerically expressed by a is numerically expressed by ab in reference to the square of the unit line as unit surface.

Since a parallelogram is double a triangle upon the same base and between the same parallels applying this to the estimated parallelogram and the estimated triangle in corresponding cases the parallelogram is expressed numerically by $2ab$ in reference to the unit triangle and the triangle by $\frac{1}{2}ab$ in reference to the unit parallelogram or applying the proposition to the unit parallelogram and the unit triangle in the like cases the parallelogram is expressed numerically by $(ab)\times 2$ in reference to the unit triangle and the triangle by $(ab)\times\frac{1}{2}$ in reference to the unit parallelogram.

Let $x'Ox$ be a straight line stretching away endlessly on each side of a point O and let OB be a straight line drawn from O to a point B not in $x'Ox$. In $x'Ox$ take a point A anywhere but at O and join AB so that OAB is a triangle. Let OU OV be equal straight lines in the same straight lines and upon the same sides of O as Ox OB severally and join UV so that OUV is a triangle. If in reference to OU or OV as unit line b express numerically OB and a express algebraically the distance from O in the direction Ox of A the quantitative element of ab expresses numerically in reference to the triangle OUV as unit surface the triangle OAB and the designative element of ab is the same as that of a. But the algebraic sign of a marks whether A be on the x or the x' side of O and OAB is upon the same side of OB as A. Therefore the algebraic sign of ab marks whether the triangle OAB is upon the same side of OB as Ox or as Ox'.

Also a parallelogram having the sides OA OB and expressed numerically by ab's absolute value in reference to the parallelogram as unit surface having the sides OU OV is upon the Ox or the Ox' side of OB according as ab's sign is + or −.

440. The matter of art. 439 may be put to such uses as the following.

(1). In a given circle whose diameter is expressed numerically by d to inscribe a rectangle having a given amount of surface expressed numerically by a in reference to the square of the unit line as unit surface.

If x y express numerically the rectangle's sides $xy=a$. And since an angle in a circular segment is a right angle precisely when the circle's center is in the segment's base each of the rectangle's diagonals is a diameter of the circle and therefore $x^2+y^2=d^2$. Hence x y can only be either of them $\frac{1}{2}\{\sqrt{d^2+2a})+\sqrt{(d^2-2a)}\}$ and the other $\frac{1}{2}\{\sqrt{(d^2+2a)}-\sqrt{(d^2-2a)}\}$. The greatest value that a can have is $\frac{1}{2}d^2$ and then $x=y$ so that the greatest rectangle that can be inscribed in a given circle is the inscribed square.

(2). About a given circle whose diameter is expressed numerically by d to circumscribe a parallelogram having a given amount of surface expressed numerically by a in reference to the square of the unit line as unit.

Since a straight line drawn through the circle's center perpendicular to the straight line of any side of the circumscribed parallelogram cuts perpendicularly the straight line of the opposite side and meets each of these lines at the point of contact the perpendicular distance between each two opposite sides of the parallelogram is equal to the diameter of the circle. Therefore if A express numerically any angle of the parallelogram $\frac{1}{\sin A}d$ expresses numerically every side and therefore in reference to the square of the unit line as unit surface $d(\sin A)^{-1}d$ or the equivalent $(\sin A)^{-1}d^2$ expresses numerically the parallelogram. Hence the parallelogram is expressed numerically by a precisely when $(\sin A)^{-1} = \frac{a}{d^2}$ and therefore precisely when each side of the parallelogram is expressed numerically by $\frac{a}{d^2}d$ or the equivalent $\frac{a}{d}$. Since the least value that $(\sin A)^{-1}$ can have is 1 which happens precisely when A expresses numerically a right angle the least parallelogram that can be circumscribed about a given circle is the circumscribed square.

If $a = b^2$ the following construction may be used. Let BKD be the circle C the center BCD the diameter at whose ends B D two opposite sides are to touch and BN DG the straight lines of those sides. If $b = d$ about BKD circumscribe a square of which BN DG are side lines. But otherwise from B as center at a distance expressed numerically by b describe a circle and let E be one of the points on opposite sides of D where $\because$ $b > d$ DG cuts this circle. Join BE and because E is on one side of BD BDE is a triangle. From E draw on that side of EB on which is the triangle BDE a straight line EF at right angles to EB. Because the triangle BDE is right angled at D the angle EBD is acute the straight lines EF BD therefore make with EB interior angles BEF EBD upon one side that are together less than a hemiperigon and therefore meet on that side if far enough produced. Let F be the point where EF BD so meet and then EBF is a triangle. Since the angle BEF is right the perpendicular ED to the side BF cuts the triangle FBE into the triangular parts EBD FED similar to itself $\therefore$ $BF : BE = BE : BD$ and therefore BF is expressed numerically by $\frac{b}{d}b$ and is therefore greater than BD. From B as center then at the distance BF describe a circle and let G be the point on one side of D where the straight line DG because it passes through the point D within cuts the circle. Join BG and since BG is expressed numerically by $\frac{b}{d}b$ the $(\sin)^{-1}$ of the angle BGD is $\frac{\frac{b}{d}b}{d}$ which $= \left(\frac{b}{d}\right)^2 = \frac{b^2}{d^2}$ and therefore the angle BGD is equal to one angle of the circumscribed parallelogram. Further since the triangle BDG is right angled at D the angle DBG is acute. Through C draw a straight line $LCHK$ cutting perpendicularly in H BG and let K L be the points on the same side of C as H and on the opposite side severally where $LCHK$ because it passes through the center cuts the circumference of BKD. Because the perpendicular CH is on that side of *CB* on which is the acute angle *CBG* the straight lines *BCD* *KCL* do not lie wholly together they therefore meet at *C* only and there cut one another. Join *DK*. The points *K* *H* being in *LCHK* on the same side of *C* are on the same side of *BCD*. But the straight line *DK* is on the same side of *BCD* as *K* and *BG* and therefore too *DG* is on the same side of *BCD* as *H*. Therefore *DK* is on the same side of *BCD* as *DG*. Moreover since *K* is in the circumference of *BKD* and not at *D* the straight lines *DK* *DB* are on the same side of *DG* as *BKD*. Since then from one end and on one side of the straight line *DB* a straight line *DG* is drawn and from the same end and on the same side of *DB* a third straight line *DK* not the same as *DG* is drawn and moreover *DK* is on the same side of *DG* as *DB* it follows that *DK* is on the same side of each of the straight lines *DB* *DG* as the other that *DK* is between *DB* and *DG* that *DK* cuts atwo the angle *BDG* that *DB* *DG* make angles *KDB* *KDG* with that lie upon opposite sides of *DK* and that *DB* *DG* are on opposite sides of *DK*. Through *K* draw a straight line *KM* touching *BKD*. Because *D* is elsewhere in the circumference of *BKD* than at *K* the straight lines *KM* *KD* meet and do not lie wholly together therefore *KM* *KD* meet at *K* only and there cut one another. Let then *KM* be the portion of *KM* that lies on the same side of *KD* as *DG* and therefore on the opposite side of *KD* to which *DB* lies on. Now the straight line which touches the circle *BKD* at either end *D* or *K* of the straight line *DK* in that circle makes with upon either side of *DK* an angle equal to the angle in the segment upon the other side and the angle in the segment *DBK* within which is the center *C* of the circle is less than a right angle. Therefore each of the angles *KDG* *DKM* is less than a right angle and therefore these angles are together less than a hemiperigon. The straight lines *DG* *KM* therefore which thus make with *DK* at the different points *D* *K* interior angles *KDG* *DKM* upon one side that are together less than a hemiperigon meet one another on that side if far enough produced. Let *M* be the point where they so meet and since *K* is not in *DM* *DM* *KM* meet at *M* only and there cut one another. Let too *N* be the point where *MK* because it cuts *DM* cuts the parallel *BN*. Because in *KCL* there is a point *K* in and a point *C* not in the straight line *MN*,—*C* being the center of a circle touching *MN*—, *KCL* *MN* meet and do not lie wholly together therefore *KCL* *MN* meet at *K* only and therefore *L* being in *KCL* but not at *K* is not in *MN*. Through *L* a point thus not in *MN* draw a straight line *LO* parallel to *MN* and let *O* be the point where *BN* and *P* the point where *DM* because it cuts *MN* one of the two parallels *MN LO* also cuts the other *LO*. The straight line *MP* and the line *MNOP* made up of the three straight parts *MN NO OP* ending at *M P* and only there meeting one another shut in a portion of plane surface and since *MP* is parallel to *NO* and *MN* to *PO* the portion shut in is a parallelogram. Since *MN* touches the circle *BKD* the straight line *LCK* drawn through the center *C* to the contact *K* is perpendicular to *MN* and since the parallels *MN PO* make with *LCK* equal alternate angles *PO* is therefore at right angles to *LCK* and therefore touches *BKD* at *L*. Since then the perpendiculars *CD CL CB CK* drawn from *C* to the side lines of the parallelogram are all equal the foot of the perpendicular to any side line is on the same side of either adjoining side line as *C*. But *C* is between the ends of each of the straight lines *BD KL* joining opposite contacts and therefore is on the same side of each of two opposite side lines as the other. Wherefore the circle's contact with any side line is on the same side of each of the two adjoining side lines as the other and therefore is between the ends of the side. The parallelogram is therefore circumscribed about the circle. Lastly because in the triangle *CBH* the right angle *CHB* is greater than the acute angle *CBH* the side *CB* is greater than the side *CH* therefore *CK* an equal to *CB* is greater than *CH* therefore of the unequal straight lines *CH CK* having a common end *C* and lying in the same straight line upon the same side of *C* the less *CH* is a part of the greater *CK* and therefore the parts *HC HK* of *CK* are upon opposite sides of *H* and therefore on opposite sides of *BG* that cuts *CK* at *H*. But *BG NM* because they cut at right angles *CK* at the different points *H K* are parallel and therefore *GM* is on the same side of *BG* as *HK* on that side to wit whereon is the parallel *NM*. Moreover *GD* is on the same side of *BG* as *HC* on that side to wit whereon is the straight line *BCD*. Therefore *GD GM* are on opposite sides of *BG* and therefore of the angles which the parallels *BG NM* make with the straight line *DGM* cut by them at the different points *G M DGB* is an exterior and *GMN* or *PMN* is the opposite interior upon the same side to wit upon that side of *DGM* whereon is the parallel *BN*. The angle *DGB* is therefore equal to the angle *PMN* and therefore of the parallelogram *PMNO* the angle *PMN* the equal opposite angle *PON* and each of the other supplementary angles *MPO MNO* has a $(\sin)^{-1}$ equal to $\frac{b^2}{d^2}$. By doing with the point on the other side of *D* where *DG* cuts the circle described from *B* as center at the distance *BF* the same as is here done with *G* another parallelogram circumscribed about the circle *BKD* is got each of whose angles has a $(\sin)^{-1}$ equal to $\frac{b^2}{d^2}$ and this and *MNOP* are the only parallelograms that can be circumscribed about *BKD* having side lines lying along *BN DG* and a surface expressed numerically by b^2 in reference to the square of the unit line as unit surface.

(3). To describe a parallelogram equiangular and equal to a given parallelogram and having a side in one of the side lines of a given triangle and the ends of the side opposite in the other two.

Let *ABC* be the given triangle and *BC* the side in the same straight line with which the sought parallelogram's side is to be. Produce *BC* endlessly both ways and toward *y* beyond *C*. Through *A* draw an endless straight line *x'Ax* cutting *By* in *D* at angles equal severally to the four angles of the given parallelogram and so that *D* is on the *x* side of *A*. Let *a h* be the numerical expressions in reference to a common unit line of *BC AD* severally and *ka* the numerical expression of the given parallelogram in reference to a parallelogram as unit surface equiangular thereto and having each of its sides equal to the unit line so that if *r s* be the numerical expressions of the given parallelogram's sides $ka = rs$ $\frac{ka}{sa} = \frac{rs}{as}$ $\frac{k}{s} = \frac{r}{a}$ and therefore *k* is the numerical expression of a first proportionate to the given parallelogram's sides and *BC*. Through any point *G* in *x'Ax* draw an endless straight line *Gy* condirectionate with *By* and let *E F* be the points where the several straight lines of *AB AC* because they as well as *x'Ax* cut *By* cut *Gy*. When *E F* are neither at *A* nor in *By* let *EFHK* be the parallelogram with sides *EK FH* parallel to *AD* and the side *KH* opposite to *EF* in *By*. This *EFHK* parallelogram then is just any parallelogram equiangular to the given parallelogram and with a side *KH* in *By* and the side *EF* opposite ending in the straight lines of *AB AC*. Let *x* be the algebraic expression of *G*'s distance from *A* in the direction *Ax*. Since from the similar triangles *EFA BCA EF* : *BC*=*FA* : *CA* and

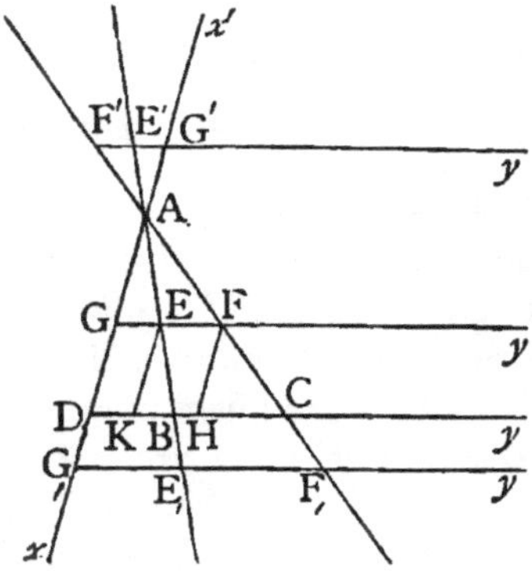

either from the similar triangles FAG CAD or because when D is C G is F $FA:CA=AG:AD$ $EF:BC=AG:AD$. Therefore and because the straight lines of AB AC cut one another at A and $x'Ax$ either cuts each of these straight lines at A or lies wholly together with one of them and cuts at A the other the distance from E in the direction Ey of F is expressed algebraically by $\frac{x}{h}a$. Besides G's distance from D in the direction Dx' is expressed algebraically by $h-x$. Hence in reference to the unit sided parallelogram equiangular to the given parallelogram as unit surface $EFHK$ is expressed numerically by the absolute value of $(h-x)\frac{x}{h}a$. Moreover $(h-x)\frac{x}{h}a$ is plus or minus according as G is or is not in the AD portion of $x'Ax$ so that the algebraic sign of $(h-x)\frac{x}{h}a$ marks if G be in Ax whether $EFHK$ is on the DA or on the Dx side of By and if G be in Dx' whether $EFHK$ is on the same side of EK as Ey or on the opposite. The parallelogram $EFHK$ is therefore the one sought precisely when $(h-x)\frac{x}{h}a=\pm ka$ and this is precisely when severally

$$(h-x)x=\pm kh \quad x=\frac{1}{2}h+\left\{\left(\frac{1}{2}h\right)^2\mp kh\right\}^{\frac{1}{2}}.$$

Of all the parallelograms then having G between A and D there is none equal to the given parallelogram if this be greater than the one having G at AD's middle if the given parallelogram be equal to the parallelogram having G at AD's middle it is equal to no other and if the given parallelogram be less than the one with G at AD's middle there are precisely two equal to it having their G points equidistant from AD's middle. But of all the parallelograms having G not between A and D there are precisely two with their Gs equidistant from AD's middle point that are equal to any given parallelogram.

Since in reference to an equilateral parallelogram upon the unit line as unit surface two parallelograms each equiangular to that parallelogram and severally with sides numerically expressed in reference to the unit line by $h-x$ x and by k h are severally expressed numerically by $(h-x)x$ and by kh the equation $(h-x)x=kh$ determines how to cut atwo AD so that two equiangular parallelograms may be equal of which one has adjoining sides equal severally to the parts and the other adjoining sides numerically expressed by k h. Likewise the equation $(h-x)x=-kh$,—being equivalent both to $(-h+x)x=kh$ and to $\{h+(-x)\}(-x)=kh$—, determines how to produce AD both beyond D and beyond A so that two equiangular parallelograms may be equal having the one adjoining sides equal severally to the whole produced line and the produced part and the other adjoining sides numerically expressed by k h.

If c stand for $\sqrt{kh}$ $(h-x)x=kh$ or c^2 precisely when $\frac{h-x}{c}=\frac{c}{x}$

$$(-h+x)x=c^2 \text{ precisely when } \frac{-h+x}{c}=\frac{c}{x}$$

$$\text{and } \{h+(-x)\}(-x)=c^2 \text{ precisely when } \frac{h+(-x)}{c}=\frac{c}{-x}$$

and therefore the distances from a straight line's ends of a point in the endless straight line in which the ended straight line lies have a given mean proportionate precisely when any parallelogram with adjoining sides equal severally to those distances is equal to a parallelogram equiangular thereto with each side equal to the mean proportionate. If in particular the parallelograms be right angled the distances from two points of a third point in one straight line with the two have a certain mean proportionate precisely when the rectangle of the distances is equal to the square of the mean proportionate.

(4). Two endless straight lines that cut one another being given and a point not in either of them through the given point to draw a straight line so as with the given straight lines to shut in a triangle having a given amount of surface.

Let $x'Ox$ $y'Oy$ be the cutting endless straight lines and P the point in neither. Of the portions Oy Oy' of yy' stretching away endlessly from upon opposite sides of O and therefore lying on opposite sides of xx' let it be Oy that lies on the same side of xx' as P and of the portions Ox Ox' of xx' on opposite sides of yy' let it be Ox that lies on the same side of yy' as P. From P not in xx' draw a straight line PM parallel to xx' and let M be the point where yy' because it cuts xx' cuts PM. Through P and a point R taken anywhere in yy' but at O and M draw an endless straight line PR. Because R is in yy' but not at M the only point common to yy' and PM R is not in PM therefore PR meets and does

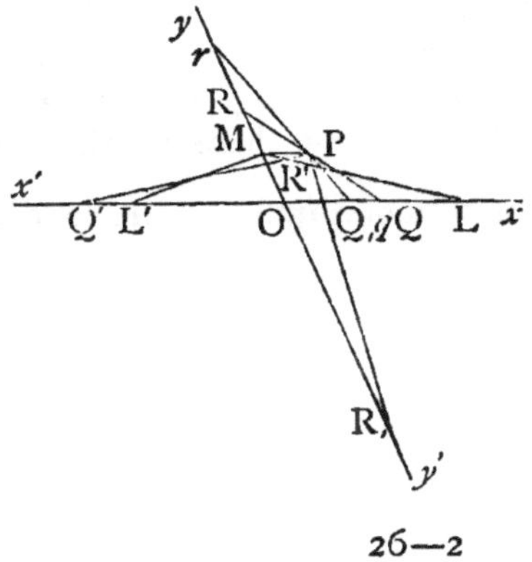

not lie wholly together with PM therefore PR meets at P only and there cuts PM and therefore PR cuts the parallel xx' to PM. Let the point where PR so cuts xx' be Q which cannot be O else two straight lines PR yy' could meet in more points R O than one and yet since P is not in yy' not lie wholly together. Then O Q R being three points not in one straight line OQR is a triangle. Let PM MO be expressed numerically by a b severally and the distances from O in the direction Ox of Q and from O in the direction Oy of R algebraically by x y severally all in reference to a common unit line so that a b being the coordinates of P relative to the axes $x'x$ $y'y$ (art. 438 (13)) $\frac{a}{x}+\frac{b}{y}=1$. Moreover in reference to a triangle as unit surface having each of two sides equal to the unit line and the angle bounded by those sides equal to the angle xOy let bc express numerically the given amount of surface so that if a straight line OL be taken along Ox and a straight line OL' along Ox' each expressed numerically by c in reference to the unit line and ML ML' be joined both OLM and $OL'M$ is a triangle equal to the given amount of surface. Then xy has an absolute value expressing numerically the triangle OQR in reference to the unit surface triangle and is plus or minus according as OQR is if OQ be in Ox on the Oy or the Oy' side of $x'x$ and if OR be in Oy on the Ox or the Ox' side of $y'y$ and hence OQR's surface is equal to the given amount precisely when $xy=\pm bc$. Since a b c are numerical quantities each else than 0 $\frac{a}{x}+\frac{b}{y}=1$ and $xy=bc$ precisely when $ay+bx=xy=bc$ and therefore precisely when $\frac{a}{x}+\frac{b}{y}=1$ and $x(c-x)=ac$ also $\frac{a}{x}+\frac{b}{y}=1$ and $xy=-bc$ precisely when $\frac{a}{x}+\frac{b}{y}=1$ and $x(c+x)=ac$. Wherefore OQR has the given amount of surface precisely when Q either so cuts atwo OL that OQ QL or so lies in OL' or $L'O$ produced that OQ QL' have the same mean proportionate as MP OL. And there are therefore precisely 4 3 or 2 triangles that can be found fulfilling the conditions according as $ac \lesseqgtr \left(\frac{1}{2}c\right)^2$.

CHAPTER VI

PARANOMIC PROCESSES

441. AN expression such as $\frac{1}{1-2x+x^2}(3-x)\ \frac{2-3x+4x^2}{7+5x}$ or $\sqrt{(1+x)}$ which has no polynomic equivalent may as in arts. 188, 189, 192, 315, 319, be operationally equivalent to an expression of the shape $A_0+A_1x+A_2x^2+\cdots+A_{i-1}x^{i-1}+Kx^i$ wherein $A_0\ A_1\ A_2 \ldots A_{i-1}$ are quantities independent of x i is any whole number chosen at will and K is a quantity dependent on x which by taking x endlessly near to 0 is made endlessly near to some fixed quantity independent of x. Inasmuch as i is any whole number whatever

$$A_0+A_1x+A_2x^2+\cdots+A_{i-1}x^{i-1}+Kx^i$$

$$=A_0+A_1x+\cdots+A_{i-1}x^{i-1}+A_ix^i+A_{i+1}x^{i+1}+\cdots+_{i+i'-1}x^{i+i'-1}+K'x^{i+i'}$$

$$=A_0+A_1x+\cdots+A_{i-1}x^{i-1}+\{A_i+x(A_{i+1}+A_{i+2}x+\cdots+A_{i+i'-1}x^{i'-2}+K'x^{i'-1})\}x^i$$

i' being any whole number greater than 0 and K' a quantity like K dependent on x which as x becomes endlessly near to 0 becomes endlessly near to a fixed quantity independent of x.

$$\therefore K=A_i+x(A_{i+1}+A_{i+2}x+\cdots+A_{i+i'-1}x^{i'-2}+K'x^{i'-1}).$$

Now if G be the greatest absolute value that $A_{i+1}+\cdots+K'x^{i'-1}$ has while x passes continuously from any given value else than 0 toward and to 0 x has only to be taken of an absolute value less than $\frac{\delta}{G}$ for $x(A_{i+1}+\cdots+K'x^{i'-1})$ to have an absolute value less than any given numerical quantity δ and if OP PQ be the directed distances severally expressed by two quantities u v one or both of them ditensive the directed distance OQ expressed by $u+v$ becomes endlessly nearly the directed distance OP as v becomes in whatever way endlessly near to 0. Hence it is the fixed quantity A_i independent of x to which K becomes endlessly near as x becomes endlessly near to 0. The symbol $A_i|$ then may fitly stand for K.

Def. If i be any whole number chosen at will $A_0\, A_1\, A_2 \dots A_{i-1}\, A_i$ quantities independent of x and $A_i\rfloor$ a quantity dependent on x that becomes endlessly near to A_i as x becomes endlessly near to 0 $A_0+A_1x+A_2x^2+\cdots+A_{i-1}x^{i-1}+A_i\rfloor x^i$ or the operational equivalent $A_0+xA_1+x^2A_2+\cdots+x^iA_i\rfloor$ is called a PARANOMIC EXPRESSION or PARANOME relative to in reference to or in x.

442. If for ever greater and greater values of n starting from a great enough value $A_0+A_1x+A_2x^2+\cdots+A_{n-1}x^{n-1}$ becomes ever nearer and nearer to a certain quantity and at length when the value of n is great enough becomes nearer thereto than by any given difference,— and it is then only and thus only that $A_0+A_1x+A_2x^2+\cdots$ taken as a result of endless successive additions has meaning—, since $A_0+A_1x+\cdots+A_ix^i+\cdots+A_{i+i'-1}x^{i+i'-1}$

$$=A_0+A_1x+\cdots+A_{i-1}x^{i-1}+\{A_i+x(A_{i+1}+A_{i+2}x+\cdots+A_{i+i'-1}x^{i'-2})\}x^i$$

and however great a given whole number i' is $A_i+x(A_{i+1}+A_{i+2}x+\cdots+A_{i+i'-1}x^{i'-2})$ may by taking x near enough to 0 be made more nearly equal to A_i than by any given difference the quantity which $A_0+A_1x+A_2x^2+\cdots+A_{n-1}x^{n-1}$ endlessly nears as n becomes endlessly great

$$=A_0+A_1x+A_2x^2+\cdots+A_{i-1}x^{i-1}+A_i\rfloor x^i.$$

443. If $A_0\, A_1\, A_2 \dots$ be an infinite series of which every term after a certain one A_{r-1} has an absolute value not greater than a certain numerical quantity A

$$A_0+A_1x+\cdots+A_{r-1}x^{r-1}+A_rx^r+A_{r+1}x^{r+1}+\cdots+A_{r+i-1}x^{r+i-1}$$
$$=A_0+A_1x+\cdots+A_{r-1}x^{r-1}+(A_r+A_{r+1}x+\cdots+A_{r+i-1}x^{i-1})x^r$$

and since if $\Theta_{\omega_1}c_1\ \Theta_{\omega_2}c_2 \dots \Theta_{\omega_i}c_i$ be any quantities protensive or ditensive having absolute values $c_1\, c_2 \dots c_i$ the absolute value of

$$\Theta_{\omega_1}c_1+\Theta_{\omega_2}c_2=\sqrt{[\{(\cos\omega_1)c_1+(\cos\omega_2)c_2\}^2+\{(\sin\omega_1)c_1+(\sin\omega_2)c_2\}^2]}$$
$$=\sqrt{[c_1^2+c_2^2+2\{\cos(-\omega_1+\omega_2)\}c_1c_2]}$$

which is not greater than c_1+c_2 and the absolute value of $\Theta_{\omega_1}c_1+\Theta_{\omega_2}c_2+\cdots+\Theta_{\omega_i}c_i$ likewise not> $c_1+c_2+\cdots+c_i$ the absolute value of $A_r+A_{r+1}x+\cdots+A_{r+i-1}x^{i-1}$ is not greater than that of $A+Ax+Ax^2+\cdots+Ax^{i-1}$ or of this's operational equivalent $A\dfrac{1-x^i}{1-x}$ when x is not 1. Besides if x's absolute value be less than 1 $-x^i$'s absolute value is less than 1 for any value but 0 of i. Hence if x have an absolute value less than 1

$$A_0+A_1x+\cdots+A_{r+i-1}x^{r+i-1}=A_0+A_1x+\cdots+A_{r-1}x^{r-1}+Hx^r$$

where H is some quantity having an absolute value less than $A\frac{2}{1-x}$ has. In like way

$$A_0+A_1x+\cdots+A_{r+i+i'-1}x^{r+i+i'-1}=A_0+A_1x+\cdots+A_{r+i-1}x^{r+i-1}+Hx^{r+i}$$

and i may be taken so great as to give x^{r+i} (arts. 173, 300) and therefore too Hx^{r+i} a less than any given absolute value. For any given value of x then that has an absolute value less than 1 however many be taken of the terms in order of the endless series A_0 A_1x A_2x^2 - - - their sum's absolute value is never greater than some definite numerical quantity and so great a number of those terms in order may be taken that their sum is more nearly equal to the sum of any greater number of them in order than by any given difference. This could not be unless the sum of endlessly more and more of the terms in order of the series became endlessly near to a certain quantity.

444. If v_0 v_1 v_2 ... be an endless series of quantities such that r being some whole number each of the quotients $\frac{v_{r+1}}{v_r}$ $\frac{v_{r+2}}{v_{r+1}}$ $\frac{v_{r+3}}{v_{r+2}}$ - - - has an absolute value not greater than a certain numerical quantity c less than 1

$$v_0+v_1+\cdots+v_r+\cdots+v_{r+i-1}$$

$$=v_0+v_1+\cdots+v_{r-1}+\left(1+\frac{v_{r+1}}{v_r}+\frac{v_{r+2}}{v_{r+1}}\frac{v_{r+1}}{v_r}+\cdots+\frac{v_{r+i-1}}{v_{r+i-2}}\cdots\frac{v_{r+2}}{v_{r+1}}\frac{v_{r+1}}{v_r}\right)v_r$$

and the absolute value of

$1+\frac{v_{r+1}}{v_r}+\cdots+\frac{v_{r+i-1}}{v_{r+i-2}}\cdots\frac{v_{r+2}}{v_{r+1}}\frac{v_{r+1}}{v_r}$ not> $1+c+c^2+\cdots+c^{i-1}$ or the operational equivalent $\frac{1-c^i}{1-c}$ and $\therefore <\frac{1}{1-c}$ whatever whole number i may be. Therefore

$$v_0+v_1+v_2+\cdots+v_{r+i-1}=v_0+v_1+v_2+\cdots+v_{r-1}+Qv_r$$

where Q is some quantity with an absolute value less than $\frac{1}{1-c}$. Likewise and because v_{r+i} has an absolute value not greater than the absolute value of c^iv_r

$$v_0+v_1+v_2+\cdots+v_{r+i+i'-1}=v_0+v_1+v_2+\cdots+v_{r+i-1}+Qc^{i'}v_r$$

and by taking i' great enough the absolute value of $Qc_{.}^{i'}v_r$ may be made less than any given numerical quantity. Hence $v_0+v_1+\cdots+v_{r+i-1}$ has never an absolute value greater than some definite numerical quantity however great i may be and i may be taken so great that $v_0+v_1+\cdots+v_{r+i-1}$ is more nearly equal to $v_0+v_1+\cdots+v_{r+i+i'-1}$ however great i' may be than by any given difference. As i then becomes endlessly great $v_0+v_1+\cdots+v_{r+i}$ becomes endlessly near to some certain quantity.

If a be not a whole number the series $a_{|0}\ a_{|1}x\ a_{|2}x^2\ldots$ is endless where as in art. 184 $a_{|0}$ is 1 and $a_{|1}\ a_{|2}\ a_{|3}\ldots$ follow therefrom in order after the law $a_{|i+1}=\frac{a-i}{i+1}a_{|i}$ and the quotient got from dividing by the ith term the $(i+1)$th $=\frac{a-i+1}{i}x=\left(\frac{a+1}{i}-1\right)x$. If a be a numerical quantity else than a whole number $\frac{a+1}{i}-1$ is minus but of an absolute value less than 1 whenever $i>a+1$ and then x has only to be taken of an absolute value not greater than c for the absolute value of $\left(\frac{a+1}{i}-1\right)x$ to be not greater than c for all values of i greater than $a+1$. If a be a minus quantity either $-a$ not> 1 and then abs. val. of $\frac{a+1}{i}-1$ not> 1 always or $-a>1$ and then although abs. val. $\left(\frac{a+1}{i}-1\right)>1$ yet for abs. val. $\left(\frac{a+1}{i}-1\right)$ to be less than $1+\kappa$ however small a given numerical quantity is κ i has only to be taken greater than $\frac{1}{\kappa}(-1-a)$ wherefore when $-a$ not> 1 abs. val. $\left(\frac{a+1}{i}-1\right)x$ not> c for all values of i if only abs. val. x not> c and when $-a>1$ abs. val. $\left(\frac{a+1}{i}-1\right)x$ not> c for any value of i greater than $\frac{1}{\kappa}(-1-a)$ if abs. val. x not> $\frac{1}{1+\kappa}c$. If a be a ditensive quantity $p+(-)^{\frac{1}{2}}q$ p q being protensives abs. val. $\left(\frac{a+1}{i}-1\right)$

$$=\sqrt{\left\{\left(\frac{p+1}{i}-1\right)^2+\left(\frac{q}{i}\right)^2\right\}}=\sqrt{\left\{1-2\frac{p+1-\frac{1}{2}\frac{(p+1)^2+q^2}{i}}{i}\right\}}$$

$$\text{and}\ \therefore\ <1-\frac{p+1-\frac{1}{2}\frac{(p+1)^2+q^2}{i}}{i}$$

which for all great enough values of i either is not greater than 1 and then x has only to be taken of an absolute value not greater than c for abs. val. $\left(\frac{a+1}{i}-1\right)x$ to be not greater than c or although greater than 1 may yet be made less than any given numerical quantity $1+\kappa$ greater by however little than 1 and then if x be given an absolute value not greater than $\frac{1}{1+\kappa}c$ abs. val. $\left(\frac{a+1}{i}-1\right)x$ not$>c$.

445. If $A_0+A_1x+\cdots+A_{\alpha-1}x^{\alpha-1}+\underline{A_\alpha|}x^\alpha=B_0+B_1x+\cdots+B_{\beta-1}x^{\beta-1}+\underline{B_\beta|}x^\beta$ for all values of x $A_i=B_i$ whatever whole number i is. For if $A_0+A_1x+\cdots+\underline{A_\alpha|}x^\alpha=B_0+B_1x+\cdots+\underline{B_\beta|}x^\beta$

$$A_0-B_0+(A_1-B_1)x+\cdots+(A_{i-1}-B_{i-1})x^{i-1}+\underline{A_i-B_i|}x^i=0$$

and $A_0-B_0+\underline{A_1-B_1|}x=0$ when x is 0 only if $A_0-B_0=0$ or $A_0=B_0$. Therefore

$$\{A_1-B_1+(A_2-B_2)x+\cdots+(A_{i-1}-B_{i-1})x^{i-2}+\underline{A_i-B_i|}x^{i-1}\}x=0$$

for all values of x and therefore when x is not 0 therefore

$$A_1-B_1+(A_2-B_2)x+\cdots+(A_{i-1}-B_{i-1})x^{i-2}+\underline{A_i-B_i|}x^{i-1}=0$$

for all values of x however near to 0 if not 0. Since then $A_1-B_1+\underline{A_2-B_2|}\,x=0$ however near to 0 x may be if not 0 x may be taken so near to 0 as to make A_1-B_1 if not 0 nearer to 0 than by any given difference. But A_1-B_1 is nowise dependent on x and hence A_1-B_1 cannot be else than 0 therefore $A_1=B_1$. Therefore as before $A_2-B_2+\underline{A_3-B_3}\,x=0$ for all values of x however near to if not 0 and since as before this makes A_2-B_2 if not 0 dependent on x $A_2=B_2$ and so on.

446. If a b be any quantities

$$(b_{|0}+b_{|1}x+\cdots+b_{|i-1}x^{i-1}+\underline{b_{|i}|}x^i)(a_{|0}+a_{|1}x+\cdots+a_{|i-1}x^{i-1}+\underline{a_{|i}|}x^i)$$
$$=b_{|0}a_{|0}+(b_{|0}a_{|1}+b_{|1}a_{|0})x+\cdots+(b_{|0}a_{|i-1}+b_{|1}a_{|i-2}+\cdots+b_{|i-1}a_{|0})x^{i-1}$$
$$+\left\{\begin{array}{l}b_{|0}\underline{a_{|i}|}+b_{|1}(a_{|i-1}+\underline{a_{|i}|}x)+b_{|2}(a_{|i-2}+a_{|i-1}x+\underline{a_{|i}|}x^2)+\\ \qquad\cdots+\underline{b_{|i}|}(a_{|0}+a_{|1}x+\cdots+\underline{a_{|i}|}x^i)\end{array}\right\}x^i$$
$$=b_{|0}a_{|0}+(b_{|0}a_{|1}+b_{|1}a_{|0})x+\cdots+\underline{b_{|0}a_{|i}+b_{|1}a_{|i-1}+\cdots b_{|i}a_{|0}|}x^i.$$

Now if a b were any whole numbers each of these operational equivalents (arts. 184, 186, 300, 354, 398) $=(1+x)^b(1+x)^a$ $\therefore$ $=(1+x)^{a+b}$

and $\therefore =(a+b)|_0+(a+b)|_1 x+\cdots+(a+b)|_{i-1}x^{i-1}+\underline{(a+b)|_i}\rfloor x^i$ wherefore (art. 445) if a b be any whole numbers $(a+b)|_i = b|_0 a|_i + b|_1 a|_{i-1}+\cdots+b|_i a|_0$. But the Laws of Operational Equivalence and the Tests of Equality on which alone this rests are the same for fractional numbers as for whole are the same for incommensurable numerical quantities as for commensurable are the same for minus quantities as for plus and are the same for ditensive quantities as for protensive. Therefore whether a b be numbers whole or fractional numerical quantities commensurable or incommensurable protensive quantities plus or minus or algebraic quantities protensive or ditensive still

$$(a+b)|_i = b|_0 a|_i + b|_1 a|_{i-1} + b|_2 a|_{i-2} + \cdots + b|_i a|_0.$$

Hence too if a b be any quantities whatever

$$(b|_0+b|_1 x+\cdots+b|_{i-1}x^{i-1}+\underline{b|_i}\rfloor x^i)(a|_0+a|_1 x+\cdots+a|_{i-1}x^{i-1}+\underline{a|_i}\rfloor x^i)$$
$$= (a+b)|_0+(a+b)|_1 x+\cdots+(a+b)|_{i-1}x^{i-1}+\underline{(a+b)|_i}\rfloor x^i.$$

The operation then which performed on a gives the result $a|_0+a|_1 x+\cdots+a|_{i-1}x^{i-1}+\underline{a|_i}\rfloor x^i$ is an operation which if symbolized by ϕ fulfils the operational equivalence

$$(\phi b)\phi a = \phi(a+b).$$

447. What now can be gathered about an operation ϕ such that $(\phi b)\phi a = \phi(a+b)$?

First $(\phi 0)\phi a = \phi(a+0) = \phi a$ and this is precisely when $(\phi 0 - 1)\phi a = 0$. Therefore either ϕ is an operation which when performed on any quantity a makes the result ϕa always 0 or $\phi 0 = 1$.

Next a b $c \ldots f$ g being any quantities

$$(\phi g)(\phi f)\ldots(\phi c)(\phi b)\phi a = (\phi g)\ldots(\phi c)\phi(a+b) = (\phi g)\ldots(\phi d)\phi(a+b+c)$$
$$= - - - = \phi(a+b+c+\cdots+f+g)$$

and therefore in particular if $a=b=c=- - - =f=g$ and there be n of them

$$(\phi a)^n = \phi n a.$$

Whence $\left(\phi\frac{1}{n}a\right)^n = \phi n\frac{1}{n}a = \phi\left(n\frac{1}{n}\right)a = \phi a$ and $\therefore$ $(\phi a)^{\frac{1}{n}} = (+)^{\frac{1}{n}}\phi\frac{1}{n}a.$

Hence again if m be any whole number

$$(\phi a)^{\frac{m}{n}} = \{(\phi a)^{\frac{1}{n}}\}^m = \{(+)^{\frac{1}{n}}\phi\frac{1}{n}a\}^m = (+)^{\frac{m}{n}}\phi m\frac{1}{n}a = (+)^{\frac{m}{n}}\phi\left(m\frac{1}{n}\right)a = (+)^{\frac{m}{n}}\phi\frac{m}{n}a.$$

Therefore and because an incommensurable numerical quantity is only known through its commensurable approximates if α be any numerical quantity commensurable or incommensurable

$$(\phi a)^{\alpha} = (+)^{\alpha}\phi\alpha a.$$

Lastly if the ϕ operation do not make ϕa always 0 and be therefore such that $\phi 0 = 1$

$$\{\phi(-a)\}\,\phi a = \phi(a-a) = \phi 0 = 1 \quad \text{and} \therefore \phi(-a) = \frac{1}{\phi a}$$

$$\therefore (\phi a)^{-\alpha} = \left(\frac{1}{\phi a}\right)^{\alpha} = \{\phi(-a)\}^{\alpha} = (+)^{\alpha}\phi\alpha(-a) = (+)^{\alpha}\phi(-\alpha a) = (+)^{-\alpha}\phi(-\alpha)a.$$

On the whole then either ϕa is always 0 or when a is any protensive quantity ϕa is equal to a value of $(\phi 1)^{a}$.

448. But $1|_0 + 1|_1 x + \cdots + 1|_{i-1} x^{i-1} + \underline{1|_i}\rfloor x^i = 1 + x$ and $a|_0 + a|_1 x + \cdots + a|_{i-1} x^{i-1} + \underline{a|_i}\rfloor x^i$ partly for that reason is not always equal to 0. Therefore (arts. 446, 447) if a be any protensive quantity $a|_0 + \cdots + \underline{a|_i}\rfloor x^i$ is the paranomic expression operationally equivalent to a value of $(1+x)^{a}$ and therefore

$$(1+x)^{a} = (+)^{a}(a|_0 + a|_1 x + a|_2 x^2 + \cdots + a|_{i-1} x^{i-1} + \underline{a|_i}\rfloor x^i).$$

Thus $(1+x)^{-1} = 1 - x + x^2 - x^3 + \cdots + (-)^{i-1}x^{i-1} + (-)^{i}\underline{1}\rfloor x^i$

$$(1-x)^{-2} = 1 + 2x + 3x^2 + \cdots + ix^{i-1} + \underline{i+1}\rfloor x^i$$

and generally if n be any whole number

$$(1-x)^{-n} = 1 + \frac{n}{\lfloor 1}x + \frac{(n+1)n}{\lfloor 2}x^2 + \frac{(n+2)(n+1)n}{\lfloor 3}x^3 + \cdots + \underline{\frac{\{n+(i-1)\}\ldots(n+2)(n+1)n}{\lfloor i}}\rfloor x^i$$

thus too $(1+x)^{\frac{1}{2}} =$

$$(+)^{\frac{1}{2}}\left\{1 + \frac{1}{2}x - \frac{1}{2.4}x^2 + \frac{3.1}{2.4.6}x^3 - \frac{5.3.1}{2.4.6.8}x^4 + \cdots + (-)^{i-1}\underline{\frac{(i\times 2-3)\ldots 5.3.1}{2.4\ldots(i\times 2-2).i\times 2}}\rfloor x^i\right\}$$

$$(1-x)^{-\frac{1}{2}} = (+)^{\frac{1}{2}}\left\{1 + \frac{1}{2}x + \frac{3.1}{2.4}x^2 + \frac{5.3.1}{2.4.6}x^3 + \cdots + \underline{\frac{(i\times 2-1)\ldots 5.3.1}{2.4\ldots(i\times 2-2).i\times 2}}\rfloor x^i\right\}$$

$$(w+x)^{-\frac{2}{3}} = w^{-\frac{2}{3}}\left[\begin{array}{l} 1 - \frac{2}{3}w^{-1}x + \frac{5.2}{3.6}w^{-2}x^2 - \frac{8.5.2}{3.6.9}w^{-3}x^3 + \\ \qquad \cdots + (-)^{i}\underline{\frac{\{2+(i-1)\times 3\}\ldots 8.5.2}{3.6\ldots\{(i-1)\times 3\}i\times 3}}\rfloor w^{-i}x^i \end{array}\right].$$

449. Since putting $\Sigma_r i$ for the sum of the products of every r of the i whole numbers 1 2 3...i and $\Sigma_0 i$ for 1 after the law (art. 183) $\Sigma_{r+1}(i+1)=\Sigma_{r+1}i+(i+1)\Sigma_r i$ stretched to the case of r being 0

$$\{a-(i-1)\}\ldots(a-2)(a-1)a$$
$$=a^i\Sigma_0(i-1)-a^{i-1}\Sigma_1(i-1)+a^{i-2}\Sigma_2(i-1)-\cdots+(-)^{i-1}a\Sigma_{i-1}(i-1)$$

it follows that a being any quantity

$$a|_0+a|_1x+a|_2x^2+\cdots+a|_{i-1}x^{i-1}+\underline{a|_i}|x^i$$
$$=1+a\left\{\frac{\Sigma_0 0}{\lfloor 1}x-\frac{\Sigma_1 1}{\lfloor 2}x^2+\frac{\Sigma_2 2}{\lfloor 3}x^3-\cdots+(-)^{i-1}\underline{\frac{\Sigma_{i-1}(i-1)}{\lfloor i}\Big|}x^i\right\}$$
$$+a^2\left\{\frac{\Sigma_0 1}{\lfloor 2}x^2-\frac{\Sigma_1 2}{\lfloor 3}x^3+\cdots+(-)^i\underline{\frac{\Sigma_{i-2}(i-1)}{\lfloor i}\Big|}x^i\right\}$$
$$+a^3\left\{\frac{\Sigma_0 2}{\lfloor 3}x^3-\cdots+(-)^{i-1}\underline{\frac{\Sigma_{i-3}(i-1)}{\lfloor i}\Big|}x^i\right\}$$
$$+\ -\ -\ -\ -$$
$$+a^i\underline{\frac{\Sigma_0(i-1)}{\lfloor i}}x^i$$

$$\text{and } \therefore\ =1+a(X+aX')$$

where X stands for $\frac{\Sigma_0 0}{\lfloor 1}x-\frac{\Sigma_1 1}{\lfloor 2}x^2+\cdots+(-)^{i-1}\underline{\frac{\Sigma_{i-1}(i-1)}{\lfloor i}\Big|}x^i$ or the equal $\frac{1}{1}x-\frac{1}{2}x^2+\frac{1}{3}x^3-\cdots+(-)^{i-1}\underline{\frac{1}{i}}\Big|x^i$ and X' for a quantity which may be made more nearly equal to $\frac{\Sigma_0 1}{\lfloor 2}x^2-\frac{\Sigma_1 2}{\lfloor 3}x^3+\cdots+(-)^i\underline{\frac{\Sigma_{i-2}(i-1)}{\lfloor i}\Big|}x^i$ than by any given difference by taking a near enough to 0.

450. Now because when u v are whole numbers $\{(1+x)^v\}^u=(1+x)^{uv}$ and

$$(1+x)^v=v|_0+v|_1x+\cdots+\underline{v|_i}|x^i=1+(v|_1x+v|_2x^2+\cdots+v|_{i-1}x^{i-1}+\underline{v|_i}|x^i)$$
$$u|_0+u|_1(v|_1x+v|_2x^2+\cdots+\underline{v|_i}|x^i)+u|_2(v|_1x+\cdots+\underline{v|_i}|x^i)^2+$$
$$\cdots+\underline{u|_{i'}}|(v|_1x+\cdots+\underline{v|_i}|x^i)^{i'}$$
$$=(uv)|_0+(uv)|_1x+\cdots+(uv)|_{i''-1}x^{i''-1}+\underline{(uv)|_{i''}}|x^{i''}$$

not only when u v are whole numbers but when u v are any quantities whatever for the laws of operational equivalence and the tests of equality whereon the proposition can anywise hang are the same for quantities generally as for whole numbers.

451. Making then $u\frac{a}{v}$ so that $\frac{a}{v}v = a$ whatever quantity v may be else than, however near to, 0 and $v_{|1}x+v_{|2}x^2+\cdots+v_{|i}x^i$ the operationally equivalent $v(X+vX')$

$$a_{|0}+a_{|1}x+\cdots+a_{|i-1}x^{i-1}+a_{|i}x^i = 1+\frac{\left(\frac{a}{v}\right)}{\lfloor 1}v(X+vX')+\frac{\left(\frac{a}{v}-1\right)\frac{a}{v}}{\lfloor 2}\{v(X+vX')\}^2+$$

$$\cdots+\frac{\left\{\frac{a}{v}-(i-1)\right\}\cdots\frac{a}{v}}{\lfloor i}\{v(X+vX')\}^i$$

$$= 1+\frac{a}{\lfloor 1}(X+vX')+\frac{(a-v)a}{\lfloor 2}(X+vX')^2+$$

$$\cdots+\frac{\{a-(i-1)v\}\ldots(a-2v)(a-v)a}{\lfloor i}(X+vX')^i$$

wherefore and because $\frac{\{a-(i-1)v\}\ldots(a-v)a}{\lfloor i}(X+vX')^i$ endlessly nears $\frac{a^i}{\lfloor i}X^i$ as v endlessly nears without being 0

$$a_{|0}+a_{|1}x+a_{|2}x^2+\cdots+a_{|i}x^i = 1+\frac{a}{\lfloor 1}X+\frac{a^2}{\lfloor 2}X^2+\cdots+\frac{a^i}{\lfloor i}X^i$$

or $1+\frac{a}{\lfloor 1}X+\frac{a^2}{\lfloor 2}X^2+\frac{a^3}{3}X^3+\cdots$ since $\frac{\frac{a^i}{\lfloor i}X^i}{\frac{a^{i-1}}{\lfloor i-1}X^{i-1}} = \frac{a}{i}X$ which whatever a X are has an absolute value less than 1 when $i >$ abs. val. Xa.

Hence if a be any protensive quantity

$$(1+x)^a = (+)^a(a_{|0}+a_{|1}x+a_{|2}x^2+\cdots+a_{|i}x^i) = (+)^a\left(1+\frac{a}{\lfloor 1}X+\frac{a^2}{\lfloor 2}X^2+\cdots\right)$$

and in particular $\quad 1+x = 1+\frac{1}{\lfloor 1}X+\frac{1}{\lfloor 2}X^2+\cdots.$

Hence too (art. 446) if a b be any quantities

$$\left(1+\frac{b}{\lfloor 1}X+\frac{b^2}{\lfloor 2}X^2+\cdots\right)\left(1+\frac{a}{\lfloor 1}X+\frac{a^2}{\lfloor 2}X^2+\cdots\right) = 1+\frac{a+b}{\lfloor 1}X+\frac{(a+b)^2}{\lfloor 2}X^2+\cdots.$$

452. If $1+x$ be a numerical quantity and therefore X a protensive quantity

$$\overline{(1+x)^{\frac{1}{x}}} = 1+\frac{\left(\frac{1}{X}\right)}{\lfloor 1}X+\frac{\left(\frac{1}{X}\right)^2}{\lfloor 2}X^2+\cdots = 1+\frac{1}{\lfloor 1}+\frac{1}{\lfloor 2}+\cdots$$

or (art. 368) the base ϵ of the Napier logarithms $\therefore 1+x=\underline{\epsilon^X}$ or $X=\log_\epsilon(1+x)$. Hence

$$\log_\epsilon(1+x)=\frac{1}{1}x-\frac{1}{2}x^2+\frac{1}{3}x^3-\frac{1}{4}x^4+\cdots+(-)^{i-1}\frac{1}{i}\Big|x^i.$$

For the special value ϵ of $1+x$ X is 1 and therefore a being any protensive quantity

$$\underline{\epsilon^a}=1+\frac{a}{\lfloor 1}+\frac{a^2}{\lfloor 2}+\frac{a^3}{\lfloor 3}+\cdots.$$

453. If $1+x$ $1-x$ be numerical quantities

$$\log_\epsilon\frac{1+x}{1-x}=-\log_\epsilon(1-x)+\log_\epsilon(1+x)$$

$$=-\left(-\frac{1}{1}x-\frac{1}{2}x^2-\cdots-\frac{1}{i\times 2+1}\Big|x^{i\times 2+1}\right)+\left(\frac{1}{1}x-\frac{1}{2}x^2+\cdots+\frac{1}{i\times 2+1}\Big|x^{i\times 2+1}\right)$$

$$=2\left(\frac{1}{1}x+\frac{1}{3}x^3+\frac{1}{5}x^5+\cdots+\frac{1}{i\times 2+1}\Big|x^{i\times 2+1}\right)$$

which if abs. val. $x<1$ (art. 443) $=2\left(\frac{1}{1}x+\frac{1}{3}x^3+\frac{1}{5}x^5+\cdots\right)$.

Thus if $\frac{1+x}{1-x}=\frac{w+h}{w}$ w being a numerical quantity and h either a plus quantity or a minus quantity having an absolute value less than w $x=\frac{h}{2w+h}$ and

$$\log_\epsilon(w+h)=\log_\epsilon w+2\left\{\frac{1}{1}\frac{h}{2w+h}+\frac{1}{3}\left(\frac{h}{2w+h}\right)^3+\frac{1}{5}\left(\frac{h}{2w+h}\right)^5+\cdots\right\}.$$

Making h 1 and w in turn 1 2 4 6...

$$\log_\epsilon 2=2\left\{\frac{1}{3}+\frac{1}{3}\left(\frac{1}{3}\right)^3+\frac{1}{5}\left(\frac{1}{3}\right)^5+\cdots\right\} \quad \log_\epsilon 3=\log_\epsilon 2+2\left\{\frac{1}{5}+\frac{1}{3}\left(\frac{1}{5}\right)^3+\cdots\right\}$$

$$\log_\epsilon 5=2\log_\epsilon 2+2\left\{\frac{1}{9}+\frac{1}{3}\left(\frac{1}{9}\right)^3+\cdots\right\}$$

$$\log_\epsilon 7=\log_\epsilon 2+\log_\epsilon 3+2\left\{\frac{1}{13}+\frac{1}{3}\left(\frac{1}{13}\right)^3+\cdots\right\}$$

and so on. Whence a set of Napierian logarithms may be found to any sought degree of nearness. From $\log_\epsilon 10$ so found,— which $= \log_\epsilon 2\times 5 = \log_\epsilon 5+\log_\epsilon 2$—, $\log_{10}\epsilon$ which (art. 366) $= \dfrac{1}{\log_\epsilon 10}$ may be found. As far as the 7th decimal place $\log_{10}\epsilon = 0{\cdot}4342945$. Then $\because \log_{10} w = (\log_\epsilon w)\log_{10}\epsilon$

$$\log_{10} 10001 = \log_{10} 10000+2\left\{\frac{1}{20001}+\frac{1}{3}\left(\frac{1}{20001}\right)^3+\cdots\right\}\log_{10}\epsilon$$

and since making w n^2-1 and h 1

$$\log_\epsilon n^2 = \log_\epsilon (n^2-1)+2\left\{\frac{1}{2n^2-1}+\frac{1}{3}\left(\frac{1}{2n^2-1}\right)^3+\cdots\right\}$$

$$-\log_{10} n+\log_{10}(n+1)$$

$$= -\log_{10}(n-1)+\log_{10} n-2\left\{\frac{1}{2n^2-1}+\frac{1}{3}\left(\frac{1}{2n^2-1}\right)^3+\cdots\right\}\log_{10}\epsilon$$

whence making n in turn 10001 10002 10003 . . . the common logarithms of 10002 10003 . . . may be readily approximated to. Moreover if $\dfrac{h}{w}$ be of a less absolute value than 1

$$-\log_{10} w+\log_{10}(w+h)$$

$$= \log_{10}\left(1+\frac{h}{w}\right) = \left\{\frac{1}{1}\frac{h}{w}-\frac{1}{2}\left(\frac{h}{w}\right)^2+\frac{1}{3}\left(\frac{h}{w}\right)^3-\frac{1}{4}\left(\frac{h}{w}\right)^4+\cdots\right\}\log_{10}\epsilon$$

and therefore if n be any one of the numbers 10000 10001 10002 . . . and abs. val. h not> 1

$$-\log_{10} n+\log_{10}(n+h) = h\frac{1}{n}\log_{10}\epsilon = h\{-\log_{10} n+\log_{10}(n+1)\}$$

as far as the eighth place of decimals.

CHAPTER VII

DITENSIVE GENERALIZATIONS

454. IF $1+x$ be any algebraic and e any numerical quantity then (art. 451) a perigon being the unit angle and X symbolizing

$$\frac{1}{1}x-\frac{1}{2}x^2+\cdots+(-)^i\frac{1}{i-1}x^{i-1}+(-)^{i-1}\frac{1}{i}\Big|_{x^i}$$

$$(1+x)^e=\Theta_{ei}\left(1+\frac{e}{\lfloor 1}X+\frac{e^2}{\lfloor 2}X^2+\cdots\right)$$

and $$(1+x)^{-e}=\Theta_{ei}\left\{1+\frac{-e}{\lfloor 1}X+\frac{(-e)^2}{\lfloor 2}X^2+\cdots\right\}.$$

On this ground is set up by the Symbolization Extension Principle (art. 325) a ditensive indexed power's

Def. If ω be the algebraic expression of any angle in reference to a perigon as unit $(1+x)^{\Theta_\omega e}$ is understood to symbolize what is such that

$$(1+x)^{\Theta_\omega e}=\Theta_{ei}\left\{1+\frac{\Theta_\omega e}{\lfloor 1}X+\frac{(\Theta_\omega e)^2}{\lfloor 2}X^2+\cdots\right\}.$$

Since in particular

$$(1+x)^{\Theta_\omega}=1+\frac{\Theta_\omega}{\lfloor 1}X+\frac{\Theta_\omega^{\,2}}{\lfloor 2}X^2+\cdots=1+\frac{1}{\lfloor 1}\Theta_\omega X+\frac{1}{\lfloor 2}(\Theta_\omega X)^2+\cdots$$

$$\{(1+x)^{\Theta_\omega}\}^e=\Theta_{ei}\left\{1+\frac{e}{\lfloor 1}\Theta_\omega X+\frac{e^2}{\lfloor 2}(\Theta_\omega X)^2+\cdots\right\}$$

$$=\Theta_{ei}\left\{1+\frac{e\Theta_\omega}{\lfloor 1}X+\frac{(e\Theta_\omega)^2}{\lfloor 2}X^2+\cdots\right\}$$

$$=(1+x)^{e\Theta_\omega}=(1+x)^{\Theta_\omega e}.$$

The full consequences of the definition however can only be got at through the straight line equal to the circumference of a circle.

455. Inasmuch as all notion of geometric equality is rooted in the Euclidic postulate "That exact fitters upon one another are equal" and a curve line is a line of which no part is straight it is not at once clear how a curve line can be equal to a straight line. The circumference of a circle is a curve line because in a straight line passing through any two points in the circumference every point between the points is within the circle and every point beyond them without.

Let *ABCDEF* be any line not straight ending at two points *A F* and draw the straight line *AF*. Since *ABCDEF* is not straight and *AF* is straight it cannot be that every point in *AF* is a point in *ABCDEF* let then *G* be a point in *AF* between *A* and *F* that is not in *ABCDEF*. From *A* as center at the distance *AG* describe a sphere. Because *AF* is greater than its own part *AG F* is at a greater distance from *A* the center of the sphere than *G* a point in the surface and therefore than any point in the surface therefore *F* is without the sphere therefore and because the center *A* is within the sphere *ABCDEF* is a line ending at two points *A F* on opposite sides of the spheric surface and therefore *ABCDEF* is cut somewhere between *A* and *F* by that surface. Let then *C* be a point in *ABCDEF* between *A* and *F* that is in the spheric surface and that is not *G* since *G* is not in *ABCDEF* and let *ABC CDEF* be the two parts into which *C* cuts *ABCDEF*. Join *AC FC*. If *C* be in the same straight line with *AF* it can only be the point where that straight line through the center *A* cuts the surface of the sphere on the side of *A* opposite to *G* and then *CF* is greater than its own part *GF* but if *C* be not in the *AF* straight line *CAF* is a triangle of which therefore the two sides *AC CF* are then together greater than the third side *AF* whence taking severally the equals *AC AG CF* is greater than *GF*. From *F* as center at the distance *FG* describe another sphere. Since *C* is at a greater distance from *F* the center of this sphere than *G* a point in the surface and therefore than any point in

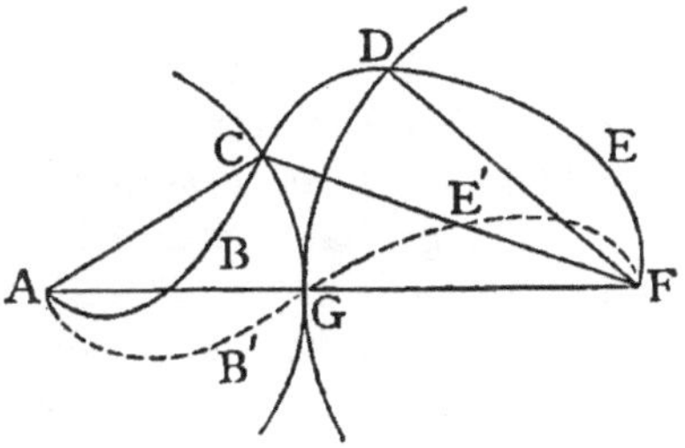

the surface C is without this sphere wherefore and because F is within the line $CDEF$ ends at two points C F on opposite sides of the spheric surface whose center is F and therefore $CDEF$ is cut somewhere between C and F by this surface. Let D be a point in $CDEF$ between C and F common to $CDEF$ and the F centered spheric surface and let CD DEF be the two parts into which D cuts $CDEF$. Join DF. The straight line AC may be exactly fitted upon the equal straight line AG and DF upon the equal GF when AC is so fitted upon AG let $AB'G$ be a line upon which ABC borne along with its fellow line AC exactly fits and when DF is so fitted upon GF let $GE'F$ be a line upon which DEF borne along with its fellow line DF exactly fits so that $AB'GE'F$ is a line joining A F. Because $CDEF$ is greater than its own part DEF and $GE'F = DEF$ $CDEF > GE'F$ therefore to the equals ABC $AB'G$ putting severally those unequals $ABCDEF > AB'GE'F$. Thus than any line $ABCDEF$ which is not straight joining two points A F there is always a less or shorter line $AB'GE'F$ joining the same points. Hence too if there be any line joining two points than which there is no shorter line joining them that line can be no other than the straight line joining the points. All this, however little it of itself help toward, must yet be taken account of in, settling what is meant by the equality of a curve line and a straight line.

Def. A straight line passing through a point in a curve line is said to TOUCH, or be TOUCHED by, the curve line at the point when of straight lines drawn each through the point and severally through points taken ever nearer and nearer thereto in some arc of the curve ending thereat the portions upon the same side of the point as the several points make ever less and less angles with the portion of the straight line upon one side of the point and at length after the points are taken near enough to the point make with that portion angles less than any given angle however small.

Let an arc ABC of a plane curve touch at the ends A C two adjoining sides AD CD of a parallelogram $ADCE$ lie wholly on one side of every straight line touching it either at or anywhere between A C and have a portion close to A on the same side of the diagonal AC as AD. Were ABC to meet AD at any other point X than A a straight line $A\alpha$ might be drawn from A to a point α in ABC so near to A as by the definition of TOUCHING to make

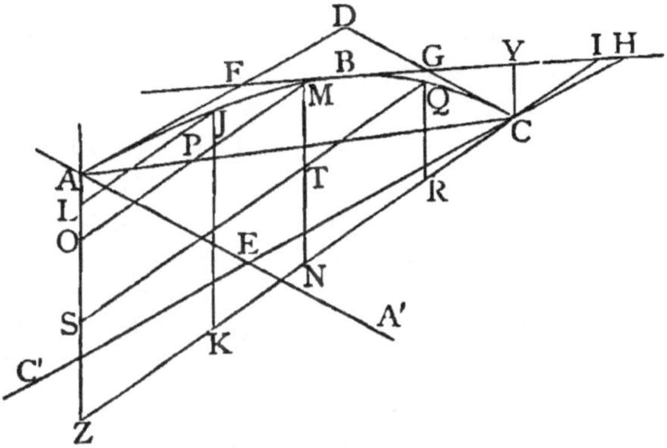

with AD an angle $DA\alpha$ less than any given angle also α might be taken so near to A that αA should make with $\alpha\beta$ the portion on the same side of αA as AD or as the small arc between A and α of a straight line $\gamma\alpha\beta$ touching ABC at α an angle less than any given angle hence α might be taken so near to A that AD $\alpha\beta$ should make with $A\alpha$ at the different points A α interior angles αAD $A\alpha\beta$ upon one side together less than a hemiperigon or than any given angle and therefore meet at some point β on that side then would $\beta A\alpha$ be a triangle with the side $A\beta$ produced to D therefore the exterior angle $D\beta\alpha$ would be equal to the off lying interior angles $\beta A\alpha$ $\beta\alpha A$ together hence α might be taken so near to A as to make $D\beta\alpha$ less than any given angle and therefore the supplementary angle $A\beta\alpha$ greater than any given angle less than a hemiperigon so therefore that the angle $A\alpha\beta$ should be less than $A\beta\alpha$ and therefore the side $A\beta$ of the triangle $A\alpha\beta$ less than the side $A\alpha$ but α might be taken so near to A as to make $A\alpha$ less than any given straight line much more therefore might α be taken so near to A as to make $A\beta$ less than any given straight line less therefore than AX and then A X would be on opposite sides of β and therefore on opposite sides of $\beta\gamma$ which meets and does not lie wholly together with and therefore cuts AX therefore portions of ABC close to A X would lie on opposite sides of $\beta\gamma$ and hence ABC cannot meet AD elsewhere than at A. Next ABC cannot but lie wholly on that side of AC whereon lie AD CD the point D and the triangle DAC for a portion of ABC ending at any point but A in the portion close to A and at a point on the other side of AC would be cut somewhere between those points by the AC straight line and the straight line touching ABC at a point X in the cutting either would be itself the AC straight line and so have the points on opposite sides of it or would cut the AC straight line and then since ABC is wholly on one side of AD and wholly on one side of CD X would be on the C side of A and on the A side of C would therefore be between A and C and therefore the portions of ABC close to A and C would be on opposite sides of the toucher at X. Again ABC thus wholly on the D side of AC could only meet AC between A and C at a point X by there touching AC and then as before portions of ABC close to A X would be on opposite sides of some straight line touching ABC between A and X and portions close to X C on opposite sides of some straight line touching ABC between X and C wherefore ABC can meet AC only at A and C. Further since the portion of ABC close to C is on the same side of AC as DC ABC as before cannot meet DC elsewhere than at C. Thus ABC is wholly on the same side of each of the sides DA AC CD of the triangle DAC as DAC and meets them at A C only any point B therefore in ABC between A and C is within the triangle DAC and therefore a straight line FBG touching ABC at B cuts the boundary of DAC twice once on each side of B but ABC because wholly on one side of FBG can as before meet FBG at B only and therefore A C and all points in AC between A and C are on one side of FG therefore the points where FG cuts DAC's boundary can only be somewhere between A and C in the line ADC made up of the straight parts AD DC moreover neither of these points can be at D since then FG would meet at two points and yet not lie wholly together with AD or CD therefore FG cuts AD at some point F between A and D and CD at some point G between C and D. Because FG cuts AD one of two parallels AD EC it cuts the straight line of the other EC in some point H and because in DC D C are on opposite sides of G the parallels AD EC are on opposite sides of a parallel to either through G therefore F H are on opposite sides of this parallel through G and therefore F H are on opposite sides of G. Produce AE endlessly to A' and CE endlessly to C' take a point Z anywhere on the same side of AA' as EC' and on the same side of CC' as EA' and therefore on the opposite side of AA' to EC and on the opposite side of CC' to EA and from Z through C draw a straight line ZCI. Because Z and C are on opposite sides of AA' Z and DC are therefore ZI meets at C and does not lie wholly together with and therefore cuts DC and Z because on the same side of AA' as EC' is on the same side of DC as CC' therefore CZ CC' are both on one side of DC and CI CH both on the other moreover because Z is on the opposite side of CC' to EA and therefore to *CD ZI* cuts *C'H* and *CD CI* lie on the same side of *C'H* hence from one end *C* and on one side of *CD CH* is drawn and from the same end and on the same side of *CD CI* is drawn and *CI* is on the same side of *CH* as *CD* therefore *CI* is on the same side of each of the straight lines *CD CH* as the other *CI* is between *CD* and *CH CI* cuts atwo the angle *DCH CD CH* make angles *ICD ICH* with which lie upon opposite sides of *CI* and *CD CH* are on opposite sides of *CI* therefore *G H* are on opposite sides of *CI* and therefore *GH* is cut in some point *I* between *G* and *H* by *CI*. Likewise from *Z* through *A* draw a straight line *ZAi* and the *EA* straight line cuts *FG* in a point *h* on the side of *A* opposite to *E* and on the side of *F* opposite to *G* and *Zi* cuts *FG* at a point *i* between *F* and *h*. From *C* not in *AZ* draw *CY* parallel to *AZ* and on the opposite side of *AC* to *AZ* or on the same side of *AC* as *CI* then because the angle *ACY* is equal to the alternate angle *CAZ* and the angle *ACD* to the alternate angle *CAE* which is a part of and therefore less than *CAZ* the angle *ACD* is less than *ACY* but these unequal angles are at a common point *C* have a common bounding line *CA* ending at *C* lie in the same plane and begin together upon the same side of *CA* therefore *ACD* is a part of *ACY* and this can be only by *CY* being on the opposite side of *CD* to *CA CE* or *CZ* and therefore on the same side of *CD* as *CI* again because the side *ZC* of the triangle *ZCA* is produced to *I* the exterior angle *ACI* is greater than the off lying interior angle *CAZ* and therefore than the equal angle *ACY* but the angles *ACI ACY* are at a common point *C* have a common bounder *CA* ending at *C* lie in the same plane and begin together upon the same side of *CA* therefore the less *ACY* is a part of the greater *ACI* and this can only be by *CY* being on the same side of *CI* as *CA* or *CD* hence from one end and on one side of *CD CI* is drawn and from the same end and on the same side of *CD CY* is drawn and *CY* is on the same side of *CI* as *CD* therefore among other things *CD CI* are on opposite sides of *CY* so therefore are *G I* and therefore the line *GI* ending at *G I* on opposite sides of *CY* is cut at some point *Y* between *G* and *I* by *CY*. Likewise a straight line *Ay* drawn from *A* parallel to *CZ* on the side of *AC* opposite to *CZ* or on the same side of *AC* as *Ai* cuts *Fi* at some point *y* between *F* and *i*. Now because in the straight line *FG i Y* are on opposite sides of *B* a parallel *BΔΛ* through *B* to *AZ* or *CY* cuts *AC* at a point Δ between *A* and *C* and *ZC* at a point Λ between *Z* and *C* and because *A C* are thus on opposite sides of

$B\Lambda$ ABC is cut between A and C by $B\Lambda$. If either of the two parts AB BC into which B cuts ABC met $B\Delta$ anywhere between B and Δ the straight line touching ABC there either would be $B\Delta$ and so have on opposite sides of it ABC's portions close to A C or would cut $B\Delta$ and so have on opposite sides of it the portion of ABC close to upon either side of B and at least one of ABC's portions close to A C wherefore every point but B in AB is on one side of $B\Lambda$ and every point but B in BC on the other. Also a parallel $B\delta\lambda$ through B to CZ or Ay cuts AC at a point δ between A and C and ZA at a point λ between Z and A and every point but B in AB is on one side of $B\lambda$ and every point but B in BC on the other. Since then every point but B in BC is on the same side of $B\Lambda$ as CY and since every point but C in BC is on the same side of CD and therefore of CY as A any point Φ in BC between B and C is on the CY side of $B\Lambda$ and on the $B\Lambda$ side of CY therefore a parallel $\Phi\Psi$ through Φ to $B\Lambda$ or CY lies between $B\Lambda$ and CY therefore has $B\Lambda$ CY on opposite sides of it and therefore cuts BY in a point Ψ between B and Y and ΛC in a point Ω between Λ and C. In like ways of points in BC every one but B is on the CZ side of $B\lambda$ and every one but C on the $B\lambda$ side of CZ a parallel $\Phi\omega$ through Φ to $B\lambda$ or CZ therefore is between $B\lambda$ and CZ therefore has $B\lambda$ CZ on opposite sides of it and therefore cuts λZ in a point ω between λ and Z and $B\Lambda$ in a point V between B and Λ. Because too of each of the parallelograms $\lambda B\Lambda Z$ $V\Phi\Omega\Lambda$ $\lambda BV\omega$ $\omega\Phi\Omega Z$ any opposite two sides are equal $\lambda B = Z\Lambda$ $V\Phi = \Lambda\Omega$ $BV = \lambda\omega$ $\Phi\Omega = \omega Z$. Thus on the whole if ABC be cut into any parts AJ JM MQ QC by first cutting ABC into any two parts AJ JC next JC into any two parts JM MC then MC into any two MQ QC and so on parallels JK JL to AZ CZ severally drawn through J cut atwo severally ZC in a point K and AZ in a point L a parallel MN through M to AZ cuts atwo KC in a point N and a parallel MO through M to CZ cuts atwo LZ in a point O and JK in a point P a parallel QR through Q to AZ cuts atwo NC in a point R and a parallel QS through Q to CZ cuts atwo OZ in a point S and MN in a point T and so on so that ZC is cut into as many parts ZK KN NR RC and AZ into as many AL LO OS SZ as ABC moreover LJ PM TQ are equal severally to ZK KN NR and JP MT QR severally to LO OS SZ and therefore the zigzag line $ALJPMTQRC$ made up of the straight parts AL LJ JP PM MT TQ QR RC is equal to the line AZC made up of the straight parts AZ ZC. The points A J M Q C that cut ABC into parts may be everywhere taken following one another so near as to have each straight part of every two straight parted line ending at the ends of a part less than any given straight line and therefore so that ABC and the zigzag may everywhere be at a less than any given distance from one another along a straight line parallel to either AZ or CZ. The endless zigzags that can be thus drawn each everywhere endlessly near to ABC are all equal to the two straight parted AZC and therefore to one another. By taking another point Z' than Z on the EC' side of AA' and on the EA' side of CC' other endless zigzags may be got in the same way each everywhere endlessly near to ABC each equal to the two straight parted line $AZ'C$ and therefore all equal to one another and this line $AZ'C$ need not be equal to the line AZC. Hence two zigzag lines may be each everywhere endlessly near to the same curve line and yet be not equal. Here however the endlessly near coincidence anywhere of the zigzag and the curve line is only as to position while nearness of coincidence anywhere as to direction is not even tried at.

The direction at a point of a curve line's portion ending at the point can only be taken to be the direction of that straight line drawn from the point to which straight lines drawn from the point through points in the portion become endlessly near as the points become endlessly near to the point and this straight line is the straight line touching the portion at the point. Hence if an ended curve line be wholly cut into any arcs none of them greater than half the curve line and the line be drawn made up of the chords of all the arcs each of the arcs be wholly cut into any arcs none greater than half the arc and the line drawn made up of the chords of all these arcs into which the curve is then wholly cut each of these arcs again wholly cut into any arcs none greater than the half and the line drawn made up of the chords of all the new arcs into which then the curve is wholly cut and so on the lines so drawn become at length endlessly near in direction to the curve line at every point of cutting. Moreover by making every point in the curve line when plane a point of cutting where either the portions close to upon opposite sides of the point lie on opposite sides of the toucher or the portions close to the point have touchers not in one straight line the curve may be wholly cut into such arcs that each of them lies wholly on one side of every straight line touching it either at its ends or anywhere between and has the portion close to one end on the same side of the chord as the toucher at that end therefore zigzag lines may then be drawn everywhere endlessly near to the curve and therefore much more the lines made up of chords. Every condition of endlessly near positional and directional coincidence is fulfilled by the

Def. A curve line is said to be EQUAL to a straight line when while the chords of any arcs into which the curve line may be wholly cut are together less than the straight line yet of a great enough number of small enough arcs which make up as parts the curve line the chords are together greater than any given straight line less than the straight line.

If a line *ABCDEF* be wholly made up of straight parts *AB BC CD DE EF* whereof no adjoining two are in one straight line and *AC AD AE AF* be joined the line *ABC* made up of two sides of a triangle is greater than the third side *AC* therefore to each putting *CD* $ABCD > ACD$ but *ACD* either is *AD* to wit if *AC CD* be in one straight line and upon opposite sides of *C* or is greater than *AD* to wit if *AC CD* be either in the same straight line and upon the same side of *C* or two sides of a triangle whereof the third side is *AD* $\therefore$ $ABCD > AD$ whence in the same way $ABCDE > AE$ $ABCDEF > AF$. Hence of any arcs into which a curve line joining two points may be wholly cut the chords are together greater than the straight line joining the points and much more therefore by the definition of the equality of a curve line and a straight line is the curve line greater than the straight line. Hence further of all lines joining the same two points the straight line is less than any other.

456. An arc *ABCDE* of a plane curve if it touch at the ends *A E* two sides *AF EF* of a triangle *AEF* lie wholly on one side of every straight line that anywhere touches it and have some portion close to one end *A* on the same side of the triangle's third side *AE* as the triangle is less than the touched sides *AF EF* together. For as shown in art. 455 every point in *ABCDE* but *A* and *E* is within the triangle *FAE* and a straight line *GCH* drawn touching *ABCDE* at any point *C* taken in *ABCDE* between *A* and *E* cuts *AF* at some point *G* be-

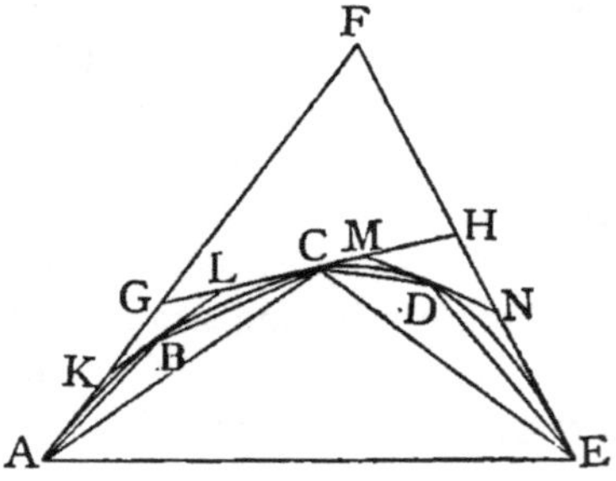

tween A and F and EF at some point H between E and F and shuts in with GF HF a triangle FGH. If then the chords AC CE be drawn of the arcs ABC CDE whereinto C cuts $ABCDE$ CAE CAG CEH are triangles therefore from the triangle ACE $(AC,CE) > AE$ and from the triangles AGC CHE $(AG,GC) > AC$ $(CH,HE) > CE$ and $\therefore$ $(AG,GH,HE) > (AC,CE)$. Moreover since from the triangle FGH $(GF,FH) > GH$ by putting to (AG,HE) these several unequals $(AF,FE) > (AG,GH,HE)$. Again the arc ABC touches at the ends A C the two sides AG CG of the triangle GAC lies wholly on one side of every straight line anywhere touching it and has a portion close to the end A on the same side of the triangle GAC's third side AC as GAC therefore as in art. 455 every point in ABC but A and C is within GAC a toucher KBL to ABC at any point B between A and C cuts AG at a point K between A and G and GC at a point L between G and C and shuts in with GA GC a triangle GKL and if the chords AB BC be drawn of the arcs whereinto B cuts ABC BAC BAK BCL are triangles. Likewise any point D in the arc CDE between C and E is within the triangle HCE a toucher to CDE at D cuts CH at a point M between C and H and HE at a point N between H and E and if the chords CD DE be drawn of the arcs whereinto D cuts CDE DCE DCM DEN and HMN are triangles. Hence as before AC (AB,BC) (AK,KL,LC) (AG,GC) are in ascending order of greatness CE (CD,DE) (CM,MN,NE) (CH,HE) are in ascending order of greatness and therefore to the first four unequals putting severally the other four (AC,CE) (AB,BC,CD,DE) (AK,KL,LM,MN,NE) (AG,GH,HE) are in ascending order of greatness. Each of the arcs into which $ABCDE$ is cut by B C D has every point in it but its ends within the triangle bounded by its chord and the touchers at its ends the same process may therefore be gone through with these arcs as was gone through with the arcs ABC CDE thus giving rise to a greater number of smaller arcs making up $ABCDE$ the aggregate of whose chords is greater than (AB,BC,CD,DE) the aggregate of whose end touchers is less than (AK,KL,LM,MN,NE) and such that the former aggregate is less than the latter the same process may then for the like reason be gone through with the arcs last got and so on for ever. Let C_1 C_2 C_3... be severally the chord AE and the ever greater and greater chord aggregates (AC,CE) (AB,BC,CD,DE) ... and T_1 T_2 T_3 ... the several ever lessening toucher aggregates (AF,FE) (AG,GH,HE) (AK,KL,LM,MN,NE) ... which yet are greater severally than

$C_1\ C_2\ C_3 \ldots$ then if $ABCDE$ were not less than T_1 it would be greater than T_2 therefore by the definition (art. 455) of the equality of a curve line and a straight line some term C_i of the series $C_1\ C_2\ C_3 \ldots$ would were the arcs taken small enough at length be come to greater than T_2 much greater therefore would C_i be than T_i. Wherefore $ABCDE$ cannot but be less than T_1.

457. If the circumference of a circle be cut at points $A\ B\ C\ D\ E$ into more than two arcs each less than the semicircumference and the chords $AB\ BC\ CD\ DE\ EA$ be drawn the two arcs $AEDCB$ AB into which $A\ B$ cut the circumference lie wholly on opposite sides of, and have no point but their ends $A\ B$ in common with, the straight line through $A\ B$ therefore the chord AE meets at A only, the chord BC at B only, and the chords $ED\ DC$ nowhere, that straight line therefore the chord AB and the line made up of the chords $AE\ ED\ DC$ CB end at the same two points and only there meet one another therefore they shut in a portion of surface or bound a figure hence and since any other neighbouring two than $A\ B$ might have been taken of the points $A\ B\ C\ D\ E$ the chords AB $BC\ CD\ DE\ EA$ are the sides of a polygon lying wholly upon one side of each of those sides and upon the opposite side of each to the circular segment whose arc is less than a semicircumference.

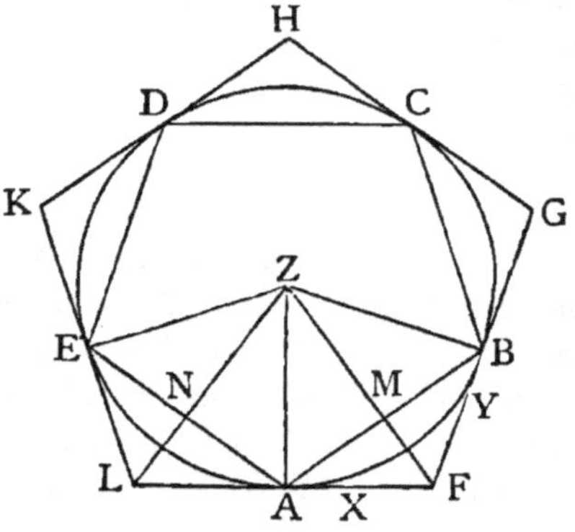

If $AX\ BY$ be those portions of the straight lines touching the circle at $A\ B$ severally which are on the same side of the chord AB as the arc AB each of the angles $BAX\ ABY$ is equal to the angle in the circular segment upon the other side of the chord AB but because the arc AB is less than a semicircumference and therefore the arc $AEDCB$ greater the angle at the circle's center standing on the arc AB is less and the angle at the circle's center standing on the arc $AEDCB$ is greater than a hemiperigon therefore the circle's centre is within the circular segment $AEDCB$ and therefore the angle in this segment is less than a right angle hence the angles $BAX\ ABY$ are each less than a right angle and therefore together less than a hemiperigon therefore *AX BY* meet one another at some point *F* on the same side of the chord *AB* as the arc *AB* is on and then *FAB* is a triangle upon the opposite side of the chord *AB* to the polygon *ABCDE*. If likewise the touchers at *B C* meet one another at *G* those at *C D* at *H* those at *D E* at *K* and those at *E A* at *L GBC HCD KDE LEA* are triangles upon the opposite sides of the several chords *BC CD DE EA* to the polygon *ABCDE*. The polygon *ABCDE* and the triangle *ABF* because upon, and upon opposite sides of, the chord *AB* make up as parts a whole polygon *AFBCDEA* then because the portions *BF BG* of the toucher at, close to, and upon opposite sides of, *B* are on opposite sides of the chord *BC* the polygon *AFBCDEA* and the triangle *BCG* are upon opposite sides of the chord *BC* and therefore make up as parts a whole polygon *AFGCDEA* then likewise because the portions *CG CH* of *GH* are on opposite sides of the chord *CD* the polygon *AFGCDEA* and the triangle *CDH* because upon opposite sides of the chord *CD* make up as parts a whole polygon *AFGHDEA* then likewise the polygon *AFGHDEA* and the triangle *DEK* upon opposite sides of the chord *DE* make up a whole polygon *AFGHKEA* and at last this polygon and the triangle *EAL* upon opposite sides of the chord *EA* make up a whole polygon *FGHKL*. Of the polygon *FGHKL* thus circumscribed about the circle the sides are as many as the points where they severally touch therefore as many as first ends of the arcs in order into which the circumference is wholly cut or as last ends of those arcs therefore as many as the arcs and therefore as many as the chords of the arcs or as the sides of the inscribed polygon *ABCDE*.

Since the arc *AB* touches at the ends two sides *AF BF* of the triangle *FAB* lies wholly on one side of every straight line touching it and has the portion close to *A* on the same side of the third side the chord *AB* as the triangle (art. 456) $(AF, FB) >$ arc *AB*. Likewise $(BG, GC) >$ arc *BC* $(CH, HD) >$ arc *CD* $(DK, KE) >$ arc *DE* $(EL, LA) >$ arc *EA*. Therefore the perimeter of the circumscribed polygon *FGHKL* made up of the greaters of those several pairs of unequals is greater than the circumference made up of the lesses. The circumference of the circle by the definition of the equality of a curve line and a straight (art. 455) is greater than the perimeter of the inscribed polygon *ABCDE*.

If the arcs *AB BC CD DE EA* be all equal it is enough that there be more than two of them (art. 92) for each to be less than the semicircumference then because equal arcs of equal circles have equal chords all the sides of the polygon *ABCDE* are equal moreover putting to the equal arcs *BC EA* the arc *CDE* the arc *BCDE* is equal to the arc *AEDC* and therefore the angles *BAE ABC* at the circumference standing on these equal arcs are equal so any other neighbouring two angles are equal of the polygon *ABCDE* and therefore all *ABCDE*'s angles are equal. Hence the inscribed polygon *ABCDE* is regular.

Find *Z* the center of the circle which because as above shown on the same side of every side of the polygon *ABCDE* as the polygon *ABCDE* is within that polygon and join *ZE ZA ZB ZL ZF*. The angles *BAF ABF* of the triangle *FAB* are equal because each equal to the angle in the circular segment upon the other side of the chord *AB* to that upon which they are therefore the sides *BF AF* of that triangle severally over against them are equal and *ZB ZA* are equal because straight lines drawn from the center to the circumference hence the points *B A* are equidistant from each of the points *F Z* and therefore are on opposite sides of the straight line *FZ* therefore *BFZ* is a triangle upon one side of *FZ* and *AFZ* a triangle upon the other. Since the straight line *FG* touches the circle the straight line *ZB* drawn from the center *Z* to *B* the point of touching is perpendicular to *FG* and since then the angle *FBZ* of the triangle *BFZ* is a right angle each of the other angles *ZFB FZB* is less than a right angle in the same way the angle *FAZ* of the triangle *AFZ* is a right angle and each of the other two angles *ZFA FZA* less than a right angle hence the angles *ZFB ZFA* upon opposite sides of *FZ* and each less than a right angle make up the whole angle *BFA* and also the angles *FZB FZA* make up the angle *BZA*. Moreover the triangles *AZF BZF* have the two sides *ZA AF* equal severally to the two sides *ZB BF* and the right angles *ZAF ZBF* equal therefore the bases are equal the triangles are equal and of the other angles $ZFA = ZFB$ and $FZA = FZB$. Hence *ZF* bisects each of the angles *AZB AFB*. In like way *ELZ ALZ* are equal and similar right angled triangles upon opposite sides of *ZL* and *ZL* bisects each of the angles *AZE ALE*. Now since the angles *AZE AZB* at the center *Z* standing on the equal arcs *AE AB* are equal their halves *AZL AZF* are equal and since *ZA* is perpendicular to *LF* the angles *ZAL ZAF* are equal the triangles *LZA FZA* then have the two angles *AZL ZAL* equal severally to the two angles *AZF ZAF* and the side *ZA* between those angles is common to the triangles therefore $AL = AF$ $ZL = ZF$ and the third angle *ZLA* is equal to the third angle *ZFA*. Of the equal angles *ZLA ZFA* then the doubles *KLF LFG* are equal and in the same way any other neighbouring two angles of the polygon *FGHKL* are equal therefore the angles of *FGHKL* are all equal. For the like reason too that *FL* is bisected in *A* is *FG* bisected in *B* and then of the equals *FA FB* the doubles *FL FG* are equal in the same way are any other adjoining two sides of *FGHKL* equal and therefore all the sides of *FGHKL* are equal.

Let *M* be the point where *ZF* because ending at two points *Z F* on opposite sides of the chord *AB*'s straight line is cut between *Z* and *F* by that straight line and therefore since the straight lines of *ZF* and the chord *AB* meet and do not lie wholly together also the point where the chord *AB* ending at two points *A B* on opposite sides of the *ZF* straight line is cut between *A* and *B* by the *ZF* straight line and let *N* be the point where in like manner each of the straight lines *ZL* and the chord *AE* is cut between its ends by the other. The triangles *ZAM ZBM* have the two sides *AZ ZM* equal severally to the two sides *BZ ZM* and the angles *AZM BZM* equal therefore the bases *AM BM* are equal the triangles are equal and of the other angles each two are equal that are over against equal sides and particularly *AMZ BMZ* wherefore the chord *AB* is bisected at *M* and *ZM* is perpendicular to the chord *AB*. Likewise *ZN* bisects perpendicularly the chord *AE*. Hence the triangles *ZMA ZNA* have the angles at *Z M* equal severally to the angles at *Z N* and the side *ZA* in common over against the equal angles at *M N* therefore among other things $ZM = ZN$. In like ways the perpendiculars are equal drawn from *Z* to any other adjoining two sides of the polygon *ABCDE* and therefore all the perpendiculars from *Z* to the sides of *ABCDE* are equal. Since *BM* is the perpendicular from the right angle of the triangle *BZF* to the side opposite the triangles *ZBF ZMB* are similar $\therefore ZB : ZM = BF : MB$ therefore and $\because AB = 2(MB)$ and $FG = 2(BF)$ $ZB : ZM = FG : AB$ but because *FGHKL*'s equal sides are just as many as *ABCDE*'s *FGHKL*'s perimeter is the same multiple of *FG* as *ABCDE*'s of *AB* and therefore *FGHKL*'s perimeter has to *ABCDE*'s perimeter the same ratio as the radius of the circle to the perpendicular drawn from the center to any side of *ABCDE*.

458. Two regular polygons of the same number of sides may hence be the one inscribed in and the other circumscribed about a given circle *ABC* whose perimeters differ from one another by less

than any given straight line *MN*. Draw a diameter *AB* of the circle produce *MN* endlessly to *P* from *NP* cut off close to *N NP* equal to 4(*AB*) find the center *O* of the circle or the middle of *AB* from the *OA* straight line cut off close to *O OD* equal to a fourth proportionate to *MP NP OA* and therefore inasmuch as *MP* is greater than its own part *NP* less than *OA* since *OD OA* then are unequal straight lines having a common end *O* and lying in the same straight line upon the same side of *O OD* is a part of *OA* and since the point *D* is at a less distance from the center *O* than *A* a point in the circumference and therefore than any point in the circumference *D* is within the circle through *D* draw a straight line *FDE* cutting at right angles *AB* and let *F E* be the points on opposite sides of *D* where *FDE* because passing through the point *D* within the circle cuts the circumference since *FDE* cuts *AB F E* are on opposite sides of *AB* let *ACB* be the semicircumference ending at *A B* that lies on the same side of *AB* as *DE* and therefore *E* is a point in *ACB* between *A* and *B* bisect the semicircumference *ACB* in *C* and join *OE* because *O D E* are three points not in one straight line *ODE* is a triangle and because this triangle's angle *ODE* is a right angle the angle *DOE* is less than a right angle therefore the arc *AE* whereon this angle at the center stands is less than one-fourth of the circumference and therefore less than the arc *AC* bisect *AC* in *G AG* in *H* and so on then since from the arc *ACB* there is taken *CB* not less than *ACB*'s half from the remainder *AC GC* not less than *AC*'s half from the new remainder *AG HG* not less than *AG*'s half and so on there is (art. 232) at length left a remainder less than *AE* let the remainder so left be the arc *AH* and join *OH* with *OA* make at the end *O* an angle *AOK* equal to the angle *AOH* and beginning upon the same side of *OA* as the semicircumference *AFB* or upon the opposite side to *ACB* and let *K* be the point where *OK* a straight line drawn from *O* a point within the circle cuts the circumference the arcs *AK AH* are equal whereon stand the equal angles *AOK AOH* at the

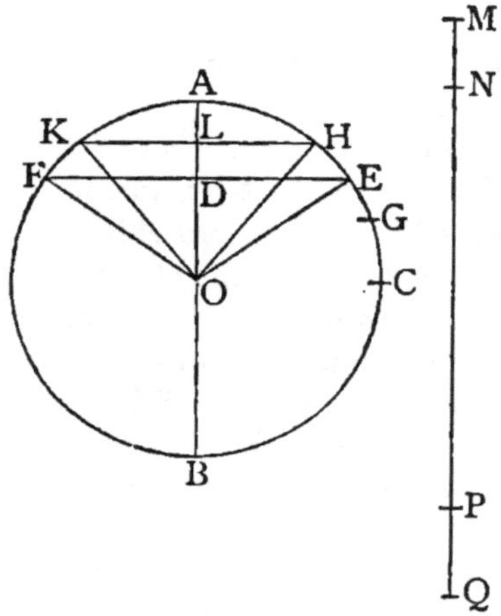

center O but AH is a part of and therefore less than the semicircumference ACB therefore AK is less than the semicircumference AFB moreover AK AFB have a common end A lie in the same line the circle's circumference and begin together upon the same side of their common end wherefore AK cannot but be a part of AFB and therefore K is in AFB between A and B join HK and since H K are on opposite sides of the AB straight line being in between the ends of the several semicircumferences ACB AFB let L be the point between H and K where HK is cut by the AB straight line L cannot be at O for then would the equal angles AOH AOK at the center be the angles ALH ALK that LA makes with upon one side of HK and therefore the arcs AH AK would be quarter circumferences therefore OLH OLK are triangles in these triangles the two sides LO OH are equal severally to the two sides LO OK and the angles LOH LOK are equal because the same either as AOH AOK or as the supplements BOH BOK therefore the bases LH LK are equal the triangles are equal and of the other angles $OLH = OLK$ now $\because$ $AH < AE$ $AOH < AOE$ but AOE is less than a right angle much more therefore is AOH less than a right angle therefore the supplement BOH is greater than a right angle and $\therefore$ $AOH < BOH$ therefore the perpendicular HL from H to the AB straight line is on that side of OH whereon is the less angle AOH or L is on the A side of O moreover L because a point in HK between H and K is within the circle and therefore on the O side of A hence L is between A and O produce MN endlessly to Q and from NQ cut off close to N NQ equal to a fourth proportionate to AL LO MN join OF as before ODF is a triangle the straight line OD passing through the center of the circle because it cuts at right angles the straight line EF in the circle not passing through the center bisects EF the triangles ODF ODE then have the two sides DO OF equal severally to the two sides DO OE and the bases DF DE equal therefore the angle DOF is equal to the angle DOE so that the angle EOF is double of AOE but angle $AOE >$ angle AOH $\therefore$ angle $EOF >$ angle HOK since then the triangles OEF OHK have EO OF equal severally to HO OK and the angle EOF greater than HOK $EF > HK$ of these unequal straight lines in the circle then the greater is nearer to the center than the less or $OD < OL$ and taking each from OA $AD > AL$ $\therefore$ (art. 263) $AD : DO > AL : DO$ and $AL : DO > AL : LO$ $\therefore$ (art. 264) $AD : DO > AL : LO$ and $\therefore$ (art. 262) $> MN : NQ$ but $\because$ $MP : NP = AO : DO$ disjointly (art. 285) $MN : NP = AD : DO$

$\therefore$ (art. 262) $MN : NP > MN : NQ$ and $\therefore$ (art. 265) $NP < NQ$ $\therefore$ (art. 9) $NQ > 4(AB)$. Now since the semicircumference ACB is a multiple of the arc AH the whole circumference is (art. 248) the same multiple of AH's double the arc KAH besides the parts each equal to AH that wholly make up the semicircumference are at least four therefore the circle's circumference may be wholly cut into more than two equal arcs whereof the arc KAH is one therefore (art. 457) a regular polygon whereof KH is one side may be inscribed in the circle and a regular polygon of just as many sides may be circumscribed about the circle with two adjoining sides touching the circle at K H and if R R' be the several perimeters of these polygons $R' : R = OA : OL$ wherefore disjointly $R'|R : R = AL : LO$ and $\therefore = MN : NQ$ but if the circumference be halved by the ends A B of a diameter AB and the halves severally halved by the ends of a diameter at right angles to AB the polygon whose sides touch the circle at the four diameter ends is a circumscribed square whereof each side is equal to the diameter and the perimeter therefore is $4(AB)$ the circumference therefore $< 4(AB)$ moreover $R <$ the circumference and $4(AB) < NQ$ $\therefore$ $R < NQ$ and $\therefore$ (art. 276) $R'|R < MN$.

459. Let AB be a side of a regular polygon inscribed in a circle ABC by drawing the chords of more than two equal arcs into which the circumference is wholly cut through the circle's center O which is not in AB because AB is not a diameter and AB's middle point E which is within the circle because in between the ends of AB draw a straight line and let C D be the several points where this straight line cuts the circumference on the opposite side of E to O and on the same side of E as O this straight line $CEOD$ too because it meets and does not lie wholly together with AB meets at E only and there cuts AB so that D is on the same side of AB as O and C on the opposite side it is therefore the arc ACB which is one of the equal arcs making up the whole circumference and the arc ADB which is made up of all the rest join OA OB since O is not in AB OEA OEB are triangles and OEA OEB have the two sides EO OA equal severally to the two EO OB and the bases EA EB equal therefore the angles EOA

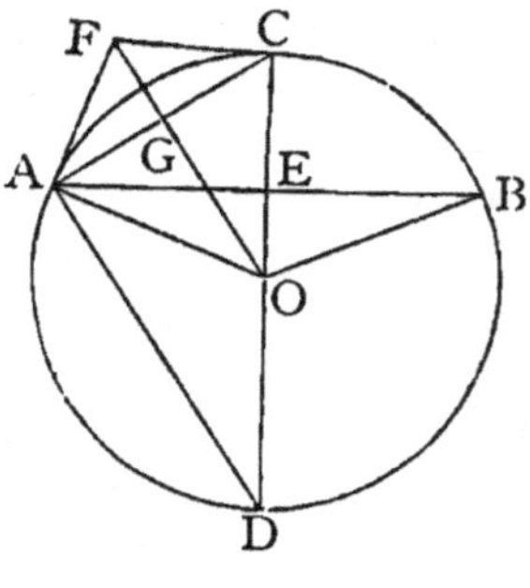

EOB are equal and therefore the arcs *CA CB* are equal whereon these equal angles at the center stand if then the chord *AC* be drawn and n be the number of equal arcs that make up the whole circumference or the number of sides of the inscribed regular polygon by bisecting each of the $n-1$ other arcs as well as *ACB* the halves of the n equal arcs are all equal and the circumference is thus wholly cut into $n\times 2$ equal arcs hence (art. 457) *AC* is one side of a regular $n\times 2$ sided polygon that may be inscribed in *ABC* by drawing the chords of the $n\times 2$ arcs and two straight lines *AF CF* may be drawn touching *ABC* at *A C* shutting in with the chord *AC* an isosceles triangle *FAC* upon the opposite side of the chord *AC* to *O* and each a half side of a regular $n\times 2$ sided polygon that may be circumscribed about *ABC* with the sides touching *ABC* at the $n\times 2$ arc ends. In reference to the diameter of the circle *ABC* as unit let σ express numerically *AB* s a side of the regular $n\times 2$ sided inscribed polygon and s' a side of the regular $n\times 2$ sided circumscribed polygon. Join *AD* and express *AD* numerically by x in reference to *ABC*'s diameter as unit. The angle *CAD* because in a semicircle is a right angle the straight line *CD* passing through the center because it bisects the straight line *AB* in the circle not passing through the center cuts at right angles *AB* therefore *AE* is the perpendicular from the right angle *CAD* of the right angled triangle *ACD* to the side *CD* opposite therefore *ADC EAC* are similar triangles so that $AD : DC = EA : AC \therefore x = \dfrac{\frac{1}{2}\sigma}{s}$ or $2xs = \sigma$ moreover from the right angled triangle *CAD* (art. 382) $x^2+s^2 = 1$. Since the angle *AOB* is less than a hemiperigon *AOC* is less than a right angle therefore *AOD* is greater than a right angle and therefore greater than *AOC* the triangles *AOD AOC* then have *AO OD* equal severally to *AO OC* and the angle *AOD* greater than *AOC* $\therefore AD > AO$. Hence

$$s = \frac{1}{2}\{\sqrt{(1+\sigma)}-\sqrt{(1-\sigma)}\} \qquad x = \frac{1}{2}\{\sqrt{(1+\sigma)}+\sqrt{(1-\sigma)}\}.$$

Join *OF* and let *G* be the point where (art. 457) *AC* cuts *OF* between *O* and *F* and *OF* bisects at right angles *AC* then *AD GO* are parallel because they make with *AC* at *A G* interior angles *GAD AGO* upon the side whereon *COD* lies that are together equal to a hemiperigon therefore and because *AD GO* cut *CD* at *D O* the exterior angle *GOC* is equal to the off lying interior angle *ADC* upon the side of *CD* whereon *CGA* lies moreover the right angles FCO CAD are equal hence the triangles FCO CAD are similar $\therefore FC : CO = CA : AD$ $2(FC) : CD = 2(CA) : 2(AD)$ $s' = \frac{2s}{2x}$

$$s' = \frac{\sqrt{(1+\sigma)}-\sqrt{(1-\sigma)}}{\sqrt{(1+\sigma)}+\sqrt{(1-\sigma)}}.$$

If upon, and upon each side of, a radius an equilateral triangle be described the ends of those three diameters whose halves are the common side and the other sides ending at the center of the triangles cut the circumference into six equal arcs the chords of these arcs therefore are the sides of an inscribed regular hexagon and each chord is $\frac{1}{2}$ a diameter. For σ then putting 0·5 the sides of regular 6×2 or 12 sided polygons inscribed in and circumscribed about a circle are severally expressed numerically by $\frac{1}{2}(\sqrt{1\cdot5}-\sqrt{0\cdot5})$ and $\frac{\sqrt{1\cdot5}-\sqrt{0\cdot5}}{\sqrt{1\cdot5}+\sqrt{0\cdot5}}$ in reference to the diameter as unit and the perimeters by 12 fold these severally for σ putting $\frac{1}{2}(\sqrt{1\cdot5}-\sqrt{0\cdot5})$ the numerical expressions are got in like manner of the sides and the perimeters of regular 12×2 or 24 sided polygons inscribed in and circumscribed about the circle and so on. Thus in reference to a circle's diameter as unit the under named regular polygons have the numerically expressed perimeters severally written against them right as far as the 8th decimal place

Sided	Inscribed	Circumscribed
12	3·10582854	3·21539031
24	- - 3262861	- 15965994
48	- - 3935020	- - 4608622
96	- - 4103195	- - - 271460
192	- - - 45247	- - - 187305
384	- - - 55761	- - - 66275
768	- - - - 8389	- - - 61018
1536	- - - - 9046	- - - 59703
3072	- - - - - 211	- - - - - 375
6144	- - - - - 52	- - - - - 293
12288	- - - - - 62	- - - - - - 72
24576	- - - - - - 5	- - - - - - 67
49152	- - - - - - - -	- - - - - - - 6
98304	- - - - - - - -	- - - - - - - 5

Wherefore of the 98304 sided regular inscribed and circumscribed polygons the perimeters differ from one another by less than $\frac{1}{2}\left(\frac{1}{10}\right)^8$ of a diameter and the circle's circumference because greater (art. 455) than the perimeter of the inscribed polygon and less (art. 457) than the perimeter of the circumscribed differs from either perimeter still more by less than this submultiple of the diameter and is expressed numerically as far as the 8th decimal place in reference to the diameter as unit by 3·14159265.

There is clearly no end to the degree of nearness to which in this way the numerical quantity may be found expressing the circumference of a circle in reference to the diameter as unit. That the same numerical quantity should express the circumference in reference to the diameter as unit of any circle whatever can only be because the ratio of the circumference to the diameter is the same for all circles.

460. *Def.* The numerical quantity expressing the circumference of a circle in reference to the diameter as unit or representing the ratio of the circumference to the diameter is symbolized by the Greek letter π.

If then r be the numerical expression in reference to any unit line of a circle's radius the numerical expression in reference to the same unit line of that circle's circumference is $\pi \times 2r$ and therefore too $2\pi r$ or $(2\pi)r$.

Hence if γ be the numerical expression in reference to a perigon as unit of an angle at the circle's center since angles at the center of a circle are proportionate to the arcs whereon they stand the arc whereon the angle stands is expressed numerically in reference to the unit line by $\gamma(2\pi)r$.

461. If from the point as center where an angle A is a circle of any radius be described and r a be severally the numerical expressions in reference to a common unit line of the radius and the arc whereon A stands since angles at a circle's center are proportionate to the arcs they stand on A is expressed numerically in reference to a perigon as unit by $\frac{a}{(2\pi)r}$ or the operational equivalent $\frac{a}{r}\frac{1}{2\pi}$ and therefore in reference to that angle as unit whose numerical expression is $\frac{1}{2\pi}$ in reference to a perigon as unit by $\frac{a}{r}$. The angle

which $\frac{1}{2\pi}$ expresses numerically in reference to a perigon as unit is the angle that A would be were a equal to r.

Def. An angle which at the center of a circle stands on an arc equal to the radius is called the CIRCULAR UNIT ANGLE.

All angles are equal to one another which at the centers of circles stand on arcs equal severally to the radii because each is expressed by the same numerical quantity $\frac{1}{2\pi}$ in reference to a perigon as unit.

If θ express numerically A in reference to the circular unit angle $\theta r = a$. The numerical expression in reference to the circular unit angle of a perigon is 2π of a hemiperigon π and of a right angle $\frac{1}{2}\pi$.

462. Let AOB be any angle less than a right angle with OA make at O an angle AOC equal to AOB and upon the opposite side of OA to AOB in OA take a point A anywhere but at O from O as center at the distance OA describe a circle and let B C be the points where OB OC drawn from the center severally cut the circumference join BC and let D be the point where BC ending at B C on opposite sides of OA is cut by the straight line of OA between B and C because the straight line of BC passes through the two points B C in the circumference D in that straight line between B and C is within the circle and therefore is on the O side of A again because BOA is less than a right angle and the equal AOC therefore less than a right angle the whole angle BOC made up of these is less than a hemiperigon therefore BOA BOC are angles upon the same side of OB with the bounding lines OA OC on that side of OB therefore every point but O in OA is on the same side of OB as every point but B in BC and therefore D is on the A side of O hence D is between O and A and therefore O A are on opposite sides of D and therefore of BC ODB ODC then are triangles and in these DO OB are equal severally to DO OC and the angles DOB DOC are equal therefore $DB = DC$ the triangles are equal

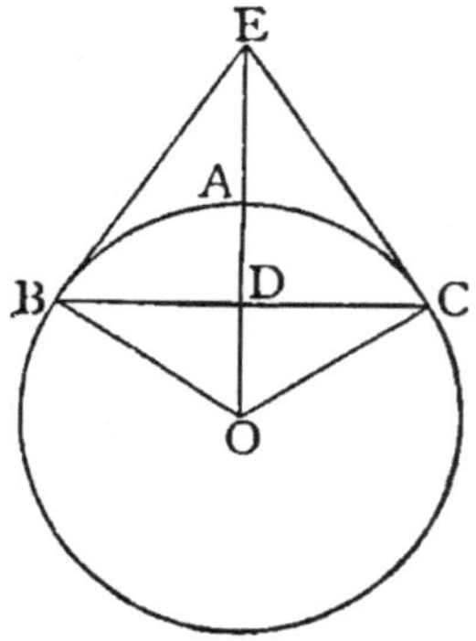

and of the other angles $BDO = CDO$ or OD is perpendicular to BC. From B draw on that side of BO whereon OD is BE at right angles to BO and therefore touching the circle at B because BE OD make with BO at the different points B O interior angles upon one side OBE BOE the one a right angle and the other less than a right angle and therefore together less than a hemiperigon BE OD meet on that side if far enough produced let E be the point where they so meet and join CE because E is not at O the only point common to OE OC EOC is a triangle then since the triangles EOB EOC have EO OB equal severally to EO OC and the angles EOB EOC equal $EB \doteq EC$ the triangles are equal and of the other angles $OBE = OCE$ OCE therefore is equal to and is therefore itself a right angle and therefore CE touches the circle at C. Now the arc BAC (art. 455) is greater than its chord BC and BC is $2(BD)$ and because the arcs AB AC are equal whereon the equal angles AOB AOC at the center stand BAC is $2(BA)$ $\therefore BA > BD$. Again the arc BAC which is $2(BA)$ because touching at the ends B C the two sides BE CE of the triangle EBC lying wholly on one side of every straight line that anywhere touches it and having the portion close to B on the same side of EBC's third side BC as EBC is (art. 456) less than (BE, EC) or $2(BE)$ and $\therefore BA < BE$. Hence if α be the numerical expression of AOB in reference to the circular unit angle and r the numerical expression of OA in reference to the unit straight line so that in reference to this unit straight line BA is expressed numerically by αr BD by $(\sin\alpha)r$ and since $\frac{1}{2}\pi - \alpha$ is the angle OEB's numerical expression in reference to the circular unit angle BE by $\frac{\sin\alpha}{\sin\left(\frac{1}{2}\pi-\alpha\right)}r$ or $\frac{\sin\alpha}{\cos\alpha}r$ the three numerical quantities $(\sin\alpha)r$ αr $\frac{\sin\alpha}{\cos\alpha}r$ are in ascending order of greatness and therefore so are $\sin\alpha$ α $\frac{\sin\alpha}{\cos\alpha}$.

$$\therefore \frac{1}{\alpha}\sin\alpha < 1 \text{ and } > \cos\alpha.$$

Therefore too and $\because$ $\sin-\alpha = -\sin\alpha$ and $\cos-\alpha = \cos\alpha$ $\frac{1}{-\alpha}\sin-\alpha < 1$ and $> \cos-\alpha$. Hence if θ be the algebraic expression in reference to the circular unit angle of any angle less than a right angle

$$\theta^{-1}\sin\theta < 1 \text{ and } > \cos\theta.$$

But by taking θ near enough to 0 $\cos\theta$ may be made although ever less than 1 yet nearer to 1 than by any given difference. Much more therefore may θ be taken so near to 0 as to make $\theta^{-1}\sin\theta$ although ever less than 1 yet more nearly equal to 1 than by any given difference.

463. If θ be the algebraic expression of any angle in reference to the circular unit angle and n be any whole number but 0

$$\left(\cos\frac{1}{n}\theta\right)^n=\left\{1-2\left(\sin\frac{1}{2}\frac{1}{n}\theta\right)^2\right\}^n=1-\underline{n}\rfloor\times 2\left(\sin\frac{1}{2}\frac{1}{n}\theta\right)^2$$

$$=1-\underline{n}\rfloor\left(\frac{1}{n}\theta\frac{1}{\frac{1}{n}\theta}\right)\frac{1}{\left(\frac{1}{2}\right)}\left(\sin\frac{1}{2}\frac{1}{n}\theta\right)^2=1-\underline{1}\rfloor\theta\left(\frac{1}{\frac{1}{2}\frac{1}{n}\theta}\sin\frac{1}{2}\frac{1}{n}\theta\right)\sin\frac{1}{2}\frac{1}{n}\theta$$

and n may be taken so great as to make $\frac{1}{2}\frac{1}{n}\theta$ nearer to 0 than by any given difference and therefore $\frac{1}{\frac{1}{2}\frac{1}{n}\theta}\sin\frac{1}{2}\frac{1}{n}\theta$ nearer to 1 and $\sin\frac{1}{2}\frac{1}{n}\theta$ nearer to 0 than by any given difference. Hence n may be taken so great as to make $\left(\cos\frac{1}{n}\theta\right)^n$ although always less than 1 yet more nearly equal to 1 than by any given difference.

464. If $\omega_1\ \omega_2 \ldots \omega_n$ be the algebraic expressions in reference to a common unit angle of any angles and x be any algebraic quantity (arts. 185, 300, 354, 398)

$$(\cos\omega_n+x\sin\omega_n)\ldots(\cos\omega_2+x\sin\omega_2)(\cos\omega_1+x\sin\omega_1)$$
$$=\sigma_n+x\sigma_{n-1}+x^2\sigma_{n-2}+\cdots+x^n\sigma_0$$

where σ_i stands for the sum of all the products made by multiplying together the cosines of every i of the n algebraically expressed angles and the sines of the $n-i$ others. Whence by making x $(-)^{\frac{1}{2}}1$ there comes (art. 393) the operational equivalence

$$\Theta_{\omega_1+\omega_2+\cdots+\omega_n}$$

$$=\sigma_n-\sigma_{n-2}+\sigma_{n-4}-\cdots+\begin{cases}(-)^{\frac{n}{2}}\sigma_0\\(-)^{\frac{n-1}{2}}\sigma_1\end{cases}+(-)^{\frac{1}{2}}\left[\sigma_{n-1}-\sigma_{n-3}+\sigma_{n-5}-\cdots+\begin{cases}(-)^{\frac{n}{2}-1}\sigma_1\\(-)^{\frac{n-1}{2}}\sigma_0\end{cases}\right]$$

the upper or the lower of the bracketed alternatives being taken according as n is even or odd. Therefore (art. 375)

$$\cos(\omega_1+\omega_2+\cdots+\omega_n) = \sigma_n-\sigma_{n-2}+\sigma_{n-4}-\cdots+\begin{cases}(-)^{\frac{n}{2}}\sigma_0 \text{ if } n \text{ be even}\\ (-)^{\frac{n-1}{2}}\sigma_1 \text{ if } n \text{ be odd}\end{cases}$$

$$\sin(\omega_1+\omega_2+\cdots+\omega_n) = \sigma_{n-1}-\sigma_{n-3}+\sigma_{n-5}-\cdots+\begin{cases}(-)^{\frac{n}{2}-1}\sigma_1 \text{ if } n \text{ be even}\\ (-)^{\frac{n-1}{2}}\sigma_0 \text{ if } n \text{ be odd.}\end{cases}$$

If $\omega_1=\omega_2=---=\omega_n$ and each $=\omega$ (art. 184)

$$\cos n\omega = n_{|0}(\cos\omega)^n - n_{|2}(\cos\omega)^{n-2}(\sin\omega)^2+n_{|4}(\cos\omega)^{n-4}(\sin\omega)^4-$$
$$\cdots+\begin{cases}(-)^{\frac{n}{2}}n_{|n}(\sin\omega)^n\\ (-)^{\frac{n-1}{2}}n_{|n-1}(\cos\omega)(\sin\omega)^{n-1}\end{cases}$$

$$\sin n\omega = n_{|1}(\cos\omega)^{n-1}\sin\omega - n_{|3}(\cos\omega)^{n-3}(\sin\omega)^3+n_{|5}(\cos\omega)^{n-5}(\sin\omega)^5-$$
$$\cdots+\begin{cases}(-)^{\frac{n}{2}-1}n_{|n-1}(\cos\omega)(\sin\omega)^{n-1}\\ (-)^{\frac{n-1}{2}}n_{|n}(\sin\omega)^n.\end{cases}$$

Hence if θ be the algebraic expression of any angle in reference to the circular unit angle i being any whole number chosen at will so as only to make no index minus

$$\cos\theta = \cos n\frac{1}{n}\theta = n_{|0}\left(\cos\frac{1}{n}\theta\right)^n - \cdots + (-)^i n_{|i\times2}\left(\cos\frac{1}{n}\theta\right)^{n-i\times2}\left(\sin\frac{1}{n}\theta\right)^{i\times2}$$

$$\sin\theta = n_{|1}\left(\cos\frac{1}{n}\theta\right)^{n-1}\sin\frac{1}{n}\theta - \cdots + (-)^i n_{|i\times2+1}\left(\cos\frac{1}{n}\theta\right)^{n-(i\times2+1)}\left(\sin\frac{1}{n}\theta\right)^{i\times2+1}$$

But $n_{|i}\left(\cos\frac{1}{n}\theta\right)^{n-i}\left(\sin\frac{1}{n}\theta\right)^i$

$$=\frac{\left(1-\frac{i-1}{n}\right)\cdots\left(1-\frac{2}{n}\right)\left(1-\frac{1}{n}\right)}{\underline{|i}}\left(\frac{1}{\cos\frac{1}{n}\theta}\right)^i\left(\cos\frac{1}{n}\theta\right)^n\theta^i\left(\frac{1}{\frac{1}{n}\theta}\sin\frac{1}{n}\theta\right)^i$$

which (arts. 462, 463) when n is taken endlessly great but with i kept always the same becomes endlessly near to $\frac{1}{\underline{|i}}\theta^i$. Therefore

$$\cos\theta = 1 - \cdots + (-)^i \frac{1}{\underline{|i\times 2}} \theta^{i\times 2} \qquad \sin\theta = \frac{1}{\underline{|1}}\theta - \cdots + (-)^i \frac{1}{\underline{|i\times 2+1}} \theta^{i\times 2+1}$$

or (art. 444)

$$\cos\theta = 1 - \frac{1}{\underline{|2}}\theta^2 + \frac{1}{\underline{|4}}\theta^4 + \cdots \qquad \sin\theta = \frac{1}{\underline{|1}}\theta - \frac{1}{\underline{|3}}\theta^3 + \frac{1}{\underline{|5}}\theta^5 - \cdots$$

Wherefore too

$$\cos\theta + (-)^{\frac{1}{2}}\sin\theta = 1 + \frac{1}{\underline{|1}}(-)^{\frac{1}{2}}\theta + \frac{1}{\underline{|2}}\{(-)^{\frac{1}{2}}\theta\}^2 + \frac{1}{\underline{|3}}\{(-)^{\frac{1}{2}}\theta\}^3 + \cdots.$$

465. If then θ be the algebraic expression of an angle in reference to the circular unit angle and c be a numerical quantity,—

$$\because \text{(arts. 451, 452)} \quad c = 1 + \frac{1}{\underline{|1}}\log_e c + \frac{1}{\underline{|2}}(\log_e c)^2 + \cdots,$$

$$\Theta_\theta c = \left[1 + \frac{1}{\underline{|1}}(-)^{\frac{1}{2}}\theta + \frac{1}{\underline{|2}}\{(-)^{\frac{1}{2}}\theta\}^2 + \cdots\right]\left\{1 + \frac{1}{\underline{|1}}\log_e c + \frac{1}{\underline{|2}}(\log_e c)^2 + \cdots\right\}$$

$$= \left[1 + \frac{(-)^{\frac{1}{2}}\theta}{\underline{|1}} + \frac{\{(-)^{\frac{1}{2}}\theta\}^2}{\underline{|2}} + \cdots\right]\left\{1 + \frac{\log_e c}{\underline{|1}} + \frac{(\log_e c)^2}{\underline{|2}} + \cdots\right\}$$

and $\therefore$ (arts. 446, 451)

$$= 1 + \frac{\log_e c + (-)^{\frac{1}{2}}\theta}{1} + \frac{\{\log_e c + (-)^{\frac{1}{2}}\theta\}^2}{\underline{|2}} + \cdots$$

$$= 1 + \frac{1}{\underline{|1}}\{\log_e c + (-)^{\frac{1}{2}}\theta\} + \frac{1}{\underline{|2}}\{\log_e c + (-)^{\frac{1}{2}}\theta\}^2 + \cdots.$$

Hence (art. 454) ω being the algebraic expression of an angle in reference to the circular unit angle and e a numerical quantity

$$(\Theta_\theta c)^{\Theta_\omega e} = \Theta_{ei\times 2\pi}\left[1 + \frac{\Theta_\omega e}{\underline{|1}}\{\log_e c + (-)^{\frac{1}{2}}\theta\} + \frac{(\Theta_\omega e)^2}{\underline{|2}}\{\log_e c + (-)^{\frac{1}{2}}\theta\}^2 + \cdots\right]$$

$$= 1 + \frac{(\Theta_\omega e)\{\log_e c + (-)^{\frac{1}{2}}\theta\} + (-)^{\frac{1}{2}} ei\times 2\pi}{\underline{|1}}$$

$$+ \frac{[(\Theta_\omega e)\{\log_e c + (-)^{\frac{1}{2}}\theta\} + (-)^{\frac{1}{2}} ci\times 2\pi]^2}{\underline{|2}} + \cdots$$

$$= 1 + \frac{\{(\cos\omega)e\}\log_e c - \{(\sin\omega)e\}\theta + (-)^{\frac{1}{2}}[\{(\cos\omega)e\}\theta + \{(\sin\omega)e\}\log_e c + ei\times 2\pi]}{\underline{|1}} + \cdots$$

$$= \left[\begin{matrix}1 \\ +\frac{1}{\lfloor 1}(-)^{\frac{1}{2}}\{(\cos\omega)e\theta+(\sin\omega)e\log_e c+ei\times 2\pi\} \\ +\cdots\end{matrix}\right]\left[\begin{matrix}1 \\ +\frac{1}{\lfloor 1}\{(\cos\omega)e\log_e c-(\sin\omega)e\theta\} \\ +\cdots\end{matrix}\right]$$

$$= \Theta_{e\{(\cos\omega)\theta+(\sin\omega)\log_e c+i\times 2\pi\}}\,\overline{\underline{\epsilon^{-(\sin\omega)e\theta}}}\,\overline{\underline{c^{(\cos\omega)e}}}.$$

466. Therefore touching the Laws of such powers as have some of their indices ditensive quantities if c d e be numerical quantities and θ ϕ ω algebraic expressions of angles in reference to the circular unit angle

$$(\Theta_\phi d)^{\Theta_\omega e}(\Theta_\theta c)^{\Theta_\omega e}$$

$$= \Theta_{e\{(\cos\omega)(\theta+\phi)+(\sin\omega)\log_e dc+(i+i')\times 2\pi\}}\,\overline{\underline{\epsilon^{e\{(\cos\omega)\log_e dc-(\sin\omega)(\theta+\phi)\}}}}$$

$$= \{(\Theta_\phi d)\Theta_\theta c\}^{\Theta_\omega e}$$

$$\frac{(\Theta_\theta c)^{\Theta_\omega e}}{(\Theta_\phi d)^{\Theta_\omega e}}$$

$$= \Theta_{e\{(\cos\omega)(-\phi+\theta)+(\sin\omega)\log_e\frac{c}{d}+(-i'+i)\times 2\pi\}}\,\overline{\underline{\epsilon^{e\{(\cos\omega)\log_e\frac{c}{d}-(\sin\omega)(-\phi+\theta)\}}}}$$

$$= \left(\frac{\Theta_\theta c}{\Theta_\phi d}\right)^{\Theta_\omega e}$$

$$(\Theta_\theta c)^{\Theta_\omega e}(\Theta_\theta c)^{\Theta_\phi d}$$

$$= \Theta_{\left[\begin{matrix}\{(\cos\phi)d+(\cos\omega)e\}\theta+\{(\sin\phi)d+(\sin\omega)e\}\log_e c \\ +(di+ei')\times 2\pi\end{matrix}\right]}\,\overline{\underline{\epsilon^{\left[\begin{matrix}\{(\cos\phi)d+(\cos\omega)e\}\log_e c \\ -\{(\sin\phi)d+(\sin\omega)e\}\theta\end{matrix}\right]}}}$$

Now $\Theta_\phi d+\Theta_\omega e=\Theta_\alpha g$ if g be a numerical quantity such that $(\cos\phi)d+(\cos\omega)e=(\cos\alpha)g$ $(\sin\phi)d+(\sin\omega)e=(\sin\alpha)g$ whence $g=\sqrt{[d^2+e^2+2\{\cos(-\phi+\omega)\}de]}$ moreover

$$(\Theta_\theta c)^{\Theta_\alpha g}=\Theta_{\{(\cos\alpha)g\}\theta+\{(\sin\alpha)g\}\log_e c+gi''\times 2\pi}\,\overline{\underline{\epsilon^{\{(\cos\alpha)g\}\log_e c-\{(\sin\alpha)g\}\theta}}}$$

but all the values of $di+ei'$ are not always precisely all the values of gi''. The law of operational equivalence here sought turns out then to be

$$(+)^{\sqrt{[d^2+e^2+2\{\cos(-\phi+\omega)\}de]}}(\Theta_\theta c)^{\Theta_\omega e}(\Theta_\theta c)^{\Theta_\phi d}=(+)^e(+)^d(\Theta_\theta c)^{\Theta_\phi d+\Theta_\omega e}.$$

Again $\because \left(\frac{1}{\Theta_\theta c}\right)^{\Theta_\omega e} = (\Theta_{-\theta} c^{-1})^{\Theta_\omega e}$

$$= \Theta_{e\{(\cos\omega)(-\theta)+(\sin\omega)\log_e c^{-1}+i\times 2\pi\}}\ \overline{\underline{\epsilon^{e\{(\cos\omega)\log_e c^{-1}-(\sin\omega)(-\theta)\}}}}$$

$$= (\Theta_\theta c)^{\Theta_{\omega+\pi} e} = (\Theta_\theta c)^{-\Theta_\omega e}$$

and $\frac{1}{(\Theta_\theta c)^{\Theta_\omega e}} = \{(\Theta_\theta c)^{\Theta_\omega e}\}^{-1}$

$$= \Theta_{-e\{(\cos\omega)\theta+(\sin\omega)\log_e c+i\times 2\pi\}}\ \overline{\underline{\epsilon^{-e\{(\cos\omega)\log_e c-(\sin\omega)\theta\}}}}$$

$$(+)^{\sqrt{[d^2+e^2-2\{\cos(-\phi+\omega)\}de]}}\left(\frac{1}{\Theta_\theta c}\right)^{\Theta_\omega e}(\Theta_\theta c)^{\Theta_\phi d} = \begin{cases}(+)^e(+)^d(\Theta_\theta c)^{\Theta_\phi d-\Theta_\omega e}\\ (+)^e(+)^d\left(\frac{1}{\Theta_\theta c}\right)^{-\Theta_\phi d+\Theta_\omega e}\end{cases}$$

$$(+)^{\sqrt{[d^2+e^2-2\{\cos(-\phi+\omega)\}de]}}\frac{1}{(\Theta_\theta c)^{\Theta_\omega e}}(\Theta_\theta c)^{\Theta_\phi d} = \begin{cases}(+)^e(+)^d(\Theta_\theta c)^{\Theta_\phi d-\Theta_\omega e}\\ (+)^e(+)^d\frac{1}{(\Theta_\theta c)^{\Theta_\omega e-\Theta_\phi d}}\end{cases}$$

$$\left.\begin{array}{l}(+)^{\sqrt{[e^2+d^2-2\{\cos(-\omega+\phi)\}ed]}}\frac{(\Theta_\theta c)^{\Theta_\phi d}}{(\Theta_\theta c)^{\Theta_\omega e}}\\ (+)^{\sqrt{[e^2+d^2-2\{\cos(-\omega+\phi)\}ed]}}(\Theta_\theta c)^{\Theta_\phi d}\frac{1}{(\Theta_\theta c)^{\Theta_\omega e}}\end{array}\right\} = \begin{cases}(+)^d(+)^e(\Theta_\theta c)^{-\Theta_\omega e+\Theta_\phi d}\\ (+)^d(+)^e\frac{1}{(\Theta_\theta c)^{-\Theta_\phi d+\Theta_\omega e}}\end{cases}$$

$$(+)^{\sqrt{[e^2+d^2-2\{\cos(-\omega+\phi)\}ed]}}(\Theta_\theta c)^{\Theta_\phi d}\left(\frac{1}{\Theta_\theta c}\right)^{\Theta_\omega e} = \begin{cases}(+)^d(+)^e(\Theta_\theta c)^{-\Theta_\omega e+\Theta_\phi d}\\ (+)^d(+)^e\left(\frac{1}{\Theta_\theta c}\right)^{\Theta_\omega e-\Theta_\phi d}.\end{cases}$$

Lastly

$$(\Theta_{\theta}c)^{(\Theta_{\phi}d)\Theta_{\omega}e} = (\Theta_{\theta}c)^{\Theta_{\omega+\phi}de}$$

$$= \Theta_{(de)[\{\cos(\omega+\phi)\}\theta+\{\sin(\omega+\phi)\}\log_e c+i''\times 2\pi]}\,\overline{\underline{\epsilon^{(de)[\{\cos(\omega+\phi)\}\log_e c-\{\sin(\omega+\phi)\}\theta]}}}$$

$$\text{and } \Theta^{\Theta_{\phi}d}_{ei\times 2\pi} = \Theta_{d\{(\cos\phi)ei\times 2\pi+i'\times 2\pi\}}\,\overline{\underline{\epsilon^{-d(\sin\phi)ei\times 2\pi}}}$$

$$\therefore\ (+)^{de}\{(\Theta_{\theta}c)^{\Theta_{\omega}e}\}^{\Theta_{\phi}d} = \{(+)^{e}\}^{\Theta_{\phi}d}(\Theta_{\theta}c)^{(\Theta_{\phi}d)\Theta_{\omega}c}.$$

www.ingramcontent.com/pod-product-compliance
Lightning Source LLC
Chambersburg PA
CBHW020906210326
41598CB00018B/1788